The Solidarity Approach in Geography

Kavita Arora • Rashmi Rani Anand
Editors

The Solidarity Approach in Geography

Cases from the Indian Subcontinent

Springer

Editors
Kavita Arora
Shaheed Bhagat Singh College
University of Delhi
New Delhi, India

Rashmi Rani Anand
Centre for African Studies
Jawaharlal Nehru University
New Delhi, India

ISBN 978-3-031-86752-1 ISBN 978-3-031-86753-8 (eBook)
https://doi.org/10.1007/978-3-031-86753-8

This Springer imprint is published by the registered company Springer Nature Switzerland AG
The registered company address is: Gewerbestrasse 11, 6330 Cham, Switzerland

Acknowledgment

The journey of compiling and editing this book has been both rewarding and challenging. This volume would not have been possible without the contributions and support of many individuals and organizations, to whom we are deeply grateful.

First and foremost, we wish to express our sincere gratitude to all the contributing authors who generously shared their expertise and insights. Their dedication and hard work have greatly enriched the content of this book. Their meticulous attention to detail and insightful feedback have significantly enhanced the quality of the work. We would like to thank Prof Elena dell' Agnese for writing a chapter for this book which is included in the introduction part and worth reading for all who are interested in knowing the linkages between geography and solidarity. The inception of the idea of this book is an International Conference organized on the same theme on the 5th and 6th of April 2021 by the Department of Geography, Shaheed Bhagat Singh College, University of Delhi. The conference was organized during the pandemic and Professor Agnese that time too delivered a very insightful talk as a chief guest.

We would also like to thank the reviewers from Springer Nature, whose critical assessments and constructive suggestions have greatly improved the manuscript. Their expertise and time are highly appreciated. We are greatly thankful for their exceptional encouragement and patience.

We are indebted to the University of Delhi (DU) and Jawahar Lal Nehru University (JNU) for providing the necessary resources and support to undertake this project. The collaborative environment fostered by the School of International Studies, JNU, and the Department of Geography, Shaheed Bhagat Singh College, University of Delhi has been conducive to the successful completion of this book.

Lastly, we would like to thank our family and friends for their unwavering support and understanding while editing this volume. Their patience and encouragement have been a constant source of strength.

To all those who have contributed, directly or indirectly, to the realization of this book, we extend our heartfelt thanks.

Contents

The Solidarity Approach in Geography: Cases from the Indian Subcontinent (About the Book)

The book, *The Solidarity Approach in Geography: Cases from the Indian Subcontinent*, examines the application of the solidarity approach in different branches of geographical studies focusing on case studies from the Indian subcontinent. Solidarity, as a concept rooted in collective action and mutual support, is analyzed within geographical research and its implications for the region's social, economic, and environmental dynamics. Solidarity is related to values that enable individuals in society to live with equality and dignity and ensure larger unity. Modern society's extreme specialization and competitiveness make it more prone to the erosion of ethical values. Having a sense of solidarity with humans improves ethical decisions by imbibing values like empathy, tolerance, and compassion. Therefore, it allows society to function with empathy and care and holds society together.

While geographers have long considered solidarity important in understanding and analyzing earth and nature, direct mention of the solidarity approach has remained largely absent in geographical knowledge. Hence, the inclusion of the solidarity approach in geography can serve as a new paradigm. The geographies of solidarity can emerge as spaces of care and responsibility. It has significant scope in geography, offering immense potential for reconceptualizing solidarity in diversity and rethinking the relational aspects of space and place, highlighting the importance of the solidarity approach in addressing global challenges.

Disciplinary knowledge in any field depends on its research approaches to global problems. Geography is unique in its subject matter and methodology, yet it maintains a multidisciplinary approach that intersects with other fields. This intersection allows geography to incorporate diverse perspectives and methodologies. Using the solidarity approach in geography can enrich geographical research in both human and physical geography. Geography derives its database from all the natural and social sciences and attempts their synthesis. It is a discipline that is a perfect amalgamation of natural and human sciences. It can help human geographers reflect on the relevance of the solidarity approach and enrich the subject matter, pushing the discipline's boundaries further. Traditionally, geography is studied under the purview of concepts of space, place, and time. Worldwide we see spatial differentiation in communities based on identity, location, territory, and their associations. However, there is growing acceptance of the relational understanding of space and place, suggesting that community-building processes can transcend social, cultural, and economic differences within a place and be defined by networks linking that place to others.

Incorporating this concept in human geography involves studying solidarity-based identities and considering scale and territory as spatial dimensions for organizing solidarity. Likewise, economic solidarity addresses issues like unemployment, poverty, provision of food, shelter, education, and institutional frameworks. By integrating solidarity, geographers can provide new insights and solutions for these pressing social and economic challenges. It can combine spatial analysis with social values to address economic disparities and promote sustainable development. Community-based economic models can be developed for globalization and local economies.

In political geography, solidarity refers to unity and shared purpose among a group of people, often based on common interests, values, or experiences. It encompasses shared identity, common interests, collective actions, and international cooperation. This solidarity can manifest in various forms, such as social movements, political activism, trade unions, NGOs, and diplomacy. It extends beyond national borders as countries and international organizations collaborate to address global challenges like climate change, poverty, health, and conflicts.

Solidarity is crucial for tackling issues that require coordinated global action. The roles and influences of international organizations are evolving, making networked support and solidarity essential for the success of global policies on climate change and natural resources. This cooperation helps build realistic solutions and large coalitions, fostering global integration.

In neoliberal globalization, support networks and solidarity are vital for the success of international agreements and global policies. The politics surrounding climate change and natural resources also depend on global solidarity and shared responsibility. It can also be a critical area of analysis looking at ongoing world conflicts like the Israel-Gaza war, the Russia-Ukraine war, and many more conflicts in Africa and other states. It can help in building realistic solutions and large coalitions, fostering global integration with changing incentives.

The solidarity approach has significant applications in physical geography and it offers a powerful framework. Physical geography studies mountains, physical features, atmospheric processes, geomorphological processes, and their effects on people. So, physical geography is the study of all natural phenomena taking place over the earth's surface. By integrating physical geography's insights with the principles of solidarity, communities can foster cooperation, enhance resilience, and work towards sustainable and equitable development. This multidisciplinary approach ensures that environmental and societal challenges are addressed comprehensively, benefiting all members of society. The study of physical geography along with Remote Sensing and GIS emerging is a strong tool for enhancing the solidarity approach. Physical geography assesses land suitability for development, while GIS helps in planning infrastructure, analyzes population density, and ensures access to services, aiding in decision-making. Solidarity is crucial in disaster response and management also, where communities support each other. Physical geography provides an understanding of natural hazards such as earthquakes, floods, and hurricanes. GIS maps of vulnerable areas predict disaster impacts and may assist in coordinating relief efforts. For example, during a natural disaster, GIS can identify the most affected regions and optimize aid distribution, demonstrating how solidarity and technology save lives. For example, in urban planning, solidarity ensures that development benefits all community members, especially marginalized groups, it can be achieved by the proper planning of cities by using the GIS.

By combining physical geography, social responsibility, spatial understanding, and technological tools, we can create sustainable and equitable solutions for a better future.

Understanding Solidarity

Historically, various scholars have attempted to define the concept of solidarity, each contributing to its understanding across different disciplines.

The etymological source of solidarity can be traced in the Roman law concept of *in solidum*. A similar concept was found in the law of classical Greece, where citizens within (*oikos*) a household, that generally has extended families—were jointly responsible for paying the debts that any of them had incurred. In the Western political domain, the discussion of the concept of solidarity started in the late eighteenth century and became widespread during the French Revolution. A political idea of fraternity or brotherhood developed during the French Revolution, and France was the birthplace of the term solidarity as well. The social and political unrest and the development of capitalism and liberalism encouraged French scholars to develop the concept of cohesion and solidarity for achieving individual rights.

In Germany, it developed later for the requirement of cohesion in the working class and labor. A prominent sociologist Émile Durkheim (1858–1917) in his theory of the development of societies, The Division of Labour in Society, introduced the concepts of mechanical and organic solidarity. Mechanical solidarity refers to the social cohesion in pre-modern societies where individuals perform similar tasks and share common values and beliefs, characterized by a collective consciousness. In contrast, organic solidarity describes the social cohesion in modern, industrial societies where individuals perform specialized tasks and roles, relying on the interdependence of individuals.

Weber (1864–1920) did not use the term solidarity but his idea of class consciousness indicates class solidarity. Marx also uses this term very less, but he developed a theory of working-class solidarity.

Karl Marx (1818–1883), a key figure in political economy, emphasized class solidarity among working-class members (proletariat) against the ruling class (bourgeoisie). He saw this solidarity as crucial for the collective action needed to overthrow capitalist systems and achieve a classless society. Marks talks about two different types of solidarity one was the working-class solidarity and another was solidarity in a postcapitalist society, under communism which he called the ideal solidarity. The advent of capitalism and associated problems compelled these scholars to find a solution.

Solidarity was not significantly used in the Leninist school of thought but Georg Lukacs (1885–1971) did develop a Leninist idea of solidarity in his book *History and Class Consciousness* and disapproved of the freedom found in capitalist society.

In political theory, Antonio Gramsci (1891–1937) extended the concept of solidarity to hegemony, where the dominant class forms alliances with subordinate classes to maintain control. For Gramsci, solidarity involves creating alliances and consensus within a social group to achieve common goals. Gramsci augmented Marxist theory with his emphasis on culture, moral understanding, and psychology. He used the term solidarity more often than the other Leninists analyzed here. Later, Mao Zedong (1893–1976) a Chinese politician emphasizes the solidarity of communist party members and between the party and the people. He did not use this idea in his theoretical deliberations but frequently used it in his speeches.

Jürgen Habermas (1929–) is known for his theory, which he presented in "The Theory of Communicative Action" (1981). The essential concern of this work is the excavating crisis of advanced capitalist societies. Habermas in philosophy and sociology focused on communicative solidarity, emphasizing the importance of communication and discourse in building solidarity. He argued that rational communication and dialogue lead to mutual understanding and agreement, fostering solidarity.

Nancy Fraser (1947–), a political philosopher, discussed solidarity in terms of social justice, involving both economic redistribution and cultural recognition. For Fraser, solidarity means supporting marginalized groups through both material and symbolic means. For Fraser, the idea of the "solidarity of the shaken" was a way of connecting at the moment of a transformation of the global order.

Alain Touraine (1925–2023) was a French sociologist best known as the originator of the phrase "Post Industrialist society," and for his studies of social movement, he linked solidarity to social movements, seeing it as a unifying force that brings individuals together to achieve social change. In this context, solidarity is the collective commitment to a cause. Iris Marion Young proposed a social connection model of solidarity based on interdependence and shared responsibility, emphasizing the need to work together to address injustices. The 1983 book *Solidarity, The Analysis of a Social Movement Poland 1980–1981* records a fascinating analysis of the Solidarity movement in Poland. He suggests an understanding of the place of social movements in present-day society and a renewed means of analyzing them through "sociological intervention."

There are many important philosophers and theologians, from various philosophical and religious traditions, engaged in an ongoing discussion and debate about the meaning, purpose, and limits of human rights language. One of them is the prominent Roman Catholic moral theologian, David Hollenbach, S.J. (1942–) viewed solidarity as a moral virtue involving a commitment to the common good and the well-being of others, particularly the marginalized.

Iris Marion Young (1949–2006) was a world-renowned feminist moral and political philosopher whose many books and articles spanned more than three decades. She explored issues of social justice and oppression theory, the phenomenology of women's bodies, deliberative democracy and questions of terrorism, violence, international law, and the role of the national security state. Iris Marion Young proposed a social connection model of solidarity based on interdependence and shared responsibility, emphasizing the need to work together to address injustices.

Amartya Sen (1933–) is a world-renowned economist, philosopher, and author. He linked solidarity to his capability approach, focusing on enhancing individuals' abilities to achieve their full potential. Solidarity, in this sense, supports others in expanding their capabilities.

Barbara Prainsack (1975–) is a professor at the Department of Political Science at the University of Vienna and at the Department of Global Health and Social Medicine at King's College London. Her work explores the social, regulatory, and ethical dimensions of biomedicine and bioscience. She along with her coauthor Alena Buyx (1977–), a Professor of Ethics in Medicine and Health Technologies and Director of the Institute for History and Ethics in Medicine, Technical University Munich, wrote a book in 2017 *Solidarity in biomedicine and Beyond* which explores solidarity-based approaches to governing biomedicine. Health databases, personalized medicine, and organ donation serve as extended case studies for applying principles of solidarity to policymaking in science and healthcare. The book also explains what solidarity can do to tackle current transboundary problems such as health crises, climate change, and democratic backsliding.

More than 40 years span of his career, Walzer has written on questions of, inter alia, democracy, obligation, social criticism, distributive justice, and war. Michael Walzer (1935–) discussed solidarity in the context of complex equality, where different spheres of justice require different forms of solidarity. He argued that solidarity involves respecting and supporting others' rights and needs across various domains of life. Complex equality is a theory of Justice given by Michael Walzer in his 1983 work *Sphere of Justice*.

In the global south anti-colonial movements in Asia, Africa, and Latin America were intrinsically tied to the concept of solidarity, as they often relied on collaboration and mutual support to achieve their goals of liberation and self-determination. Throughout the twentieth century, anti-colonial movements in Asia, Africa, and Latin America played a pivotal role in dismantling colonial rule and establishing independent nations. These movements fostered transnational solidarity, with leaders like Ho Chi Minh, Kwame Nkrumah, and Patrice Lumumba inspiring struggles for liberation across the globe. Che Guevara (1928–1967) was one of the most important innovators of Marxism in Latin America and a leading theorist of decolonial Marxism in world politics generally. Che precisely defined his revolutionary thesis for the third world and emphasized the solidarity of the struggle for national liberation.

Some prominent thinkers and geographers who have contributed to the solidarity approach in the Global South include:

Born on the island of Martinique under French colonial rule, Frantz Omar Fanon (1925–1961) was one of the most important writers of black Atlantic theory in an age of anti-colonial liberation struggle. Frantz Fanon: A psychiatrist and revolutionary from Martinique, Fanon's influential works like *The Wretched of the Earth* and *Black Skin, White Masks* explored the psychological and social effects of colonialism, highlighting the importance of decolonization and self-determination for the liberation of oppressed peoples. The colonial discriminatory practices form the basis of Fanon's distinctive account of Third World political solidarity as a form of intercontinental populism between formerly colonized nations that could obliterate the bonds of racialized fraternity that defined colonial relations and continued to define core and periphery relations in the postcolonial era. In his work, Fanon displays that race conciliates how social solidarity was expressed and practiced in such a way that colonial subject-citizens were at the same time contained within and let off from republican citizenship.

Amílcar Lopes Cabral (1924–1973) was a revolutionary leader and theorist from Guinea-Bissau and Cape Verde, Cabral's writings emphasized the importance of national liberation movements being rooted in the culture and history of their people. He advocated for a "unity of theory and practice" that would enable the masses to actively participate in the liberation struggle. Cabral was an internationalist who engaged in solidarity with anti-colonial struggles throughout the world while at the same time seeking to learn from those many diverse experiences. One of Cabral's significant ideological contributions was his critique of Marx's *Communist Manifesto*. While largely supportive of its main message, Cabral saw this as Eurocentric without sufficient recognition of the existing reality in much of the world particularly in Africa and other territories occupied by colonialism.

Eduardo Hughes Galeano (1940–2015) was a Uruguayan journalist, writer and novelist considered, among other things, a literary figure of Latin America. Galeano's most famous work, *Open Veins of Latin America*, provided a scathing critique of Western powers' exploitation of Latin America. He called for a new form of development that would prioritize social justice and environmental sustainability. He prefers solidarity over charity and mentions "I don't believe in charity. I believe in solidarity. Charity is so vertical. It goes from the top to the bottom. Solidarity is horizontal. It respects the other person. I have a lot to learn from other people."

Raewyn Connell (1944–) is an Australian sociologist. Connell's work has focused on the intersections of gender, race, and class in global power relations. She has challenged the dominance of Western knowledge systems and advocated for a more inclusive approach to development that recognizes the diversity of experiences and perspectives in the Global South.

Vandana Shiva (1952–) is an Indian scholar and environmental activist. She has been a vocal critic of globalization and its negative impacts on the Global South. She advocates for a return to traditional knowledge systems and sustainable agricultural practices, emphasizing the importance of food sovereignty and local economies. Dr Shiva Vandana is a philosopher, environmental activist, eco-feminist, and author of several books. She participated in the non-violent Chipko movement during the 1970s. The movement, in which predominantly women participated, formed human circles around trees to protect them from ruthless cutting. She is a renowned figure of the global solidarity movement known as the alter-globalization movement.

This discussion reflects the diverse ways in which scholars have understood and conceptualized solidarity. While the specific emphases and contexts may differ, a common thread across these thoughts is the idea of mutual support, shared goals, and collective responsibility. Despite its widespread use in disciplines such as sociology, political science, international relations, and economics, "solidarity" has never been a prominent concept in geography but still, the notion of solidarity has been explored and applied in various ways to understand and address socio-economic, cultural, and environmental challenges.

Its application to core geographical concepts like space, place, scale, landscape, mobility, and nature remains limited. However, a few geographers have also made significant contributions to the conceptualization of solidarity, particularly by exploring spatial relationships, place-based identities, and environmental concerns. Their work expands the traditional understanding of solidarity by integrating geographical concepts that highlight the importance of location and environment. They are generally concerned with studying the nature of solidarity in different branches of geography. For example, tribal populations, rural areas, and self-help groups are closely knitted and solidarity has always existed from the ages. Geographers have tackled the topic of solidarity from a variety of angles, highlighting the significance of comprehending the socio-economic dynamics inherent in the world's major capitalist economies. Through the Solidarity and Cooperation Economy (SCE), they have investigated the significance of solidarity in bridging geographically separated communities and promoting alternative geopolitics. The historical geographies of solidarity have also been studied by geographers, who have drawn attention to the importance of mobility in creating varied solidarities across social divides as well as the risk of entrenching power disparities. In addition, geographers have put forth creative ideas for cost allocation that both encourage individual reason and make up for unavoidable costs, such as the solidarity egalitarian method. Together, these many viewpoints add to a thorough comprehension of solidarity in certain geographic circumstances.

Looking into geographers using "Solidarity," let us see the work of Jacques Élisée (1830–1905). He did not see humanity as competing with nature but rather wrote that humanity's development is most intimately connected with the nature surrounding it. In 1905 in L'Homme et La Terre, I, Les ancetres, he also stressed human interdependence. He understood human civilization as a shared achievement, insisting that there is no longer any possibility of progress, other than for the world as a whole. In 1908 in L'Homme et La Terre, IV, histoire contemporaine, he mentioned that the lessons of geography should advance the cause of human solidarity. In its essence, human progress consists of the common cause being found among all peoples. This was very different from geography in the service of the Empire as propounded by many of that time scholars.

David Harvey (1935–) is a notable Marxist geographer, who talked about the idea of Spatial Solidarity. He emphasized the importance of spatial relationships in understanding social processes. He defines solidarity in terms of spatial justice, advocating for the equitable distribution of resources and opportunities across different geographic areas. He talked about how spatial solidarity and efforts to bridge the urban-rural divide and create inclusive public spaces. In his studies, he highlighted equitable development and ensured that both urban and rural populations benefit from economic growth and infrastructural advancements. By examining urban planning projects, rural development schemes, and public space enhancements, we reveal how spatial solidarity can contribute to a more balanced and cohesive society. Therefore, we can say that Harvey's work of solidarity involves recognizing and addressing spatial inequalities and fostering a collective responsibility towards disadvantaged regions.

Doreen Massey (1944–2016), known for her work on space, place, and power, discussed relational solidarity. Relational solidarity focuses on the interconnectedness of economies, cultures, and communities. She views solidarity as a relational concept deeply tied to the interconnectedness of places and people. Massey argues that solidarity involves understanding and acting upon the interdependencies between different places and communities, highlighting the importance of considering global-local linkages and the impacts of local actions on global scales. Through inter-state economic partnerships and cultural festivals, we demonstrate how relational solidarity can strengthen social and economic ties, promoting a more integrated and harmonious growth of the region.

Gillian Hart (1946–) is best known for her books *Rethinking the South African Crisis*, *Disabling Globalization*, and *Power, Labor, and Livelihood*. She emphasized the significance of scale in understanding solidarity. She contends that solidarity must be built across different scales, from local to global, to address issues such as economic inequality, social justice, and environmental sustainability. Hart's perspective on solidarity involves multi-scalar networks of support and activism. She argued that how scale-focused solidarity plays a crucial role in addressing local issues while connecting to broader frameworks.

Noel Castree (1968–) focused on environmental issues, defining solidarity in terms of collective action towards environmental sustainability. He underscores the need for solidarity in addressing global environmental challenges like climate change, biodiversity loss, and pollution. Environmental solidarity, according to Castree, involves a shared commitment to protecting the environment and ensuring equitable access to natural resources.

Michael Watts (1951–), renowned for his work on political ecology, views solidarity as a form of collective resistance against oppressive systems and practices. He highlights the role of solidarity in social movements and struggles for justice, particularly in the context of environmental and economic exploitation. Watts emphasizes building alliances and networks to resist and challenge dominant power structures. He emphasized on the interconnectedness of economic systems and the importance of fostering cooperative economic relationships to promote social and environmental justice.

Eric Sheppard (1950–), from the Department of Geography, UCLA a key figure in economic geography, defines solidarity as a means to challenge neoliberal globalization and promote alternative economic practices. His work highlights how economic processes are not isolated but are deeply embedded in broader social and spatial contexts. Sheppard advocates for economic systems that prioritize social justice and equity. He critiques neoliberal economic policies that often exacerbate inequalities and environmental degradation. Instead, he calls for economic solidarity that supports fair trade, equitable resource distribution, and sustainable development.

Jennifer Hyndman (1964–), known for her work on migration and humanitarianism, defines solidarity in terms of humanitarian action and support for displaced populations. She emphasizes the importance of solidarity in responding to crises such as refugee movements and natural disasters, advocating for compassionate and equitable approaches to humanitarian aid. Her perspective on solidarity emphasizes the importance of care, empathy, and practical support in addressing human suffering. She emphasizes the need for a compassionate response to human suffering, advocating for policies and actions that prioritize the dignity and well-being of displaced and marginalized populations.

Diarmaid Kelliher is a historical geographer who focuses on the spatial politics of labor and social movements. He has been mainly interested in developing theoretical ideas around solidarity and agency. In his book *Making Cultures of Solidarity: London and the 1984-5 Miners' Strike* he highlighted how the solidarity of 1984–5 was rooted in the development of mutual relationships of support between the coalfields and the capital since the late 1960s. He emphasizes that struggle brings people together and develops a culture of solidarity and the legacies of support movement create long-term solidarity of people. History provides different persuasive examples of the development of solidarity across geographical and social boundaries.

Going through these thoughts of geographers, we can say geographers' perspective of solidarity often emphasizes the spatial dimensions of social relationships and the importance of addressing inequalities and injustices across different scales and places. Whether through the lens of spatial justice, environmental sustainability, economic alternatives, or humanitarian action, geographers highlight the interconnectedness of people and places and the need for collective action to build a more equitable and sustainable world.

These geographers collectively expand the understanding of solidarity by incorporating diverse spatial, social, and environmental dimensions, offering valuable insights for addressing contemporary global challenges.

Solidarity and Indian Subcontinent

Solidarity in geography within the Indian subcontinent can be examined through various lenses, drawing on the perspectives of prominent geographers and the region's unique socio-economic and cultural contexts. The book delves into solidarity and examines cases from the Indian subcontinent, highlighting the political, economic, social, and cultural cooperation among the people, states, and countries in the region, providing a comprehensive understanding of how solidarity can address contemporary global challenges.

India's vision of global solidarity is deeply rooted in *Vasudhaiva Kutumbakam*. It views the world as an interconnected and interdependent entity, where the well-being of one nation is inextricably linked to the well-being of others. The Sanskrit phrase *Vasudhaiva Kutumbakam*, literally translates to "the whole world is one family." It reflects an ancient Indian philosophy that emphasizes the interconnectedness of all beings and promotes a sense of global kinship and *Sarve Bhavantu Sukhinaḥ*. It advocates for universal brotherhood, compassion, empathy, and respect for all, regardless of nationality, race, or religion. It encourages cooperation, understanding, and peaceful coexistence among nations.

India's perspective on global solidarity and *Vasudhaiva Kutumbakam* is deeply intertwined and forms a core tenet of its foreign policy and engagement with the world. India actively promotes South-South cooperation, recognizing the shared challenges and aspirations of developing countries. It champions initiatives that empower developing nations and enhance their capacities. It has consistently extended humanitarian assistance and disaster relief to countries in need, reflecting its commitment to global solidarity and compassion.

Solidarity in the Indian subcontinent encompasses various dimensions, including political, economic, social, and cultural cooperation among the countries in the region. It reflects the region's complexity and the diverse ways in which communities come together to address common challenges and aspirations. In terms of political solidarity, regional organizations like SAARC (South Asian Association for Regional Cooperation), established in 1985, promote regional integration and cooperation among member countries such as Afghanistan, Bangladesh, Bhutan, India, Maldives, Nepal, Pakistan, and Sri Lanka. SAARC aims to foster mutual understanding, good neighborly relations, and collaboration in various fields. Diplomatic engagement through bilateral and multilateral dialogues, such as India-Pakistan bilateral talks, is essential for resolving conflicts, building trust, and maintaining peace and stability in the region.

Economic solidarity is facilitated through trade and investment initiatives like SAFTA (South Asian Free Trade Area), which aims to reduce trade barriers and promote intra-regional trade. For example, the reduction of tariffs on goods traded between India and Bangladesh enhances market access and economic growth. Additionally, the BCIM (Bangladesh-China-India-Myanmar) Economic Corridor seeks to improve connectivity and economic cooperation among the involved countries, boosting trade and development through infrastructure projects like highways and railways. Regional projects such as the development of regional transportation networks, exemplified by the Trans-Asian Railway network, facilitate the smoother movement of goods and people. Joint energy projects, like hydroelectric power plants in Bhutan supplying energy to India, address the region's energy needs effectively.

Social solidarity is evident in collaborative efforts in health and education. Cross-border healthcare initiatives, such as joint efforts to combat pandemics like COVID-19 and improving healthcare infrastructure, are crucial. Educational exchange programs, including scholarships for South Asian students to study in India, foster mutual understanding and knowledge sharing. Additionally, joint efforts in poverty reduction, exemplified by SAARC Development Fund projects, and coordinated disaster management and relief efforts, such as the joint response to earthquakes in Nepal, improve the quality of life and resilience in the region.

Cultural solidarity is promoted through cultural exchanges and the preservation of shared heritage. Cultural festivals, art exhibitions, and sporting events like the SAARC Film Festival promote cross-cultural understanding and strengthen bonds among South Asian countries. Joint efforts to preserve and promote the region's rich cultural heritage, such as the restoration of ancient sites like Nalanda University, enhance a sense of common identity and pride. People-to-people connections are fostered through tourism initiatives, such as Buddhist circuit tourism in Nepal and India, which enhance mutual understanding and provide economic benefits. Engaging the South Asian diaspora through events like India's Pravasi Bharatiya Divas promotes the region's interests globally.

Security cooperation is vital for regional stability, involving counterterrorism and border security efforts. Collaborative intelligence-sharing mechanisms, such as India-Bangladesh intelligence cooperation on terrorism, and joint border management efforts, like India-Nepal border security coordination, help reduce cross-border crimes and enhance regional security. These multifaceted solidarity efforts illustrate the diverse and interconnected ways in which the countries of the Indian subcontinent collaborate to address shared challenges and promote mutual growth and stability.

While the potential benefits of solidarity in South Asia are significant, several challenges need to be addressed. Political tensions, such as the historical and ongoing conflicts between India and Pakistan, can hinder regional cooperation. Economic disparities among South Asian countries also pose challenges to equitable collaboration, as varying levels of development can create imbalances in cooperation efforts. Additionally, the region's cultural and linguistic diversity, although a strength, can create barriers to mutual understanding and solidarity.

Despite these challenges, fostering solidarity in the Indian subcontinent is crucial for the region's peace, stability, and development. By promoting political, economic, social, cultural, and security cooperation, South Asian countries can address common challenges more effectively and work towards a prosperous and harmonious future.

Introduction to the Parts in the Book

This book delves into the application of the solidarity approach through diverse cases from the Indian subcontinent, showcasing how different forms of solidarity can foster inclusive development and resilience. This book underscores the diverse ways in which the solidarity approach can be applied to address various challenges in the Indian subcontinent. By fostering unity, cooperation, and mutual support, solidarity has the potential to drive inclusive development, resilience, and social justice. Through these examples, this book aims to inspire further exploration and application of solidarity in geography and beyond, contributing to a more equitable and sustainable world. The book is arranged into eight parts. The first part is Introduction, which consists of Chapter 1, Geography and Solidarity. What geography should (really) be... by Elena dell' Agnese. In this chapter, the author mentions that the importance of geography's military role is well known. In times of war, when knowledge of terrain becomes crucial, geography courses multiply and academic geographers themselves often become "spies, teachers, and occupiers" (Barnes, 2016). Once the war is over, geographical knowledge is or can be used in diplomacy and civil affairs for the reorganization of the country. But geography can play a much more important role in educating citizens for peace. It can teach people to see the other as human beings like themselves, it can help them to put themselves in the other's shoes, and it can teach solidarity. Moreover, the tools of geographical analysis can help to understand how solidarity can be exercised in a system that is not necessarily limited to the local, but which, precisely because place is a complex system of relations (Massey 2004), extends or can extend beyond the nation-state in the construction of transnational networks.

The second part is titled as Globalization, Regional Solidarity, Urbanization, and Economic Development. The era of "globalization" conveys the notion of an interconnected world. Globalization refers to the phenomenon when people from around the globe march towards a uniform goal. Hypothetically, a globalized harmonious world can be developed through regional solidarity, urbanization, and economic development. Regional solidarity goals of joint socio-economic development and international solidarity can be achieved by international assistance and cooperation, sustainability in international relations, the peaceful coexistence of all members of the international community, equal partnerships, and the equitable sharing of benefits and burdens. International solidarity also refrains from harming or posing obstacles to the greater well-being of others, which includes the global economic system and our typical ecological habitat. This part is focused on all the above aspects, how the solidarity approach is the answer for a more just and equitable globalized world, and the challenges to achieving the goals, as many research studies, argue globalization as a world economic system is more favorable for capitalism.

Globalization, regional solidarity, urbanization, and economic development are all deeply interconnected forces that have profoundly shaped the Indian subcontinent trajectory in recent decades. It has been a major catalyst for economic growth and development. Opening up to international trade, investment, and technology flows has led to increased productivity, expanded markets, and greater integration into global value chains. This has created millions of jobs, boosted exports, and raised living standards for many.

However, it has to be noted that globalization and regional solidarity are interconnected and can be seen as opposing forces as regional solidarity focuses on balanced regional inclusive development. The relationship is dynamic and evolving, and the outcome depends on various factors, including economic policies, cultural attitudes, and political will. While globalization can pose challenges to regional solidarity, it also creates opportunities for regions to cooperate and strengthen their bonds based on shared interests and values. Regional cooperation initiatives play a vital role in promoting equitable growth, addressing regional disparities, and fostering a sense of national unity. This includes collaboration on infrastructure projects, sharing best practices in governance, and promoting inter-regional trade and investment. By working together, states can leverage their strengths, mitigate weaknesses, and ensure that the benefits of globalization are more evenly distributed across the Indian subcontinent.

Regional solidarity and globalization are interconnected forces that shape contemporary societies in complex ways and have an impact on urbanization. However, it has to be noted that globalization, regional solidarity, urbanization, and economic development have challenges. Increased competition has displaced some domestic industries, while uneven growth has exacerbated existing inequalities. There are concerns about environmental sustainability, cultural homogenization, and the vulnerability of the Indian economy to external shocks. These forces are not isolated phenomena but interconnected forces that shape the Indian subcontinent's present and future. By navigating these forces strategically, India can harness the benefits of globalization, promote regional balance, manage urbanization sustainably, and achieve its goal of inclusive and equitable economic development.

This part includes five chapters. It begins with Chap. 2 titled Balancing Act: Role of India in Geopolitics of Global South. This chapter explores the complex dynamics of India's involvement in the Geopolitics of the Global South, emphasizing its diverse contributions, obstacles, and strategic necessities. Chapter 3 is Unlocking Potential: Transboundary Rivers as Drivers of Regional Solidarity in South Asia investigates the potential for using water as an instrument to strengthen solidarity rather than creating conflict in South Asia. Chapter 4 Application of Solidarity Approach to Space Identity Crisis: Issues of Muslim Pockets in Kolkata is an attempt to study the mental maps as perceived by the residents of Kolkata city in India away from Muslim localities to identify and analyze the factors responsible for creating a false identity of these spaces, to compare and contrast the perceived scenario of these spaces with other parts of Kolkata, and finally to apply the solidarity approach to deal with these misconceptions. Chapter 5 Lost in the City: A Study of Homelessness in Chandigarh, India investigates the social, economic, and administrative challenges faced by the homeless population in the planned city of Chandigarh, one of the most progressive and capital cities of India. The chapter explored the care dimension of the solidarity approach to provide long-term and sustainable solutions to the problem faced by the homeless population there. Chapter

6 Sustaining Solidarity: Dharavi's Socio-Economic Landscape in *Rediscovering Dharavi* by Kalpana Sharma. This chapter deals with the pertinent issue of the urban poor in the Indian Subcontinent, with Dharavi as the primary focus. Dharavi, one of the largest slums of Asia in Mumbai, is home to more than a million people and is characterized by social diversity and economic vitality. This chapter, by using the application of the Solidarity Approach within the framework of Human Geography, analyzes the subculture of Solidarity and Resistance developed in Dharavi. A crucial aspect of the discussion is to explore literary text as an alternative form of modeling human cohesion through Kalpana Sharma's *Rediscovering Dharavi* where the slum dwellers' uphill battle in building their identities is brought to life through a series of spellbinding stories. The focus is on concrete solidarity performances manifested through a sense of collective responsibility and agency.

The third part is about the Sustainable Development of Rural Societies and the Indigenous Knowledge Base. Ensuring no one is left behind in the developmental planning and implementation has captured the world's attention and gravitated towards the Sustainable Developmental Goals (SDG) and Targets. The focus on the people-centric approach and principles of accountability, transparency, and participation ensures SDGs are a near-perfect holistic development model. SDG implementation talks about the inclusion and participation of the last person to implement this global development program successfully. However, the biggest challenge is to guarantee the development measures reach these marginalized rural communities and Indigenous people. The discrimination based on work and descent (DWD) failed to cut SDGs and Targets. The SDGs indirectly influence the various concerns of the excluded communities. At the same time, it could not recognize the DWD communities or caste or ethnicity as an essential determinant of poverty and underdevelopment. Increasing inequality and no specific attempt to reduce it results in political and economic instability within the communities. SDG is based on the solidarity approach, which upholds the principle of good governance, transparency, accountability, and participation in implementing the goals and targets. This part looks into the development of rural societies and creative communities through the lens of SDG's principle of solidarity, "Leaving no one behind." This part includes seven chapters that cover various dimensions of rural life in the Indian subcontinent. Chapter 7 is Socio-economic Development Through Ecological and Social Solidarity: A Case of NABARD Springshed-based participatory watershed development program in the North East Region. This chapter focuses on establishing a participatory and collaborative model with institutional support to achieve ecologically balanced sustainable growth, thereby eradicating poverty by creating a robust multidimensional structural framework. The Springshed-based Watershed Development Program implemented by NABARD (National Bank of Agriculture and Rural Development) in the North East region of the country would further promote biodiversity conservation and accelerate the socio-economic growth process. The community's involvement in the development, monitoring, and maintenance of Springshed is paramount for its effective implementation which develops solidarity and supports the solidarity economy in the region. Chapter 8 is Development Under the Aegis of MGNREGA: A Case Study of Riyali Panchayat of District Kangra,

Himachal Pradesh. Rural societies in India have their own social and economic problems. Growing population, lack of employment opportunities, underdeveloped health, and transportation infrastructure, and declining water resources are to name a few. To address the majority of these issues, the Government of India implemented the Mahatma Gandhi National Rural Employment Guarantee Act (MGNREGA). This chapter tries to examine the impact of MGNREGA at a minute scale so that problems, strengths, and changes can be understood at the root level. Also, the study attempts to understand how the scheme is further enhancing the solidarity of people belonging to different religious and cultural backgrounds as they work in harmony under the MGNREGA scheme. Chapter 9 is Enhancing Indigenous Knowledge through Community Solidarity and Sustainable Development in Rural Society. This chapter describes through community solidarity indigenous knowledge can be enhanced which further facilitates sustainable development in rural communities. It stresses the importance of sustainable management of natural resources. Chapter 10 is Food Insecurity and Emotional Experiences in the Tribal Community of Vagad region, Rajasthan, India. This chapter highlights human attachment to food, nutrition, and diet. Food nurtures emotions and emotions nurtures food. Food insecure households generally hold negative emotions like sadness, anger, demoralization, etc. This chapter analyzes the human emotions linked to food insecurity. Food insecurity further led to distress and conflict in society. The chapter is based on extensive fieldwork done in the Vagad region of Rajasthan by the authors. Chapter 11 is Corresponding Mental Health and *Prakriti* in Indian *Ayurvedic* Literature. Life experiences and individual factors govern the mental health of every single person differently. In continuation of individuality, *Ayurveda* has a principal notion of *Prakriti* that expounds the unique peculiarity of a person. The manuscript shows how Mental Health and *Prakriti* correlate with each other. The chapter also proposed that society should emphasize achieving social and environmental solidarity. A positive environment with social warmth will help in reducing mental health distress. Chapter 12 is Towards Rural Social Sustainability: An Index to Assess the Sustenance Potentialities. This study presents a Social Sustainability Index (SSI) that helps to assess the potentiality of present-day Indian villages to sustain future challenges while identifying the existing development gaps. This chapter examines the solidarity approach as a potential alignment with social sustainability in Indian villages to build self-reliant and locally managed communities. Sustainable villages should instill a sense of social acceptability, well-being, and security among rural communities while ensuring equity in resource allocation and distribution. To create a more equal and balanced future for India's rural landscapes, it is essential to embrace social sustainability through the lens of solidarity.

Part fourth of the book is Geopolitics, International Diplomacy, and Political Movements within Nation-States. Geopolitics refers to studying the geographical factors in world politics and inter-state relations. It emphasizes explicitly regional strategic links, location, resources, and accessibility. The impact of the COVID-19 pandemic has been devastating for the entire world and humanity. Developed countries, too, struggled to mitigate and manage the welfare and safety of their people. COVID-19 is a crisis for the world's balanced development. To tackle this unique complex health crisis, various forms of solidarity approaches have been incorporated into public policy, international diplomacy, and political movements within the states. Solidarity bears two different meanings for domestic and external use. In contrast, the former focuses on resilience-building; the latter is a tool for standard action in the form of welfare and safety. Geopolitics, international diplomacy, political movements within nation-states, and global solidarity are interconnected forces that shape the contemporary world order. By understanding their complex relationships, we can better navigate the challenges and opportunities of the twenty-first century and build a more just, equitable, and sustainable global community.

This part describes international solidarity, underlying the right to development and the key to its realization, the impetus for collective responses to interconnected challenges in an interdependent world, and the shared responsibilities. This part also looks into international commitments and organizations and the workings of a broad range of stakeholders and the global civil society in reiterating the significance of international solidarity for the expected future. Effective diplomacy can build trust among nations, creating a foundation for cooperation on global issues and strengthening solidarity. This part includes four chapters.

Chapter 13 is India's Quest for Geopolitical Gains Through Metamorphosing NAM Solidarity into Global South Solidarity. This chapter identifies the specific factors and components that have led to the transformation of NAM (Non-Aligned Movement) solidarity into Global South Solidarity. This study aims to elucidate the current realignment of India and solidarity with struggling nations, focusing particularly on its competition with China for influence in the Indian Ocean Region and Indian subcontinent. The chapter also traces India's historical geopolitical goals and its evolution to its current status through the solidarity approach. This research essentially investigates India's efforts to revive solidarity within the NAM through a fresh approach known as Global South Solidarity and explores the strategies and mechanisms employed by India in pursuit of these goals.

Chapter 14 of this part is Strategic Location of Siliguri Corridor and Its Issues. The chapter discusses that the Siliguri Corridor, which links India with the Northeastern States, is a strategic region in the Himalayan foothills. It is often referred to as Northeast India's (NE) entrance. It establishes a land connection between India and its three neighbors: Nepal, Bhutan, and Bangladesh. The corridor significantly contributes to India's progress. On the one hand, it can create opportunities due to its strategic, geoeconomics, and geopolitical significance in the region and through its connectivity enhance regional solidarity. On the other hand, it can create challenges due to geographical proximity to neighboring countries. That is why the main focus of the present study is on the strategic location of the corridor and its issues. Chapter 15 is Enhancing Regional Solidarity Through HADR: A Case Study of Indo-Nepal Disaster Response. This chapter is an attempt to evaluate the distinctive yet intertwined dimensions of disaster diplomacy and also help to understand the contribution of India to the improvement of regional solidarity through humanitarian assistance. In particular, this study examines the focal point of India's disaster response policy in the context of HADR at the time of the Earthquake and Flood incidences in Nepal. Chapter 16 is about Sikhism, Ad-Dharm

Movement, and Ravidasi Identity in the Doaba region of Punjab. This study focuses on the process of change of the "Ravidasi" community in the Doaba province of Punjab with the influence of the Ad-Dharm movement. In this chapter, the author looks at the growth of caste distinctions among Sikhs in many different processes. While the Sikh religious scriptures thoroughly reject all types of discrimination in society, the practice of untouchability exists among Sikhs.

The Ad-Dharm movement and the rise of caste consciousness in East Punjab and the Doaba region during the 1920s, a well-known movement called Ad-dharma started by Mangoo Ram was successful in mobilizing the large majority of Chamars from (Indian Punjab districts named Jalandhar, Hoshiarpur, Kapurthala, and Shaheed Bhagat Singh Nagar, and develop solidarity to achieve inclusion and equal social status.

Part fifth is Climate Change, Sustainable Development, and International Solidarity. International solidarity is the need of the world today. We need a framework of ethics and an ethically informed practice today to fight climate change. Recent climate change is a human-induced phenomenon of the world that is deeply divided by policies for the developed and developing world. The impact and effect of climate change further intensify and multiply divisions and vulnerabilities that already exist. Subnational levels of marginalized social groups and communities in the poorer nations are much more vulnerable to the risks of climate change. These developing nations also find it challenging to adopt measures for climate change, as they have limited means and resources to do so. Developing countries are thus dependent on assistance for adaptation of the developed nation. In the context of climate change, sustainable development, and international solidarity, it is a call for unity, identification, compassion, support, and assistance. This solidarity principle is for wider target audiences in the fight against climate change, which results in extreme weather events (floods, hurricanes, droughts, forest fires).

The Indian subcontinent is highly vulnerable to climate change. Rising temperatures, erratic rainfall patterns, extreme weather events (like floods, droughts, and cyclones), and sea-level rise pose significant threats to agriculture, water resources, infrastructure, livelihoods, and human health. These impacts disproportionately affect the most vulnerable populations, exacerbating existing inequalities. Given the region's vulnerability, sustainable development is not merely a choice but an imperative.

Adopting practices and policies that promote economic growth, social well-being, and environmental protection in a balanced and integrated manner is essential for sustainable development. Key areas of focus include clean energy transition, climate-resilient agriculture, water resource management, sustainable urbanization, and ecosystem conservation. Shifting away from fossil fuels towards renewable energy sources like solar and wind power is crucial for reducing greenhouse gas emissions and mitigating climate change. Developing and implementing agricultural practices that can withstand climate variability and extreme events is essential for food security and livelihoods. Implementing efficient irrigation techniques, rainwater harvesting, and wastewater treatment can help address water scarcity and ensure equitable access to this vital resource. Designing energy-efficient cities, promoting public transportation, and incorporating green spaces can help reduce emissions and enhance resilience. Protecting forests, wetlands, and other natural habitats can help sequester carbon, regulate water cycles, and provide livelihoods for local communities.

Addressing climate change and achieving sustainable development in the Indian subcontinent requires international solidarity. This involves providing financial and technological support, as developed countries have a historical responsibility for climate change due to their past emissions. They need to assist developing countries in the region in transitioning to low-carbon economies and adapting to climate impacts. Additionally, knowledge sharing and capacity building through collaboration on research, technology transfer, and training can enhance the region's ability to tackle climate change and implement sustainable solutions. Regional cooperation is also vital, as countries in the Indian subcontinent can benefit from sharing knowledge, resources, and best practices in areas like disaster risk reduction, early warning systems, and climate-smart agriculture.

This part of the book looks into the practice of humanitarian aid that works under such extreme conditions, and their mobility within states or internationally. Also, the part will look into the broader aspect of human solidarity, the interdependence of humans on one another, and ecosystems for survival and flourishing. This part includes three chapters. Chapter 17 of this part is International Solidarity to Address Climate Change Through Conservation of Natural Resources: Problems, Prospects, and Indian Perspective. This chapter focuses on the importance of global unity in the fight against climate change through resource conservation, identifying the challenges, opportunities, and viewpoints from India. Due to the global nature of climate change, resource conservation is necessary to mitigate its effects and strengthen resilience. Nevertheless, economic differences, competing interests, and governance problems frequently obstruct the road to conservation. The chapter underlines the importance of international solidarity in promoting sustainable resource management and transformation to a conservation-based model. Chapter 18 is Climate Change Impacts Mitigation, and Adaptation through Solidarity Strategies in the Indian Subcontinent. Climate change denotes alterations in temperature, precipitation, and weather patterns that occur over a large time and space. It is mainly caused by the emission of greenhouse gases (GHGs), especially carbon dioxide (CO_2), into the troposphere due to anthropogenic activities. It has significant impacts on our planet, including increasing frequency of disasters, loss of ecosystem services and biodiversity, infestation of pests and diseases, degradation of natural resources, economic impacts, water scarcity, human health, etc. These impacts serve as a hurdle towards attaining the Sustainable Development Goals (UN SDGs) set by the United Nations. In this chapter, the authors discuss various solitary approaches adopted in the Indian subcontinent to achieve the objective of sustainable development. Chapter 19 of this part is Climate Change, Urban Challenges, and Collaborative Solutions: Addressing Heat Islands, Urbanization, and Water Crises in South Asian Countries. South Asian countries face significant urban challenges as they experience rapid urbanization, including urban heat islands, infrastructure inadequacies, and water crises. This chapter examines these interconnected issues across the diverse climatic and geographical landscape of South Asia, encompassing countries like India,

Pakistan, Bangladesh, and Nepal. South Asian countries have implemented various policies and collaborative initiatives through organizations like the South Asian Association for Regional Cooperation (SAARC). These efforts focus on sustainable urban development, including green building codes, urban water management, and climate resilience strategies. This chapter concludes that while progress has been made, continued regional cooperation and knowledge sharing are crucial for developing effective, locally appropriate solutions. It emphasizes the need for a solidarity approach to tackle the complex urban challenges faced by South Asian countries, allowing for more efficient resource use and fostering innovation in sustainable urban development.

Part six of the book is Geographies of Care, Responsibility, and Social Mix. These concepts explore how social relationships, spatial patterns, and power dynamics shape the distribution of resources, access to services, and overall well-being. Women disproportionately bear the burden of care work in the Indian subcontinent, including childcare, eldercare, and household chores. This unpaid labor is often undervalued and unrecognized, limiting women's opportunities for education, employment, and social mobility. The availability of affordable and accessible childcare, healthcare, and eldercare services is crucial for supporting caregivers and promoting social well-being. However, such infrastructure is often lacking, particularly in rural areas and marginalized communities. The Indian subcontinent is characterized by significant spatial segregation along lines of caste, class, religion, and ethnicity. This segregation can perpetuate inequality, limit social mobility, and exacerbate social tensions.

Various international summits and declarations aim for a common goal of "a society for all." Similarly, SDG 16 also envisions a world free of inequality and aims for the common goal of "leaving no one behind." This inclusive approach and concept emerged from the geographies of care, responsibilities, and social mix. Geographies of care and human geography recently grabbed the attention of many researchers. The issue of social inclusion not only aims to eradicate poverty, but also aims to achieve universal education, gender equality, reducing child mortality, and improving maternal health, etc. There is an urgent need to pay closer attention to the issue of social inclusion of the marginalized and vulnerable section of society, who continue to be excluded from the progress made so far. The successful implementation of inclusive care policies depends on the inter-relationships between various players.

This part examines and analyzes the inter-relationships, illustrated by various policymakers and other stakeholders of the society, and how voluntary support is affected by differing local contexts of care that impact the marginalized community.

By recognizing the interconnectedness of care, responsibility, and social mix, we can create more inclusive, equitable, and sustainable societies in the Indian subcontinent. This part includes six chapters. Chapter 20 is Solidarity Through Sports: A Geographical Reading. The geographical concepts of space, scale, and territory are intertwined with sports. Sport serves as a platform for cultural expression for individuals, regions, and the nation-state. The purpose of this chapter is to expand the dialogue on the solidarity approach in human geography through sports. The chapter focuses on analyzing the content of sporting events reported in print and social media to substantiate the connections between solidarity, space, and sporting culture. The chapter claims that sportsmanship and international cooperation embody the essence of solidarity.

Chapter 21 is A Study of Socio-cultural Aspects of Tribal Existence in Association with the Sense of Care and Responsibility. This study is an effort to trace how geography in the current pervasive nature of the unemployment prevailing in society and the impact of the same on poverty, inner city decline, and environmental destruction while strengthening the solidarity of people, places, and care ethics can create new relations, institutions, and integrated actions to promote well-being.

Chapter 22 of this part is the Role of Social Networks to Increase Socio-economic Solidarity for Transformative Participatory Development in the Mewat Region, India. This study interrogates the unexplored role of social networks in facilitating transformative participatory development outcomes in the Mewat region of India. Despite its proximity to Gurugram and its location within the National Capital Territory of Delhi, the area is still experiencing significant levels of underdevelopment. Identifying research gaps and underpinning the conceptual framework from existing literature, this research focuses on assessing how "community" and households at the unit level promote participatory development through a solidarity approach using social networks.

Chapter 23 is- Where and in Which Aspect to Care? Insights from Spatial Variation and Levels of Human Development in Bardhaman District of West Bengal. This chapter is an attempt to evaluate HDI at the intra-district level of the then-undivided Bardhaman district of West Bengal. Results point out that the eastern part of the district has a better outcome of human development. The analysis also states that the HDI tends to be better in those parts of the district close to Purba Bardhaman's headquarters-Burdwan. Solidarity within the geographical scholarship is related to forms of a shared connection to address varied forms of inequality with a strong feeling of shared territory, which in this case is at the CD block level representing the stark inequality in the HDI between the eastern and western part of the district. Therefore, concentrating and caring for the developmental outcomes in the CD blocks with low HDI should be a primary step in the policy response.

Chapter 24 is Unpaid Domestic and Caregiving Duties in India and Conflict in Solidarity of Society: An Assessment Through SDG 5.4. This chapter discusses that the successful functioning of a society is a function of solidarity of roles performed by the individuals in a society. Solidarity is necessary but solidarity with the compromise of independence creates a conflict in solidarity. This chapter studies how the solidarity of gender roles and conventions has created a gender disparity in society. The chapter examines SDG 5.4, which illustrates the gender differences in time spent on unpaid domestic and care labor, using the Indian Time Use Data 2019. Research finding indicates that, in India, men typically devote less than an hour a day to unpaid domestic and caregiving duties, whereas women typically devote over 6 h. Based on the findings, public policy recommendations are made about the significance of time-use surveys and how solidarity in society can be maintained without hampering individual growth, social status, and economic opportunities for women.

Part seven of this book is Hazard, Disaster, and Integrated Crisis Management. Rapid shifts in climate patterns have marginalized the livelihoods of vulnerable indigenous populations globally, many of whom are predominantly smallholder subsistence farmers relying on rain-fed agriculture. Hazard, disaster, and integrated crisis management are crucial for fostering solidarity and resilience in the Indian subcontinent, as the region is highly prone to various natural and human-induced hazards. Effective management of these crises can significantly reduce their impact and promote a sense of shared responsibility and collective action. The Indigenous people, the vulnerable communities, are geographically located in environmentally fragile regions exposed to temperature and rainfall variability and a variety of climate risks.

Various research and reports indicate that disasters hit the vulnerable sections of society the most, such as women, a person with disabilities, and groups that are socially excluded and discriminated against based on their identity emanating from caste, religion, or ethnicity. This is also evident from the various case studies observed in different parts of the world that most of the vulnerable communities live on the outskirts of the village, in low land, with poor habitation facilities, which makes them vulnerable in the first stage. Poor drainage, inaccessible drinking and safe water, poverty, and malnutrition make them prone to suffering the outbreak of diseases in times of disaster. Lack of information and early warning issued by state machinery does not reach vulnerable communities, as most live outside the mainland, as hamlets, as ghettos. Discriminatory evacuation practices, death, and loss of infants during floods are unrecognized, as most vulnerable communities' birth registration is not maintained. During a crisis, the degree of solidarity is the basis of the endurance of any society. Social solidarity can act as a cementing force in reducing the social distance that arises during any disaster. This part elaborates that solidarity plays a pivotal role in bridging the social distance by developing a collective consciousness in society during any disaster.

By prioritizing hazard, disaster, and integrated crisis management, and fostering a spirit of solidarity, the Indian subcontinent can build a more resilient and sustainable future for its people. This part consists of four chapters Chap. 25 is Creative Leadership and Solidarity in the Management of the COVID-Related Lockdown Situation in a Grama Panchayat. This chapter studies the COVID-19 resistance Activities of a grama panchayat in Kerala during the lockdown period of COVID-19. The focus is on the creative and democratic leadership of the Panchayat President who succeeded in gaining vertical and horizontal support emanating from a strong foundational history of solidarity among the citizens in the area under his jurisdiction. Sensitivity to the needs of the vulnerable, acknowledging and utilizing the talents and resources of the citizens to fulfill the needs, the planning and synergistic coordination of activities to confront a crisis of unprecedented nature offer lessons for leadership across situations. Chapter 26 of this part is Use of ICT for Disaster Relief Distribution Across Hostile Terrain. This chapter discusses how the use of ICT can foster solidarity with disaster-affected people. The basic premise of any

Disaster Risk Management plan is the provision of rescue and essentials despite scarcity or even the absence of known means of conventional transportation and scarcity of communication resources/channels. It becomes very difficult for communities to come together in solidarity with the affected people due to a lack of coordination in adverse circumstances and the ICT has been a savior to this cause, bringing rescue and relief teams closer to each other over technological advancements. The relief distribution applications being discussed in this study are contextually relevant to the underlying assumption that most people in the rescue and relief teams shall be equipped with a basic smartphone running Android, and there is a provision for Unmanned Aerial Vehicles (UAVs), popularly known as drones, capable of delivering small loads over short distances. Chapter 27 is from Adversity to Unity: Women, Pandemic and Social Solidarity. This study delves into the multifaceted challenges confronted by women during the pandemic and explores the role of social solidarity in empowering them to navigate these hardships. Drawing on existing studies and research, the research sheds light on critical issues such as education dropout, economic crises, direct and indirect health concerns (with a significant impact on maternal health), food scarcity, increased workload, and various forms of violence against women. Notably, violence against women emerged as a parallel pandemic, exacerbating the challenges faced by women during this tumultuous time. Chapter 28 in this part is Geography of Disaster, Coordination for Relief and Conflict Resolution: A Case of Kashmir Earthquake 2005. Disaster creates a context and circumstance in which the two estranged communities or nations sit together and coordinate. The chapter highlights the prospect of cooperation and peacebuilding after the earthquake occurred on either side of Kashmir on 8th October 2005. The chapter analyzes the modalities of relief and rehabilitation that allow two nations to sit together and resolve problems. It also discusses how the geography of disaster plays a vital role in bringing states together, coordinating, fostering solidarity, and thus finally leading to conflict resolution.

Part eight of the book is Environmental Solidarity and Natural Resource Management. Our relationship to the natural environment is one of the great issues of our age. The region's diverse ecosystems and rich natural resources are essential for the livelihoods and well-being of millions of people. However, these resources are under increasing pressure due to climate change, population growth, and unsustainable consumption patterns.

It is also one of the most controversial issues of the hour because of the driving forces like economy vs ecology, urbanization vs land use change, globalization vs local issues, dependency vs self-reliance, and development vs disasters. Environmental degradation has generated problems for poor environmental refugees in the present decade. Environmental solidarity suggests that human beings are related to each other and the natural world. It is bound to shape both our science and politics at present and in the future. The virtue of solidarity commits nations and individuals to work for the common good and to sacrifice for the sake of others.

The Indian subcontinent faces numerous challenges in achieving environmental solidarity and sustainable natural resource management, including Poverty and Inequality, Political Will and Governance, and Climate Change Impacts. The Indian subcontinent possesses abundant natural resources, including solar and wind energy, fertile land, vast biodiversity, and diverse ecosystems. These resources can be harnessed to promote sustainable development and improve livelihoods.

Limited resources and unequal power dynamics can hinder equitable access to natural resources and participation in decision-making processes. Weak governance and competing interests can pose obstacles to implementing effective environmental policies and regulations. However, there are also significant opportunities as the region has a wealth of traditional knowledge and practices related to natural resource management. Integrating this knowledge with modern science can lead to innovative and sustainable solutions. Therefore, *Solidarity* forms the basis of environmental ethics in the present environmentally degraded world, where all the countries are competing for their most developed stature. The environment is the global public good and is required for all living organisms. However, the capacity of the environment is limited and a shared matter for the whole of mankind is to conserve this limited global environment so that all people on earth (and coming generations) can enjoy a healthy life. The right to a healthy and sustainable environment is getting utmost importance in the present context of climate change concerns. With fast-changing geopolitical situations and ever-increasing conflicts, the conservation of natural resources has become a critical factor for building and maintaining peace, reducing poverty, maintaining and improving health, and fostering sustainable development. So, it is important to adopt environmental solidarity worldwide to address the problems of environmental pollution, temperature rise, changing weather patterns, depletion of the ozone layer, rise in sea level, outbreak of diseases, species extinction, and natural resource management.

There is growing recognition of the need for regional cooperation on environmental issues. So, this part aims to study, compare, and qualify the "environmental solidarity" and explore the several dimensions affecting uneven development vs environment debate as an Action Research Agenda, and how environmental solidarity is important for solving environmental problems and natural Resource Management. This last part of the book includes five chapters. Chapter 29 in this part is Commercialization of Water and Shift from Traditional to Modern Technologies: Evidence from Water Insecure Western Rajasthan. This chapter discusses that when in the arid regions of western Rajasthan, the traditional water harvesting systems managed by the local communities are being replaced either by public piped tap connections or by private water vendors. This process has brought about the commercialization of water where villagers are now paying to buy drinking water from private water tankers, hitherto unknown to this waterscape. The access to water technology brought a process of socio-cultural change where there is a gradual transformation of village solidarity-based water resource management and conservation systems to commercialized "pay as you use" water use systems. The study finds that tanker water technology is leading to intragender, intra-caste, and intervillage socioecological scarcity exuberated by the existing social hierarchy and economic disparities. However, in the context of climate change adaptation and growing awareness to foster community resilience to climate change impacts, there is a renewed interest in going back to the "traditional" and "local". The chapter concludes that it is time to reconceptualize a hybridized form of water system that harmonizes these two systems together. Chapter 30 is Prevention of Hydrological Disasters Cantered on Hydro-Solidarity: Estimation of Sediment Yield in Bhagirathi Basin and Prioritization. A basic objective of hydro-solidarity is the integrated, unified management of shared water resources. As water flows down a gradient across the landscape, the relationship between upstream and downstream entities and the connection between terrestrial (land-based) conditions and water resources must be considered for effective water management. These interlinkages required coordinated management efforts within watershed or river systems not only for the efficient use and conservation of the resources but also to avoid or resolve water-related conflicts and hydrological disasters. Assessment of sediment yield from reach and prioritization of watershed is one of the major concerns in mountainous regions for watershed management and prevention and mitigation of hydrological disaster. In this study, Soil and Water Assessment Tool (SWAT) model is operated to quantify the sediment yield from the Bhagirathi basin and prioritize the sub-basin of high risk. Chapter 31 in this part is the Ecology-Social and Solidarity Economy Model (ESSE) for Sandstone Mining in Mirzapur, Uttar Pradesh. The Ahraura town, municipal board, and Marihan tehsils of Mirzapur district are well known for sandstone mining. In a District Survey Report published by the Uttar Pradesh Geology and Mining Department in 2018, an area stretch of 674.37 hectares was used for mining activities to bring out sandstones in the financial year 2017–2018 Mirzapur. Though these quarries are crucial for the district as they provide employment and revenue these activities have destructed once dense forests in these two tehsils that were connected to the Chandraprabha Wildlife Sanctuary which is situated in the adjoining Chandauli district. Animals like panthers and bears were once found in these forests but mining dispersed once flourished wildlife from this area. The land is now a barren wasteland. Quarries in the Marihan and Ahraura tehsils of the Mirzapur district, where the sandstone is mined and delivered all over the place, have left its rural inhabitants with severe health issues and breathing disorders including asthma, silicosis, and tuberculosis. In this background, this chapter presents an alternative analytical approach by integrating social and environmental values in the solidarity economy model. For centuries, these ESSE (Ecology-Social and Solidarity Economy) principles have been imbibed in the life of rural India and need to be used to advance the sustainable development of the mining sector.

Chapter 32 is Enhancing Solidarity through a Comparative Analysis of NDVI and SAVI Vegetation Analysis: A Geospatial Approach for Sustainable Land Use and Environmental Management. The comparative analysis of the Normalized Difference Vegetation Index (NDVI) and Soil Adjusted Vegetation Index (SAVI) with the aid of geospatial technology has emerged as a powerful approach to increase ecological, social, economic, and agricultural solidarity while facilitating the development of environmental solidarity to improve the sustainability of land use practice. This study explores the application of geospatial technologies in the analyzing NDVI and SAVI data of LANDSAT series for the years 1975, 1979–80, 1989–90, 2000, 2010, and 2019–20 of pre- and post-monsoon to promote solidarity in vegetation change analysis for land use sustainability in Central Haryana. NDVI and SAVI models are applied to get the results with the help of ERDAS and ARC GIS software.

Chapter 33 of this part is Fostering Environmental Solidarity Through the Assessment of the Hydrogeomorphic Problems of the Shiddhi Hill Region for the Sustainable Development of Shirur town in Pune District, Maharashtra. Urbanization is a persistent global trend. Land use changes in India over the last ten years have converted into urban settlements due to population growth and metropolitan expansion. In the Siddhi Hill region of Shirur, development causes expansion of the urban area foot of the hill, groundwater level rise due to recharge to aquifer system is the effect of high rainfall resulting in a myriad of problems including basement water. Gravity water is a serious problem in the basement of the toe of the hill there, where with good planning and by using suitable engineering techniques this problem can be solved also by these techniques society property loss will be reduced and the basement will be better used for commercial purposes. Hence, fostering environmental solidarity and implementing comprehensive strategies, Shirur town can address the hydrogeomorphic problems of the Shiddhi hill region. Sustainable development efforts will protect the natural resources as well as also enhance the well-being and resilience of the local community.

At present issues of poverty, education, zero hunger, health and well-being, gender equality, clean water, affordable and clean energy, decent work and economic growth, industry, innovation and infrastructure, reduced inequality, sustainable cities and communities, climate actions, life below water, life on land, peace and justice, partnership to achieve goals, etc. have acquired central significance in the world today. These issues transcend the boundaries of sovereignty, given the transnational character of the problem itself. And therefore, they require transnational solutions. These are the global challenges and all the countries are targeting to solve these through achieving SDGs by 2030.

Solidarity can play a crucial role in the realization of the 2030 Agenda and the 17 SDGs and 169 targets by promoting inclusiveness of all the territories where no one is left behind. Attaining Sustainable Development Goals is the central challenge of our times.

Therefore, the book aims to explore and develop an understanding of the application of the solidarity approach in solving the problems of trans-regional characters which is the essence of the various existing branches of the discipline of geography.

Geographers critically examine the power dynamics inherent in research and knowledge production, challenge one-dimensional perspectives, and amplify marginalized voices. This involves acknowledging the diversity of knowledge systems and recognizing the agency of local communities. It also adopts a relational perspective that emphasizes the interconnectedness of people, places, and environments. This approach challenges individualistic and competitive frameworks and highlights the importance of cooperation and mutual support. It also incorporate care ethics into their analysis, recognizing the importance of care work, interdependence, and social responsibility. This approach emphasizes the ethical dimensions of geographical relationships and challenges the dominance of economic rationality. This involves analyzing the spatial dynamics of social movements, their strategies, and their impact on social change. Geographers can study social movements that challenge injustice and promote solidarity, such as environmental activism, labor movements, and anti-racist struggles. It also examine the experiences of migrants and diaspora communities, highlighting the importance of solidarity networks and transnational connections. This involves investigating the social, economic, and political factors that shape migration patterns and the experiences of migrants.

Hence, there is a huge scope for a solidarity approach in geographical research. It is a powerful concept that embodies unity and mutual support within and across communities, transcending individual interests for collective well-being. Studying geographical aspects through the solidarity approach gives a new meaning to places and spaces. Likewise, solidarity is deeply rooted in our day-to-day places and spaces such as workplaces, occupational or wider class identities, and divisions across other social cleavages, such as gender, sexuality, and race, are crucial in elaborating a more nuanced understanding of solidarity. In the case of mobility, solidarity plays a major role in building relationships between particular individuals and groups with locals in foreign lands. This book is an attempt to compile thoughts on the Application of the Solidarity Approach in different domains of human and physical geography and its subfields. It is an effort to deepen the understanding of changing ideas and innovative methods of the Solidarity Approach in Geography.

Shaheed Bhagat Singh College, Kavita Arora
University of Delhi
New Delhi, India

Centre for African Studies, Rashmi Rani Anand
Jawaharlal Nehru University
New Delhi, India

Editors and Contributors

About the Editors

Kavita Arora is currently working as a Professor at the Department of Geography, Shaheed Bhagat Singh College, University of Delhi, India. She has more than 15 years of experience in teaching and research. She earned her Ph.D. in Political Geography from Jawaharlal Nehru University, New Delhi. She has published her research in many national and international Journals. Her areas of interest are political geography, regional planning, development, environmental discourse, disaster management, and Indigenous peoples. She was awarded a postdoctoral fellowship by the Indian Council of Social Science Research, New Delhi. She has also developed content and recorded various modules for ePG Pathshala and Massive Open Online Courses (MOOCs) under the Ministry of Human Resource Development. She has authored and translated various modules on geography, economics, and sociology for the Indira Gandhi National Open University, New Delhi. She has authored a book *Indigenous Forest Management in the Andaman and Nicobar Islands, India*, published by Springer and co-authored a book *Urban Green Space, Health Economics and Air Pollution in Delhi*, published by Routledge India.

Rashmi Rani Anand is an assistant professor at the School of International Studies, JNU, New Delhi. Before this, she also served in the Department of Geography at Shaheed Bhagat Singh College, University of Delhi for 10 years. Her areas of interest are African Studies, Geospatial Technologies, Political Geography, and Developmental Issues. She holds an M.A. from the Department of Geography, Delhi School of Economics, University of Delhi, and an M.Phil. and Ph.D. degrees from the Centre for African Studies, School of International Studies, JNU. She is a trained geographer and has published widely in international and national journals. She has also created content and recorded modules for e-PG Pathshala and Massive Open Online Courses (MOOC) under the Ministry of Human Resource Development.

About the Contributors

Arnav Anshuman is a Ph.D. Research Scholar at the Department of Geography, Delhi School of Economics, University of Delhi. He is working on the topic "Sports and Space: Exploring the Spatialities of Cricket in Delhi." He looks to make meaningful contributions to the field of geography through his innovative and critical approach.

Anindita S. Chaudhuri is an Indian geographer specializing in sustainable development, GIS, and remote sensing applications. She earned her Ph.D. from Amity University, focusing on urbanization's impact on hydrological regimes. With over 6 years of academic experience, she has contributed to multiple research publications, book chapters, and conferences on environmental sustainability, urban heat islands, and groundwater management. She has served as an Assistant Professor at Indraprastha College for Women, University of Delhi.

Gursimran Cheema is a Student at Fraser Valley India (Chandigarh), University of Fraser Valley, British Columbia, Canada.

Seema Choudhary is currently working as an Assistant Professor in the Geography Department of Himachal Pradesh University, Summer Hill, Shimla. In 2010 and 2011, she earned a gold medal in the M.A. and M.Phil. programs. After passing the UGC JRF exam in 2014, she started working as an Assistant Professor at Himachal Pradesh University in 2016. She has supervised seven M.Phil. and four Ph.D. (ongoing) students and has a strong interest in both teaching and research. She has more than a dozen research publications and book chapters to her name that have been published in domestic and foreign magazines. She regularly participates in delivering talks at various venues and is a member of several academic organizations.

Nigamananda Das is a Senior Professor and Head of the Department of English at Nagaland University. Besides teaching, he shouldered several academic and administrative responsibilities as Warden of the Hostel, Head of the Department of English, Academic In-charge of the University Campus, and Dean of the School of Humanities and Education at Nagaland University, Kohima Campus, Meriema, Kohima, Nagaland, India. Commonwealth Literature, Saint Literature, Multiethnic Literature, ELT, and Contemporary Theory are his major areas of interest, he holds 31 years of teaching and 22 years of postdoctoral research experience. He has authored, edited, and co-edited 16 books on Contemporary Commonwealth Literature and Saint Literature. He has also published 80 research papers in national and international journals and anthologies and has so far completed three UGC-sponsored Minor Research projects and one ICSSR-sponsored Short-Term Empirical research project.

Moushila De is a distinguished environmental planner with over 5 years of experience, including teaching and research in urban and rural planning, disaster management, and GIS applications. She holds a gold medal in her Master's in Urban and Rural Planning from DCRUST, Murthal, and has pursued further studies with a PG Diploma in Disaster Management and a PG Certificate in Geoinformatics from IGNOU. Currently, she is completing her Ph.D. in Climate Responsive Planning for Arid Areas of Rajasthan at the School of Planning and Architecture, New Delhi. Moushila has served as a guest and visiting faculty at prominent institutions like Jamia Millia Islamia and SPA New Delhi, where she has taught and guided numerous students. Her research and professional work focus on sustainable development, climate change adaptation, and urban planning, making her a key contributor to innovative solutions in environmental sustainability and urban development.

Elena dell'Agnese has been the Director of the Centre for Visual Research at the Department of Sociology and Social Research of the University of Milano-Bicocca since 2008. She is also the former IGU Chair and Vice-President. At the same university, she also coordinates a Master's Program in Tourism and Development and teaches Political Geography, Communication Geography, and Geography of Development. She has been a Visiting Fellow at the Australian National University and the Universitas Indonesia, Jakarta. Currently, her research interests are mostly focused on popular geopolitics, critical geography, and gender issues (nation-building processes and regional iconographies in Europe and Asia; European borders and their role as identity markers; cities as symbolic landscapes and political places/spaces; and the political geography of masculinity). She is the author of a dozen books and approximately 100 chapters and articles in refereed journals. Among her most relevant publications are *Manuale di Geografia Politica* (1995, with Giacomo Corna-Pellegrini), the textbook which revived Italian political geography in the 1990s; then *Geografia e geopolitica dell'Oriente Estremo* (2001); *Geopolitiche dei Balcani* (2002, with Enrico Squarcina); *Geografia politica critica* (2005); *Mascolinità all'italiana* (2007, with Elisabetta Ruspini); and *Paesaggi ed eroi: Cinema, nazione, geopolitica* (2009). Elena dell'Agnese serves as a member of the editorial board of the book series "Spaces," Guerini, Milan, and co-director of the book series "Studi e ricerche sul territorio," Unicopli, Milan. She is also a member of the editorial board of GeoJournal, of the Bollettino della Società Geografica. Since 2008, she has been an official referee for Urban Studies.

Bratati Dey is a postdoctoral research fellow from ICSSR working on urban simulation modeling and holds a Ph.D. degree with a specialization in land cover change and prediction. Besides, Dey has five years of experience in teaching remote sensing and GIS at Swastik Edustart-Institute of GIS and Remote Sensing, Delhi. There are many research papers published by Dey on various topics ranging from resource mapping, LULC change detection and prediction, and hydrological analysis.

Teesta Dey is an Assistant Professor of Geography and has been working as the head of the department for the last 8 years at Kidderpore College, Kolkata. She stood first in Class I and received Gold Medals from the University of Calcutta twice, both in graduation and post-graduation. She received various international and national awards throughout her academic and professional life. Some of them are like Gold Medal from the University of Calcutta, the Guru Saday Dutta Memorial Prize, UGC Merit Scholarship, the Best Researcher Award in International Geographical Union, Jagiellonian University, Krakow, Poland, Young Researcher Award by Netaji Institute by Asian Studies, Young Scientist Awards by Centre for Environment and Economic Development (CEED) New Delhi, NITI Aayog, Govt. of India, etc. She has published more than 30 research papers and articles in various international and national journals. Her area of interest is various branches of human geography.

Meenakshi Dhote is a highly respected academician and practitioner in Environmental Planning, currently the Dean of Academics at the School of Planning and Architecture, New Delhi. With over 32 years of experience, she has made substantial contributions to the fields of biodiversity conservation in urban environments and hill area development. Dr. Dhote holds a Ph.D. in Urban Biodiversity and has led numerous major projects funded by national and international organizations, demonstrating her expertise in integrating ecological considerations into urban planning. Her work is widely recognized, with numerous publications in esteemed journals and presentations at prominent conferences. As a committed educator, she has mentored countless students and professionals, shaping the next generation of planners and environmentalists. Dr. Dhote's extensive experience and impactful work continue to influence the development of sustainable and resilient urban landscapes both in India and globally.

Bilal Faizal Faheem a student of business administration, is a driven and accomplished student known for his exceptional academic performance and active involvement in student activities. His passion for learning is evident in his top grades and enthusiasm for challenging coursework. In addition to his academic pursuits, he is deeply engaged in a variety of extracurricular activities. Currently, he is serving as the Vice-President of Students at the University of the Fraser Valley Student Union Society.

Ayushman Ganguly is a dedicated and curious student with a strong academic record and a profound commitment to social causes. His passion for understanding complex societal issues has driven him to excel in his studies, along with taking on projects focusing on social neglect and injustice. His interest in social causes extends beyond the classroom; he actively volunteers with local organizations and engages in community service, seeking to make a tangible impact. His goal is to channel his academic strengths and advocacy skills into a career where he can drive meaningful change and contribute to a more equitable society.

Bhoomi Gupta is a Professor in the Department of IT, MAIT, GGSIPU, Delhi. She has 21 years of experience in computer science education and is a dedicated professor of computer science. She holds a Ph.D. in computer engineering and multiple Microsoft MTA certifications that demonstrate her proficiency and knowledge in software development, database administration, and Windows server administration.

Sachin Gupta is a CSE professor at the Department of Computer Science, MVN University, Palwal, Haryana. His research tries to bridge the gap between theory and practice and to prepare employable graduates who can adapt to the evolving technological changes. He has also designed curriculum and managed faculty teams, ensuring that the academic standards and quality are maintained and enhanced. Teaching and technology are my fascinations and my mission, and I love to prove people wrong who stereotype CS faculty as possessing only theoretical knowledge.

Vipasa Gupta is currently pursuing a Ph.D. in Geography at the University of Kentucky and holds a Master's degree from Jawaharlal Nehru University, New Delhi. She completed her Bachelor's degree at Indraprastha College for Women, University of Delhi, where she received a research grant to study the social sustainability of villages. This chapter is part of her Bachelor's research conducted under that grant. As a social geographer, she explores social sustainability, community dynamics, and the complex interactions between people and their environments, aiming to contribute to innovative approaches for building more resilient and sustainable societies.

Tamali Halder is a Senior Research Fellow affiliated with the Centre for Regional Studies at the University of Hyderabad in India. As a research scholar, she is working on the educational dynamics in a mining region and its intersection with caste, class, gender, and residence. Apart from this, her interest lies in exploring the reasons behind regional disparities with specific attention to education and health, using a quantitative and ethnographic approach. She has two book reviews published in international peer-reviewed journals, one research article in a Scopus-indexed journal, and two research articles in UGC care list journals. She has been awarded the Best Presentation Award for presenting at the International Workshop on Geography and Sustainability 2023, organized by Beijing Normal University, China. She also has a profound knack for teaching geography. She has taught geography to secondary and senior secondary students at an ICSE school for 4 years. She has taught geography to UPSC aspirants as well. Her interaction with the students she tutored has motivated her toward the cause of education.

Atosü Hibo is an Assistant Professor in the Department of Commerce at Patkai Christian College (Autonomous), Nagaland. She is currently pursuing her doctoral studies at Nagaland University, Kohima Campus. Her research focuses on the institutional role of banks in the economic and rural development of Nagaland. Her areas of interest include microfinance, indigenous sustainable livelihood practices, and community development programs.

Prajakta Jadhav is a dedicated academic with a Ph.D. in Geography from the University of Mumbai, awarded in March 2018 for her research on land use planning in Purba Medinipur, West Bengal. She has extensive research experience, including a UGC-funded project on micro-level land use planning and involvement in an Emeritus project in Sindhudurg District. Dr. Jadhav has also contributed to corporate governance research at S.P. Jain Institute of Management and Research. With a solid teaching background spanning various esteemed colleges, she has guided 58 students in research projects focused on environmental issues. Her academic contributions include participation in 14 conferences and 7 publications in UGC-recognized journals. Dr. Jadhav's expertise extends to data handling, event management, and administrative activities, underpinned by strong computer skills.

Ravindra Gahininath Jaybhaye holds a distinguished academic background with an M.Sc. and Ph.D. qualifications. Currently serving as the Senior Professor, Department of Geography at Savitribai Phule Pune University, Prof. Jaybhaye brings with him a wealth of experience and expertise. Prof. Jaybhaye has a rich teaching experience, having spent 13 years at Nowrosjee Wadia College, Pune, and another 16 years at the Department of Geography, Savitribai Phule Pune University. His dedication to education has been recognized through awards, such as the "Best Teacher Award" from Maharashtra Bhoogol Parishad and the UGIT's Excellency Award 2023. In the realm of research, Prof. Jaybhaye's Ph.D. focused on "Environmental Management for Sustainable Development of Ecotourism in the Western Part of Pune district, Maharashtra." With 62 published research papers and more than 100 papers presented in international, national, and state-level conferences, Prof Jaybhaye has made significant contributions to geographical studies.

Atul M. Jethe is an Associate Professor at the Department of Geography, C.T. Bora College, Shirur District, Pune. He holds a Ph.D. in Applied Geomorphology from Savitribai Phule Pune University, Pune. He has 18 years of teaching experience at the UG level and 11 years at the PG level. Currently, he is a BOS member at Rajarshi Shahu (Autonomous) College, Latur, and working as a Ph.D. research guide at Savitribai Phule Pune University, Pune. His research interests are in fluvial geomorphology, watershed management and planning, hydrology, applied geomorphology, environmental geomorphology, remote sensing, and GIS. He has completed the NORMS-ISRO Certificate Course in Remote Sensing and GIS Technology. He has also undertaken postdoctoral research in geochemistry. Currently, he serves as the Coordinator of the IIRS-ISRO Outreach program at the college level. He has participated in various national and international seminars and conferences and has published 28 research papers in UGC-listed journals. He has authored three books, including entitled *Physical Geography* and *Geography of Maharashtra* (for SYBA and SYBSc classes). The Best Book Award 2021 was achieved for *Physical Geography*. He has been awarded the title of Paryavarndoot by the Shirur Municipal Corporation. Currently, he is an active member of the Indian Institute of Geomorphologists (IGI), Allahabad, the Institute of Indian Geographers (IIG),

Pune, International Association of Geomorphologists (IAG), Geological Society of India (GSI), and Maharashtra Bhugalshatra Parishad (MBP), Pune.

Arushi Jha has a Bachelor's degree in Life Sciences and a Master's degree in Environmental Sciences and qualified UGC-NET twice for eligibility of Assistant Professor in June 2019 and then again in December 2019 in Environmental Sciences. She completed a postgraduate diploma in remote sensing and GIS from Swastik Edustart-Institute of GIS and Remote Sensing, Delhi. She is currently pursuing a Ph.D. from Guru Gobind Singh Indraprastha University, Delhi. Her research work entails the impact of climate change on the hydrology of river basins and its future projections, hydrometeorological studies, and disaster risk assessment studies. She has worked with various hydrological models like SWAT and HEC-RAS.

Neena Joseph is a Professor at the Institute of Management in Government (Retd) Faculty, XLRI/LIPI. Her areas of interest include solidarity in management, general management, gender soft skills, and leadership. She has presented papers widely in national and international journals. She also conceived and implemented 45 training programs on "The Protection of Women from Domestic Violence Act, 2005."

Aparna Joshi is working as a Visiting Faculty in the Department of Geography at the University of the Fraser Valley, Fraser Valley, British Columbia, Canada. She was a university topper in the subject of geography in the Bachelor of Arts Program (2001) and also won the University Medal in Master of Science (Geography) in 2003. She holds a doctoral degree in geography, and her area of interest lies in regional development. She has more than 20 years of teaching experience in various colleges and universities. She has also been guiding and teaching aspirants for the prestigious Indian civil services examination for about 20 years now.

An active researcher, she has to her credit 16 articles published in various international and national journals, 6 book chapters, and 3 books (edited 2 and authored 1). She has also presented 17 papers/articles at various national and international conferences and seminars.

Nilotpal Kant is currently pursuing a Ph.D. in the Latin American Studies Program at the School of International Studies, Jawaharlal Nehru University, New Delhi, India. His research focuses on migration policies and the evolution of the culture of migration in Mexico. The title of his Ph.D. thesis is "Mexico's Policies of Migration toward the United States, 2006–2018." His areas of interest include area studies, particularly Latin America, and also the Global South. Additionally, he organizes sessions with scholars and academicians through a study circle called the "Global South Forum," where discussions and debates on issues pertinent to the Global South are conducted. He has received a National Fellowship for Other Backward Classes from the Ministry of Social Justice, Government of India. He holds a Bachelor of Arts degree in Geography (Honors) from Banaras Hindu University, Varanasi, India, and a Master of Arts degree in International Relations from Jawaharlal Nehru University, New Delhi, India.

Rajwant Kaur earned her Ph.D. from the Centre for the Study of Social Systems at Jawaharlal Nehru University, New Delhi. She received her M.Phil. from the Department of Sociology at the Delhi School of Economics, University of Delhi, and completed her Master's degree from the same centre for the Study of Social Systems at Jawaharlal Nehru University, New Delhi.

Ramanuj Kaushik serves as an Assistant Professor of Geography at D.A.V. Post Graduate College, Siwan. With a Master's degree in Geography from Patna University, an M.Phil., and a Ph.D. in Diplomacy and Disaster Relief from Jawaharlal Nehru University, New Delhi, Dr. Kaushik has cultivated extensive academic and practical expertise in natural hazards and disaster management. His Ph.D. research focused on the response to the 2004 Indian Ocean Tsunami in India and Indonesia, analyzing the geopolitical impacts of disaster relief efforts and the role of diplomacy in these distinct contexts. Additionally, Dr. Kaushik holds a postgraduate diploma from the Indian Institute of Remote Sensing in Dehradun, specializing in geoinformation science with an emphasis on natural hazards and disaster risk management. Driven by a passion for understanding the interplay between climate change and geopolitics, he continuously monitors natural hazards and related global issues, contributing valuable insights to the field.

Maya Kumari is a social worker currently pursuing her Ph.D. at the Delhi School of Social Work, University of Delhi. Her area of specialization includes gender and public health.

Pushpanjali Kumari is a Ph.D. Fellow at the Department of Sociology, Banaras Hindu University, Varanasi. She has completed her Master's degree in Social Work with a specialization in rural reconstruction and urban development from Banaras Hindu University, Varanasi.

Naresh Kumar has an M.A. and Ph.D. from Jawaharlal Nehru University (JNU), New Delhi, India. He has also conducted 2-year postdoctoral research in the Delhi NCR region, concentrating on participatory development and social capital. His broad research work in social geography, regional planning, and development studies has been published in reputed journals, and he has participated in seminars and conferences internationally. Dr. Kumar has also previously taught at Kurukshetra University in India. Currently, he is an independent researcher based in Delhi, India.

Sanjeev Kumar is a Research Assistant at the Indian Council of Social Sciences Research (ICSSR). He is a graduate of the Department of Geography, Shaheed Bhagat Singh College, University of Delhi. His area of specialization includes agriculture, food security, rural development, GIS, remote sensing, and data analysis.

Surender Kumar is an Assistant Professor of Geography at Dayanand College, Hisar. Dr. Kumar published several papers in preferred journals and chapters in books. Dr. Kumar received his postgraduate degree, an M.Sc. in Geography, from Kurukshetra University, Kurukshetra, and an M.Tech. in Geoinformatics from Guru

Jambheshwar University of Science and Technology, Hisar. He obtained his PhD on the topic "Geospatial Application for Waterlogging and Salinity Assessment: A Case Study of Central Haryana" from Lovely Professional University in Phagwara, Punjab. He presented various academic as well as research-based papers at several national and international conferences. He specializes in the area of geospatial technologies.

Rahul Mahadeo Lad holds a position as an Assistant Professor in the Geography Department, School of Liberal Arts and Social Science, Faculty of Education and Humanities, JSPM University, Pune, India. His academic journey culminated in a Doctorate focused on the "Geostrategic Assessment of Transboundary River Water Sharing in the South Asia Region." Beyond academia, he contributes to the intellectual discourse as a member of the prestigious Bangalore-based think tank, The Takshashila Institution. Within Takshashila, he serves as a Network for Advanced Study of Pakistan (NASP) Fellow, benefiting from the guidance of Dr. TCA Raghavan, a distinguished former Indian Ambassador.

His scholarly insights extend beyond traditional academic circles, with his work prominently featured in esteemed international policy publications, such as the London School of Economics and Political Science, South Asian Voices, The Diplomat, Global Water Forum, Modern Diplomacy, The Geopolitics, Down To Earth, and *South Asia Journal*.

A.K. Mishra is currently working as a Professor at the Department of Commerce, Nagaland University, which is a Central University and possesses about 28 years of postgraduate teaching and research as well as academic and administrative experiences. His area of research is public experience in economic development, sustainable economic development, banking, management of public enterprises, self-help groups, etc. He has successfully guided five doctoral research studies for the award of a Doctorate of Commerce. He has completed a research project funded by ICSSR as a Project Coordinator. The title of the project is "Micro Enterprise Development Education in Nagaland." He has presented several research papers in national as well as international Seminars. He was a member of various academic decision-making bodies at other universities.

Juveriya Naaz has completed her Bachelor's and Master's degree with first division in Geography from Aligarh Muslim University, Aligarh, India. She has qualified for the UGC-NET, making her eligible for the position of Assistant Professor, which she achieved on 2nd December 2019 in Geography. Additionally, she has a postgraduate diploma in remote sensing and GIS from Swastik Edustart-Institute of GIS and Remote Sensing, Delhi. She has developed a keen interest in the application of remote sensing and GIS to study various geographical phenomena, such as climate change, disasters, and impacts of anthropogenic activities on the physical environment. She also holds 1 year of experience working as a GIS consultant and currently a faculty member at Swastik Edustart-Institute of GIS and Remote Sensing, Delhi, India.

Tirth Pahouja is a dedicated business administration student specializing in sales, currently based in Abbotsford, British Columbia. Before moving to Canada, he was actively engaged in addressing social issues in India, focusing on the challenges faced by marginalized communities. Now, despite everything being new in Canada, Tirth is committed to becoming an active participant in giving back to society. He is currently working with HUB International, an insurance brokerage known for its strong culture of social responsibility. Through HUB's dedicated program, HUB Gives, Tirth is finding meaningful ways to continue his passion for social issues and community service, striving to make a positive impact in his new environment.

Fidel Rahmati was a Lecturer at Parwan University in Afghanistan and holds a Ph.D. in Geopolitics from Osmania University, India. Dr. Rahmati's research focuses on South Asian Geopolitics and Afghanistan, where he has over a decade of experience in both teaching and research. He is also an active journalist with Khaama Press News Agency. His research interests are geography, international relations, international relations theory, geopolitics, and China studies.

Swati Rajput is a Professor at the Department of Geography, Shaheed Bhagat Singh College, University of Delhi. She has 23 years of experience in teaching and research. She holds her M.Phil. and Ph.D. degrees from the Department of Geography at the Delhi School of Economics, University of Delhi, India. She has been recognized as a National Resource Person in the Social Sciences by UNFPA and MHRD. She has been a rank holder throughout her academic journey. She received the Delhi University Gold Medal and the NS Pradhan Award. In 2020, she received the Venus International Women's Award for her commendable research in the field of gender studies. She is an expert in the fields of health and well-being geography, food and nutrition security, agricultural geography, and urban ecology. She has published numerous research papers in national and international journals, as well as chapters in renowned books. She has supervised and examined M. Phil and Ph.D. research in the sub-fields of agricultural geography, food and nutrition security, and tourism geography. Her book *Urban Green Space, Health Economics and Air Pollution in Delhi* (2021) has been published by Routledge, Taylor and Francis Group, UK. The book excerpts have been included in the "Sustainable Development Goals online collection." The book received the Honor Award in the Green Literature Festival of India 2023–2024. She has also edited a book on *Sustainable Smart Cities of India*, published by Springer, which has been acknowledged on various international platforms. It was among the top 25% most downloaded e-books in 2019. She has also published a book *Geography of Health* in 2024. She has also convened, participated in, and presented research papers at many national and international conferences. She has also been acknowledged as a guest speaker, expert, and keynote speaker in various workshops and conferences. Besides that, she has been associated with E-learning modules and MHRD projects of ePathshala and IGNOU. She has completed various research projects related to green spaces, land use-land cover, food and nutrition security, Yamuna water treatment, etc., sponsored by the University of Delhi and ICSSR.

Ravi Shankar Raj is a Senior Research Fellow at the Department of Political Science, Central University of Haryana, Mahendragarh, India. He holds a UGC-NET JRF in Political Science and has a strong academic background that includes an M.A. from Banaras Hindu University and a B.A. (Hons) in Political Science from Shaheed Bhagat Singh College, University of Delhi. His research interests lie in various aspects of international relations, particularly in areas such as Indian Foreign Policy, IR theories, geopolitical issues, and the soft power dynamics of India in global politics. His passion for research and teaching is evident through his active participation in numerous seminars, conferences, and workshops, where he has presented thought-provoking papers on topics ranging from geopolitical issues, India's soft power, and cultural diplomacy to its foreign policy strategies. His scholarly contributions extend beyond presentations to publications, with several UGC CARE, peer-reviewed articles, both at the national and international levels, and a book chapter.

Robin Raj is a Research Scholar at the Indian Institute of Technology Bombay. Raj has been awarded a Junior Research Fellowship by the University Grants Commission (UGC), India. Raj's area of interest lies in demography, development economics, and sustainable development.

Deeksha Rana is currently pursuing her Ph.D. in the Department of Geography at Himachal Pradesh University. She is currently working on sustainable development in her Ph.D. program. She has completed her M.Sc. and M.Phil. in Geography in First Division. During her M.Phil. program, she has worked on drinking water and sanitation in Shimla city. She cleared NTA NET-JRF and HP SET in 2019 and 2020, respectively. She regularly takes part in various conferences and seminars. She has published four research papers and three chapters to her name.

Shweta Rana is currently pursuing her Ph.D. in the Department of Geography at Himachal Pradesh University, Summer Hill, Shimla. She is currently working on the impacts of ecotourism in her Ph.D. program. She has scored second position in B.A. and M.Sc. Geography in the state. She has completed her M.Phil. studying the impact of MGNREGA on regional development. She has also cleared UGC-NET JRF and HP SET in 2018. She regularly takes part in various conferences and seminars. She has also published four research papers and two book chapters.

Chadetrik Rout is currently working as an Associate Professor in the Department of Civil Engineering, Maharishi Markandeshwar (Deemed to be University), Mullana, Ambala, Haryana. He has more than 16 years of teaching and research experience in the field of Environmental Science and Engineering (PG and UG level). His broad research areas include water resource management, environmental impact assessment studies, bioremediation, air and water quality assessment and modeling, solid waste management, etc. He has guided 12 M.Tech. students in their dissertation work and one Ph.D. scholar in their thesis work. Dr. Chadetrik Rout has published 3 books and 8 book chapters and more than 60 research and review papers in different national and international journals and conferences. He currently serves as a reviewer for more than 15 internationally reputed journals and has reviewed over a hundred journal papers.

Kasturi Saikia is a Ph.D. scholar in the Department of English at Nagaland University, India. Her research focuses on gender, diaspora, subaltern, and urban studies. She previously earned her M.Phil. in English from the same department and holds an MA in English from Tezpur University, India. Kasturi's work explores marginalized voices and urban narratives, aiming to shed light on overlooked experiences within literature. Her research contributes to a deeper understanding of how identity, displacement, and power dynamics are portrayed in contemporary literary and cultural contexts.

Armaan Singh Sandhu is a dedicated and hardworking student with a strong commitment to academic excellence and personal growth. Throughout his educational journey, he has consistently demonstrated a deep passion for learning and an unwavering determination to succeed. Beyond academics, he balances a rigorous schedule that includes leadership roles in student organizations, student company programs, and volunteering in community service projects. These experiences have honed his organizational and teamwork skills, further fueling his drive to achieve his goals. His perseverance and work ethic are at the core of his approach to both his studies and extracurricular endeavors, as he strives to make a positive impact and continuously improve.

Shubham Kumar Sanu is an Assistant Professor in the Department of Geography at Indraprastha College for Women, University of Delhi. With over 35 research papers published in prestigious journals and more than 50 presentations at national and international conferences, his expertise encompasses disaster management, environmental studies, and urban space.

Anindita Sarkar is a Senior Researcher at the Department of Ecology and Natural Resources Management at the Center for Development Research (ZEF), University of Bonn, Germany. She is a Professor at the Department of Geography at Miranda House, University of Delhi. Her research is anchored at the intersection of development and environment, and her work shows how technologies and strategies of resource use and access shape spaces. Her published work focuses on how policies facilitate or impede water access through institutions influencing food security, water security, and livelihood security of communities. She has been associated with ACIAR, World Bank, IWMI, NCERT, ICSSR, and IGNOU in various capacities. She has worked on research projects in Kenya, Bangladesh, and India. She received the UGC Research Award in 2016 and was conferred the Delhi University Excellence Award in 2023 for her outstanding contributions to research and academic activities.

Soma Sarkar is an Assistant Professor at the School of Global Affairs, Dr. B.R. Ambedkar University, Delhi. With a Ph.D. in Geography from the University of Delhi, her expertise spans regional planning, disaster management, and advanced geospatial technologies. Dr. Sarkar's extensive experience includes faculty positions at various colleges under the University of Delhi and serving as a Geospatial Consultant for the DST Climate Change Programme (CCP) project at the ICMR-National Institute of Malaria Research, Delhi. A prolific researcher, Dr. Sarkar has published over a dozen research articles in prestigious international journals, such as *Environmental Modelling and Software*, *Geocarto International*, *Natural Hazards*, *Geospatial Health*, and *Malaria Journal*. Her work is highly regarded in the academic community, with several of her publications being widely cited. In addition to her research, Dr. Sarkar actively participates in the academic community, having delivered numerous invited lectures and presented her research at various national and international conferences.

Pratham Seth a 21-year-old psychology major at the University of the Fraser Valley (UFV), exemplifies the integration of academic excellence and a deep-seated passion for helping others. With a holistic perspective, he engages actively in both co-curricular activities and rigorous academic pursuits, allowing him to approach complex social issues with a well-rounded and empathetic understanding. Pratham's commitment to addressing societal challenges is evident in his research, where he explores the multifaceted problem of homelessness. His work aims to uncover the underlying causes of this issue and contribute to the development of sustainable solutions. By combining his academic knowledge with a genuine concern for social welfare, Pratham aspires to make a significant impact in the field of psychology and beyond. His dedication to both scholarship and service positions him as a promising advocate for change in addressing the pressing social issues of our time.

Arabinda Sharma Associate Professor and HoD at Gangadhar Meher University, boasts an impressive academic record with a Ph.D. from IIT Kharagpur and an M.Tech from Birla Institute of Technology. His 15+ years of experience encompass teaching, research, and administration. With expertise in geoinformatics, water resources, and environmental fields, he has authored 85 publications with an h-index of 13 and edited 6 reference books. Dr. Sharma actively mentors students (16 M.Tech./M.Phil., 1 Ph.D., and 6 ongoing Ph.D.), serves on editorial boards of prestigious journals, and organizes academic events. Recognized for his excellence, he has featured in "Marquis Who's Who in the World" and received national awards like the Jawaharlal Nehru Outstanding Doctoral Thesis Award 2012 from ICAR, New Delhi, and Young Achiever Award (2013) from the Indian Society of Geomatics, Ahmedabad.

Shivam is pursuing a bachelor's degree in liberal arts at the University of the Fraser Valley.

Kameshwer Nath Singh a seasoned academic and expert in Rachana Sharir, has over 23 years of experience in teaching. He joined the Department of Rachana Sharir at IMS BHU in 2002 as its first faculty member. Dr. Singh holds a B.A.M.S., M.D. (Ayurveda), and Ph.D. in Rachana Sharir from BHU. He has guided 12 M.D. and 10 Ph.D. students, published over 75 papers, and presented over 80 papers at conferences. His editorial work includes proceedings on Tantra Sharir and Srotas Sharir. Among his accolades are the Sharir Shrestha Susruta Award and the VD GOOD International Scientist Lifetime Achievement Award. Dr. Singh also serves as a life member of several prestigious scientific bodies and as Vice President of the PG (Ayurveda) Association's Varanasi Branch. His contributions continue to shape the future of Ayurvedic education and research.

Ripudaman Singh is a Professor of Geography at Amity Institute of Social Sciences, Amity University, Noida, India. He specializes in the fields of regional development, urban and regional planning, and remote sensing applications. He has been working on the regional disparities in India and analyzing how these regional disparities have evolved and how their nature, extent, and intensity have been varying over time and space. Apart from his books on *Regional Disparities in Pre and Post-Reform India*, *Globalization and Regional Disparities in India*, and *Changing Geopolitical Contours in the Twenty-First Century World* (Saptrishi, Chandigarh, 2016; 2017), two edited volumes on *Re-envisioning Remote Sensing Applications: Perspectives from Developing Countries* (CRC Press, 2021) and *Re-envisioning Advances in Remote Sensing: Urbanization, Disasters and Planning* (CRC Press, 2022) have been published by Taylor & Francis Group. Another book, *Twenty-First Century India: Land, People and Economy*, is in the process of publication by Springer.

Shantesh Kumar Singh is currently serving as a Professor at the Centre for International Politics, Organization and Disarmament (CIPOD) and Director of the Human Rights Studies Programme (HRSP), School of International Studies (SIS), Jawaharlal Nehru University, New Delhi. He also served as a Professor in the Department of Political Science at the Central University of Haryana, Mahendragarh, India. His academic journey has also taken him to international platforms, including his role as a Postdoctoral Fellow in Global Health at the International Institute for Global Health (IIGH), United Nations University, Kuala Lumpur, Malaysia. Prof. Singh's academic credentials include an M.Phil. and Ph.D. from the School of International Studies, Jawaharlal Nehru University, New Delhi, India. In addition to his academic pursuits, Prof. Singh serves as the Managing Editor of the *Journal for National Security and Terrorism*, NCTN-IACSP, based in Helsinki, Finland, contributing to the dissemination of cutting-edge research and analysis in the field of national security and terrorism studies. With a rich and diverse academic background, Prof. Singh has made significant contributions to the fields of political science, international relations, and global health governance. He has a significant amount of experience working in the academic and research fields in a variety of highly regarded universities and organizations both in India and abroad.

Gopal Sonkar is a scholar who specializes in interdisciplinary fields. He holds an honors degree in Geography from Banaras Hindu University. He pursued his master's degree in Environmental Studies at the University of Delhi. He obtained his M.Phil. degree in Development Studies from Jamia Millia Islamia. The Department of Geography at the University of Delhi awarded him with the Doctor of Philosophy degree.

He has specialized in the areas of biomass energy, clean mechanisms, sustainable development, and climate change. He has published 17 research publications in the form of articles and chapters. He has delivered over 20 papers at both national and international academic conferences. He was employed as a research associate at the Organization of Unearthing Suppressed Realities of Hindustani History in New Delhi, where he collaborated with renowned Indian historian, Professor Kapil Kumar. He also worked as a research associate at the Centre for Urban Studies at the Indian Institute of Public Administration in New Delhi, under the supervision of Dr. Sachin Chowdhry. He has published his most noteworthy research paper, "Environmental Activism: Historical Review, Development, and Expansion," in Springer, and "Assessment of Arsenic Contamination in the Indian Rivers." He was awarded the Best Reader Award 2022–2023 from Nehru Memorial Museum and Library, New Delhi.

A diligent researcher, with a clear academic vision, Dr. Gopal Sonkar is desirous of making a mark in the fields of renewable energy, renewable energy governance, regional development, climate change governance, and many other aspects of environmental and geographical issues.

Anjali Srivastav is currently working as a Research Associate at India Child Protection. She has a Master's in Population Studies, which she completed at the International Institute for Population Sciences, Mumbai, India, in 2023. She has also qualified for UGC-NET in Population Studies. Her research interests include migration and urbanization, women and health, aging, and gender issues.

Smrity Upadhyay is a Senior Research Fellow at the Delhi School of Economics, University of Delhi, with a distinguished focus on disaster management, disaster diplomacy, human security, and research ethics. Her work is characterized by a deep commitment to understanding and addressing the complex challenges within these fields. S. Upadhyay is also an active member of the Brahmavarta Geographical Society of India.

Archana Verma is a Project Scientist at UP Remote Sensing Application Centre, Lucknow. She works on the Disaster Management and GIS project. Her areas of interest are natural and human resources, disaster management, and urban studies.

Jyoti Yadav is an emerging scholar in the field of geography. She laid the foundation of her academic career at Kendriya Vidyalaya, Mughalsarai, where she developed an early interest in spatial sciences. She pursued higher education at Banaras Hindu University (BHU), earning a Bachelor of Arts and a Master of Arts in

Geography. Her academic rigor and passion for research led her to the Institute of Medical Sciences (IMS) at BHU, where she completed her Ph.D. in geography, specializing in medical and social geography. Currently, she is engaged in postdoctoral research at Jawaharlal Nehru University (JNU), where she continues to explore advanced topics in geography. Her work is characterized by a multidisciplinary approach, integrating traditional geographical methodologies with contemporary theoretical frameworks. Dr. Yadav is poised to make significant contributions to academic discourse and the broader field of geographical studies.

Part I
Introduction

Geography and Solidarity? What Geography Should (Really) Be…

Elena dell'Agnese

Introduction

The importance of geography's military role is well known. In times of war, when knowledge of terrain becomes crucial, geography courses multiply and academic geographers themselves often become "spies, teachers, and occupiers" (Barnes, 2016). Once the war is over, geographical knowledge is, or can be, used in diplomacy and civil affairs for the reorganization of the country. But geography can play a much more important role in educating citizens for peace. It can teach people to see the other as human beings like themselves, it can help them to put themselves in the other's shoes, and it can teach solidarity. Moreover, the tools of geographical analysis can help to understand how solidarity can be exercised in a system that is not necessarily limited to the local, but which, precisely because place is a complex system of relations (Massey, 2004a, b), extends or can extend beyond the nation-state in the construction of transnational networks. After mentioning in the first paragraph the role of British and American geographers in the First and Second World Wars, the chapter will present the positions of some geographers who thought of geography as an instrument of peace and will conclude by mentioning networks of solidarity articulated at different scales. Finally, special mention will be made of interspecific forms of solidarity.

E. dell'Agnese (✉)
Department of Sociology and Social Research, Università di Milano-Bicocca, Milan, Italy
e-mail: elena.dellagnese@unimib.it

K. Arora, R. R. Anand (eds.), *The Solidarity Approach in Geography*,
https://doi.org/10.1007/978-3-031-86753-8_1

Is Geography All About Making War?

Everyone remembers the title of a famous booklet by Yves Lacoste, *La géographie ça sert d'abord à faire la guerre* ("*Geography is all about making war*"), a classic of radical geography from 1976. Investigating US warfare techniques in Vietnam, Lacoste demonstrated that their military successes were due to a thorough knowledge of the terrain. So, he declared that studying geography is now, and it has always been, about warfare. His cry of accusation had great resonance at the time (his work was translated into different languages), but he was not discovering anything new.

"It is often remarked that war stimulates interest in geography, and there can be no doubt that the Franco-Prussian War of 1870–1871 had this effect" (Hudson, 1977, p. 13). Indeed, as remarked by J. Scott Keltie, in a report about geographical education edited by the Royal Geographical Society (Keltie, 1885, pp. 36–37), "Chief Justice Daly, in his address to the American Geographical Society, draws attention to the Franco-German War as a war fought as much by maps as by weapons' and attributes the result to 'skillful military movement', performed by an army thoroughly acquainted with all the geographical features of the country over which it was moved." Indeed, Keltie adds, the Franco-Prussian war had taught that "if the fate of a nation may depend upon a battle, a battle may depend on a knowledge of geography."

In this respect, the establishment of the so-called new geography (Stoddart, 1992) in the British Empire at the end of the nineteenth century was closely linked to the realization that in-depth geographical knowledge could serve national interests. As underlined by Hudson (1977, p. 12), at the time, "the study and teaching of the new geography at an advanced level were vigorously promoted, largely, if not mainly, to serve the interests of imperialism in its various aspects including territorial acquisition, economic exploitation, militarism and the practice of class and race domination."

It soon became clear that university education would also have to be expanded to train the military. In 1906, Halford Mackinder, then Rector of the London School of Economics, introduced a special course for military personnel. On the other side of the Atlantic, at the United States Military Academy at West Point, a Department of Geography, History, and Ethics had begun teaching geography, history, and ethics as early as 1818. The teaching of geography, which for a few decades was abandoned along with that of history and ethics because it seemed to divert attention from the supposedly more important courses in mathematics, science, and engineering, was reactivated at the turn of the century (Finnegan, 2004), and even today the Academy has a Department of Geography and Environmental Engineering within it.

During the years of the First World War, a debate arose within the Royal Geographical Society about the strategic role of geographical knowledge and the potential role of eminent geographers as spies. At that time, Albrecht Penck, who was arguably Germany's leading university geographer, and had also received a medal from the Royal Geographical Society, was arrested by the British authorities as suspected of being "a German special agent who had exploited the innocent hospitality of unsuspecting Britons" (Heffernan, 2000, p. 271). Probably, Penck was not a spy; however, back at home, he felt morally committed to making presentations on behalf of the war and even published an article with the title *Der Krieg und das Studium der Geographie* (Penck, 1916) (war and the study of geography) (Fahlbusch et al., 1989).

Again, during the First World War, there was the example of American geographers working as propagandists, to promote, in the American public, the idea of the United States' entry into the war on the side of the Allies. For instance, Douglas W. Johnson, in 1917a published *The Peril of Prussianism*, a booklet where he discusses the idea of the divine right of the state as central to the doctrine of Prussianism, then explains that such an idea justifies the most horrible war crimes and, on this basis, adds that "The Prussian ideal must perish, the American ideal must live" (Johnson 1917b, p. 53). In the same year, he published *Topography and Strategy in the War* and, to bring the sub-discipline of military geography "out of the military academy and into the civilian world and the university" (Ginsburger, 2005, p. 270), he began lecturing on the subject at the American Geographical Society.

The military commitment of geographers did not die out with the end of the First World War. To stay with the example of the United States, during the Second World War, American geographers were even more important and numerous within military intelligence, especially in one branch of US intelligence, named the Office of Strategic Services (OSS) (Barnes & Crampton, 2016), which became one of the main employers of geographers during this period. At its height, 129 were working there (Barnes, 2016, p. 545). One of them was Richard Hartshorne.

If geography related to geostrategy, propaganda, and intelligence activities can be considered potentially warmongering, also the so-called geography of professors (Lacoste, 1976), albeit apparently innocent, may have questionable implications. Indeed, as critical geopoliticians and, more generally, all those engaged in postcolonial/decolonial studies have well explained, "imaginative geographies as representations of the Other can easily be accused of emanating from the existing power relations often associated with colonialism or imperialism" (Shimazu, 2015).

Geography and Peace

Such reflections could lead to erasing geography, as "a quintessential science of empire" (Bryan, 2016, p. 507), from any contemporary school curriculum. But indeed, this is just half of the story. Luckily, geography and geographers can also become instrumental in making peace.

In fact, "geography and geographers have been mobilized in attempts to build a world of mutual understanding and peace since the second half of the nineteenth century" (Shimazu, 2015, p. 98). For instance, the first International Geographical Congress was held in Antwerp in 1871. Again, the Franco-Prussian War had a role, since the congress was organized after its end and was intended to function as a peace-oriented international assembly; so, among its aims, it had that of overcoming "distances, difference in nationality, and sometimes even hostile prejudices" (Shimazu, 2015, p. 99).

In addition to these early attempts at "scientific diplomacy," geographers were acting as real diplomats. The same Johnson, who acted as a propagandist during the war, and was a major with the American Army's Intelligence Division (Ginsburger, 2005), was also part of *The Inquiry*, the research team created by President Wilson to make recommendations for post-war revisions and adjustments by collecting geographical and ethnographic information. *The Inquiry* was based at the American Geographical Society headquarters, in New York, and led by its director, Isaiah Bowman. Together with Mark Jefferson, who was the head of the cartographic program of *The Inquiry*, Johnson, and Bowman were part of the American delegation participating in the Paris Peace Conference in 1919 (Bowman later became also US President Franklin D. Roosevelt's advisor during Second World War) (Smith, 2003). However, despite Wilson's good intentions, the deployment of geographical knowledge aimed at bringing peace to Europe did not work. The politics of self-determination and the Fourteen Points "were under constant interpretative pressure," because, due to the extremely difficult-to-define ethnic factors, strategic considerations were often added. So, even if the conference was supposed to bring peace to Europe, it just opened up a series of territorial issues bound to create the conditions for a new war.

During the Second World War, though, many geographers were commissioned by President Roosevelt to draw up the so-called *M-Project* (where M stands for migration), a survey developed to understand how to resettle European refugees, including Jews, at the end of the war; the resettlement plan was not implemented, but according to Henry Field, one of the members of the research team, it was the basis of the famous *Point Four Program* of assistance to "developing countries" announced by US President Harry S. Truman in his inaugural address (20 January 1949) (Martin, 2015).

Geography, however, is not only a science *ad usum delphini*, as suggested by the students of the Italian geographer Lucio Gambi in the years of "contestation" (Gambi, 1968) (dell'Agnese et al., 2021). It can be much more. There are several geographers who, instead of thinking of geography as an instrument of war, have realized that it can be a tool to promote mutual understanding, reciprocity, and solidarity.

The Russian anarchist geographer Petr A. Kropotkin and his pedagogical manifesto, *What Geography Ought to Be*, should be cited first and foremost in this regard. Written by Kropotkin while in prison in France in 1885, *What Geography Ought to Be* is quite famous, but it still deserves to be quoted directly:

> *Geography … must teach us, from our earliest childhood, that we are all brethren, whatever our nationality. In our time of wars, of national self-conceit, of national jealousies and hatreds ably nourished by people who pursue their own egotistic, personal, or class interests, geography must be … a means of dissipating these prejudices and of creating other feelings more worthy of humanity. It must show that each nationality brings its precious building stone for the general development of the commonwealth and that only small parts of each nation are interested in maintaining national hatred and jealousy. … there is [another task], perhaps still greater; that of dissipating the prejudices in which we are reared about the so-called 'lower … ".*

In the same years, Kropotkin was working on his theory of "mutual aid" (Kinna, 1995), a thesis that was intended to be both a scientific explanation of the change, identifying cooperation as the dominant factor in evolution rather than Darwinian competition, and a didactic tract (since cooperation is supposed to be present not only among animals but also in human societies). Mutual aid (Kropotkin, 1902) was to become a fundamental tenet of anarchist communism, and it is still a very important principle in the creation of "social solidarity." Indeed, survival is for the fittest, but the fittest are, in Kropotkin's terms.

> *not the physically strongest, nor the cunningest, but those who learn to combine so as mutually to support each other, strong and weak alike, for the welfare of the community* (Kropotkin, 1890, p. 338)

Similarly, Elisée Reclus, an equally famous anarchist geographer, believed in the possibility of making a "free society founded on natural and moral laws rather than the oppressive laws of the state..., a form of social organization that arises out of mutual aid, cooperation, solidarity, and compassion" (Clark, 2003).

The anarchists were not alone in believing that geography was a useful tool for mutual understanding. Beyond the world of anarchism, another geographer, politically and ideologically distant from them but with highly innovative ideas about geography as a tool for mutual understanding, was Halford Mackinder. Mackinder already quoted here as the Rector of the London School of Economics who introduced geography courses for the military deserves to be mentioned for something else. In collaboration with the Colonial Office Visual Instruction Committee (COVIC 1902), he wanted to organize lantern slide lectures in which a series of photographs of the different regions of the British Empire would be projected. The general idea was to combine visual technologies with geographical instruction to teach children throughout the Empire "an equal form of belonging across borders, class divisions, and ethnic backgrounds." To foster feelings of sympathy and unity in viewers, and the "idea of a common but equal status," Mackinder sent a photographer and illustrator, Hugh Fisher, across the Empire, who took over 6000 photographs (Moser, 2017). COVIC intended that colonial schoolchildren would receive sets of slides about the "Mother Country," while British schoolchildren would see slides about the colonies. This method of organization created a dual audience, each with their own set of expectations and lived experiences of empire. But Mackinder's imagination went further, believing that

> *The effort on the part of English children to imagine themselves in the position of Indian children should tend to arouse and impress a valuable feeling of political sympathy* (Mackinder, 1909, p. VI).

So, he believed that pictures could allow "viewers to not only learn about the wider world but to imagine what it might be like to occupy another subject position and to look back at representations of themselves from this outside perspective" (Moser, 2017, pp. 200–201).

Another author worth mentioning for his ideas on geography and peace is the Italian Arcangelo Ghisleri, a socialist and freemason who was strongly influenced by the ideas of Elisée Reclus. Like Kropotkin, Reclus, and Mackinder, Ghisleri was convinced of the importance of geographic education for everyone and dedicated a good part of his scholarly work to making geography an instrument for increasing the capacity for action of the popular classes. He recognized the power of geography but wanted it to be a power shared by the people. To this end, he devoted himself passionately to teaching and launched several journals, including *Geografia per tutti* (*Geography for everyone*).

For Ghisleri, geography not only had to function as an instrument of reciprocal knowledge (as for Kropotkin), or even as a field for building the capacity to identify with the other (as for Mackinder), but it also had to be a discipline capable of demonstrating the scientific inconsistency of racism and colonialism. So, in another journal directed by him, *Cuore e critica* (Heart and Criticism), Ghisleri published numerous articles aimed at refuting the racist theories of his contemporaries and even opened a column, entitled *Razze spregiate* (Despised races), where the qualities of individuals belonging to this or that discriminated category (including women) were presented to the readers. In 1896, he published a study on colonialism and the "rights of the people," in which he tried, with a critical approach, to dispel the "naturalising" assumptions of the colonialist discourse:

> *...the theories of those who pretend to justify with Science the arrogance of a race against the others, have no basis other than in a bilious ignorance...* (Ghisleri, 1896, p. 73).

Later on, in 1935, the president of the Associations of American Geographers, Wallace Atwood, stressed the responsibility of geography to "overcome ignorance among people about other people in the world, and remove the leftovers of hatred ..." (Atwood, 1935, p. 15). In the years immediately after the Second World War, also the Australian geographer T. Griffith Taylor (1946) took up the same theme, arguing that geographical education not only helped to overcome prejudices and stereotypes but that in this perspective geopolitics could be transformed in meaning and become *geopacifics*.

Geography and Solidarity

The idea that geography can be a means of mutual understanding, and thus an instrument of solidarity, is not new, as can be seen from this very brief historical excursus. By helping people to understand how we can be similar while being different, and how to put ourselves in the shoes of others, as Mackinder wanted to teach us, geography helps people to build solidarity not only within their community, with those similar to them, but also outside it, in difference (Oosterlynck et al., 2017). Moreover, it offers conceptual instruments, such as the relational concept of place (Massey, 2004a, b), which can help people understand that places are sites of heterogeneity and interaction, on many different scales. Moreover, political geography may also help politicians to get out of the "territorial trap" (Agnew, 1994), and to stop thinking that solidarity applies, as a "taken for granted," only inside national communities.

As Kelliher (2018) notes, there are many different and sometimes only loosely related ways in which the concept of solidarity can be used. It has a history in Christian thought as well as a strong association with the political left. It is a form of support offered by people who share a similar idea or problem, or simply feel close to the pain of others (in this perspective, solidarity is "about communal identification with the other through understanding the perspective of the other") (Cojocaru & Cochrane, 2023).

Solidarity is also closely associated with forms and relationships of activism. For example, the definition offered by the Collins English Dictionary emphasizes the political meaning of the term, saying that "when a group of people shows solidarity, they show support for each other or another group, especially in political or international affairs."

Solidarity can be expressed through a gesture or even just a symbol (for example, flying the flag of Ukraine or Palestine to show concern for political realities that are suffering from territorial aggression.) (Militz et al., 2022). Providing an even more political definition of the term ("a relation forged through political struggle which seeks to challenge forms of oppression", Featherstone, 2012, p. 5), David Featherstone emphasizes that solidarity is not about similarity, but is a form of a relationship built from the bottom up that can unite different places, activists, and struggles. Moreover, "solidarities are constructed through uneven power relations and geographies" (p. 6). They also may represent "new ways of relating," because solidarity is "inventive."

Solidarity is changing. Thus, in addition to the structures and manifestations of solidarity of the past (e.g., trade unions, fundraising, public displays of symbols, etc.), new forms of solidarity are emerging, linked to the possibilities offered by networked communication technologies and social media. Examples include global justice networks and mutual aid networks.

Global justice networks may involve different types of political actors, build alliances of different groups, connect different place-based movements, negotiate linkages, and create spaces of resistance, with a mixture of wider and more local social relations (Massey, 2004a, b). Thanks to the possibility of entering the global arena, global justice networks allow activists to broaden the horizons of their territorial struggle and develop the potential to build alternative politics from the bottom up, as they can provide them with resources (material and intellectual) that can be used within their national society (Routledge & Cumbers, 2009). For example, the global solidarity network provides the Zapatista movement with discursive and material resources at the transnational level for a struggle that remains ultimately local.

Mutual aid is a very broad term, covering a wide range of actions such as community response, neighborhood assistance, or collective action. However, unlike other manifestations of political solidarity, it is a form of direct social action, i.e., it does not focus on demanding something from those in power but on directly changing some specific aspects of society by acting (Zamponi, 2023). For example, after a disaster, cross-border mutual aid networks can be set up to help people cope with the aftermath (see the Mutual Aid Disaster Relief and the Common Ground sites set up after Hurricane Katrina in New Orleans in 2005, which provided health clinics, a legal clinic, community gardens, a women's shelter, a radio station, and also community media centers). Mutual aid networks have also been set up in several countries to overcome the difficulties associated with the COVID epidemic (2020–2021), using the Internet to meet not only people's material needs but also their desire to stay together when social distance would otherwise be imposed (Zamponi, 2023).

So far, we have considered solidarity as a relationship between people, near or far, similar or different, united by a common cause, or simply sympathetic. But there is more that geography can teach us. As remarked by William Bunge:

> *Geographers must labor to make sure that the earth's surface truly becomes the home of man [sic], that mankind [sic] comes to some peace, some rest, some harmony, and ultimate unity with nature... As a first step, it must be seen that man [sic] and nature are one. The seeming separation is a measure of the degree of the unnaturalness of man himself [sic]* (Bunge, 1973, pp. 290–291).

Bunge used the term "man" because he was writing in the 1970s (today we would say "human being"), but his message is clear. With this phrase, he addresses the need to overcome the binarisms typical of Western thought (nature–culture, human–animal, etc.), because, to ensure "human survival" (the title of the article), it is not enough to be in solidarity with one another, we must be at one with nature itself. So, solidarity must be rethought in a non-anthropocentric way. The idea that non-human animals are fellow members of multi-species communities suggests a more comprehensive approach to the concept (although it is very easy to feel compassion for our pets, it is less easy to feel compassion for wild animals and is even more difficult to feel compassion for those we habitually eat, for which a kind of cognitive dissonance comes into play) (Rothgerber, 2020).

For example, the all-volunteer global movement, Food Not Bombs, collects surplus food that would otherwise go to waste, prepares and serves community meals, and distributes free food to the hungry and homeless. Ideologically based on anarchist principles of mutual aid, FNB differs from traditional forms of charity, firstly because it serves only vegetarian and vegan food (solidarity with the environment and with non-human animals), but also because it does so in public places, outdoors, to make poverty visible. It wants to expose hunger. It, therefore, chooses not only public spaces but also particularly visible (or symbolically important) public spaces. In this way, FNB can be seen as a form of geographical activism that "transforms the city, not just because a table is set up and free food is shared in a place like Copley Square, Boston, or Athens, Georgia, but also because it shows how spatially connected networks of committed individuals emerge and recognize themselves as a force of resistance" (Heynen, 2010) and, of course, a force of solidarity.

References

Agnew, J. (1994). The territorial trap: The geographical assumptions of international relations theory. *Review of International Political Economy, 1*(1), 53–80.

Atwood, W. W. (1935). The increasing significance of geographic conditions in the growth of nation-states. *Annals of the Association of American Geographers, 25*(1), 1–16.

Barnes, T. J. (2016). American geographers and world war II: Spies, teachers, and occupiers. *Annals of the American Association of Geographers, 106*(3), 543–550.
Barnes, T. J., & Crampton, J. (2016). Mapping intelligence: American geographers and the office of strategic services and GHQ/SCAP (Tokyo). In S. Kirsch & C. Flint (Eds.), *Reconstructing conflict* (pp. 227–252). Routledge.
Bryan, J. (2016). Geography and the military: Notes for a debate. *Annals of the American Association of Geographers, 106*(3), 506–512.
Bunge, W. W. (1973). The geography of human survival. *Annals of the Association of American Geographers, 63*(3), 275–295.
Clark, J. (2003). Epilogue: Humanity, nature, and the anarchist ideal. In J. Clark & C. Martin (Eds.), *A voyage to New Orleans: Anarchist impressions of the old south* (pp. 85–97). Glad Day Books.
Cojocaru, M. D., & Cochrane, A. (2023). Solidarity with wild animals. *Ethics, Policy & Environment, 26*(2), 198–216.
dell'Agnese, E., Minca, C., & Schmidt, M. (2021). Italian critical geographies: A historical perspective. In L. D. Berg, U. Best, M. Gilmartin, & H. G. Larsen (Eds.), *Placing critical geography: International histories of critical geographies* (pp. 202–222). Routledge.
Fahlbusch, M., Rössler, M., & Siegrist, D. (1989). Conservatism, ideology and geography in Germany 1920–1950. *Political Geography Quarterly, 8*(4), 353–367.
Featherstone, D. (2012). *Solidarity: Hidden histories and geographies of internationalism.* Zed Book.
Finnegan, P. (2004). The study of law as a foundation of leadership and command: The history of law instruction at the United States Military Academy at West Point. *Military Law Review, 181*, 112.
Gambi, L. (1968). *Geografia e contestazione*. Fratelli Lega.
Ghisleri, A. (1896). *Le razze umane e il diritto nella questione coloniale*. Istituto Italiano d'Arti Grafiche.
Ginsburger, N. (2005). An American geographer between science and diplomacy: The mission of Douglas W. Johnson in Europe, May–November 1918. In P. Purseigle (Ed.), *Warfare and belligerence: Perspectives in First World War studies* (pp. 265–294). Brill.
Heffernan, M. (2000). Professor Penck's Bluff: Geography, espionage and hysteria in World War I. *Scottish Geographical Journal, 116*(4), 267–282.
Heynen, N. (2010). Cooking up non-violent civil-disobedient direct action for the hungry: 'Food not bombs' and the resurgence of radical democracy in the US. *Urban Studies, 47*(6), 1225–1240.
Hudson, B. (1977). The new geography and the new imperialism: 1870–1918. *Antipode, 9*(2), 12–19.
Johnson, D. W. (1917a). *The peril of Prussianism*. J.P. Putnam's Sons.
Johnson, D. W. (1917b). *Topography and strategy*. Legare Street Press.
Kelliher, D. (2018). Historicising geographies of solidarity. *Geography Compass, 12*(9), e12399.
Keltie, J. S. (1885). *Geographical education: Report to the Council of the Royal Geographical Society*. J. Murray.
Kinna, R. (1995). Kropotkin's theory of mutual aid in historical context. *International Review of Social History, 40*(2), 259–283.
Kropotkin, P. A. (1885). What geography ought to be. *The Nineteenth Century, 18*, 940–956.
Kropotkin, P. A. (1890). Mutual aid among animals. *The Nineteenth Century*, 337–354.
Kropotkin, P. A. (1902). *Mutual aid: A factor of evolution*. Heinemann.
Lacoste, Y. (1976). *La géographie, ça sert, d'abord, à faire la guerre*. Maspero.
Mackinder, H. J. (1909). *Lantern lectures on the United Kingdom: Seven lectures on the United Kingdom for use in India (reissued for use in the United Kingdom)*. Visual Instruction Committee of the Colonial Office/Waterlow & Sons Limited.
Martin, G. J. (2015). *American geography and geographers: Toward geographical science*. Oxford University Press.

Massey, D. (2004a). The political challenge of relational space: Introduction to the Vega symposium. *Geografiska Annaler, Series B, Human Geography, 86*(1), 3.

Massey, D. (2004b). Geographies of responsibility. *Geografiska Annaler, Series B, Human Geography, 86*(1), 5–18.

Militz, E., Ruppert, L., & Schurr, C. (2022). Solidarity, the national flag, and embodied differences. *Political Geography, 99*, 102738.

Moser, G. (2017). Photographing imperial citizenship: The Colonial Office Visual Instruction Committee's Lanternslide Lectures, 1900–1945. *Journal of Visual Culture, 16*(2), 190–224.

Oosterlynck, S., Schuermans, N., & Loopmans, M. (Eds.). (2017). *Place, diversity and solidarity*. Routledge.

Penck, A. (1916). Der Krieg und das Studium der Geographie. In *Zeitschrif der Gesellschaft für Erdkunde zu Berlin* (pp. 158–176, 222–248).

Rothgerber, H. (2020). Meat-related cognitive dissonance: A conceptual framework for understanding how meat eaters reduce negative arousal from eating animals. *Appetite, 146*, 104511.

Routledge, P., & Cumbers, A. (2009). *Global justice networks: Geographies of transnational solidarity*. Manchester University Press.

Shimazu, T. (2015). War, peace, and a geographical internationalism: The 1871 Antwerp International Geographical Congress. *Geographical Reports of Tokyo Metropolitan University, 50*, 97–105.

Smith, N. (2003). *American empire: Roosevelt's geographer and the prelude to globalization*. University of California Press.

Stoddart, D. R. (1992). Geography and war: The 'new geography' and the 'new army ' in England, 1899–1914. *Political Geography, 11*(1), 87–99.

Taylor, T. G. (1946). *Our evolving civilization: An introduction to geopacifics: Geographical aspects of the path toward world peace*. University of Toronto Press.

Zamponi, L. (2023). Mutual aid and solidarity politics in times of emergency: Direct social action and temporality in Italy during the COVID-19 pandemic. *Social Movement Studies, 23*, 1–20.

Part II
Globalisation, Regional Solidarity, Urbanisation, and Economic Development

India's Balancing Act in the Global South: Strategic Imperatives Amidst the Resurgence of Geopolitics

Shantesh Kumar Singh and Ravi Shankar Raj

Introduction

Nations play an essential role in international relations and geopolitical dynamics based on historical trajectory, economic imperatives, and strategic goals. Amidst an ever-evolving geopolitical landscape, Latin American, African, and Asian developing nations have increasingly become important actors in international politics. This region displays a unique mix of socio-economic and politico-economic dynamics which has undergone rapid transformations in the post-Cold War years. Global South, which is the successor term to 'Third World', encompasses many nations characterized as developing or emerging economies. It is important to recognize the diversity of nations within this region. It is also important to acknowledge that, despite their diversity, these countries share a common set of developmental challenges and have had similar historical trajectories. However, 'Global South' does not simply refer to a geographical region. It also has political and economic ramifications. The historically marginalized nations of the Global South attempted to create their own paths in a West-dominated world order. In the post-Cold War period characterized by regional power, multipolarity, and the emergence of regional

S. K. Singh
Centre for International Politics, Organisation and Disarmament, Jawaharlal Nehru University, New Delhi, India
e-mail: shantesh@jnu.ac.in

R. S. Raj (✉)
Department of Political Science, Central University of Haryana, Mahendergarh, Haryana, India

Research Analyst, ROGER HANSEN INSTITUTE for Global Interchange of Perspectives, Netherlands, Netherlands

K. Arora, R. R. Anand (eds.), *The Solidarity Approach in Geography*,
https://doi.org/10.1007/978-3-031-86753-8_2

actors, a geopolitical imbalance shift has been observed. India, with its deep historical legacy and increasing economic prowess along with resilient democratic structures as well as distinct geopolitical positions, is an important player in global politics. India, the largest democracy in the World, holds a significant place within the Global South. The fact that India is the largest democracy in the world highlights not only its regional ambitions and strategic moves within international politics but also its global aspirations. India is a prominent player that occupies an important position in the international arena, strategically situated at the intersection of regional dynamics, historical legacies, and global aspirations. The primary objective of this chapter is to provide a comprehensive assessment of India's significant influence on the geopolitical scene of the Global South.

Geopolitics has undergone dramatic change over the past century. This change was mostly driven by the rising influence of emerging economies of the Global South. India is blessed with an exceptional cultural legacy, historical significance, and geographical advantage—factors which have all helped expand its international influence while strengthening its soft power. India has become an ever-increasing factor in global South geopolitics since climate change and economic inequality became global challenges. India is uniquely situated, from its northern Himalayan foothills all the way southward to strategic Indian Ocean maritime regions making it a key player in both maritime and continental geopolitics. Indian history plays an integral part in understanding its present geopolitical standing. Since independence in 1947, India has sought to craft a foreign policy which protects both its sovereignty and strategic autonomy. In order to be effective at this endeavour India has focused on treading a fine line between global powers while emphasizing solidarity among developing nations. The Non-Aligned Movement (NAM), established by leaders like India's Jawaharlal Nehru, became emblematic of this approach, standing for collective neutrality during the Cold War. India's legacy and colonial experience have enabled it to foster bonds of solidarity with nations of the Global South. India's shared history has allowed it to foster collaborative ties based on mutual respect, cooperation, and shared developmental goals. Even as the Cold War fades from memory, India remains committed to non-alignment and south-south cooperation as part of its foreign policy, though with adjustments tailored specifically for current global circumstances. Though NAM has become controversial over time, its core principles—non-alignment and strategic autonomy—remain influential in shaping India's foreign policy approach and shaping India's diplomatic strategies. These values remain deeply embedded into its diplomatic strategies.

Current geopolitics is marked by a transition from unipolarity to multipolarism, with emerging powers seeking to assert their influence and redefine global norms. In this context, India's role becomes increasingly relevant. On the one hand, its aspirations for a more prominent role in institutions like the United Nations Security Council underscore its global ambitions. On the other, its active participation in platforms like BRICS, IBSA, SCO, etc. showcases its commitment to fostering a collective voice for the Global South. Economically, India's trajectory within the Global South presents a tale of transformation. It has metamorphosed from a largely agrarian economy into a major global player with significant technological, and manufacturing sectors. Its economic growth, while impressive, has also been accompanied by challenges—economic disparities, infrastructure bottlenecks, and the need for sustainable development. The manner in which India navigates its domestic economic challenges while simultaneously advancing its global economic aspirations entails a complex interplay of strategic considerations and pragmatic approaches. Geopolitics in the Global South extends far beyond economic power alone, taking in regional conflicts, historical grievances, and strategic rivalries across Africa, Latin America, and Asia. India plays an instrumental role in this geopolitical landscape through its relationships with its neighbours, involvement in regional conflicts, pursuit of energy security goals, and role in international organizations like the UN—all factors which play into its geopolitical calculus. These multifaceted elements collectively contribute to India's strategic decision-making process within the realm of geopolitics.

India has to perform a difficult balance act. India's attempts to maintain its strategic independence, while also building relationships with other major powers and addressing regional issues at the same show how complex diplomacy can become. India's role as a geopolitical player in the Global South is complex, entangled in a web of regional rivalries and historical legacies, for example, India is confronted with a myriad of geopolitical challenges stemming from its on-going rivalry with China, historical contentions with Pakistan, and apprehensions surrounding the emergence of a new Cold War paradigm. The bilateral relationship between India and China serves as a compelling illustration of the intricate and multifaceted dynamics inherent in the geopolitical manoeuvrings of the Global South. The dynamic as mentioned above is distinguished by a combination of cooperative efforts on a global scale and a drive for competition within regional arenas. Moreover, the international community places significant expectations upon India to serve as a leading advocate for democratic principles and humanitarian values within its region. This, in turn, imposes additional layers of responsibility upon the country. India's role as a geopolitical player in the Global South is fundamentally a study of balancing acts—between history and modernity; idealism and realism; regional obligations and global ambitions.

Nurturing Solidarity: Tracing Historical Context of India's Engagement in Global South

Throughout history, India's connections with countries in the Global South have evolved significantly. With its rooted historical heritage, India has a long history of engaging in cultural, economic, and political exchanges with various regions worldwide. In the twenty-first century, India's interactions with Global South nations have been gaining momentum influencing its foreign policy goals noticeably. India aims to position itself as a 'Vishwa Guru' showcasing its leadership ambitions.

Colonial Legacy and Anti-colonial Struggles

India's engagement with countries, of the Global South can be understood by looking at its interactions during the colonial period. The historical backdrop of modern India is closely associated with its experience of colonial rule under the British Empire, i.e. Modern India's connections to other parts of the world including Africa and Southeast Asia may be traced back to when it was a colony of the British Empire. Despite the stifling constraints of colonialism, these interactions laid the groundwork for future connections by facilitating cultural, economic, and political exchanges. Throughout the era of British colonialism, the Indian subcontinent experienced a multitude of manifestations of exploitation, economic subjugation, and political marginalization. The aforementioned experience played a pivotal role in shaping India's commitment to global anti-colonial initiatives. India, having successfully liberated itself from the burdensome shackles of colonial rule, naturally developed a profound sense of empathy towards other nations grappling with identical anti-colonial struggles. This common ground served as a foundation for India's solidarity with other recently decolonized nations and as a basis for India's commitment to anti-colonial efforts on a global scale (Raghavan et al., 2022). As a prominent actor on the global stage, India has undeniably exerted considerable influence in championing the cause of decolonization movements, with a particular focus on the African and Asian regions. India, guided by influential figures like Mahatma Gandhi and Jawaharlal Nehru, played a pivotal role in advocating for self-rule and independence, forging strong alliances with recently emancipated nations, and shaping its engagement with the emerging countries of the Global South.

Non-Alignment Movement

India's active participation in the Non-Alignment Movement (NAM) constitutes a significant and noteworthy episode in the timeline of India's modern history. Throughout the historical period known as the Cold War, Western World witnessed a notable division, characterized by a clash of ideologies and military rivalries between two dominant global powers, namely the United States and the Soviet Union. India along with the newly independent countries of Asia and Africa adopted the policy of Non-alignment. India's engagement with NAM was based on two key objectives—firstly, NAM will be helpful in establishing a multipolar global governance challenging the western dominance and secondly, it will act as a platform to develop closer relations with the newly independent countries. India has been actively working to challenge the dominance of Western countries and has strongly supported the rights and needs of nations, in the Global South. India's strong support for Global South nations is based on promoting ideas like sovereignty, self-determination, and equality, between nations (Abraham, 2008). The Bandung Conference in 1955 which preceded the establishment of NAM showcased India's dedication to anti-colonialism and fostering cooperation among countries from the Global South. Through participation in NAM, India played a dynamic role enabling itself to actively participate in global political matters while staying true to its unique beliefs and values (Singh, 2017). India through its leadership at NAM has successfully developed diplomatic ties with countries from the global south. These diplomatic engagements helped India strengthen its position to emerge as a leader of the newly independent nations that were undergoing the process of decolonization.

Cultural Influence

India has huge cultural influences through its Bollywood, delicious cuisine, profound yoga traditions, and spiritual practices all over the globe that include not only the developed Western countries but also the nations comprising the Global South (Rathore, n.d.). Bollywood movies have captivated a range of viewers surpassing borders. The rich culinary traditions of India known for their variety of flavours and spices have gained recognition globally. Additionally, it is interesting to observe the increasing appreciation, for practices like yoga and meditation which have roots, in India's ancient heritage. These cultural exports from India to different parts of the world have helped not only in shaping a positive perception of India but have also helped to attract more tourists and facilitating interpersonal engagements.

India's Soft Power and Development Assistance

India has risen as a significant player, on the global stage showcasing not just impressive economic growth and military strength but also standing out as a significant source of soft power. Soft power refers to a nation's ability to sway others using non-coercive means like cultural appeal and diplomatic efforts and offering foreign aid. India's rich cultural heritage, development assistance initiatives, and diplomatic activities have greatly contributed to enhancing its expanding soft power impact in regions, like the Global South. India's soft power influence has been greatly shaped by its efforts in supporting countries in the Global South particularly through its aid initiatives. When it comes to education, India provides scholarships, training programs, and capacity-building schemes to people from developing nations. These actions are made possible through institutions like the Indian Council for Cultural Relations and the Indian Technical and Economic Cooperation program. The aforementioned endeavours have effectively bolstered the interconnections and amicable relations between India and its partner nations, concurrently fostering the development of individual skills and capabilities. India, as a reliable and strong ally, has emerged as a praiseworthy partner in the developmental odysseys of nations within the Global South. This is primarily attributable to its noteworthy contributions in the domains of healthcare, agriculture, and the transfer of technology. The best example is India's Development Initiatives in Afghanistan. Its commitment to Afghanistan's reconstruction is reflected in its investments in various infrastructure projects like the Salma Dam and the Afghan Parliament Building (Raj & Singh, 2022). In addition, India's support, during the conflict in Afghanistan after 2021, demonstrates the nation's involvement in maintaining stability within the region. The aid and rebuilding initiatives provided by India in Afghanistan indicate its dedication to promoting progress and peace in a country ravaged by war reflecting its interactions, with the Global South.

India's involvement, in the Global South is based on a story that includes a colonial past mutual development challenges and the spirit of solidarity among the Global South nations. This involvement reflects India's desire for a world order, with multiple power centres coexisting and exerting influence. It also highlights India's role as a stabilizing force aiming to uphold balance and stability amid the complex web of international relations (Khayat, 2024). India faces challenges in maintaining its influence and meeting its objectives in the Global South due to increasing competition from China and internal constraints. Its interactions with Global South nations are not just based on historical factors but also on evolving geopolitical aspirations, security needs, and economic priorities. India's diverse engagement reflects its aim of solidarity within the Global South, against the Western hegemony.

The Return of Geopolitics in International Relations: Implications in South Asia and the Role of India

In recent years, there has been an increase in geopolitical activities globally marking a shift from the era of globalization and multilateralism following the Cold War. The resurgence of geopolitics has had impacts in regions like South Asia. India has emerged as a key player in geopolitics due to major changes offering the potential to wield substantial influence on both regional and global interactions. Various factors have contributed to this resurgence, such as the diminishing power of the United States, the growing prominence of emerging players, and the revival of national interests. With the decline of unipolarity, there is now an evolving power dynamic where global and regional players are vying to assert their influence and safeguard their concerns. Consequently, considerations related to territory, resources, and strategic positioning have regained importance in international relations as a recognized field. Examples include issues like Taiwan's status, Indo-China relations, and conflicts such as Russia-Ukraine.

Return of Geopolitics and Its Implications for South Asia

The South Asian area is widely known for its population and diverse mix of religions, cultures, and languages. It has been a hotspot, for geopolitical conflicts, territorial disputes, religious rivalries, religious extremism, power struggles between countries, like India and its neighbours, and the looming threat of nuclear war. These longstanding tensions have continued into the day contributing to instability in the region. In the twenty-first century, geopolitical considerations have resurfaced, exacerbating existing tensions and introducing complexities to the security landscape of South Asia. The resurgence of geopolitics has added another layer of complexity as different regional players pursue their goals. Notable issues include:

China's Expanding Footprint in South Asia

In South Asia, a blend of cultures diverse, rich history, and modern-day challenges are experiencing changes, in the realm of geopolitics. The People's Republic of China plays a key role in driving this transformation reshaping established alliances and altering the region's strategic landscape. This development carries implications, for both South Asia and the global community. Through its ambitious infrastructure and connectivity project BRI, China has expanded its presence in South and Southeast Asia. The investments made as part of the BRI not only focus on infrastructure development but also demonstrate China's strategic objectives in terms of its economic and geopolitical ambitions demonstrating an increase in Chinese influence across trade, investment, and strategic partnerships, unlike anything seen before. China's increasing presence presents advantages and challenges, for nations in South Asia. While it brings investment and financial support, it also pulls these countries towards its sphere of influence possibly reducing their independence. In essence, China's Belt and Road Initiative goes beyond building infrastructure; it showcases its goals, in South Asia with its important ports and transit hubs being crucial (Mishra & Mishra, 2021). China's BRI promises growth, development, and connectivity, yet these investments also raise concerns over debt traps, sovereignty issues, and long-term political influence.

China–Pakistan Economic Corridor (CPEC) is one of several initiatives within the Belt and Road Initiative (BRI). China–Pakistan Economic Corridor (CPEC), estimated at over USD 10 billion, represents an ambitious project which seeks to provide a link between Xinjiang province in China and Gwadar port in Pakistan. China–Pakistan Economic Corridor (CPEC) holds enormous promise of spurring significant economic development, strengthening energy security and driving infrastructural expansion. China's strategic interests are served by providing a direct maritime pathway to the Arabian Sea, thus bypassing the Malacca Straits as a geopolitical bottleneck. Economic implications of the CPEC are evident, with the potential to spur industrial development and improve infrastructure within Pakistan. However, it is essential to acknowledge its strategic underpinnings. India's opposition to CPEC arises from its proposed passage through Pakistani-occupied Kashmir, an area subject to longstanding territorial dispute. India views this development as an attack on her sovereignty and territorial integrity (Rajagopalan, 2022).

China's increased economic engagements in nations such as Sri Lanka, Nepal, and Bangladesh are indicative of a strategic endeavour to encircle India, commonly known as the 'String of Pearls' approach. The 99-year lease of the Hambantota Port in Sri Lanka by China, prompted by Sri Lanka's debt distress, resulted in concerns within India. In a similar vein, the close relationships that Beijing has nurtured with Kathmandu and Dhaka, primarily through the implementation of infrastructural projects and the provision of economic aid, suggest a notable departure from established patterns of alliances (Paul, 2018). In the realm of geopolitical dynamics, it is noteworthy to observe China's assertive stance in the South China Sea, which has led to territorial claims and exertion of influence over Southeast Asian nations. This development has inevitably brought China into direct confrontation with India's Act East policy, as both nations are competing for regional influence and strategic positioning.

India-Pakistan Rivalry: An Enduring Geopolitical Flashpoint

The geopolitical chessboard has become an arena of power politics where nations flex their muscles and engage in strategic manoeuvres. In the realm of tensions, South Asia stands out prominently particularly due to the rivalry between India and Pakistan which significantly influences the regional geopolitical landscape. The historical religious and territorial disputes have fuelled competition between these two nations since their partition from British India, in 1947. Over the years, they have engaged in conflicts often centred around Kashmir as a matter of national significance. This prolonged discord is rooted in identity-driven sentiments that are deeply ingrained within the fabric of both societies.

Amidst the changing landscape shaped by China's emergence, a new dynamic has emerged in South Asia. Pakistan traditionally seen as a rival, to India in the region has been aligning itself with China to counterbalance India. In this pursuit of equilibrium, Pakistan has found a partner in China. Through its might and strategic goals, Beijing has cultivated an alliance with Islamabad to offset New Delhi's influence in South Asia. The China–Pakistan Economic Corridor (CPEC) an initiative under China Belt and Road project symbolizes the depth of this collaboration by offering advantages to Pakistan while granting China direct access to the Arabian Sea. By backing Pakistan, China aims to establish a presence that challenges India's dominance in the region (USIP China-South Asia Senior Study Group, 2020). The bond between China and Pakistan presents challenges for India that extend beyond concerns into realms such as economics, technology, and soft power.

India and Pakistan are still embroiled in a conflict, over Kashmir with several wars fought over this mountainous region in the past few years. Both countries currently have a military presence in the area making it one of the most heavily militarized zones globally. The escalating geopolitical situation has heightened tensions.

Solidified hardline stances from both sides as external forces increasingly take positions on either side. The conflict in Kashmir goes beyond territorial disputes. It reflects deep-rooted distrust and historical tensions between two nuclear-armed neighbouring countries. Each event in Kashmir has repercussions on the international stage posing threats to stability. Despite attempts at discussions and peace initiatives, Kashmir continues to be an issue. India's decision to revoke Article 370 in 2019, thereby removing the region's special status introduced another level of complication leading to reactions from Pakistan and raising concerns, among the international community (Ahlawat & Izarali, 2020).

The tensions between the two countries go beyond disputes over land. The issue of cross-border terrorism often linked to groups operating from Pakistan but not officially affiliated with the state has been a significant barrier to achieving peace. The tragic assaults on soldiers in Uri and Pulwama serve as evidence of this on-going challenge. India consistently accuses Pakistan of backing these attacks, an allegation that Islamabad vehemently denies. These incidents not only raise tensions, between the two nations. Also, pose a risk of escalating into larger military conflicts, if any such cross border terrorism activity in future is going to take place, keeping aside their nuclear capabilities. Pakistan's suspected involvement, in supporting terrorism is more than a matter between two nations; it poses a challenge for the region. The consequences of these actions could lead to a larger conflict drawing in regional and global actors. Moreover, given the events in Afghanistan there are concerns that extremist organizations could seek refuge there adding complexity to the security situation, in South Asia.

Afghanistan's Geopolitical Uncertainty: A New Regional Power Play

In the changing South Asian geopolitical scene, characterized by instability and unpredictability, there are significant shifts taking place. Afghanistan, known as the graveyard of empires, for nations, has once more resisted a dominant force. With the US troops leaving Afghanistan, the situation in South Asia has become more precarious. The rapid rise of the Taliban resulted in Kabul's fall and their control, over the nation. This resurgence carries consequences for neighbouring nations. Now that the Taliban has taken control, there is growing worry about the rights of women and minorities and the possible reversal of democratic principles that have been established over time (Amnesty International, 2023). The swift change in leadership has led to a situation in South Asia with far-reaching implications for countries like India, Pakistan, and even China. India, given its cultural connections, to Afghanistan is now grappling with a challenging dilemma (Shekhawat, 2023). The Taliban's control of Afghanistan significantly diminishes India's influence in the country, cultivated over two decades through diplomacy, trade, and significant infrastructural investments. The Indian government, which had robust relations with the ousted Afghan administration, is now grappling with the challenge of navigating ties with a Taliban-led regime, which might be more inclined to side with Pakistan, India's arch-rival.

Pakistan's historical ties and support to the Taliban are no secret. Islamabad has viewed the Taliban as a strategic asset to counterbalance Indian influence in Afghanistan. With the Taliban now in power, Pakistan might have increased leverage in Kabul (Threlkeld & Easterly, 2021). This relationship not only challenges India's diplomatic standing in Afghanistan but also raises concerns about Pakistan's influence on the Taliban's policies, particularly those affecting India and the wider region.

The vacuum left by the US's withdrawal and the subsequent power shift has allowed China to make inroads into Afghanistan and thus marks a new chapter in the geopolitical landscape of South Asia. Beijing's primary interest is to ensure that Afghanistan does not become a breeding ground for Uighur militants who could threaten its Xinjiang province. Moreover, the vast mineral resources in Afghanistan provide an economic incentive for Chinese investments. However, China's engagement, particularly its economic interests and potential infrastructure projects, may further complicate the geopolitical matrix (Wani, 2023). India will have to tread cautiously, given its on-going border disputes and broader strategic rivalry with China. This growing Chinese presence could challenge India's strategic interests in the region, particularly if Beijing and Islamabad work together to curtail New Delhi's influence.

The Taliban's ascent poses direct security challenges for India. Historically, India has maintained cordial ties with Afghanistan, investing significantly in its infrastructure, education, and health sectors. With these investments and diplomatic efforts, in place the reemergence of the Taliban is concerning, especially considering their hostility towards India. History has shown that when left unchecked militant groups can directly threaten India's security as evidenced by the attacks in the late 1990s and early 2000s. India has key issues to deal with in Afghanistan after the withdrawal of US troops, to deal with China's growing presence in Afghanistan along with the Taliban's resurgence and Pakistan's strategic influence presents a challenge for India. To navigate this situation effectively, India should try to adjust its strategies by engaging in talks with the Taliban and strengthening relationships with other countries, in Central Asia (Wani, 2022). There is a possibility in future that both India and Taliban led Afghanistan can come closer to each other, if any geopolitical changes happens in South Asia and can counter China's and Pakistan's present status of close proximity with Afghanistan.

The resurgence of geopolitical dynamics has ranging implications, for the entire South Asian region. It raises security concerns escalating the potential for conflicts and contributes to the proliferation of arms race. On-going border disputes such as the India-Pakistan conflict over Kashmir and the India-China border tensions serve as areas of concern. The region's economic progress and growth have faced obstacles due to geopolitical rivalries and cross-border terrorism. Geopolitics brings about both opportunities and challenges. Enhancing connectivity developing infrastructure and fostering collaboration can facilitate regional integration and prosperity. Initiatives like the Bay of Bengal Initiative for Multi-Sectoral Technical and Economic Cooperation (BIMSTEC) offer avenues for working on shared issues such as climate change or maritime security concerns.

India's Role in the Changing Geopolitical Landscape

South Asia, an important region in terms of both its location and population size, has experienced a profound transformation in its geopolitical dynamics over recent decades. South Asia has witnessed centuries of political upheavals, colonial rule, wars, and tenuous diplomatic relationships. At present, as global power shifts are realigning, South Asia finds itself at the crossroads of significant strategic interests primarily between India, China, and the US. India, with its vast territory, diverse history, and robust economy has long occupied an integral place within South Asian geopolitics. From its days as part of the Non-Aligned Movement through today's global realpolitik, India's involvement has been distinguished by strategic interests, regional stability, and engagement with major global powers. Due to changes in geopolitics and changing geopolitical dynamics, its role and involvement have grown increasingly significant in regional politics. Here are key aspects of India's role:

India understands the importance of building alliances and strategic relationships to navigate geopolitics. With global power dynamics changing and regional tensions rising, India is actively participating in multiple platforms to address these challenges.

India and the Quad

India's strategic partnerships include its association with the US, Japan, and Australia in the Quadrilateral Security Dialogue (Quad). Its primary aim is to promote an open Indo-Pacific; especially as China asserts itself more in the region, this forum seeks to counterbalance Beijing's influence by emphasizing rule of law principles such as freedom of navigation as well as upholding territorial integrity and sovereignty.

Security Cooperation: The Quad nations hold regular joint naval exercises, most notably the Malabar exercise. These exercises aim to foster interoperability between member countries as well as build mutual understanding among them while preparing them for potential security threats.

Balance China's Influence: China has long been seen with suspicion by many nations in Asia, as its rise as both an economic and military power. Though not explicitly an anti-China coalition, QUAD strives to balance out China's influence in Asia ensuring no single power dominates it unilaterally.

Economic Integration: Beyond military considerations, the Quad serves as an avenue for economic integration. Due to scepticism regarding China's Belt and Road Initiative (BRI), there has been an effort among Quad nations to offer alternatives for infrastructure development and investment in Indo-Pacific regions.

Shared Democratic Values: QUAD members all hail from democracies with shared democratic values of freedom, human rights and the rule of law—providing them a common platform both strategically and for upholding global values.

Promoting an Open and Free Indo-Pacific: The Indo-Pacific region spanning both the Indian and Pacific oceans is a key trade corridor, so India and its QUAD partners seek to ensure it remains free, open, and inclusive while encouraging transparent infrastructure investment and exchanges of trade between its member nations. Respect for international laws and norms is of utmost importance.

BRICS and SCO

India has made considerable strides towards global leadership through participation in multilateral forums like BRICS and SCO, beyond its core involvement with Quad. India seeks to further show its strategic interests while contributing towards multipolarism in world governance and fulfilling its ambition of playing an instrumental role.

BRICS: Brazil, Russia, India, China, and South Africa form the BRICS grouping, an alliance of major emerging economies. While initially intended as an economic alliance, over time BRICS has evolved to address political and security-related matters; India sees it as an avenue for South-South cooperation, reform of global financial institutions, and multipolarity promotion.

Shanghai Cooperation Organization (SCO): While BRICS is more economically oriented, SCO, which includes countries like Russia, China, Kazakhstan, Kyrgyzstan, Tajikistan, Uzbekistan, Pakistan, Iran and India, focuses on political, defence, and security-related cooperation. For India, joining SCO was a way to engage with its immediate neighbours and major Eurasian powers, especially in combating terrorism, enhancing regional security, and promoting trade and connectivity. As a full member since 2017, India has leveraged the SCO to enhance its connectivity with Central Asia and engage with major players like Russia and China on security and defence issues and also to maintain dialogue with adversaries like China and Pakistan, given their membership.

Regional Groupings: From SAARC to BIMSTEC

India plays an instrumental role in regional groupings like the South Asian Association for Regional Cooperation (SAARC) and the Bay of Bengal Initiative for Multi-Sectoral Technical and Economic Cooperation (BIMSTEC). While SAARC has faced challenges due to India-Pakistan rifts, BIMSTEC is emerging as a promising platform for regional cooperation, with India taking the lead in various initiatives.

South Asian Association for Regional Cooperation (SAARC) was established in 1985 and brought together eight nations with the aim of promoting peace, stability, development, and prosperity in South Asia. India, given its size and strategic position, has been the driving force behind many SAARC initiatives. New Delhi-based South Asian University attracts students from across the South Asian Association of Regional Cooperation nations, serving as a testament to India's dedication to regional integration and knowledge-sharing. However, SAARC's potential has often been undermined by political rivalries between India and Pakistan; their ongoing tensions often overshadow SAARC. Meetings have been delayed or cancelled and initiatives shelved, creating the perception of SAARC being a somewhat ineffectual body. India-Pakistan tension has not only hindered cooperation within SAARC but has also had an adverse effect on its international standing. India has always championed regional cooperation despite these difficulties, often seeking to resolve bilateral differences for the greater regional good. But for SAARC to be truly effective again, its member nations must put aside bilateral issues and focus on shared regional aspirations instead.

Established in 1997, the Bay of Bengal Initiative for Multi-Sector Technical and Economic Cooperation (BIMSTEC) now comprises seven member states from Asia; these are India, Bangladesh, Bhutan, Nepal, and Sri Lanka from South Asia; Myanmar and Thailand are from Southeast Asia. This grouping's primary purpose is to connect South Asia and Southeast Asia in terms of physical connectivity projects as well as economic and cultural exchanges; India plays an essential role in this regard. It acknowledges BIMSTEC's potential as an alternative to SAARC where shared goals can be accomplished without being hindered by India-Pakistan antagonism. BIMSTEC's focus areas, spanning trade and tourism to technology, culture, and environment, fit perfectly with India's 'Act East' policy and serve as a great vehicle for regional outreach. India's participation in BIMSTEC is both strategic and practical in nature. India is asserting its leadership of various initiatives by not only reinforcing its place as a regional powerhouse but also offering an alternative narrative for South Asian cooperation, free from longstanding political differences. Given shared regional challenges such as climate change and security threats, its involvement in BIMSTEC could open up opportunities for more regional cooperation in future years.

India and China

China's rise as a global superpower has significantly shaped South Asian geopolitics, particularly within South Asian states that have adopted China's Belt and Road Initiative (BRI). BRI offers new dimensions of engagement as well as competition. China has undertaken an ambitious initiative known as the Belt and Road Initiative, or BRI, which seeks to connect Asia, Europe, Africa, and beyond through infrastructure projects. South Asian nations such as Pakistan, Sri Lanka, Bangladesh, and Nepal have demonstrated immense enthusiasm for the Belt and Road Initiative

(BRI), seeing it as an invaluable opportunity for infrastructure development and economic expansion. India recognizes that China's BRI presents significant geostrategic and strategic ramifications, as its influence grows more notable within India's territory and threatens its sovereignty. India has been concerned about China's increasing influence in their backyard through ports like Gwadar in Pakistan and Hambantota in Sri Lanka. Gwadar Port in Pakistan's Baluchistan province, funded and run by China, serves as a strategic foothold in the Arabian Sea for Beijing. India views any possible Chinese naval presence at Gwadar as a potential threat (Basrur et al., n.d.). Meanwhile, Sri Lankan Hambantota Port was taken over by China under a 99-year lease agreement due to its inability to repay Chinese loans. As mentioned above, this development has raised alarms in India about China's 'Debt-trap Diplomacy' and its potential military implications. These ports, combined with other strategic points across the Indian Ocean, give rise to the String of Pearls' theory, a perceived strategy by China to encircle India through a series of naval bases and commercial ports.

Balancing Act with Major Powers

India's strategic partnership with the United States, bolstered by shared democratic values and concerns about China's assertiveness, has grown. As the world's two largest democracies, both nations cherish the ideals of freedom, human rights, and the rule of law. This shared democratic ethos provides a foundational bond. The rise of China and its increasingly assertive posture in the South China Sea, the Indian Ocean, and along the India-China border have caused concerns in both New Delhi and Washington. This has led to closer strategic and defence cooperation between India and the United States, as exemplified by the Quadrilateral Security Dialogue and joint naval exercises. Simultaneously, India continues to engage with Russia, especially in the defence and energy sectors. Despite global shifts, India's relationship with Russia remains steadfast. Historically, the USSR and subsequently Russia have been reliable partners for India. Russia has been India's primary defence partner for decades. The Sukhoi fighter jets, T-90 tanks, and the S-400 missile defence system exemplify this deep-rooted military cooperation. Russia plays a crucial role in India's energy security, providing nuclear technology for civilian purposes and oil and gas deals, and engaging in hydrocarbon trade. Both nations advocate a multipolar world order and frequently collaborate on global platforms like BRICS and the Shanghai Cooperation Organization (Rajagopalan, 2023). On various global platforms, including the UN, Russia has often supported India, be it in its fight against terrorism or its aspirations for a permanent seat in the UN Security Council.

India's dual engagement, with the United States on one side and Russia on the other, isn't a matter of convenience but a deliberate strategy. India is guided by principles of non-alignment in its foreign policy approach. This involves making decisions based on national interests without unduly being influenced by any major power, while still maintaining robust relations with both America and Russia—this dual engagement exemplifies India's nuanced foreign policy approach of non-alignment and strategic autonomy.

Regional Connectivity and Economic Initiatives: Soft Power and Development Diplomacy

India stands as an essential player in South Asian geopolitics, where historical animosities, military might, and new alliances continually shift the strategic landscape. India has emerged as an influential voice and key actor. India's approach towards its neighbours and wider region demonstrates its transition towards soft power and development diplomacy. India has chosen not to rely heavily on military force or coercive diplomacy when engaging other nations; rather it seeks to strengthen cultural, economic, and developmental ties instead (Xavier, 2020). Soft power stems from an understanding that economic integration and regional cooperation not only benefit trade and commerce but also foster mutual trust and guarantee regional stability. India has made considerable efforts in regional connectivity and infrastructure development across South Asia and beyond, as evidenced by projects that represent this strategy. By supporting projects such as the India–Myanmar–Thailand Trilateral Highway (IMTTH), Bangladesh, Bhutan, India, and Nepal Motor Vehicle Agreement, and India–Middle East–Europe Economic Corridor as well as Chabahar Port in Iran, India demonstrates its dedication to regional integration and collaboration thereby effectively using soft power diplomacy for development purposes.

The IMTTH is a showcase of India's regional connectivity strategy. Spanning over 1400 km, this highway links India's northeastern states with Southeast Asian nations like Myanmar and Thailand—its goal being not only trade but also cultural exchange and mutual understanding among participating nations. For India specifically, this route represents an alternate trade route which reduces dependence on traditional passageways like the Siliguri Corridor. Furthermore, it opens up India's northeastern region, traditionally viewed as peripheral, to broader economic opportunities, integrating it further with the national economic mainstream (Ghatak, 2021). Thus, it can be observed that the IMTTH Highway serves three key purposes. Firstly, it stands for economic connectivity. The project is expected to significantly bolster trade, allowing the seamless movement of goods among the three nations. Secondly, it promotes cultural exchange by facilitating tourism and interpersonal exchanges, the highway will help reinforce cultural ties and mutual understanding. Lastly, it seeks strategic depth by bypassing traditional chokepoints, India is carving an alternative path that reduces its dependence on vulnerable sea routes.

In the larger maritime landscape of the Indian Ocean Region, the Chabahar Port has emerged as a strategic pivot for India. Located in southeastern Iran, overlooking the Arabian Sea, this port offers India a direct gateway to landlocked Afghanistan and Central Asia, bypassing the traditional and often contentious route through Pakistan. The development of the Chabahar Port underscores India's intention to enhance its maritime footprint and provide an alternative to China's Belt and Road Initiative (BRI), which has been expanding across the region. By investing in Chabahar, India aims to balance the regional scales, provide Afghanistan with a viable trade route, and counterbalance China's growing influence in the region. Another dimension of investing in Chabahar, India cements its bilateral relations with Iran, crucial in light of Iran's significance in regional geopolitics and its oil reserves. Moreover, this port emphasizes India's commitment to fostering regional integration and leveraging development diplomacy. It's not just about geopolitical competition; it's about creating sustainable and mutually beneficial economic linkages.

The BBIN initiative focuses on a sub-regional cooperation mechanism to enhance road connectivity among the four nations. By fostering transportation linkages, it aims to facilitate the seamless movement of people and goods, thus boosting trade and tourism and fostering regional integration. Moreover, by bringing these nations closer, the BBIN network reinforces India's 'Neighbourhood First' policy. This regional camaraderie allows for better collective bargaining on global platforms, helps counteract any efforts to undermine regional stability, and underscores the benefits of shared growth and interdependence.

India–Middle East–Europe Economic Corridor initiative is yet another testament to India's vision of expanding its trade and investment horizons. By focusing on establishing seamless connectivity between South Asia, the Middle East, and Europe, India envisages a multi-fold increase in economic partnerships, promotes energy cooperation and cultural exchanges, and ensures maritime security. Such an initiative will not only uplift India's global economic stature but will also positively impact the regional economies involved. Against the backdrop of the evolving geopolitics involving China's Belt and Road Initiative (BRI), this corridor stands as a testimony to India's ambitions and proactive foreign policy.

The initiatives of India's regional connectivity are part of a larger strategy for making use of soft power and encouraging economic interdependence across the region. By focusing on infrastructure projects, India sends a clear message that its growth could be a path to cooperation and shared prosperity rather than hegemony and dominance. Additionally, these initiatives represent India's efforts to influence its South Asian geopolitical narrative, which emphasizes collaboration, mutual respect, and benefits shared by all. India wants to be recognized as a generous regional leader who understands the complexities that govern regional relations and is committed to forging an optimistic, brighter, and more prosperous future for everyone involved in regional cooperation.

Conclusion

India's geopolitical engagement within the Global South has undergone significant evolution in recent times. India has swiftly emerged as a prominent participant in regional dynamics, offering substantial contributions to the maintenance of regional stability, promotion of economic prosperity, and facilitation of cooperative initiatives. This has been achieved through the diligent pursuit of diplomatic, economic, and security endeavours. Given the enduring complexities that India faces, it becomes imperative to acknowledge its strategic positioning, economic prowess, and steadfast dedication to fostering South-South cooperation. The aforementioned factors serve as a robust foundation for India to maintain its active involvement and exert its influence within the realm of the Global South. The resurgence of geopolitics in the realm of international relations has sparked significant transformations in global power dynamics and the distribution of influence. These developments have had profound and wide-ranging implications, particularly for regions such as South Asia. The multifaceted and evolving role of India as a burgeoning global power necessitates a comprehensive analysis of the associated challenges and opportunities at hand. In order to thrive in this contemporary epoch, it is imperative for India to deftly navigate the delicate equilibrium between its domestic aspirations and a magnanimous and collaborative approach towards its neighbouring nations and the international community at large. India possesses the potential to make substantial contributions towards upholding stability, security, and prosperity in the South Asian region, while simultaneously exerting its influence on the development of the global order.

References

Abraham, I. (2008). From Bandung to NAM: Non-alignment and Indian Foreign Policy, 1947–65. *Commonwealth & Comparative Politics, 46*(2), 195–219. https://doi.org/10.1080/14662040801990280

Ahlawat, D., & Izarali, M. R. (2020). India's revocation of Article 370: security dilemmas and options for Pakistan. *The Round Table, 109*(6), 663–683. https://doi.org/10.1080/00358533.2020.1849495

Amnesty International. (2023). *Report 2022/2023—The state of the world's Human Rights*. Retrieved from www.amnesty.org

Basrur, R. M., Mukherjee, A., & Paul, T. V. (n.d.). *India-China maritime competition: The security dilemma at sea (First)*. Taylor and Francis.

Ghatak, S. (2021). The Indian perspective on connectivity corridors with Southeast Asia: From barrier to bridge. *The Review of International Affairs, LXXII*(1183), 5–28.

Khayat, S. (2024). *How India connects the Global South and developed world*. Retrieved March 5, 2024, from Pacific Forum website: https://pacforum.org/publications/pacnet-1-india-can-unite-the-global-south-with-the-developed-world/

Mishra, M., & Mishra, V. K. (2021). China's belt and road initiatives examining debt trap diplomacy for regional Supermacy in the context of Sri Lanka. *World Affairs, 25*(1), 122–135. Retrieved from https://www.jstor.org/stable/48622938?seq=2

Paul, T. V. (2018). When balance of power meets globalization: China, India and the small states of South Asia. *Politics, 39*(1), 50–63. https://doi.org/10.1177/0263395718779930

Raghavan, P., Bayly, M. J., Leake, E., & Paliwal, A. (2022). The limits of decolonisation in India's international thought and practice: An introduction. *The International History Review, 44*(4), 812–818. https://doi.org/10.1080/07075332.2022.2090990

Raj, R. S., & Singh, S. K. (2022). India's soft power in Afghanistan: Myth or reality. *Satraachee, 37*(25), 83–95.

Rajagopalan, R. P. (2022). *India's latest concerns with the China-Pakistan economic corridor*. ORF. Retrieved October 23, 2023, from Observer Reasearch Foundation website: https://www.orfonline.org/research/indias-latest-concerns-with-the-cpec/

Rajagopalan, R. P. (2023). *India's balancing act viewed through recent military exercises*. Retrieved October 23, 2023, from Observer Reasearch Foundation website: https://www.orfonline.org/research/indias-balancing-act-viewed-through-recent-military-exercises/

Rathore, N. (n.d.). *India's soft power and leadership in Global South*. Retrieved October 22, 2023, from Kautilya Public Policy website: https://kautilya.org.in/indias-soft-power-and-leadership-in-global-south/

Shekhawat, S. (2023). *India's Taliban dilemma: Diplomatic engagement and moral disquietness*. Retrieved March 5, 2024, from Observer Reasearch Foundation website: https://www.orfonline.org/expert-speak/indias-taliban-dilemma

Singh, B. (2017). Non-alignment movement: It's relevance in present context. *International Journal of Research—GRANTHAALAYAH, 5*(6), 86. https://doi.org/10.5281/zenodo.820965

Threlkeld, E., & Easterly, G. (2021). *Afghanistan-Pakistan ties and future stability in Afghanistan*. United States Institute of Peace. Retrieved from www.usip.org

USIP China-South Asia Senior Study Group. (2020). *China's influence on conflict dynamics in South Asia*. USIP China-South Asia Senior Study Group. Retrieved from https://www.usip.org/publications/2020/12/chinas-influence-conflict-dynamics-south-asia

Wani, A. (2022). *Taliban's Afghanistan: An emerging security challenge for India and Central Asia*. ORF.

Wani, A. (2023). *China's dichotomous ties with the Taliban and Uyghurs*. Retrieved October 23, 2023, from Observer Reasearch Foundation website: https://www.orfonline.org/expert-speak/chinas-dichotomous-ties-with-the-taliban-and-uyghurs/

Xavier, C. (2020). *Sambandh as strategy India's new approach to regional connectivity*. CSEP.

Unlocking Potential: Transboundary Rivers as Drivers of Regional Solidarity in South Asia

Rahul M. Lad and Ravindra G. Jaybhaye

Introduction

Transboundary rivers, the lifeblood of nations, traverse political boundaries and sustain the livelihoods of millions across South Asia. These majestic waterways have long played a pivotal role in shaping the socio-economic, political, and environmental landscapes of the region. South Asia, a region known for its diverse cultures, histories, and complex geopolitical relations, is home to some of the world's most significant transboundary river systems. The Indus, Ganges, Brahmaputra, and their tributaries have historically been critical channels of prosperity, connecting nations, and supporting agriculture.

For all of the continental countries of South Asia, transboundary water resources are more important (Bradnock, 2015). Most of the rivers in South Asia have an international basin because they traverse one or more political boundaries. Twenty significant rivers are shared by the South Asian countries, excluding the island nations of Sri Lanka and the Maldives (Dholakia, 2019). The Indus Basin spans China, India, Afghanistan, and Pakistan. India, Nepal, Bangladesh, and Bhutan are connected by the river basins of the Brahmaputra and Ganges. The Kosi, Gandaki, and Mahakali rivers connect India and Nepal. India and Bangladesh share 54 rivers, with the Brahmaputra, Ganges, and Teesta being the largest. The Kabul River basin connects Pakistan and Afghanistan.

The two main common river systems in South Asia are the Ganges-Brahmaputra-Meghna (GBM) and the Indus River basin. These two river basins collectively

R. M. Lad
Department of Geography, School of Liberal Arts and Social Sciences, Faculty of Education and Humanities, JSPM University, Pune, Maharashtra, India

R. G. Jaybhaye (✉)
Department of Geography, Savitribai Phule Pune University, Ganeshkhind, Pune, Maharashtra, India

K. Arora, R. R. Anand (eds.), *The Solidarity Approach in Geography*, https://doi.org/10.1007/978-3-031-86753-8_3

support around 700 million people, and they have the potential to influence almost 100 million hectares of arable land in the region. Both river basins have been proven to have hydropower potential exceeding 100,000 MW (Desai, 2010). As a result, the water resources in this region of South Asia are incredibly significant.

Except for Sri Lanka and the Maldives, the majority of mainland South Asian nations experience a significant dependency ratio (Fig. 1). The dependency ratio describes how dependent downstream riparian nations are on water resources originating from their upstream counterparts, especially when it comes to transboundary rivers. The shared water resources in transboundary river basins call for careful management, given that the utilization of these resources by one nation can significantly impact water availability for downstream nations. This underscores the necessity for regional cooperation through multilateral institutions regarding cross-border river water management, as highlighted by Bradnock (2015).

However, the potential of these transboundary rivers extends beyond mere utility. They possess the inherent ability to foster regional solidarity and cooperation among the nations they traverse. By harnessing their potential through sustainable and collaborative management, South Asian countries can unlock a pathway to shared prosperity, regional stability, and environmental sustainability. This chapter endeavors to explore the untapped potential of transboundary rivers as catalysts for enhancing regional cooperation and fostering a sense of solidarity among the nations of South Asia. In the pages that follow, we will delve into the issues associated with these transboundary rivers, highlighting the symbiotic relationship between the river systems and the nations they traverse. We will examine the challenges and opportunities presented by these rivers, exploring the impact of climate change, water scarcity, and other factors on their sustainable management. Moreover, we will analyze successful examples of transboundary river cooperation from around the world, seeking to draw lessons that can be applied to the South Asian context.

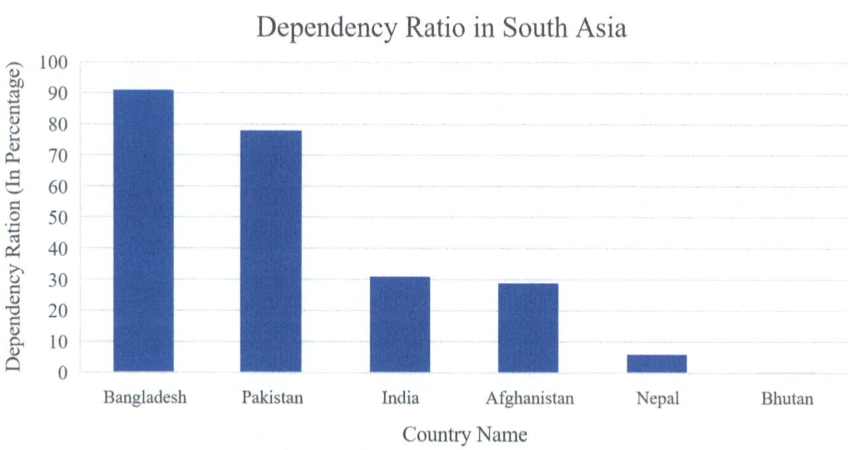

Fig. 1 Dependency ratio of Mainland South Asian Countries (in percentage). (Source: The Routledge Atlas of South Asian Affairs)

By understanding various dimensions of transboundary river management, this chapter aims to provide insights into how these vital waterways can become unifying forces, fostering greater collaboration, harmony, and solidarity among the nations of South Asia. Ultimately, unlocking the potential of transboundary rivers holds the promise of a brighter, more interconnected future for the region, characterized by collective progress and sustainable development.

Literature Review

The study of transboundary rivers and their role in promoting regional solidarity is a multidisciplinary field. In the South Asian context, major rivers like the Ganges, Brahmaputra, Indus, and their tributaries traverse multiple countries, presenting complex challenges and opportunities for the nations involved. Wolf (1995) thoroughly discussed how transboundary rivers contribute to deciding political relations among the nations and aspects of conflict as well through the theoretical framework. Mirumachi (2015) examines transboundary river governance and cooperation, emphasizing the significance of power dynamics, legal frameworks, and institutional arrangements in shaping collaborative approaches. This work also provides valuable insights into the institutional and political dimensions of transboundary river management, which are crucial aspects of promoting regional solidarity in South Asia.

The scholars have focused on both aspects of environmental issues creating conflict and cooperation among the countries. The connection between environmental problems including climate change, the scarcity of renewable resources, water sharing, and natural disasters, and how these problems might spark conflict has attracted the attention of academics and decision-makers (McDonald, 2013; Van Baalen & Malin, 2017). In contrast, Ide (2019) has furnished an elaborate overview of the scholarly works concerning environmental concerns and prospects for collaboration between states. The author claims that scholars have employed a range of terminology to describe the relationships that exist between environmental problems, collaborative environmental efforts, and peacebuilding—a notion referred to as "environment-peace links." These designations include "disaster diplomacy" (Kelman, 2006), "ecological diplomacy" (Griffin & Ali, 2014), "water diplomacy" (Islam & Repella, 2015), "environmental peacebuilding" (Carius, 2006), "environmental peacemaking" (Conca, 2002), "environmental peace perspective" (Ide & Scheffran, 2014), and "peace ecology" (Amster, 2015). Even if some phrases are synonymous, they frequently refer to different literary works and theoretical systems.

According to the literature on environmental peacemaking, conflicting countries can set aside their differences and work together to address common environmental concerns, which will promote more harmonious relationships between them. Environmental peacemaking places emphasis on the fact that environmental problems, like pollution, water scarcity, and climate change, frequently cross international borders and have an impact on all nations. Understanding how their ecosystems are interconnected and how important it is for everyone to manage resources sustainably can help organizations find common ground for communication and collaboration. In the end, this mutual knowledge can promote cooperation, the development of trust, and the resolution of conflicts, resulting in a more harmonious cohabitation and long-term environmental sustainability. Therefore, environmental challenges have the potential to strengthen state-to-state regional integration, which may ultimately result in regional solidarity.

On this background, the present study attempts to explore the prospects and potential of environmental issues such as transboundary river water sharing as a catalyst to foster regional solidarity in South Asia.

Limits of Transboundary Water Cooperation in South Asia

South Asia is home to some of the world's most significant transboundary rivers, including the Ganges, Brahmaputra, and Indus. These rivers flow through multiple countries, making water sharing a critical issue for regional cooperation and stability. However, transboundary river water governance has largely remained a bilateral exercise in South Asia. Most of these bilateral regimes were enacted decades ago and lack provisions essential for addressing contemporary challenges, such as climate change. The inadequate representation of basin member countries and the rigid nature of the governance mechanisms create significant challenges in the region. For example, the Indus Water Treaty governs transboundary water management between India and Pakistan but does not include China and Afghanistan, which also share the Indus basin. Additionally, the treaty lacks mentions of critical contemporary issues such as climate change, groundwater extraction, water pollution, and population growth, all of which are pressing concerns in contemporary times. These omissions hinder the treaty's effectiveness. Similarly, in the Ganga basin, India and Bangladesh are struggling to find a solution for the diminished water availability in the river. Aggravated domestic water demand due to industrialization, irrigation, and population growth exacerbates the existing problem in the Ganga basin (Lad & Jaybhaye, 2023).

Although there are several issues regarding transboundary rivers, South Asia remains largely silent on initiating a regional effort to resolve them. The sole regional organization unifying all of South Asia, the South Asian Association for Regional Cooperation (SAARC), has abstained from addressing disputed issues among its members, such as the governance of transboundary river water (Jaybhaye & Lad, 2022). As a result, South Asia lacks a regional platform that unites all basin countries within a single organization to address these intricate issues. However, this deadlock can be resolved by prioritizing cooperation over non-traditional issues. By fostering a spirit of cooperation and mutual understanding, these initiatives can pave the way for addressing contentious matters like transboundary river water sharing.

Prioritizing Non-traditional Issues to Break the Stalemate

Target 6.5 of the Sustainable Development Goals (SDG) 6 states that the application of Integrated Water Resources Management (IWRM) should be prioritized, and it should include a transboundary component. Thus, to fully realize the 2030 Agenda for Sustainable Development, transboundary water cooperation is essential (Barua et al., 2019). In this context, "cooperation" refers to coordination between states in which they work together to accomplish shared goals and reap benefits. This implies that all of the basin countries need to collaborate to accomplish the shared objective of peacefully managing transboundary rivers. However, there are instances when nations that share rivers find it challenging to work together for a variety of reasons. However, in terms of hydro politics, some river basins around the globe, including the Jordan, the Nile, and Mekong, prove that water naturally tends to persuade even hostile co-riparians to work in tandem (Barua et al., 2019). The major driver for this kind of cooperation is to set aside the differences and formulate a common framework for optimum utilization of resources, in this case, water. In South Asia, countries like India, and Pakistan, not only share rivers but also historical animosity, mistrust, etc. But despite the presence of the Indus Water Treaty in 1960, the issues regarding water governance persist. Hence, the region-wide cooperation over transboundary rivers remains dismal in South Asia. Therefore, prioritizing the non-traditional issue to initiate the regional integration process can lead to solidarity among the nations.

In the following sections, this study discusses several non-traditional challenges that have the potential to foster cooperation and unity among countries, rather than inciting conflict. These challenges include cross-border cooperation over flood control, hydropower trade, navigation through shared rivers, river water pollution, early warning systems, and climate change.

Flood Control: A Need for "Democratization of Data"

Sharing of data among the key stakeholders holds primacy in the transboundary river water governance. It enables people and the government downstream to prepare for the upcoming danger. However, in South Asia, the exchange of water and climate-related data is not an easy practice. Most of the South Asian countries, including India and Pakistan, keep hydrological data secret/classified (Surie, 2015). While regional cooperation has already been established over trade, energy exchange, etc., these sectors are growing steadily, hydro-cooperation lags in the region. A reason for this highly securitized approach might be because of historical animosity, geopolitics, and cross-border disputes. However, this approach has hampered informed planning and management practices of shared basins in the region. In addition, the hydrological data is being used for political purposes by the countries to counter its adversaries. For example, in 2017, China didn't provide hydrological data to India, despite having a bilateral agreement (Manhas & Lad, 2024). However, the same data was being provided to Bangladesh. It happened against the backdrop of the Doklam crisis between India and China. It implies that hydrological data is being used as a weapon by countries in war or war-like situations against their perceived adversaries.

However, despite numerous obstacles to data sharing, some joint drives are currently underway. For instance, to lessen the vulnerability of impacted communities, the International Centre for Integrated Mountain Development (ICIMOD) has been collaborating with the governments of Pakistan, Bangladesh, Nepal, and Bhutan to create a Regional Flood Information System in the Hindu-Kush Himalayas (Surie, 2015). The flood data in the Ratu River, a shared river between India and Nepal, has also been exchanged since 2015 (Pyakurel, 2023). But this is not at the government level, it happens at the NGO and volunteers level. This needs to elevate at the government-to-government scale, so communities across the shared basins will be benefited.

Hydropower Trade

The South Asian region, particularly the countries of India, Nepal, and Bhutan, is endowed with significant hydropower potential due to the abundant river systems originating from the Himalayas. These rivers not only serve as crucial water sources but also present vast opportunities for hydropower development, which can act as a catalyst for regional cooperation and economic integration (Shrestha, 2018). Nepal, possessing about 83,000 MW of hydropower potential, and Bhutan, with about 30,000 MW, make both countries critical for hydropower trade in the region (Pandey, 2020; Dorji, 2019). Despite this proven potential, both countries couldn't utilize this to the fullest. It might be because of poor technology, lack of infrastructure, and difficult terrain. However, India, which lies on the southern slope of both countries, requires power to support its domestic demands. In addition, India has pledged to achieve net zero carbon emissions by 2070, an attempt to combat climate change. Since power generation is responsible for the majority of the carbon emissions, it needs to shift toward renewable sources of energy. However, India's generation capacity from renewable sources has not been satisfactory in the last few years. According to the Central Electricity Authority, as of January 2023, India's installed capacity stood at 412 GW of which 175 GW were produced using non-fossil fuels. This represents about 42% of the total installed capacity (Government of India, Ministry of Power, 2022). If this persists, it would be challenging to realize the neat zero emission in the given period.

Hence, it would be beneficial for India to import power from its Himalayan neighbors. It will be mutually beneficial for these countries as Nepal and Bhutan may earn foreign exchange through the export of hydropower to India. Lessons from the United States and Canada should be applied to this regional hydropower trade paradigm. These nations have effectively established a hydropower trade to bolster peace within the region. Large hydroelectric dams are to be constructed in accordance with the Columbia River Treaty (three in Canada and one in the United States) in order to maximize annual electricity production and lower the risk of catastrophic floods. The Canadian province of British Columbia is home to the dams. For the extra electricity generated as a result of the operation of the Canadian dams, the United States pays Canada (i.e., Canadian Entitlement) between $180 and $300 million a year. The Treaty is regarded as a prototype for regional cooperation in the domains of hydropower development and transboundary water management (Canada, 2022).

The above given model can be replicated in South Asia as India, a downstream country, may help Nepal and Bhutan financially by receiving excessive hydropower and reduced risk of floods in return. This will act as a strong cementing bloc to enhance regional harmony in the region.

Navigation through Shared Rivers

The inland water transport through rivers is a significant and environment-friendly transport system. If harnessed properly, it can enhance trade and transport in addition to regional amicability. In South Asia, rivers like Ganga, Brahmaputra, and Indus are the major transboundary rivers with potential for inland water transport. In the extensive Ganga-Brahmaputra-Meghna delta, where a wide network of waterways formerly carried over 70% of the region's trade and passengers, rivers constitute the oldest means of transportation. However, as of right now, less than 2% of goods are moved by water in the region (Stankevich et al., 2024).

Currently, countries like India and Bangladesh have developed inland waterways within their territorial limits (Stankevich et al., 2024). But, like shared rivers, the possibility of shared navigation among the countries through these rivers may be explored. The South Asian countries have already started cooperative engagement over navigation. For example, India and Bangladesh which share the Ganga delta are exploring the possibilities of reviving the defunct water channels. Both countries have proposed to initiate a transnational navigation under the comprehensive "Eastern Waterways Grid" (Law, 2023). The attempt to revitalize the economies of the eastern subcontinent through this grid poses the significant potential to benefit the region's 600 million people, most of them are among the poorest in South Asia. Moreover, India's Himalayan neighbors, Nepal and Bhutan, have started utilizing the rivers potential to facilitate their access to India and Bangladesh. Along with this, India has granted permission to Nepal to use its three major inland waterways, namely Kolkata-Kalughat, Raxaul; Kolkata-Sahebgunj, Biratnagar and Kolkata-Varanasi-Raxaul (Khanal, 2019). Furthermore, an Indian waterway in Assam facilitates trade between Bhutan and Bangladesh. The ship, carrying Bhutanese stone, set to sail from Dhubri (Assam) and will travel to Narayanganj (Bangladesh), over the Brahmaputra River and through the Indo-Bangladesh Protocol Route. Cargo transportation using this route will result in an 8–10-day reduction in journey time and

30% cost savings (Antara, 2024). This type of cooperation over transboundary rivers needs to be explored and strengthened in South Asia.

River Water Pollution

Hydropolitics primarily concentrates on ensuring sufficient water supply, with minimal emphasis placed on the waters quality. The quantity of water normally comes first when negotiating the water sharing agreement between the nations. For most of countries, however, "water quality rather than quantity lies at the crux of the problem" (Alam, 1998). According to Richards and Waterbury (1996) as well, the problem of poor water quality will make it harder to get enough water. When this contaminated water flows across the border, it may be responsible for downgrading the bilateral relations between the riparian countries. For instance, in 2019, India alleged Pakistan of releasing polluted water coming from the tanneries in Pakistan into the Sutlej river (Jagga, 2019). In 2017, the Siang river, which is the name for the Brahmaputra in the Indian states of Arunachal Pradesh, abruptly became muddy and darkened, leading to suspicions regarding Chinese activities upstream (Mohanty, 2021). Water tests revealed elevated turbidity levels, indicating an increased presence of suspended particles. Fish died in large numbers as a result of this event because the grills were clogged with suspended particles. The event had a severe impact on the fishing communities and agricultural output in the Siang Valley (Lad & Jaybhaye, 2023).

This kind of phenomenon may destabilize the bilateral relations among the riparian countries. However, the precautions by the upstream in the form of water treatment, reduced human interventions in the water channel, etc. can enhance the trust level which could lead to a better environment for future cooperation in South Asia. This type of cooperation exists in the Danube River basin through the Trans-National Monitoring Network (TNMN). This network is tasked with reporting the overall condition and long-term trends of surface water, and when necessary, the status of groundwater within a basin-wide context (ICPDR, 2018).

Early Warning System (EWS)

In South Asia, the majority of transboundary rivers originate in the Himalayan region. Since these rivers traverse through the high mountains, they receive water through rainfall, snowfall, and glacial melt. In the Monsoon season, the heavy and intense rainfall in the catchment area of these rivers creates havoc in the downstream region. In this situation, climatic cooperation in terms of advanced data sharing among the basin countries can make a considerable impact to avert the loss. According to the World Meteorological Organization, if an Early Warning System

(EWS) is implemented 24 h before floods, it can save damage and save lives (WMO, 2023). This implies that early warning systems from South Asian meteorological authorities may play a key role in lowering annual losses and damages in the region. This approach has been used in Nepal and appears to be having a favorable effect on the communities. About 13 million SMS notifications were given to populations in danger by Nepal's Department of Hydrology and Meteorology in 2022, as opposed to 3.5 million in 2019 (Uprety, 2024). These SMS alarm systems are turning into vital life-saving tools for Nepali villages impacted by floods. But, due to poor trans-border data sharing mechanism, the people living in the Indian side used to suffer as data is delayed, unclear, or in a language not familiar to local population (Sangomla, 2019). The possibility of similar incidents occurring could also be a concern in the context of transboundary rivers shared between Bhutan and India, as well as between China and India.

Climate hazards disregard national borders, making cross-border collaboration crucial for the safety and security of flood-affected communities in South Asia. Countries experiencing similar disasters, such as floods, should initiate efforts to develop a robust regional Early Warning System (EWS) to mitigate the impacts of these disasters across their borders. By sharing data, resources, and expertise, nations in South Asia can create a more resilient and responsive system that benefits all countries involved. Thus, shared water resources can become a unifying force among cross-border communities in the region, fostering cooperation and mutual support in the face of climate challenges.

Climate Change

Climate Change can be a dividing and uniting factor for the countries simultaneously. The observable impacts of it already started showing its footprints in various sectors in South Asia, water is one of them. As the Himalayan region is particularly vulnerable to climate change, the rivers originating in the region are also under dire threat of water availability (Chauhan et al., 2023). Since rivers like the Indus, largely rely on Himalayan glaciers for their flow, especially in dry season, it is critical for the survival of the people downstream. Above all, 70% of the rivers summer flow is meltwater (Adeel & Wirsing, 2016). As a result, the Indus flow regime is quite fragile given the anticipated climate predictions. A recent study conducted in the Netherlands suggests that by 2050, the Indus River flow may decrease by 8% due to glacier melting (Mandhana, 2012). This could pose a significant challenge for the extensive irrigation network sustained by the Indus River, which extends across the border.

Hence, it has become inevitable for the countries in the region to employ a cooperative approach to combat this shared adversary. However, South Asia lacks significant initiatives to counter it at the bilateral, multilateral, or institutional levels. Existing transboundary river water treaties, such as the Indus Water Treaty, the

Ganga Waters Treaty, and the Mahakali Treaty, do not include any provisions addressing this issue. The only notable achievement in this regard is the India-Bhutan agreement (The Statesman, 2024). In contrast, there are numerous examples of successful regional initiatives for climate change around the world. The governance regimes for the Danube and Mekong rivers serve as successful examples of regional cooperation.

Regarding the Danube, the ICPDR has implemented a climate change strategy for the entire basin to assist Danube countries in developing more efficient water management plans in the face of an increase in catastrophic floods and droughts. Another example of cooperation in addressing climate change can be seen in the collaborative efforts between China and its downstream riparian countries in Southeast Asia. Growing cooperation between the China-led upstream Lancang-Mekong Cooperation and the downstream Mekong River Commission has resulted from joint studies conducted in the Mekong basin to evaluate the effects of climate change (Peter, 2021). But, these kinds of multilateral governance regimes are not available in South Asia. However, South Asia needs to employ a basin-wide management approach to make the region climate resilient. To start with, finding a common ground for dialogue could initiate the process of regional cooperation. For instance, a whole-basin approach is vital in the Indus Basin to ensure that all users have access to adequate water despite more unpredictable climatic condition. To address these challenges, initiating joint studies, fostering technical cooperation, and establishing mechanisms for real-time hydrological data sharing could serve as essential first steps for the states within the Indus Basin. Specifically, Pakistan's Living Indus Initiative and India's Composite Water Management Index offer valuable frameworks that could be integrated into a collaborative effort to enhance water management practices across the basin (Rollason et al., 2024).

Furthermore, the Indus Forum, which evolved into the Indus Basin Knowledge Forum (IBKF), is an informal mechanism comprising participants from each of the four basin countries: India, Pakistan, China, and Afghanistan. Since its inception in 2013, the forum has successfully brought together a wide range of stakeholders, including academicians, scientists, media professionals, and civil society representatives from all four basin countries. The IBKF's main focus is to foster cooperation on critical issues such as climate change, data sharing, and supporting joint research initiatives. This collaborative effort aims to address the shared challenges and opportunities within the Indus Basin, promoting sustainable development and regional stability (World Bank, 2022).

The IBKF's efforts not only facilitate the sharing of knowledge and resources but also help to build mutual trust and understanding among the nations involved. Such trust is crucial for preventing conflicts and ensuring that all countries benefit equitably from the basin's water resources. This kind of approach will significantly create a conducive environment for increased cooperation among the riparian countries in South Asia.

Regional Governance Framework: Inspiration from the International Commission for the Protection of the Danube River (ICPDR)

River Danube, one of the longest rivers (2857 km) in Europe, traverses across 19 countries and its basin covers 817,000 km^2. The river supplies water to many major cities, including the capitals of four countries: Vienna, Bratislava, Budapest, and Belgrade. Furthermore, hydropower generation, agriculture, navigation, water supply, recreation, and the natural environment depend heavily on it (ICPDR, 2018).

Since the river crosses numerous political boundaries in Europe, its management of it becomes critical. However, the breakthrough was achieved in the form of the International Commission for the Protection of the Danube River (ICPDR), in 1998. Since its founding, it has offered the nations a forum to evaluate the river's condition and formulate plans for the whole basin to deal with critical issues like pollution, hydropower, navigation, and climate change adaptation. The 1992 United Nations Economic Commission for Europe (UNECE) Convention on the Protection and Use of Transboundary Watercourses and International Lakes served as the model for the ICPDR. Based on this, South Asia needs to learn from the Danube River Protection Convention, which has shown to be a successful example. This kind of initiatives will act as a common platform for the multilateral discussion over climate-induced challenges in the region. This would also promote cooperation and collaboration among countries in the region, leading to better management of shared water resources. By learning from the ICPDR's approach, South Asia can work toward a more integrated and effective framework for addressing water-related issues in the region. This kind of amicable and sustainable solution would certainly act as a confidence-building measure and ultimately strengthen the regional solidarity in South Asia.

Conclusion

Given the fact that water is an essential requirement for the livelihood of the people, its governance becomes of the utmost importance. If the river is transboundary, its management becomes more complex and complicated. Furthermore, contemporary issues like climate change are posing increasing risks to the water supply in the South Asian region. However, regional collaboration between the nations is crucial to addressing the current problems. Similar to other parts of the world, transboundary river water sharing problems also affect South Asia. However, South Asia lacks a regional or multinational framework for water governance, in contrast to other regions like Europe and South East Asia. There are several potential explanations for this lack of multilateral efforts, including persistent historical hostilities, a lack of mutual trust, rising nationalism, closed-door talks focused just on water-related issues, etc. However, this trend has the potential to seriously undermine regional harmony in the future. As a result, the nation's ought to band together to work jointly on the aforementioned issues. To start with, the study advocates prioritizing the non-traditional issues to achieve wider regional consensus. Issues like climate change, shared navigation, early warning systems, data sharing, etc. may be proven as confidence-building measures. These non-traditional issues can serve as a foundation for building trust and cooperation among South Asian countries, paving the way for addressing more complex and contentious matters in the future. By focusing on common goals and shared interests, countries in the region can work toward enhancing regional stability and solidarity.

References

Adeel, Z., & Wirsing, R. G. (Eds.). (2016). *Imagining Industan: Overcoming water insecurity in the Indus basin*. Springer.

Alam, U. Z. (1998). *Water rationality: Mediating the Indus Waters Treaty*. Doctoral dissertation, Durham University.

Amster, R. (2015). *Peace ecology*. Paradigm Publishers.

Antara. (2024). *Indian waterways boost Bhutan-Bangladesh trade*. IndBiz | Economic Diplomacy Division. Retrieved July 20, 2024, from https://indbiz.gov.in/indian-waterways-boost-bhutan-bangladesh-trade/

Barua, A., Deka, A., Gulati, V., Vij, S., Liao, X., & Qaddumi, H. M. (2019). Re-interpreting cooperation in transboundary waters: Bringing experiences from the Brahmaputra Basin. *Water, 11*, 2589. https://doi.org/10.3390/w11122589

Bradnock, R.W. (2015). The Routledge Atlas of South Asian Affairs (1st ed.). Routledge. https://doi.org/10.4324/9781315683089

Carius, A. (2006). *Environmental peacemaking: Environmental cooperation as an instrument of crisis prevention and peacebuilding: Condition for success and constraints*. Adelphi.

Central Electricity Authority, Ministry of Power, Government of India. (2022). *CEA annual report 2021–22*. Retrieved March 18, 2023, from https://cea.nic.in/wp-content/uploads/annual_reports/2022/AR_2021_22_dated_03.11.2022.pdf

Chauhan, H. K., Gallacher, D., Bhatt, A., & Bisht, A. K. (2023). The Himalayas: A climate change laboratory. *Environmental Development, 45*, 100814.

Conca, K. (2002). The case for environmental peacemaking. In K. Conca & G. Dabelko (Eds.), *Environmental peacemaking* (pp. 1–22). John Hopkins University Press.

Desai, V. V. (2010). "The Political Economy of Regional Cooperation in South Asia" (No. 54). ADB Working Paper Series on Regional Economic Integration.

Dholakia, A. (2019). Transboundary Rivers of South Asia–The Case for Regional Water Management. IMPAKTER (August 6, 2019) https://impakter.com/transboundary-rivers-of-south-asia.

Early Warning System. (2023). World Meteorological Organization. Retrieved July 17, 2024, from https://wmo.int/topics/early-warning-system

Environment and Climate Change Canada. (2022). *Canada-US Columbia River treaty*. Canada.ca. Retrieved July 18, 2024, from https://www.canada.ca/en/environment-climate-change/corporate/international-affairs/partnerships-countries-regions/north-america/canada-united-states-columbia-river.html

Griffin, P. J., & Ali, S. H. (2014). Managing transboundary wetlands: The Ramsar convention as a means of ecological diplomacy. *Journal of Environmental Studies and Sciences, 4*(3), 230–239.

Ide, T. (2019). The impact of environmental cooperation on peacemaking: Definitions, mechanisms, and empirical evidence. *International Studies Review, 21*(3), 327–346.

Ide, T., & Scheffran, J. (2014). On climate, conflict and cumulation: Suggestions for integrative cumulation of knowledge in the research on climate change and violent conflict. *Global Change, Peace & Security, 26*(3), 263–279.

International Commission for the Protection of the Danube River. (2018). *Lessons from the Danube: A world leader in transboundary river basin management*. Retrieved July 20, 2024, from https://www.icpdr.org/sites/default/files/nodes/documents/lessons-from-the-danube-a-world-leader-in-transboundary-river-basin-management.pdf

Islam, S., & Repella, A. C. (2015). Water diplomacy: A negotiated approach to manage complex water problems. *Journal of Contemporary Water Research & Education, 155*(1), 1–10.

Jagga, R. (2019, August 22). Polluted water from Pakistan tanneries adding to flood woes in Ferozepur, says DC. *The Indian Express*. Retrieved July 20, 2024, from https://indianexpress.com/article/india/polluted-water-from-pakistan-tanneries-adding-to-flood-woes-in-ferozepur-says-dc-5927062/

Jaybhaye, R., & Lad, R. (2022). Transboundary river management in South Asia: The exigency of multilateral institutional framework. *Annals of the National Association of Geographers, India, 42*(2), 209–225.

Kelman, I. (2006). Acting on disaster diplomacy. Journal of International Affairs, 59(2), 215–240.

Khanal, R. (2019). India agrees to allow Nepal to use three inland waterways. *The Kathmandu Post*. Retrieved July 20, 2024, from https://kathmandupost.com/money/2019/10/03/india-agrees-to-allow-nepal-to-use-three-inland-waterways

Lad, R., and Jaybhaye R. G. (2023). *Geo-strategic assessment of trans-boundary river water sharing in South Asia region*. Thesis, Savitribai Phule Pune University, Pune, India.

Law, B. A. (2023). India eyes 5,000 km eastern regional waterways grid connecting Bangladesh, Myanmar. *The Hindu Business Line*. Retrieved July 25, 2024, from https://www.thehindubusinessline.com/economy/logistics/india-eyes-5000-km-eastern-regional-waterways-grid-connecting-bangladesh-myanmar-sarbananda-sonowal/article67086464.ece

Mandhana, N. (2012). Water wars: Why India and Pakistan are squaring off over their rivers. *Time*. Retrieved February 1, 2022, from https://content.time.com/time/world/article/0,8599,2111601,00.html

Manhas, N. S., & Lad, R. M. (2024). China's weaponization of water in Tibet: A lesson for the lower riparian states. *Journal of Indo-Pacific Affairs, 7*(2), 211.

McDonald, M. (2013). Discourses of climate security. *Political Geography, 33*(1), 42–51.

Mirumachi, N. (2015). *Transboundary water politics in the developing world*. Routledge.

Mohanty, A. (2021). *Water as a political security tool: The Himalaya's strategic conundrum*. Institute for Security & Development Policy.

Pandey, A. (2020). Hydropower potential in Nepal: A review. *Renewable Energy Review, 55*(4), 277–290.

Peter, Z. (2021, September 24). China joins lower Mekong river countries to study impacts of climate change, dams. *Voice of America*. Retrieved July 22, 2024, from https://www.voanews.com/a/china-joins-lower-mekong-river-countries-to-study-impacts-of-climate-change-dams-/6244149.html

Pyakurel, D. (2023). *Community cooperation across Nepal-India border saves lives during floods—ICIMOD*. ICIMOD—International Centre for Integrated Mountain Development. Retrieved July 20, 2024, from https://www.icimod.org/article/community-cooperation-across-nepal-india-border-saves-lives-during-floods/

Richards, A., & Waterbury, J. (1996). *A political economy of the Middle East* (2nd ed.).

Rollason, R., Green, T., & Bhattarai, B. (2024). *Elevating river basin governance and cooperation in the HKH region: ICIMOD Summary report III, Indus River Basin*. ICIMOD.

Sangomla, A. (2019, August 13). Communication gap: Bihar floods show why India, Nepal need to get their act together. *Down to Earth*. Retrieved July 23, 2024, from https://www.downtoearth.org.in/climate-change/communication-gap-bihar-floods-show-why-india-nepal-need-to-get-their-act-together-65961

Shrestha, A. (2018). Renewable energy in South Asia: Prospects and challenges. *International Journal of Energy Research, 42*(7), 1125–1140.

Stankevich, N., Nora, E., & Bandyopadhyay, A. (2024, March 16). *Eastern Waterways Grid to benefit Bangladesh, Bhutan, India and Nepal*. World Bank Blogs. Retrieved July 20, 2024, from https://blogs.worldbank.org/en/endpovertyinsouthasia/eastern-waterways-grid-benefit-bangladesh-bhutan-india-and-nepal.

Surie, M. D., (2015, March 25). *South Asia's water crisis: A problem of scarcity amid abundance*. The Asia Foundation. Retrieved July 19, 2024, from https://asiafoundation.org/2015/03/25/south-asias-water-crisis-a-problem-of-scarcity-amid-abundance/

The Statesman. (2024). *India and Bhutan agree to work together on climate change, and air quality*. Retrieved July 22, 2024, from https://www.thestatesman.com/india/india-bhutan-agree-to-work-together-on-climate-change-air-quality-1503319749.html#google_vignette

Uprety, D. R. (2024). Early warning systems: Sharing technology to combat climate change in South Asia. *Dialogue Earth*. Retrieved July 17, 2024, from https://dialogue.earth/en/climate/early-warning-systems-sharing-technology-to-combat-climate-change-in-south-asia/

Van Baalen, S., & Malin, M. (2017). Climate change and violent conflict in East Africa: Integrating qualitative and quantitative research to probe the mechanisms. *International Studies Review, 20*(4), 547–575.

Wolf, A. T. (1995). *Hydropolitics along the Jordan River: Scarce water and its impact on the Arab-Israeli conflict* (Vol. 99). United Nations University Press.

World Bank Group. (2022). *From source to sea: 2013–2021 South Asia Water Initiative Completion Report*. Washington, DC. Retrieved July 25, 2024, from https://documents1.worldbank.org/curated/en/099312405272210424/pdf/IDU0274da53607b1704a6c0beba048461972fe3e.pdf

World Meteorological Organization (2023). Early Warning System. Retrieved July 17, 2024, from https://wmo.int/topics/earlywarning-system.

Application of Solidarity Approach to Space Identity Crisis: Issues of Muslim Pockets in Kolkata

Teesta Dey

Introduction

The multidimensional identity of geography has been experiencing some prominent theoretical and technological changes in the new millennium. From neoliberal globalization to the postmodern analysis of space, from satellite image interpretation to drone-based micro-level area study, the last two decades have witnessed enormous transformations in geographical studies worldwide. Paradigm shifts in geography are inevitable and necessary. Such paradigm shifts not only upgrade geographical research potentiality in academia but also reestablish the importance of this subject in the socio-political spheres of human lives. One such example of a paradigm shift is the application of the solidarity approach in geographical research that started viewing various environmental, human, and social issues from a different perspective. In 2007, in the presidential address of *Annals of the Association of American Geographers*, Victoria Lawson clearly illustrated the justified establishment of geographies of care and responsibilities in the broader sphere of geographical research (Lawson, 2007). This concept later on directly influenced the application of the solidarity approach in various aspects of spatial analysis (Al Emadi et al., 2016). The responsibility of geographers to care about different global and local issues became the prominent background of this new geographical perspective (Massey, 2004). However, only theoretical discussion can yield the entire essence of this paradigm. In this context, the application of the solidarity approach in geographies of care and responsibilities brings about a framework for future research in real-world scenarios (Turner, 1967). The solidarity approach imbibes empathy, tolerance, and compassion in any kind of socio-environmental issue (Kelliher, 2018). From the politics of separatism to the crises of global climate change, geographers can apply this approach to generate unity, ethics, and values for global concerns.

T. Dey (✉)
Department of Geography, Kidderpore College, University of Calcutta, Kolkata, India

K. Arora, R. R. Anand (eds.), *The Solidarity Approach in Geography*,
https://doi.org/10.1007/978-3-031-86753-8_4

Space and spatial analysis in the geographical discipline has had immense impetus since the 1960s. From Haggett's idea about spatial structure to Morrill's ideas about spatial organization, geographers have always tried to find out the specific patterns and processes that influence spatial characteristics (Bernstein, 2005). Later on, Lefebvre, David Harvey, and Soja illustrated their view on spatial functions where in each aspect society and societal functions play the dominant roles. However, there has always been a lack in the application of any specific approach that directly affects space identity-related issues. The solidarity approach in this context can be used as a unique perspective through which one can view any specific spatial issue from a more humanitarian background and can accordingly apply specific steps to deal with that issue. In this paper, one such attempt has been made to apply a solidarity approach to deal with the spatial identity crisis of the Muslim-dominated areas in the city of Kolkata to address the causes of spatial identity crisis and the methods through which such identity crisis can be reduced.

Spatial Identity and Identity Crisis

As per the Dictionary of Human Geography, spatial identity refers to how identities are constituted, articulated, and contested about space and place (Gregory et al., 2009). Deterritorialized and reterritorialized identities of space are also associated with this issue of spatial identity (Kostovicova, 2005). While some research has viewed space merely as a container for identity, or as a stage on which identities are played out, other studies particularly since the 1990s trace the mutual constitution of both space and identity and their multiple and contested articulations (Keith & Pile, 1993). This is inspired by poststructuralism, psychoanalysis, and postcolonialism. It tries to ground and locate the material spatialities of identity, which are closely bound up with ideas about situated knowledge, the politics of location, and the spatial politics of identity (Rohkrämer & Schulz, 2009). Geographical analysis of spatial identity explores the metaphorical and material relationships between space, identity, and power (Ghandour, 2018).

Postcolonial and feminist theorists have been particularly influential in theorizing spatial identity. In his classic work, *Orientalism*, Edward Said (1979) explored the imaginative geographies of 'self' and 'other' through the intensely politicized spaces of the 'Orient'. Other postcolonial theorists have explored the spatial identities bound up with migration, diaspora, and the contested politics of hybridity, multiculturalism, and cosmopolitanism (Hall, 1990).

There are two schools of thought defining space identity. According to the first school of thought, people perceive space as having certain fixed identities (Paechter, 2004). For example, people living in the suburbs feel that they have a distinct mentality and lifestyle in comparison to people living in the core city. The second school of thought emphasizes the plurality of identity of a particular space, and this viewpoint is well supported by poststructuralism, psychoanalysis, and postcolonial theories. In the Muslim-dominated areas in Kolkata, the concept of self-other-binary is appropriate where the Muslims are identified as the other by the middle-class Hindu society. In the case of these Muslim localities, the postcolonial approach is also appropriate where it is still felt that even after decolonization, there was the creation of an internally deprived group, fractured by various vertical powers. According to Soja, space identity is associated with two different aspects. These are as follows:

1. **Identity politics of space**: It refers to the concealed process of power and attendant ideological frameworks through which spaces are produced, reproduced, and also re-configured. Although initially people try to consider a space neutral, the notion of neutrality becomes faded over time from their perceptions. The construction of identity of a space depends largely upon the structure of political power where they sustain themselves on ideological concerns and it always keeps changing. In Kolkata, the power structure is mostly governed by the middle-class cultured Bengali society where there is not enough representation of the Muslim community. There has never been any planned infrastructural or societal development plan for any specific Muslim area in Kolkata city. Few political faces sprung up at different times but no consistent, proper, sufficient infrastructural plan can be observed for most of the Muslim-dominated areas. Even in the Sachar Committee Report, it was identified that although in Kolkata and West Bengal, Muslims are safe, they are still less developed in terms of the availability of education, health infrastructure, and civic amenities (Prime Minister's High Level Committee, Cabinet Secretariat, 2006). Due to religious identities and associated prejudices, the Muslim-dominated areas have long been victims of institutional negligence and the existing political structure is responsible for creating different types of false identities of the Muslim areas in Kolkata.
2. **Spatialized politics of identity**: It refers to the processes of identity formation confronted with experiences of unbelonging that propel the quest for alternate spaces, both literal and figurative, characterized by radical openness. This does not mean the formation of any kind of spatialized identity, rather this concept emphasizes on formation of a particular concept of identity irrespective of any specific space.

Such creation of false space identity often leads to the occurrence of various levels of socio-cultural tensions in certain areas. Many such places in Kolkata have been struggling with such identity crises and experiencing the politics of separatism from different layers of society. Soja illustrated the concept of the identity crisis of space from two different perspectives.

1. **Space as opaque (Illusion of opaqueness)**: It is based on the concentration of concrete forms, where space is fixed, dead, and undialectical. It ignores the deeper social origins of spatiality, its problematic productions and reproduction, and its contextualization of politics, power, and ideology (Keith & Pile, 1993). But later on, this idea was highly criticized as proper historical and societal analysis can always help to build ideas about space.
2. **Space as transparent (illusion of transparency)**: It dematerializes space and becomes an abstraction, a supposedly real representation of concrete forms.

> Spatiality is reduced to a mental construction alone, social space folds into mental space, away from materialized social realities. Here a space becomes transparent based on empirical perception-oriented viewpoints, by completely ignoring the material reality.

The Muslim localities in Kolkata city have been facing both these two types of identity crises. The majority of the residents have a strong, fixed, and static perception of the Muslim-dominated areas as notorious, criminalized, less developed, and problematic spaces. This static notion represents their illusion of opaqueness, which can be reduced by proper study of the historical background and other social aspects of these Muslim-dominated areas. When people totally ignore the material identity of any space and only perceive based on their mental maps, then these Muslim-dominated areas become distant, less developed, unknown places for a major section of residents in Kolkata city. This creates the illusion of transparency.

Research Background

Space identity is mostly associated with the perception of the majority who directly or indirectly influence that space. Specific localities within a particular city are often identified and characterized in public perception through otherizing discourses based on class, caste, religion, infrastructure, and such other considerations. Kidderpore, Rajabazar, Belgachia, and Park Circus are the most prominent Muslim-dominated areas within Kolkata. These areas have long been subjected to otherizing discourses. The Muslim communities exist in Kolkata as part of a communal geography, marked by pockets of Muslim-majority spaces subjected to relentless ascription of religious identity. According to Anusua Chatterjee, 'These musholman paras, as they are colloquially referred to, nonetheless, continue to carry enormous symbolic and social baggage, identified as they are as zones of the 'other', and hence perpetually unknown and, in the least, avoidable for the ordinary city-dweller of Kolkata'. To identify the context, she also stated, 'The stigmatization of Muslim-predominant neighbourhoods in the city, the popular cultural constructs around them, and the negative description and practiced avoidance of those who live in them can be seen as an extension of the cultural prejudice of the Hindu bhadralok which has continued to this day and which has, in its way, increasingly pushed Muslims to the margins of the city's social life'. Among all the major Muslim-dominated areas, Kidderpore is identified as the most ignored and misinterpreted area where people consider this area as the most underdeveloped, criminalized, less developed, distant part of Kolkata city. This is partly because of the lingering trauma associated with the partition and partly because of subsequent riots and attendant communal discourses which have also created the provocative misnomer, 'Mini Pakistan' about the Kidderpore area. These circumstances have led to the identity of Kidderpore as a defamiliarized space that is either considered to be unsafe or dominated by illegal practices or extremely distant from other parts of Kolkata or a hub of religious parochialism or a space of impoverishment and unsanitary ambiance (Chatterjee, 2017). The image of this urban social space is acted on not in terms of its physical reality but in terms of the mental pictures that people of distant localities have built up gradually. Such a complex multidimensional process of spatial identity crisis reflects a typical example of the localized sense of fear and distress about a particular space where cultural relativity is neglected and a distinctive demarcated otherized zone has evolved (Puniyani, 2018). In this context, an attempt has been made to study the mental maps as perceived by the residents of Kolkata about the Muslim localities like Kidderpore, to identify and analyse the factors responsible for creating such false identity of this space, to compare and contrast the perceived demographic and social scenario of Kidderpore with the real-world case and finally to apply solidarity approach through which these misconceptions may be eradicated from the minds of the city dwellers.

Materials and Methods

The entire work is divided into two broad parts. The first part deals with the perception of the residents of Kolkata city about the Muslim-dominated spaces, more specifically about the Kidderpore area. In 2019, from March to October, a perception study was conducted based on a questionnaire survey of a total 280 permanent residents of Kolkata city. There are 15 boroughs (major administrative units) in Kolkata comprising a total of 144 wards. Kidderpore and its adjacent locality are situated in Borough IX consisting of five wards, viz. ward no 76, 77, 78, 79, and 80. A systematic random sampling method has been applied here selecting 20 permanent residents from each borough. But Borough IX has been excluded from this list as people residing in this borough have a direct understanding of the Kidderpore locality. Hence 20 respondents from 14 boroughs are selected to gauge their perceptions. Here the target population includes persons who are permanent residents of Kolkata city for more than 20 years but have no previous, direct interaction with Kidderpore locality. These respondents know this Kidderpore area based on information and news they received from different sources like Mass media, books, friends, colleagues, acquaintances, and of course from different social media. They are well aware of this locality and have their own opinions about this area but have never seen this area or experienced any incidents associated with this locality. The primary reason behind selecting these types of respondents is to gauge their perception of a place in their city which is highly influenced by external sources of information. To identify the role of different sources that influence the perceptions of commoners about the Kidderpore locality, this attempt has been made. The respondents were asked to share specific ideas about this place. A mental map of the respondents about the Kidderpore locality has been created based on the words mostly given by the respondents about this area. They were also given options from which they have chosen the words which they think are most appropriate for identifying the space of Kidderpore locality. Their perception of this place reveals some social taboos that are well injected into their minds and hence directly affect their viewpoints. The respondents were also asked the reasons behind the generation of these perceptions about this specific locality of Kidderpore. They illustrated their reasons and firmly confirmed that they believed their perceptions.

The second part of this study is completely based on the secondary data source which reflects some basic demographic, social, and economic status of the residents of each ward in Kolkata Municipal Corporation (KMC). At present, the city has 144 wards but as per the available latest data of KMC (2011), data have been found for 141 wards. As the additional three wards have been attached later on, no census data is available for them. Some of the data has been collected from the Kolkata Municipal Corporation (KMC) office and the Police headquarters. The administrative boundary and the police station boundaries are different for the city of Kolkata. The Kolkata city is divided into 8 main police sub-divisions and 69 police stations which have data on different crimes occurring in this city. Hence the crime data has been collected from the Police Headquarters at Lal Bazar for the year 2012–2015 for all the 69 police station zones within Kolkata. All the demographic, social, economic, and crime data reveal the real scenario of different localities within the city at the micro level. These data also exhibit the present status of the Kidderpore locality and help to identify the degree of differences between the perceptions of the people and the reality.

Locational Identity of Kidderpore Area

Kidderpore is one of the oldest parts of the Kolkata city exhibiting the historical imprints of old Calcutta, comprising Mughal and British influences on its social, architectural, and cultural aspects. From a broader perspective, Kidderpore encompasses Kidderpore, Mominpore, Ekbalpore, Metiabruz, Wattgunge, and Garden Reach areas in Borough IX comprising wards 76–80 of Kolkata Municipal Corporation (KMC) (Fig. 1). These areas share some common features like the dominance of the Muslim population and related cultural imprints on land and adjacency to the port (Seabrook & Siddiqui, 2011). Hence a broader locational similarity is observed here. Considering its location on the cityscape, this area is in the proximity of the central part of Kolkata and is highly connected with the rest of the city. This area connects Kolkata city with the Kolkata port and has one of the prominent markets for imported products. Kidderpore also comprises a good share of mixed communities with various socio-cultural aspects.

Kidderpore has a long history of urban growth and socio-cultural transformations. It is one of the oldest parts of Calcutta which has remained as Calcutta's fringe since the colonial period. Back in the eighteenth and nineteenth century, in Alipore, Garden Reach and adjacent areas affluent company employees set up garden houses while the migrant king, Nawab Wajid Ali created 'ek doosra Lucknow' in the southwestern fringe of Calcutta, Metiaburz. From 1880 onwards, Kidderpore also began to be developed as a port by the East India Company (EEIC). In 1884, it

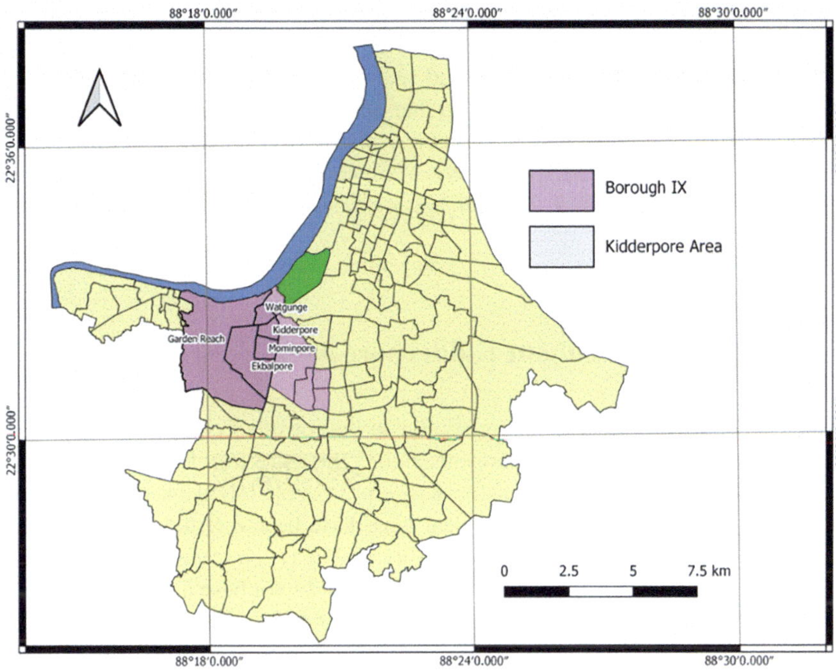

Fig. 1 Location map of Kidderpore area within Kolkata Municipal Corporation

was selected as the site for the wet docks of the Port of Calcutta and by 1892 it became functional. Along with its rich cultural leanings, Kidderpore thus developed as a busy industrial and shipping area, attracting migrant labourers from different parts of the country, having different social and cultural identities. The locality became a bustling industrial and cultural mosaic with Bengali-speaking Hindus and Muslims, Urdu/Hindi-speaking Muslims and Hindus primarily from the Purvanchal region (present eastern UP, Bihar Jharkhand), and Odisha with a small percentage of Anglo-Indian and other communities. No wonder the cosmopolitan nature of the port area of the city of Kolkata had a chequered past and is still holding on to its pluralistic present.

Amusingly enough, the multiplicity of the Kidderpore area is often mistaken to be a 'melting pot' where 'other' cultures have faded into oblivion to give way to a hegemonic, dominant Muslim culture. This assumption might be statistically true but socio-culturally anomalous. The coexistence of English, Urdu, Oriya, Hindi, and Bengali medium schools including different ethnic communities in the Kidderpore area bears out the point. Once home to three of Bengal's prominent poets, viz. Michael Madhusudan Dutta, Rangalal Bandyopadhyay, and Hemchandra Bandyopadhyay, this area champions polyphonic lifestyles. Therefore, the Kidderpore area is not a 'melting pot' but a 'salad bowl' (Zangwill, 1908), which continues to nourish Masjids and Imambaras, along with Temples and Churches.

The metanarrative of Kolkata's history has hardly included the potent mini-narratives of the richly layered Kidderpore area. Thus, the Kidderpore area has remained one of the lesser-discussed and seldom-visited places by the city's educated middle classes and elites. But researchers and academicians must re-investigate this politics of representation, to unearth the cause of this exclusionary treatment, to interrogate whether the Kidderpore area is a crowded, poor part of the city inhabited by a specific ethnic group or a proper amalgamation of everything for which the city stands.

Popular Perceptions about Kidderpore Area

Despite its landscape features being similar to that of other common localities of Kolkata city, a unique perception prevails among the residents of other parts of this city regarding the very identity of the Kidderpore area. Based on the perception survey of a total of 280 respondents, clear ideas have been registered regarding the general public idea of the Kidderpore locality (Table 1). The target group includes all possible socio-economic spheres of the city that do not have any direct connection with the Kidderpore locality. The options for broad perceptions about this locality were selected based on a pilot survey which went on to reflect the primacy of some specific factors that are responsible for creating sustained false identity about the Kidderpore locality.

All the respondents believe that Kidderpore is mostly occupied by the Muslim community. Almost 88% of people agreed with the fact that this place is an abode of various illegal and criminal activities, mostly due to the location of ports, the dominance of Muslim communities, and to concentration of poverty-stricken people. This identity of the Kidderpore area as Muslim-dominated, crime-oriented, and illegal trade-based locality has become a popular stereotype for this place (Bhattacharya, 2018). About 84% of respondents assume Kidderpore is a part of Kolkata city with a very high concentration of slums and squatter settlements

Table 1 Major perceptions about Kidderpore locality

The perceptions	Percentage (approximate) of respondents in support of the perceptions
Kidderpore as the land of Muslims	100
Kidderpore as an abode of criminal activities	88
Kidderpore as congested with slums and squatters	84
Kidderpore is identified as a place with most of the informal workers and illegal activities	80
Kidderpore as illiterate urban space	78
Kidderpore is a distant place, edge of the city, and poorly connected with the rest of the city	72
Kidderpore is the poverty-stricken part of the city	64

Source: Primary Survey, 2019

because it is a poverty-stricken area with very low socio-economic development, 78% of respondents consider that people living in this area are highly illiterate, and 80% of respondents think people living in the Kidderpore area are mostly engaged in informal economic sectors and also in different types of illegal trade-related activities. Almost 72% of people also assumed that Kidderpore is somehow a very distant place from the central part of Kolkata city and very few mass transport services are available there. The overall response reveals that the majority of the respondents feel Kidderpore locality is a highly underdeveloped area in Kolkata city, accompanied by poor infrastructural development, concentration of slums and squatter settlements and no proper so-called middle-class Bhadralok (English meaning is with decent cultural background) Bengali Hindu family would like to stay over there. According to them, Kidderpore is mostly the locality of the Muslims and only Muslims and other poor first-generation migrants stay over there. Many people also have some strong ideas about this locality, like it is a poverty-stricken, nasty, underdeveloped area where the government does not provide any support for regional development. Prevalence of such fixed notions about this locality reflects the issues of false spatial identity which is indicative of various relevant issues of space identity crisis within this city.

Reasons behind the Negative Perceptions about the Kidderpore Locality

The broad analysis of the perception study reveals that the majority of the respondents consider Kidderpore only as the most notorious place in this city. The common notions about this area are that it is the main abode of different types of crimes, including homicides, theft, robbery, dacoity, and assaults against women. As per their view, maximum criminal activities happen here. Muslim dominance is observed in other parts of the city, but the location of the Kolkata port and port-based trade-related activities are the main reasons behind the generation of such perceptions among the commoners. While interrogating the respondents regarding the reasons behind such strong perceptions of them, a few factors emerged. These factors are illustrated below.

Dominance of Muslim Community

The most prominent perception about the Kidderpore locality is the concentration of the Muslim population. Almost 100% of the respondents shared their viewpoint about this religious dominance. Although it is true to some extent, the problem associated with this fact is that the non-Muslim people are less interested in visiting this place while they are accustomed to visiting other Muslim pockets like Rajabazar,

Park Circus, Belgachia, Topsia, or Entally in this city. This is mostly because the Kidderpore area is located in the western part of the city and away from the central connecting localities within the city. However, in reality, the Kidderpore locality has a mixed demographic feature where a significant amalgamation of various religious events and structures is visible. People from different religious and cultural backgrounds live here peacefully, and a mixed cultural imprint can be observed on the cultural landscape of the Kidderpore area. Another interesting fact is that here the number of Hindu temples is higher than the number of mosques and Mazars.

Improper Transportation System

The majority of the respondents (72%) feel that the degree of connectivity of Kidderpore with other parts of the city is very poor. There is a serious lacuna in the availability of mass transport services that connect Kidderpore locality with different parts of the city. But the real scenario depicts a very strong degree of connectivity of this area with all possible parts of the city through various public transport modes. It is served by a wide range of public and private buses, auto services, and other para-transit modes. It is also well connected with the nearest five metro stations in different directions.

Influence of the Movie 'Zulfiqar'

Release of the famous movie called Zulfiqar, directed by Srijit Mukherjee in 2016, starring some famous actors of Tollywood depicted port-oriented terrorism in the Kidderpore locality. It has also attempted to negatively stereotype the Muslim community in this area creating a long-term effect on people, who later on perceived the Kidderpore area as a notorious place. In this movie, a notion of otherization is instilled among the viewers, and Kidderpore locality is identified as a 'state of exception' meaning a no man's land between the public law and political fact and between the judicial order and life—a country within a country—at odds with the majority's perception of Kolkata associated with icons like *Howrah Bridge*, *Tagore* or *Durga Puja*. It also emphasized popular misconceptions of increased criminality in port areas. The combined effect of this movie was that it created a negative image of the Kidderpore locality among the viewers which still has its imprint on them.

Socio-Political Tensions in a College

Another important reason was an incident that happened in a college named Hari Mohan Ghosh College, located in this area, where a police sub-inspector and four students were killed when clashes took place between two college unions. This incident happened in 2013 and severely impacted popular perceptions. This incident led parents to avoid this college for their wards' admission, and people were sceptical about the conditions of the higher educational institutions and the safety and security of the students in the Kidderpore area.

Murder of an IPS Officer

Although it is a very old incident but people remember the Murder of IPS officer, Deputy Commissioner of Police Vinod Mehta and his bodyguard Mokhtar Ali Khan in Garden Reach on March 18, 1984 by the prime suspect Idris Mian. Such a sensational assassination created a long-lasting effect on the commoners of Kolkata who later on avoided visiting this place.

Murder of an Underworld Don

Murder of an underworld don Uday Sing in 1997 also raised fear among the residents of Kolkata who still believe that Kidderpore is the main abode of all types of crimes happening in this city.

Socio-Economic Realities of Kidderpore

Creation and sustenance of these false perceptions and myths about the Kidderpore locality deviate largely from the actual scenario. The census-based secondary data analysis highlights some specific demographic and socio-economic conditions of Kidderpore locality in comparison to other areas within the city. These analyses reveal that though some of the perceptions are true for this area, they do not support the fact that it is a hub of criminal activities. Detailed illustration of these databases confirms that Kidderpore area is much less crime-infested than many other localities within the city, but people usually do not bother about the other localities like they do in case of Kidderpore area.

Population Density

Population density underscores the spatial preference and dependency of the city residents. Each ward has its own locational attributes which directly influence people's decisions regarding whether they want to stay in that locality or want to shift anywhere else. The notion is that the Muslim-dominated parts of the city have very high population density as they intend to stay close together and their fertility rate is also very high. But the census data reveals that the central part of the city (borough II, IV, V and VI mostly) has maximum population density. Among the 141 wards, only four wards have very low population density, among which one is the Garden Reach area (ward no 80), which is also a part of Kidderpore locality. The other wards in Kidderpore area also exhibit low-to-moderate density which nullify the perception of the common people regarding the high concentration of people in Kidderpore area (Fig. 2).

Decadal Growth Rate

The city reveals the negative population growth rate in the decade 2001–2011. The city has clear distinction between the core areas with negative population growth rate and the peripheral areas with positive population growth rate. This data clearly

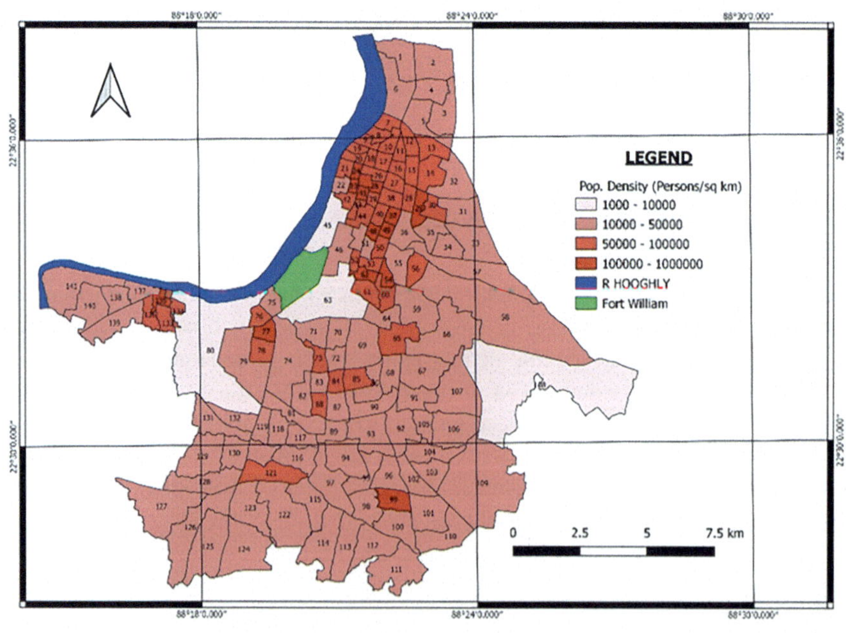

Fig. 2 Population density of KMC (2011)

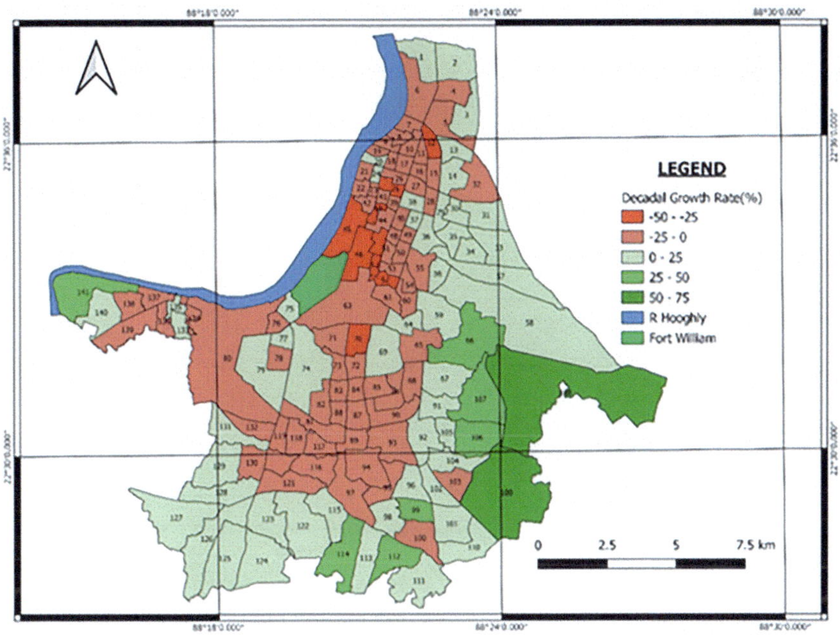

Fig. 3 Decadal growth rate of KMC (2001–2011)

exhibits the locational influence on population growth. About 81 wards among 141 wards exhibit negative population growth rate (Fig. 3). The Kidderpore area shows a mixed pattern in case of population growth rate but the overall growth rate in this area is comparatively low. The notion is that Muslim areas always have very high population growth rate and in the Kidderpore areas. Muslim population growth rate is very high in comparison to other parts of the city. But the reality reveals that ward no 77 and 79 have low positive growth rate but the rate is much lower than other wards of the city. On the other hand, Watgunge, Mominpur, and Garden Reach areas have recorded negative population growth rate.

Literacy Rates

In association with the conception of insufficient educational centres and dominance of poor people, a general idea has evolved among the respondents (78%) that Kidderpore locality has very poor literacy rates. The notion was that the poor literacy rate is one of the major reasons behind the low social development of Kidderpore area. But the census data reveals that many wards in central Kolkata have comparatively low literacy rate. Only the southern, south central, and few northern parts of Kolkata city exhibit highest level of literacy rate. Only Garden Reach area in

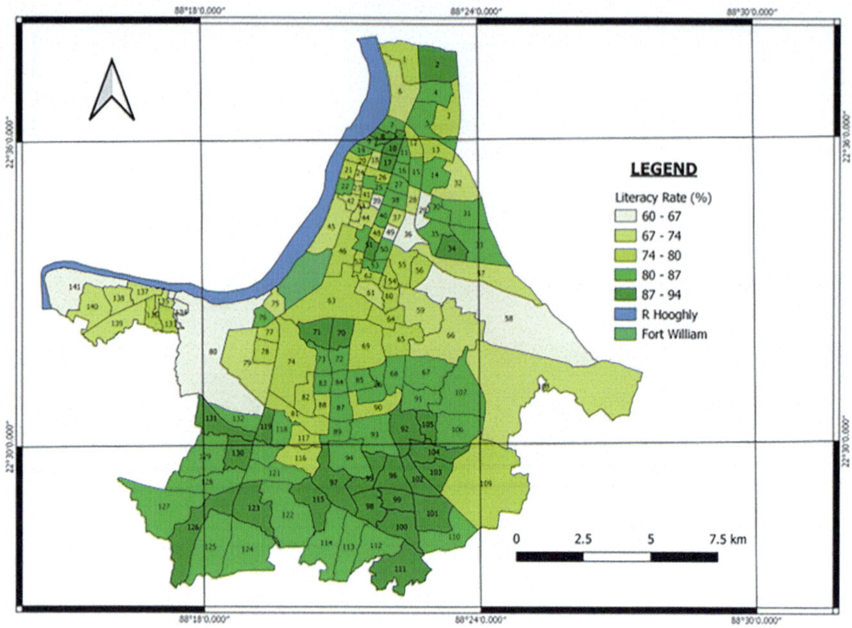

Fig. 4 Literacy rate of KMC (2011)

Kidderpore has comparatively low literacy rate. But in the rest of the Kidderpore area, literacy rate ranges between 80% and 84% (Fig. 4).

The notion of very poor literacy rate in Muslim-dominated areas cannot be applicable in Kidderpore locality. In association with all these perceived negative ideas about Kidderpore locality, a significant number of people (66%) assume that this area is devoid of any good educational institution. But this area actually consists of many reputed convent and other schools like St. Thomas School, St. George High School, St. Teresa's High School, The Khidirpur Academy, etc. and students from various parts of the city come to study in these schools.

Dominance of Informal Sector of Economy

The cumulative effect of all these perceptions leads to the general idea of prevalence of informal sector in this region. Common people have a basic notion that in this region, people are mostly engaged in informal sectors related to trade-oriented activities like goods shipments, real-estate development, and transactions of various imported materials both legally and illegally. However, while considering the employment rates in various wards of this city, it has been found that the entire southern part of this city has very low employment rate. The share of people who are directly involved in the capital generation process reflects the level of economic

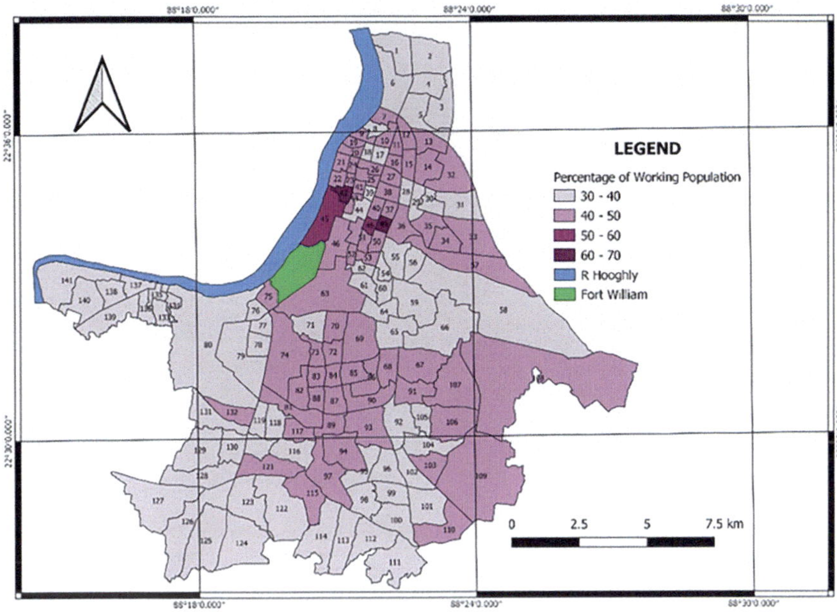

Fig. 5 Share of working population in KMC (2011)

development of a region. Noticeably, the employment rate data does not include people who are engaged in informal activities. Hence a large number of data is missing while representing the employment rate spatially. The entire northern, western, south western, southern, and eastern part of the city reveal very low share of working population as majority of the residents in these wards are involved in activities of informal sectors (Fig. 5). The over concentration of higher level of working population only in the central, south central, and south eastern part of the city is not a good reflector of economic development of this city. The entire Kidderpore locality reveals low level of share of working population.

High Rate of Poverty

Poverty level shows a great degree of variety over the city. Almost 64% of respondents think that Kidderpore area has maximum concentration of poor people in comparison to other parts of the city. But in reality, maximum concentration of poor people can be seen in central part of the city. In Kidderpore area, less than 10% people live below poverty line, whereas in many areas in central Kolkata, the value ranges from 15% to 25% (Fig. 6). The ill effects of overurbanization and imbalance between availability of job and supply of labour lead to various negative outcomes on the cityscape. Rapid immigration of people from the adjacent rural areas in

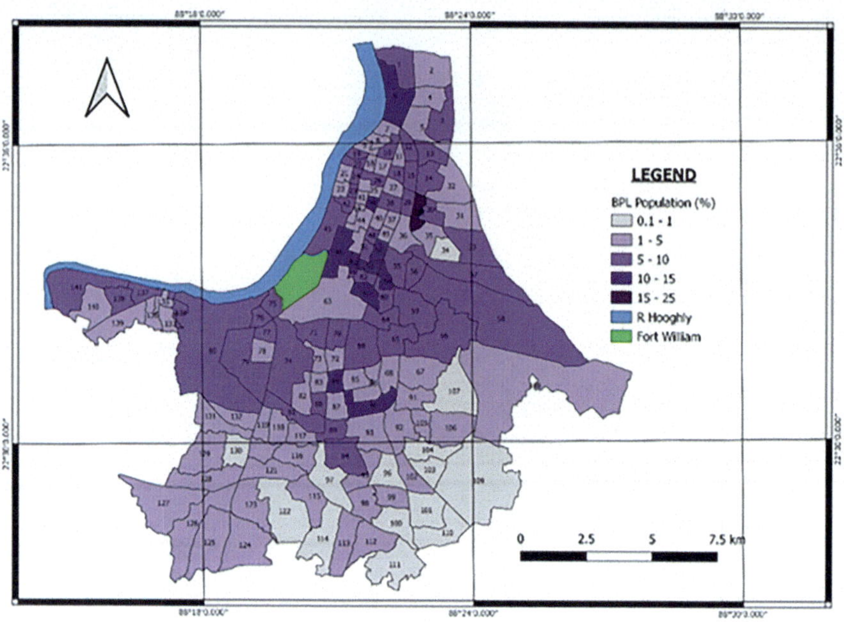

Fig. 6 Share of poor population in KMC (2011)

search for jobs aggravates the urban economic stress and increases the share of people living below poverty line. It is indeed the reflection of faulty urban development and inefficiency of urban governance. The low level of working population in the formal sector is a justified reason behind the existence of people below poverty line in Kidderpore area. In case of Mominpur, the percentage share of poor people is comparatively less than many affluent areas of the city. Comparatively, the peripheral areas of southern Kolkata have lesser share of poor people. The notion of high poverty rate and concentration of slums in Kidderpore area leads to another notion of high occurrence of criminal activities of the local people in this area. Although high poverty rate is observed in various parts of this city, undue emphasis has been put on Kidderpore locality.

Concentration of Slums

Slum concentration within a city exhibits the results of over urbanization and improper urban development process. Urban economic, social, and infrastructural issues are associated with the share of slum concentration within a city. In Kolkata,

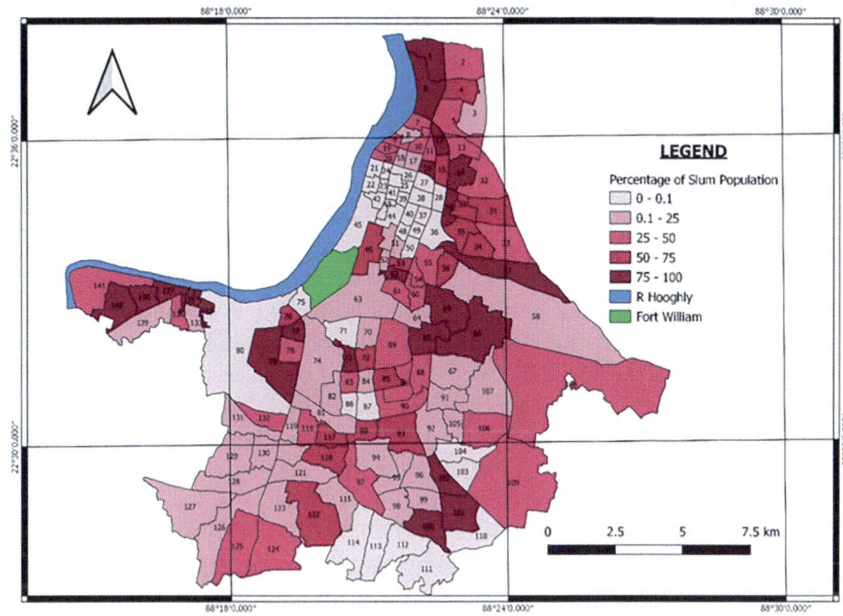

Fig. 7 Share of slum population in KMC (2011)

the slums are the inevitable results of excessive immigration from the surrounding areas, lack of job availability, higher share of informal workers, improper urban development, etc. An estimated 3.82 million people live in 3500 registered and unregistered slums in Kolkata. The spatial variations of slum population reflect concentration of slums in the peripheral areas and not in the central city. There are nine wards having more than 75% of slum population in KMC (2013); 21 wards have 50–75% of slum population. Total 39 wards did not have any slum population which are mostly located at the central part of the city; 84% of respondents believe that maximum people living in Kidderpore locality are very poor and live in slums or in squatters' colonies. They also think that here most of the area is occupied by slums and no proper local urban development has happened over here. Noticeably, it is true that the wards in the Kidderpore locality have higher percentage of slum population. Two wards in Kidderpore locality show very high concentration of slums but such condition is also applicable in another 20 wards dispersed all over the city (Fig. 7). But arguments never happened to assess the reasons of such high percentage share of slum population in those wards. Except 39 wards, all other wards in KMC have slum population. But the concentration of slum population is not specific to a particular location.

Comparison of Socio-Economic Standards of Kidderpore Locality with Other Parts of Kolkata

To assess the combined level of local urban area development, all the mentioned demographic and socio-economic indicators are clubbed and composite score is calculated based on Kendal's method. The selected six indicators reflect the urban socio-economic conditions of a place and have direct influence on the perception of the commoners about the localities. Through the analysis of the spatial data, a clear understanding of the Kidderpore locality in comparison to the other wards can be generated. In this method, the wards are ranked based on individual indicators and then all the ranks of each ward are summed up to get the combined score. Here the ranks are given based on the qualitative assessment of the dataset. Ward representing best urban socio-economic scenario is given rank 1, and the successive wards are ranked accordingly. This means that the lower the rank, higher will be the urban development standard and vice versa. The composite score reveals five different levels of urban development of the wards within KMC. There are a total of 29 wards with lowest composite score which are mostly located in the southern part of the city. These are the highly developed localities of Kolkata city. The second level has 28 wards located in south western and central part, the third level has 39 wards in the central and northern part, and the fourth level of development has 28 wards concentrated in the western and north eastern part. About 17 wards have lowest level of urban socio-economic development and are located in the entire western part and in some central pockets of the city. From this distribution and spatial variations in levels of socio-economic development, the Kidderpore area falls in the fourth level along with some other wards (Fig. 8). Low percentage of literacy, high share of workers in informal sectors, and concentration of slum population are the main reasons behind it. But there are many other wards within the city which reveal a more pathetic scenario than the Kidderpore area. Despite that, people do not have any stubbornly negative image of those localities like they have for Kidderpore area. Majority of the residents think all these poor socio-economic conditions along with few other factors are responsible for higher occurrence of crimes in Kidderpore area. Hence to assess the real-world scenario of the crime rate in Kidderpore area, crime-related data have been collected from 69 police stations of KMC and are assessed here.

The Crimescape of Kidderpore: Perception and the Reality

The Kolkata Police Jurisdiction is divided into 8 police divisions and 69 police stations. Each station has specific locality-based crime data. The data consists of homicides, dacoity and robbery, theft and allied crimes, cheating and criminal breach of trust (CBT), etc. Finally, the total incidents of criminal activities have been calculated based on these data for each police station for the year 2012–2015. Among the eight police divisional zones, the port police sub-division falls into Kidderpore area.

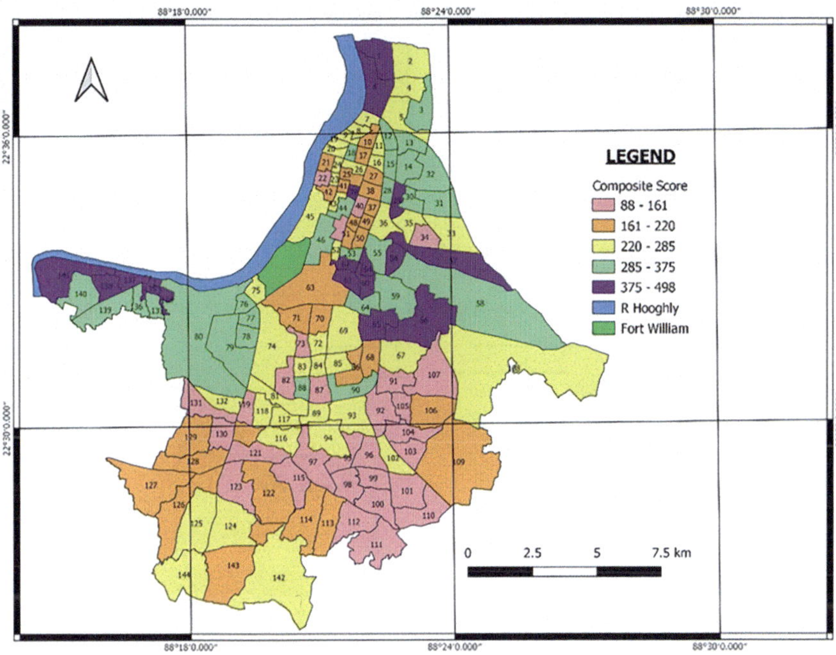

Fig. 8 Combined socio-economic status of KMC (2011). (Based on Kendal's composite score)

This police sub-division has nine police stations among which six cover almost all the areas under Kidderpore locality. These police stations are Watgunge, Garden Reach, Ekbalpore, South Port, West Port, and Metiabruz.

Approximately 88% of the residents in Kolkata perceive Kidderpore locality as a major crime hub in this city. Proximity to the port, maritime trading and associated activities, dominance of Muslim community, influence of social media, and other political reasons created the strong perception of prevalence of various criminal activities in Kidderpore locality. To compare the crimescape of Kidderpore locality with other areas in the city, each criminal activity-based spatial analysis has been conducted here.

Homicides

The homicide cases show great variations over the entire city. Higher number of homicide cases are observed in the southern part, whereas the core city exhibits comparatively lesser occurrences of homicides. While Tiljala and Jadavpur areas have very high number of homicide cases, Kidderpore area has significantly low number of cases (Fig. 9). There are no such police records that reveal the port area has higher occurrences of homicide cases. Even many affluent areas in this city like Jadavpur, Patuli, and Regent Park show higher occurrence of homicide cases than Kidderpore area.

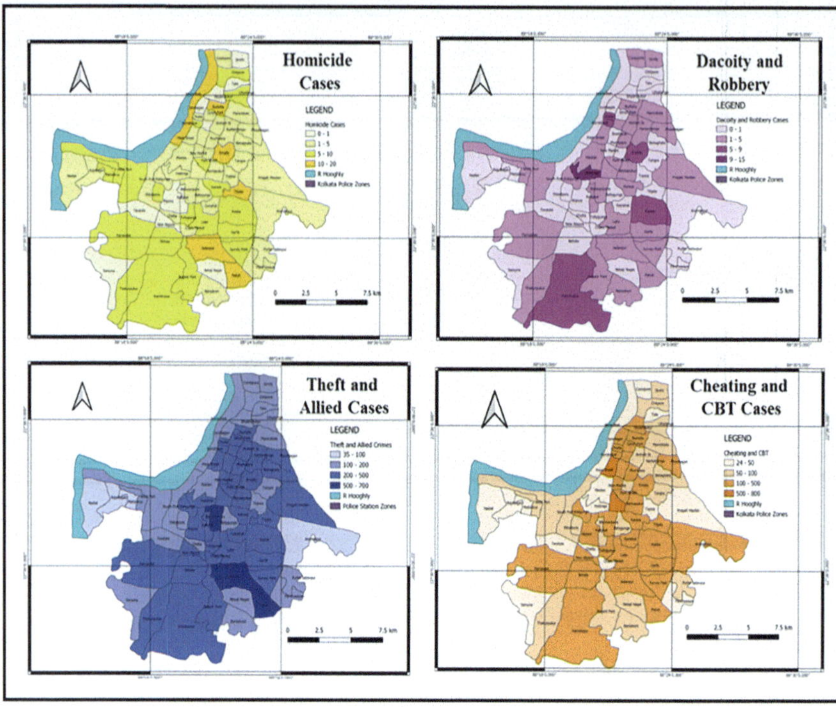

Fig. 9 Recorded criminal cases under Kolkata Police Jurisdiction (2012–2015)

Dacoity and Robbery

The cases of dacoity and robbery show an alarming concentration in the central and south central parts of Kolkata while the western part comprising Kidderpore locality and the port area have noticeably insignificant number of such cases. Within 2012–2015, Ekbalpore and South Port area did not have any such case while Metiabruz had recorded only one dacoity case (Fig. 9). Wattgunge and West Port recorded two such cases each and Garden Reach area had three such incidents. While other parts of the city have comparatively higher number of dacoity cases, Kidderpore locality does not show any such pattern.

Theft and Allied Crimes

The concentration of cases of theft and allied crimes (more than 500 cases) are observed at the central and southern part of the city, whereas the port area has less than 200 cases (Fig. 9). Maximum police stations in this city have more than 200 such cases, and the southern part shows highest occurrence of such cases, specifically in Jadavpur and Patuli areas.

Cheating and Criminal Breach of Trust (CBT)

Highest cases of Cheating and Criminal Breach of Trust are observed at the northern central part. Here the numbers of cases vary from 100 to 500 within the study period. On the contrary, Kidderpore locality, including the port area, has less than 100 such cases and thus reflects fewer criminal activities (Fig. 9).

Comparison of Criminal Activities of Kidderpore Locality with Other Parts of Kolkata

Finally, a map is prepared combining all the criminal cases in the entire city (Fig. 10). The police station wise criminal records reveal that the Kidderpore area does not have any significant criminal activities in comparison to other parts of the city. Maximum concentration of crimes is observed in south central part and also in a few other police stations in central part of the city. Total number of cases in these areas are much higher than 500. On the contrary, the port area has less than 500 crime cases and cannot be identified as a notorious zone of Kolkata with maximum crime cases. The creation of false identity among the commoners regarding the identity of Kidderpore locality as the main abode of different types of crimes must

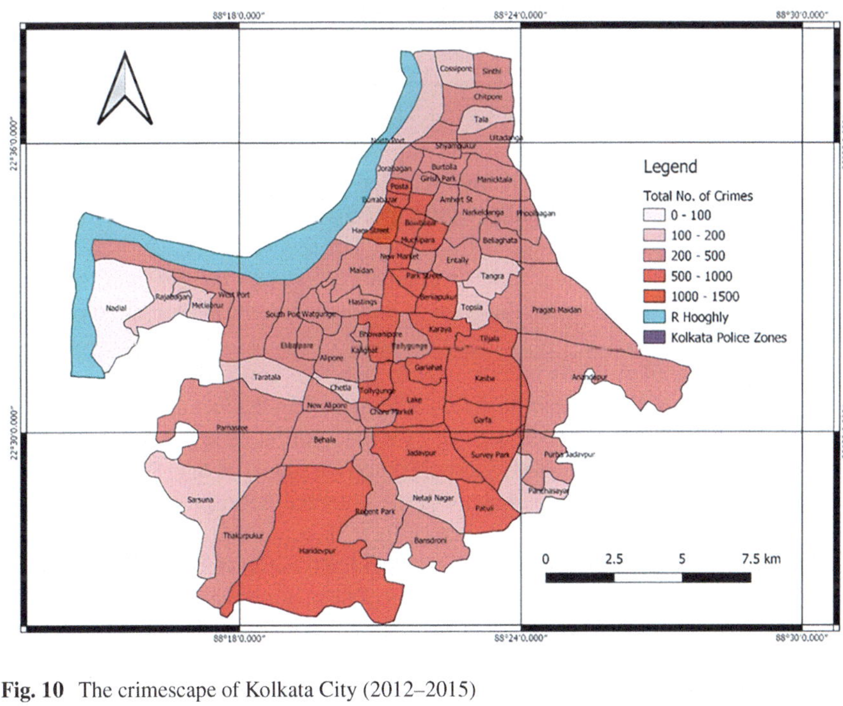

Fig. 10 The crimescape of Kolkata City (2012–2015)

be rejected based on the authentic police station-based data. Although many crime cases are not registered officially, that happens for each and every space within the city. Only the official data can give a proper idea about the overall scenario of a particular region. To justify the secondary data source, field visits and perception study of the residents in Kidderpore area have been conducted which support the authenticity of the database.

Application of Solidarity Approach to Redefine Kidderpore Locality

The difference between perception and reality about Kidderpore is the outcome of an age-old politics of space and communal issues. While this area in reality is endowed with the location of many heritage structures and is a place with multiple attributes of cultural heritage, people only identify this place as a Muslim residential area with multiple nefarious activities. From the perspectives of natural heritage like the history and importance of Adi-Ganga and role of Kolkata Port as India's first river-based port to the tangible heritage of age-old mosques, churches and temples, historical buildings, and the intangible heritage of people, culture, cuisine, literature and much else, Kidderpore area is quite unique in Kolkata city. The history of the migrant king Nawab Wajid Ali Shah, diffusion of his Awadhi culture in this area and the emergence of a Doosra Lucknow, creation of a unique Food Culture, amalgamation of various religious, ethnic, linguistic groups, and their harmonious living—all render this area unique in the bigger context of national identity (Silverman, 2011). And this should be visualized by everyone who has scant or no idea about the rich cultural heritage of the Kidderpore locality. But the spread of negative ideas about this area among the commoners plays an important role in the gradual degradation of the potentiality of this area in being identified as a heritage space. The negative perception about Kidderpore locality as the space of the other hinders the influx of tourists from national and international sources. The prejudiced perceptions regarding the Kidderpore area reflect a sheer imbalance between the potential heritage resources this area possesses and the level of maintenance and preservation of these heritage identities.

In this context, there is an ardent need to generate consciousness about these heritage structures and aspects of this area at least among the local residents. As per the perception survey of the local people of Kidderpore area regarding their ideas about the heritage potentiality of Kidderpore, surprisingly more than 84% of respondents said they do not have any idea about the potential heritage identities of their own locality. The absence of preservation of local histories by the local researchers and academicians and the lack of interest of the local authorities are the main reasons behind this space ignorance and the spread of false notions about the Kidderpore locality. In this scenario, application of solidarity approach can generate awareness among the commoners by imbibing empathy, tolerance, and compassion (Saltiel &

Strüver, 2022). Few techniques can be applied to generate some basic awareness about this space among the local people to generate the interest in Kidderpore as a heritage site and deconstruct the false identity of locality. Application of solidarity approach in this context can bring a positive change in the perception of the commoners about this locality. Fostering the communal harmony by applying solidarity approach can transform the identity of this place in the social sphere of the city. Spreading the sense of solidarity among the commoners of the city can help in debunking their false perceptions about Kidderpore locality. Various steps have been taken by the local teachers and students to unveil the harmony among the local people and heritage potentialities of this area.

To foster inter-community harmony, various steps have been taken like

- ***Hosting programmes*** like Iftar party (during Eid) like 'Dosti-ki-Iftar' by the local Hindu communities strengthen the bonding between Hindus and Muslims.
- ***Organizing blood donation camps*** for all upgrades the relations among the two communities.
- ***Performing environmental awareness programmes*** in the interior localities of Kidderpore like covering open drains, planting trees, preventing water logging in certain localities, proper garbage disposal on a regular basis by the students of different communities from local colleges create a more inclusive ambiance.
- ***Extension of free education*** to the local children of economically weaker Muslim families accelerates the rate of social cohesion in this locality.
- ***Organizing culinary fair*** by the local colleges or other academic and cultural communities where local Awadhi cuisines were showcased by the local Muslim people to the people who have not yet tasted such dishes reveals another way of spreading harmony among the two communities. The fear of touch and the sense of separatism and unbelongingness are thus gradually discarded by applying the solidarity approach in this context.

Spreading of awareness about the local heritage and tourism potentiality of Kidderpore area among the local people also helps in upgrading the identity of this area in the greater forum. The steps taken by the local teachers and students have been mentioned here

- ***Neighbourhood walking*** to visit the heritage structures and interaction with local people.
- ***Regional and local mapping of heritage*** structures of the Kidderpore area.
- ***Application of community-based participatory approach*** like the involvement of local people in organizing small heritage tours in this area, the opening of food stalls offering unique cuisine of this area, and sightseeing by cycle or toto rides for the tourists.
- ***Generation of capacity building*** among the local people especially the youth in preparation of an inventory of heritage resources through workshops and training.
- ***Establishment of systematic inventory of heritage structures*** using community network and

- ***Awareness programmes*** have made the local community aware of their heritage resources and generated a sense of pride in them.
- ***Maintenance of local resources*** prevents deterioration and safeguards the heritage structures.
- ***Formal and informal networking*** has emerged for local-level volunteers working in this field.
- ***Creation of sources of information*** like publication, documentary films have already been prepared entirely with local support.
- ***Institutionalization*** like signing MoU between the heritage commission and local academic institutions to initiate various programmes based on the local cultural heritage can accelerate the heritage development of this area as well.
- ***More involvement of the students*** through field study and interaction with local people.

These attempts result in the generation of certain important aspects for the preservation and holistic improvement of the heritage identities of the Kidderpore locality and helps in reducing the false space identities of this area.

Conclusion

The identity of any space depends on various factors. It is the society that creates the context through which the general perception of any space develops (Purakayastha, 2018). Kidderpore locality has a rich history of urban heritage which started much before the advent of British rule. The gradual urban development process, segregation of a specific community, intentional ignorance in societal and infrastructural development by the government, otherizing mentality of the political decision-makers, colloquial viewpoints, stories, and exaggerations about the local events, etc. have created a false reality about this locality for a significant time period among the common people who do not have any direct experience about this locality. These reasons are also responsible for creating a wrong perception among the common people who consciously avoid visiting this place. In this context, decoding the actual identity of this space by using the solidarity approach can generate a vast range of knowledge among the commoners. Finally, it can be said that public perception of any space depends largely on class, caste, religion, and other determinants. Kidderpore, like other 'Muslim' localities has been subjected to derogatory and communally prejudiced representations. But conscious and planned study of the local history and heritage through the application of solidarity approach can bypass these obstacles and glorify Kidderpore with all its unique spatial attributes.

References

Al Emadi, N., Leson, H., Lucas, J., Meier, P., & Borge-Holthoefer, J. (2016). Geography of solidarity: Spatial and temporal patterns. In *Procedia Engineering 159. Humanitarian Technology: Science, Systems and Global Impact 2016, HumTech 2016* (pp. 27–34). https://doi.org/10.1016/j.proeng.2016.08.060

Bernstein, M. (2005). Identity politics. *Annual Review of Sociology, 31*, 47–74.

Bhattacharya, P. (2018, October 24). *Crimescape analysis of Kolkata, the city of joy*.

Chatterjee, A. (2017). *Margins of citizenship: Muslim experiences in urban India*. Routledge & CRC Press.

Ghandour, H. (2018). A space for identity: The case of Lebanon's naturalised Palestinians. In *The politics of identity: Place, space and discourse. Manchester Scholarship Online*. Oxford Academic.

Gregory, D., Johnston, R., Pratt, G., Watts, M., & Whatmore, S. (2009). *The dictionary of human geography* (5th ed.). Wiley-Blackwell.

Hall, S. (1990). Cultural identity and diaspora. In J. Rutherford (Ed.), *Identity: Community, culture, difference* (pp. 222–237). Lawrence & Wishart.

Keith, M., & Pile, S. (1993). *Place and the politics of identity*. Taylor & Francis Group.

Kelliher, D. (2018). Historicising geographies of solidarity. *Geography Compass, 12*, e12399. https://doi.org/10.1111/gec3.12399

Kostovicova, D. (2005). *Kosovo: The politics of identity and space*. Routledge.

Lawson, V. (2007). Presidential address: Geographies of care and responsibility. *Annals of the Association of American Geographers, 97*, 1–11.

Massey, D. (2004). Geographies of responsibility. *Geografiska Annaler. Series B, Human Geography, 86*, 5–18.

Paechter, C. (2004). Space, identity and education. *Pedagogy, Culture and Society, 12*, 307–308. https://doi.org/10.1080/14681360400200204

Prime Minister's High Level Committee, Cabinet Secretariat. (2006). *Social, economic and educational status of the Muslim community in India*. Government of India.

Puniyani, R. (2018). Muslims and politics of exclusion. In A. Shaban (Ed.), *Lives of Muslims in India: Politics, exclusion and violence* (pp. 88–106). Routledge India.

Purakayastha, A. S. (2018). Communalisation of politics in West Bengal: Religion and the public sphere. *Economic and Political Weekly, 53*(16), 12–15.

Rohkrämer, T., & Schulz, F. R. (2009). Space, place and identities. *History Compass, 7*, 1338–1349. https://doi.org/10.1111/j.1478-0542.2009.00627.x

Said, E. W. (1979). *Orientalism*. Vintage Books.

Saltiel, R., & Strüver, A. (2022). Just care! Rethinking the uneven geographies of care. *Erdkunde, 76*, 161–170. https://doi.org/10.3112/erdkunde.2022.03.01

Seabrook, J., & Siddiqui, I. A. (2011). *People without history: India's Muslim ghettos*. Pluto Press.

Silverman, H. (2011). *Contested cultural heritage: Religion, nationalism, erasure, and exclusion in a global world*. Springer.

Turner, R. H. (1967). Types of solidarity in the reconstituting of groups. *The Pacific Sociological Review, 10*, 60–68. https://doi.org/10.2307/1388423

Zangwill, I. (1908). *The melting-pot*. The Macmillan Company.

Lost in the City: A Study of Homeless in Chandigarh, India

Aparna Joshi, Ayushman Ganguly, Armaan Singh Sandhu, Tirth Pahouja, Bilal Faisal, Pratham Seth, Gursimran Cheema, and Shivam

Introduction

> **No matter where we travel in the world, people living on roads, pavements, open and public spaces, is a common sight in every city and this number is constantly increasing each year.**

Homelessness has emerged as one of the biggest and most serious problems across the globe, challenging and questioning the claims of various state-run welfare programs and human rights protection. Acknowledging the seriousness of the issue, the United Nations in its 58th session of the UN Commission for Social Development agreed upon the text of the first-ever United Nations Resolution on Homelessness on 19 February 2020 (DISD, n.d.).

The United Nations has estimated that worldwide about 1.6 billion people reside in inadequate housing with another 15 million being evicted forcibly each year (Filipenco, 2023). Even the richest countries of the world have a huge section of deprived populations living with minimum resources and constantly struggling to find a permanent roof over their heads. The situation, however, is very gloomy in the developing countries of the world with a very high percentage of the homeless population to their total population. It has been estimated that the ten countries with the highest number of homeless populations in the world are located in Asia (Pakistan, Syria, Bangladesh, Afghanistan, Philippines, and Yemen), Africa (Nigeria, Egypt, and the Democratic Republic of Congo), and South America (Columbia).

A. Joshi (✉) · A. Ganguly · A. S. Sandhu · T. Pahouja · B. Faisal · P. Seth · G. Cheema · Shivam
Fraser Valley India (Chandigarh), University of Fraser Valley, Abbotsford, BC, Canada
e-mail: aparna.joshi@ufv.ca; Ayushman.Ganguly@student.ufv.ca; Armaan.Sandhu2@student.ufv.ca; Tirth.Pahouja@student.ufv.ca; Bilal.Faheem@student.ufv.ca; Pratham.Seth@student.ufv.ca; Gursimran.Cheema@student.ufv.ca

K. Arora, R. R. Anand (eds.), *The Solidarity Approach in Geography*,
https://doi.org/10.1007/978-3-031-86753-8_5

The common causes of homelessness have been identified as poverty, unemployment, eviction, and a shortage of affordable housing. Studies have shown that homelessness is closely linked to poverty and economic inequality. Many people experiencing homelessness are unable to access affordable housing or stable employment and may also face barriers to accessing healthcare, education, and other services. In particular, racial and ethnic minorities are disproportionately represented among the homeless population (OHCHR, n.d.-a, b). Mental illness and substance abuse are also common factors contributing to homelessness. Individuals with mental illness are more likely to experience homelessness than the general population and may face challenges accessing appropriate care and support (Fazel et al., 2014). Homelessness can also result from substance abuse since addicts may find it difficult to keep up relationships or jobs. Natural disasters like earthquakes and floods, as well as social and political unrest, also make people homeless.

There is no single universally accepted definition of homelessness. The countries have adopted different definitions which, at times, seem to be completely incompatible. Greece follows a very vague definition of homelessness as 'insufficient accommodation', whereas, in Russia, any person without a place to stay or without a fixed abode is termed as homeless. While in South Korea, terms such as 'vagrants' and 'rough sleepers' are prevalent, in Zimbabwe, people are considered homeless if they don't possess a house in an approved residential area.

In India, the definition provided by the Census of India 2011 is followed as that, homeless are those 'who do not live in buildings or census houses but live in the open on roadside, pavements, in hume pipes, under flyovers and staircases, or in the open in places of worship, mandaps, railway platforms, etc.' With estimates ranging from 1.77 million to over 18 million individuals, India is one of the countries with the biggest numbers of homeless people worldwide.

Homelessness is a very complex social-economic problem. A serious issue of violation of human rights, homelessness, causes deprivation of many other human rights such as the right to life, the right to the highest attainable standard of health, and the right to liberty and security. A homeless person also faces discrimination against various economic, social, and civil rights due to the lack of residential address, thus undermining the essence of social welfare schemes and benefits, participation in political affairs along depriving the person of economic gains and opportunities. Living on the streets for a long time leads to dejection, despair, and hopelessness for a better and safer future. People tend to accept their fate and life on the roads without any expectation of support from the local administration and the government.

In India, various governments and non-governmental organizations have worked to end homelessness through several initiatives and policies, including social welfare programs, offering job opportunities and vocational training, along with providing housing and basic utilities. These initiatives, however, have had a limited reach and impact, and many homeless people still lack access to basic requirements and struggle mightily to find stable housing and jobs. Providing permanent solutions for homelessness needs a long-term commitment and a multidimensional strategy that involves cooperation between the government, civic society, and other stakeholders.

It is now universally accepted that development is incomplete if it is not sustainable which can be ensured by focusing on the three pillars of sustainable development—economic viability, social equality, and environmental sustainability. The human dimension of development is core to the idea of sustainability and emphasizes poverty reduction, achieving zero hunger, good health and well-being, quality education, and employment generation along with ensuring equality for all. However, these goals will not be achieved till the mainstream population develops empathy and solidarity for the unprivileged class. Though there is solidarity among the unprivileged population, homeless people in the case of this research paper, their social conditions will remain unchanged if there is no empathy for them from society. By caring for this homeless population, not only their socioeconomic conditions can be improved, but even social integration can be strengthened by overcoming the economic, regional, and racial barriers.

In this paper, the care dimension of the solidarity approach has been explored to provide long-term and sustainable solutions to the problem faced by the homeless population in the planned city of Chandigarh, one of the most progressive and capital cities of India.

Objectives

1. To understand the problems and challenges faced by the homeless in the 'City Beautiful'.
2. To suggest effective and viable solutions to provide comfortable and affordable housing to the homeless.
3. Formulation of a long-term plan to keep the city full of opportunities.
4. To explore the care dimension of the solidarity approach for providing a long-term solution faced by the homeless in the city.

Study Area

Located on the geographical coordinates 30.7333° N and 76.7794° E, Chandigarh is a union territory in northern India, serving as the shared capital of the states of Punjab and Haryana. It is bordered by the state of Punjab on the north, west, and south, and by the state of Haryana on the east. Chandigarh is one of the earliest planned cities in post-independence India and is internationally known for its architecture and urban design.

It has also been rated as one of the 'Wealthiest Towns' of India with Chandigarh's Gross State Domestic Product (GSDP) at Rs. 436,743.4 million (US$ 6195.8 million) in 2019–2020, rising to a Compound Annual Growth Rate (CAGR) of 10.52%

Fig. 1 Make shift housing structures along the main road

between 2015–2016 and 2018–2019. Chandigarh's Net State Domestic Product (NSDP) was Rs. 3,917,275.0 million (US$ 5557.2 million) in 2019–2020 (IBEF, n.d.).

The estimated population of Chandigarh city in 2023 is 1,325,000, while Chandigarh metro population is estimated at 1,415,000. The last census was conducted in 2011, and the scheduled census for Chandigarh city in 2021 was postponed due to Covid. The current estimates of Chandigarh City are based on past growth rates (Chandigarh city population 2023). According to the findings of the first-ever Chandigarh Homeless Survey (2020) under the National Urban Livelihood Mission, there were 2064 homeless people in the city. However, over the last two years, there has been an increase in the number of homeless people in the city, primarily due to the in-migration of rural migrants from different states of North India (Figs. 1, 2, 3, and 4).

The majority of homeless people work in the informal sector and lack the required paperwork, such as voter IDs or Aadhar cards. They become 'invisible' in the eyes of the city government as a result, and their voice is ignored. According to a 2019 Municipal Corporation Chandigarh survey, 49% of the homeless did not have any form of identification. Typically, they are classified as nuisances to urban living. The topic of exclusion is not covered in the current scheme guidelines. Thus, it is imperative to guarantee inclusivity (Sheltering the urban homeless, n.d.).

Fig. 2 Shanties located in the open grounds of Sector 25

Fig. 3 Toilets installed by the Municipal Corporation in Sector 25

Fig. 4 A view of the environment in a shanty settlement in the industrial area

Data Collection and Research Methodology

The present study is based on mixed-method research to understand the issues, challenges, and aspirations of the homeless population in the 'City Beautiful'. A structured questionnaire containing both closed-ended and open-ended questions was used along with conducting semi-structured interviews for collecting data from 450 homeless people 'living rough' across the city. The respondents were living in Sector 25, Industrial Area Phase I, and Sector 35. The field survey for the primary data collection was conducted between March 2023 and May 2023 (Figs. 5 and 6).

Besides, secondary information regarding various housing schemes of the central government and Chandigarh administration had been collected from different government websites and offices.

Addressing Homelessness: Government in Action

The **Ministry of Housing and Urban Affairs** oversees creating and carrying out national housing, urban development, and urban governance policies, programs, and initiatives. The ministry develops and carries out national housing policies and programs, such as the creation of cheap housing, the renovation of slums, housing financing, and rental housing schemes. It strives to give all societal groups access to safe, decent, and cheap housing. The ministry is essential to the growth and

Fig. 5 Data collection during the field survey

Fig. 6 Data collection during the field survey

planning of urban regions. It develops rules and policies for urban planning, infrastructure construction, and urban transportation. Additionally, it encourages inclusive and sustainable urban development methods.

- **Deendayal Antyodaya Yojana-National Urban Livelihoods Mission (DAY-NULM)** was launched in September 2013 to reduce poverty and vulnerability of the urban poor households by enabling them to access gainful self-employment and skilled wage employment opportunities. The mission focused on the urban homeless and aimed to provide well-equipped shelters to them in a phased manner. Another target of the mission is to safeguard the livelihood of the street vendors by providing them access to suitable spaces, institutional credit, social security, and skills to access emerging opportunities in the market.
- Established in June 2015, the **Atal Mission for Rejuvenation and Urban Transformation**, also known as **AMRUT**, is an Indian government initiative started by the Ministry of Housing and Urban Affairs aimed at altering urban regions and raising inhabitants' quality of life through enhanced service delivery and infrastructure development.
- The Prime Minister's Housing Scheme, also known as **Pradhan Mantri Awas Yojana (PMAY)**, is a prominent housing initiative introduced by the Indian government in 2015. The program is focused on encouraging the building of residences promotes the adoption of sustainable and green building techniques and intends to serve a variety of social categories, including economically disadvantaged individuals (EWS), low-income individuals (LIG), and middle-income individuals (MIG). The main goal of PMAY is to offer qualified people in India's urban and rural areas solutions for affordable housing by providing a Credit-Linked Subsidy Scheme (CLSS) which effectively reduces the house loan repayment burden.
- **JNNURM, or Jawaharlal Nehru National Urban Renewal Mission**, was a prominent program introduced by the Indian government in 2005 to enhance urban infrastructure, living conditions of the urban poor, and administration in many cities throughout the nation. It advocated for an integrated strategy for urban development that addressed both infrastructure development and social and economic challenges. JNNURM indirectly sought to decrease homelessness and enhance the general well-being of urban inhabitants by enhancing overall living circumstances and offering better housing options. The mission's primary objective was to improve living conditions in slum communities by providing basic amenities. It encouraged slum dwellers to take part in the design and execution of slum redevelopment initiatives and promoted the building of affordable housing.
- **Chandigarh Housing Board (CHB)**, a government organization founded in 1976, is tasked with helping the city's citizens find affordable housing options by planning, developing, and implementing housing projects for the residents belonging to various socioeconomic backgrounds in the city. For various income

categories, including the economically disadvantaged sections (EWS), low-income groups (LIG), middle-income groups (MIG), and high-income groups (HIG), it builds residential complexes, apartments, and homes.

- ***Raen Basera*/Night Shelters**: *Raen Basera*, also known as common or night shelters, is an initiative that the Chandigarh Administration and various NGOs started in the winter 2022, we have seen the houses being set up. The Chandigarh Administration allotted places for night shelters (little tin shacks) to be installed in some prominent sectors of Chandigarh such as like Sector 16, 17, 22, 26, 43, etc. These were total ten in number (two permanent and eight temporary). These shacks ensured a safe and comfortable living to the homeless people in the cold winter months along with the provision of medical assistance, food, entertainment, toilets, and clean water within these shelters.

Analysis of the Field Survey

A comprehensive survey was conducted, wherein individuals from diverse backgrounds were invited to share their experiences and perspectives regarding homelessness. The survey gathered information from participants across various demographics, including age, gender, and religious affiliation.

The results revealed a compelling distribution of religious affiliations within the surveyed population. Among the 450 participants, it was found that 84% identified as Hindus, indicating a significant presence of individuals from the Hindu faith. Additionally, 5.80% of respondents identified as Muslims, highlighting the representation of the Muslim community among those experiencing homelessness (Fig. 7). The remaining participants belonged to various other religious faiths.

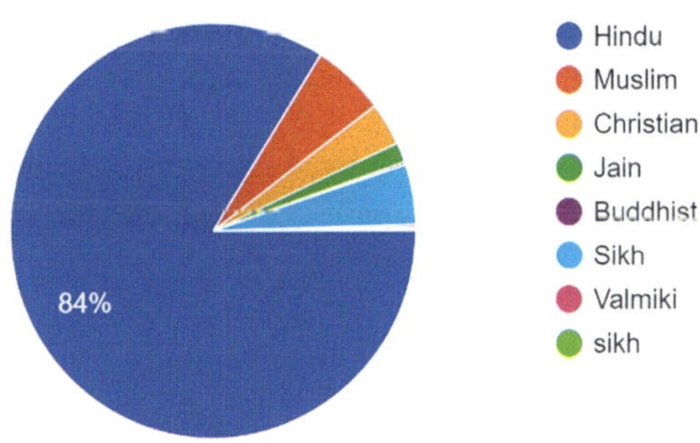

Fig. 7 Religious affiliation. (Source: Field Survey, March–May 2023)

Residence Status

The findings highlighted that nearly 60% of the individuals experiencing homelessness identified themselves as migrants. This finding brings attention to the intersectionality between migration and homelessness, emphasizing the unique challenges faced by migrant populations in securing stable housing. Factors such as language barriers, limited access to social support networks, unfamiliarity with local resources, and potential legal complexities contribute to the vulnerability of migrants to homelessness.

The high proportion of migrants among the homeless population underscores the need for culturally sensitive and inclusive strategies that address the specific needs of this group. It calls for comprehensive policies that prioritize not only affordable housing but also integration support, language assistance, employment opportunities, and accessible healthcare for migrants. By recognizing and addressing the distinct circumstances faced by migrant individuals experiencing homelessness, we can strive towards fostering a more inclusive and equitable society for all.

Proof of Identity

This survey aimed to understand the prevalence of homelessness among the economically disadvantaged population and explore the accessibility and utilization of vital identification documents such as Aadhar cards and ration cards.

Among the 450 individuals surveyed, a striking 96.20% reported possessing either an Aadhar card or a ration card. This highlights the fact that the majority of the economically disadvantaged population in this study had access to at least one essential identification document. It is worth noting that the remaining 3.80% of respondents did not possess an Aadhar card or a ration card. This minority subset could face additional barriers in terms of accessing government assistance programs, social services, and employment opportunities due to the lack of official identification.

By implementing targeted strategies to facilitate the acquisition of Aadhar cards and ration cards, we can bridge the existing gaps, empower individuals, and create pathways towards a more inclusive society

Education Status

The survey results indicate the significant educational challenges faced by economically disadvantaged individuals experiencing homelessness. Limited access to education can perpetuate the cycle of poverty and hinder opportunities for social and economic advancement.

It was found that around 10% of respondents reported being totally uneducated, indicating a lack of formal education within this group. Furthermore, a majority of participants indicated that they had not completed education beyond the ninth grade, highlighting limited educational opportunities and potential barriers to secure stable employment and improved living conditions.

Further, out of the 450 participants, 50.10% reported sending their children to school, indicating that a significant portion of the surveyed population has been able to provide access to education for their children despite their economic circumstances. This finding highlights the commitment of many parents to prioritize their children's education and the potential for breaking the cycle of poverty through education. However, it is concerning that the remaining 49.90% of respondents reported not sending their children to school, indicating a lack of access to education for a substantial number of children within this group. This underscores the need to address barriers to education faced by economically disadvantaged families.

It is crucial to address these barriers by advocating for increased educational opportunities and tailored interventions that support individuals in overcoming educational challenges. The findings indicate a concerning prevalence of low educational attainment, with a significant proportion reporting no formal education and the majority not progressing beyond the ninth grade. These results highlight the urgent need for targeted initiatives that focus on improving educational access and promoting lifelong learning for the less fortunate, with the aim of breaking the cycle of poverty and empowering individuals to secure a brighter future.

Employment Status

The survey sought to understand the connection between employment and homelessness, shedding light on the employment experiences of those who have experienced homelessness.

Out of the 450 participants, 42% reported having a job before experiencing homelessness. This finding indicates that a significant portion of the surveyed population had previous employment experience, highlighting the potential impact of economic factors on homelessness. It suggests that factors such as job loss, insufficient wages, or lack of job stability can contribute to homelessness among individuals who were previously employed (Fig. 8).

Conversely, the remaining 58% of respondents reported not having a job before experiencing homelessness. This subset underscores the need to address unemployment and the lack of economic opportunities among the economically disadvantaged population to prevent homelessness.

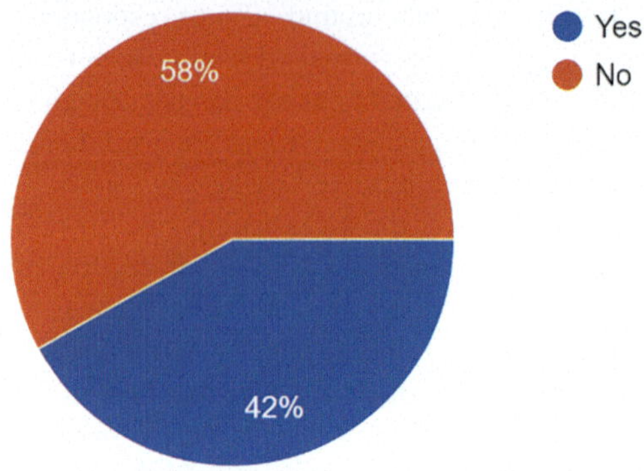

Fig. 8 Employment status. (Source: Field Survey, March–May 2023)

Duration of Being Homeless

Figure 9 reveals that over 50% of the respondents indicated that they had experienced homelessness for a period exceeding one year, and astonishingly, they still found themselves without a stable place to call home. This data sheds light on the persistent and entrenched nature of homelessness, emphasizing the urgent need for comprehensive interventions and support systems to address this pressing societal challenge. The fact that such a significant proportion of individuals remain trapped in homelessness for extended periods underscores the complex and multifaceted nature of the issue, necessitating a holistic approach that encompasses affordable housing, access to healthcare, mental health services, education, and employment opportunities.

The survey results highlight the urgency of concerted efforts from various stakeholders, including government bodies, non-profit organizations, and communities, to collaboratively work towards providing sustainable solutions that ensure the basic human right of shelter for all individuals in need.

Status of Daily Meals

When the survey delved into the daily food intake of individuals experiencing homelessness, a disheartening reality unfolded. Majority of respondents (59.10%, Fig. 10) reveals that they did not have access to three meals a day, indicating the immense challenges they face in meeting even their basic nutritional needs. This distressing finding highlights the harsh and difficult circumstances that prevent these individuals from having regular and nourishing meals.

Insufficient access to food not only compromises their physical health but also adds to the emotional and psychological strain of their already burdensome lives. It underlines the urgent need for comprehensive support systems that address food insecurity among the homeless population. Collaborative efforts from government

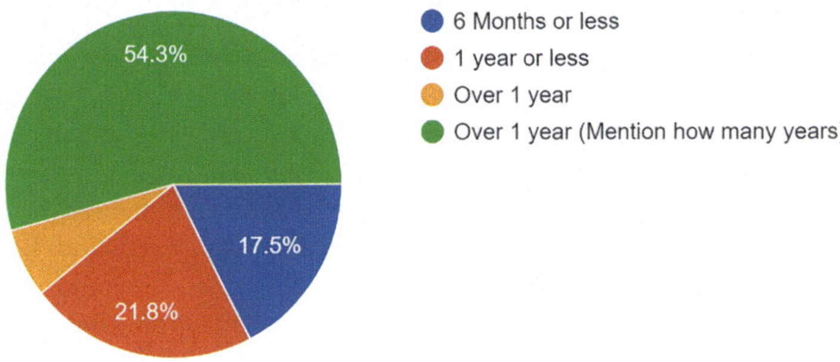

Fig. 9 Duration since being homeless. (Source: Field Survey, March–May 2023)

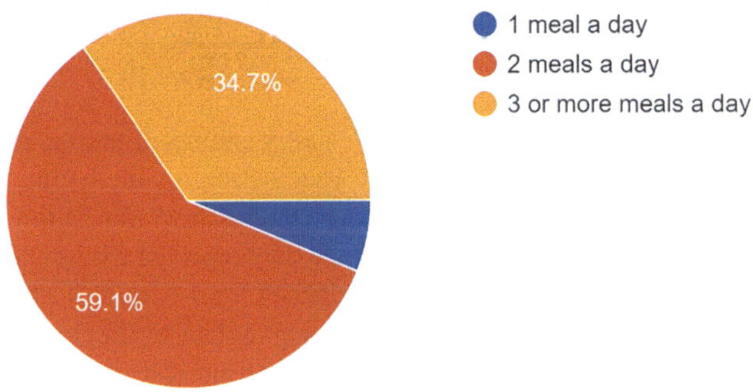

Fig. 10 Number of meals per day. (Source: Field Survey, March–May 2023)

agencies, non-profit organizations, and community initiatives are crucial in establishing sustainable solutions, such as expanding access to food banks, community kitchens, and nutrition assistance programs. By recognizing and actively addressing the issue of food insecurity, we can contribute towards alleviating the hardships faced by individuals experiencing homelessness and promoting their overall well-being.

Duration of Sleep

The survey further explored the sleeping conditions of individuals experiencing homelessness, revealing yet another disheartening reality. Shockingly, a significant number of respondents reported receiving inadequate amounts of sleep on a nightly basis. Disturbingly, some individuals reported getting less than 3 h of sleep, while the majority indicated sleeping for less than 6 h per night. These findings shed light on the immense challenges faced by those without stable housing in obtaining proper rest and the profound impact this has on their overall well-being.

Sleep deprivation not only affects their physical health but also contributes to heightened levels of stress, fatigue, and cognitive impairments, further exacerbating the difficulties they encounter in their daily lives. It emphasizes the urgent need for safe and secure sleeping arrangements for individuals experiencing homelessness, along with access to appropriate support services. Collaborative efforts are required to provide temporary shelters, affordable housing, and comprehensive social services that address the multifaceted needs of this vulnerable population. By recognizing and addressing the profound issue of sleep deprivation, we can work towards creating more compassionate and inclusive communities that prioritize the fundamental right to rest and rejuvenation for all individuals, regardless of their housing status.

Status of Vaccination Against COVID-19

This survey aimed to assess the vaccination status among economically disadvantaged individuals in order to understand the vaccination uptake within this population and its implications for public health and homelessness mitigation efforts.

Out of the 450 participants, a substantial 85.10% reported having taken the vaccination against COVID-19. This finding indicates a positive trend in vaccination uptake among the economically disadvantaged population affected by homelessness, demonstrating a significant commitment to public health measures. Conversely, the remaining 14.90% of respondents reported not having taken the vaccination (Fig. 11). This subset presents an opportunity for targeted outreach efforts to ensure equitable access to vaccination services and address potential barriers that may hinder their vaccination uptake.

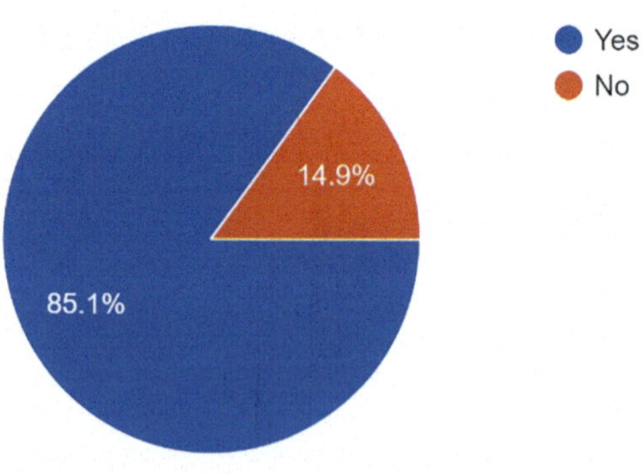

Fig. 11 Vaccinated against COVID-19. (Source: Field Survey, March–May 2023)

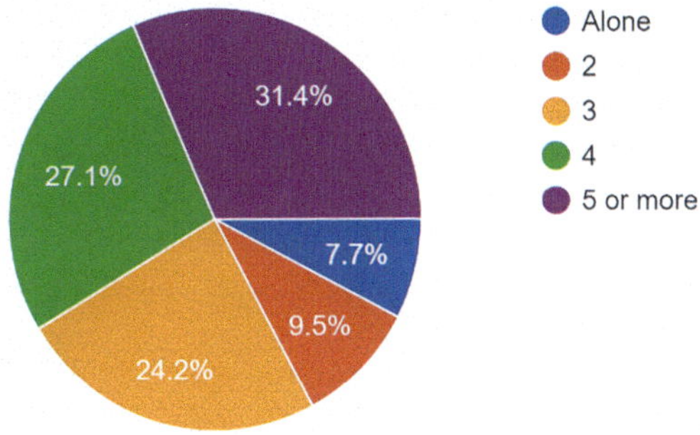

Fig. 12 Inmates in temporary accommodation. (Source: Field Survey, March–May 2023)

Housing Arrangements

The survey sought to understand the prevalence of proper housing arrangements within this population and shed light on the challenges faced by those without a stable and suitable place to stay. Out of the 450 participants, 77.40% reported having a proper place to stay, indicating some level of housing stability among the economically disadvantaged. This finding suggests that a significant portion of the surveyed population has access to suitable housing arrangements (in the form of a shack/tent fixed at a place), which can contribute to their overall well-being and quality of life though to a very limited extent. However, it is concerning that the remaining 22.60% of respondents reported not having a proper place to stay, indicating a substantial proportion of individuals facing housing insecurity and potential homelessness and frequent displacement within this group.

With regard to the number of inmates the homeless stay with, the findings from the survey revealed a significant trend of crowded living conditions among the less fortunate. It was found that the majority of respondents (82.70%), comprising a substantial percentage of the surveyed population, reported having either three or more than three individuals living in their temporary areas (Fig. 12). This indicates the prevalence of overcrowding as a challenge faced by individuals experiencing homelessness.

In addition to exploring the migration status of individuals experiencing homelessness, the survey also sought to understand their sense of safety within their current living conditions. The findings revealed a distressing reality, as an overwhelming two-thirds of respondents expressed feeling unsafe (Fig. 13).

This unsettling statistic sheds light on the inherent vulnerabilities faced by those without stable housing, further exacerbating the challenges and hardships they endure on a daily basis. The lack of safety not only perpetuates a cycle of fear and insecurity but also hampers individuals' ability to rebuild their lives and seek opportunities for upward mobility. It underscores the urgent need for interventions that prioritize not only providing shelter but also creating environments that foster a sense of security and well-being. This may include implementing measures to enhance public safety, ensuring adequate lighting in public spaces, and establishing support networks that address the unique safety concerns of those experiencing

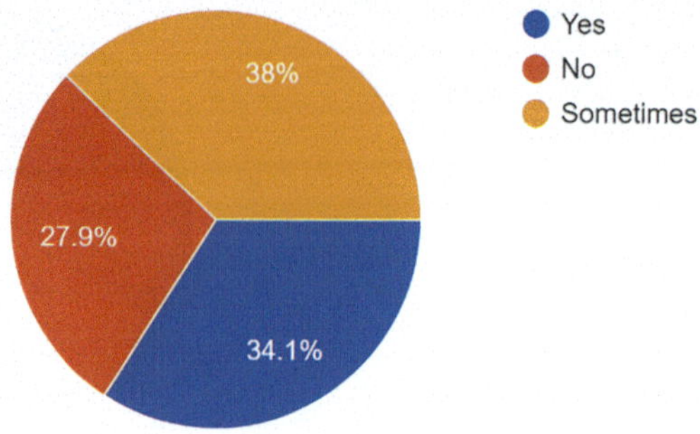

Fig. 13 Feeling of safety. (Source: Field Survey, March–May 2023)

homelessness. By acknowledging and addressing the deep-rooted issue of safety, we can work towards creating inclusive communities where everyone has the right to feel secure and protected, regardless of their housing situation.

Financial Obligations for Accessing Housing

This survey aimed to assess the financial obligations faced by individuals facing poverty and homelessness in relation to their housing arrangements. It sought to understand the prevalence of financial burdens associated with occupying a place to stay and shed light on the challenges faced by economically disadvantaged individuals in meeting such obligations.

Out of the 450 participants, 30.60% reported paying something to other people for the place they occupy to stay. This finding suggests that a significant portion of the surveyed population faces financial obligations in order to secure housing, highlighting the additional burden faced by those already experiencing poverty and homelessness.

Conversely, the remaining 69.40% of respondents reported not having to make any payments to others for their housing arrangements, indicating a level of housing stability and potentially reduced financial strain among this subset of individuals.

Harassment from Law Enforcement Bodies

The homeless population usually set up its makeshift accommodation on empty plots or vacant lands owned by the administration. This population continues to occupy the area until it is asked to vacate either for making way for some construction activity, redevelopment project, or city beautification. Sometimes, this evacuation can cause harrowing experience for the homeless if force and aggression are exercised by the law enforcement agencies.

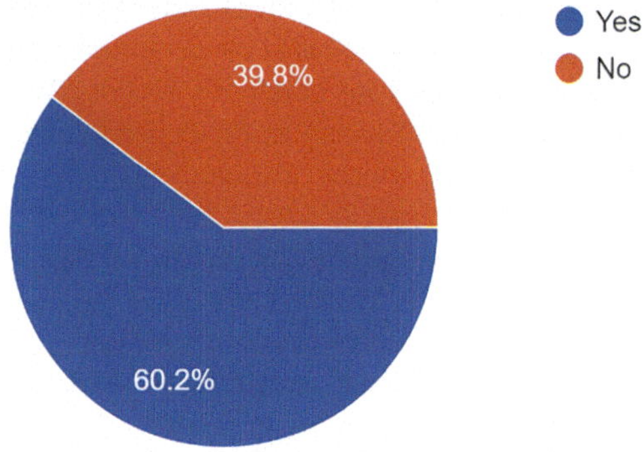

Fig. 14 Harassment faced from law enforcement bodies to vacate occupied land. (Source: Field Report, March–May 2023)

During the survey, 60.20% of the homeless respondents reported instances where law enforcement agencies troubled them to pack their belongings and move elsewhere. These findings highlight a concerning pattern of displacement faced by a significant portion of the homeless population. Conversely, 39.80% of respondents reported not encountering such troubles with law enforcement, indicating a variation in experiences within this group (Fig. 14).

The survey findings underscore the need to address the relationship between law enforcement agencies and the homeless population. It is crucial to promote fair treatment and protect the rights of individuals facing homelessness. Recommendations include providing training programs to law enforcement agencies to enhance their understanding about the homelessness, fostering partnerships with homeless advocacy organizations, establishing clear guidelines for interactions, and promoting community engagement and awareness campaigns. By implementing these recommendations, we can work towards a more empathetic and supportive society that respects the rights and dignity of those experiencing homelessness.

Challenges, Interests, and Preferences of the Homeless Population

In reverence to their interests in professions and perceptions of socioeconomic obstacles, varied and diversified replies were gathered for the analysis of this report. The information was collected through semi-structured interviews which were carried out to learn about the public thinking on job opportunities and other societal issues. The main points of discussion have been summarized as under:

1. **Job preferences**: When questioned about their ideal career, most of the survey participants indicated that working in cafes, government institutions, as a personal driver, or in the private sector were among their top choices. This suggests that people have a wide range of occupational preferences, various career objectives, and propensities for both service-oriented and private sector roles.

2. **Financial priorities**: When asked about their financial priorities, respondents named three main areas of concentration: saving money, building a decent house, and paying for their children's school. These priorities emphasize the value of education and housing as fundamental aspects of people's lives. The emphasis on preserving money also conveys a worry for future stability and financial security.
3. **Perception of society**: The words 'outrageous' and 'downtrodden' were used by the respondents to characterize the impression of the society about them. This suggests a sense of irritation or unhappiness with current social norms and the way resources are allocated. These phrases are used to express a perceived lack of fairness and equality, implying that some people feel neglected or marginalized by societal structures.
4. **Challenges of the current generation**: According to the respondents, their problems have persisted since they were born, indicating that the current problems have a generational component. This shows that socioeconomic issues have been passed down from one generation to the next and are deeply engrained. Given the fact that it is a generational vicious cycle and strong changes need to be made on the base level, like easy access to education, quick access to jobs, free healthcare, and basic financial education.
5. **Hurdles and difficulties**: The survey replies revealed a number of obstacles and difficulties faced by the respondents, such as bringing clean drinking water, doing simple tasks, and getting little sleep. These difficulties are a reflection of some of the underlying struggles people experience, such as the lack of access to necessities, employment prospects, and enough sleep.
6. **Employment challenges**: A worsening environment and a lack of employment possibilities were frequently highlighted as important obstacles, according to the research. The responders listed a few of the choices, including jobs as a car driver, maid, food server, and cleaner. These comments indicate a lack of steady, well-paying employment opportunities, which forces people to take on low-skilled positions that are frequently underpaid. The situation's deterioration suggests a declining socioeconomic climate that makes it more difficult to get work.
7. **Perception about the government programs/initiatives**: In general, respondents had conflicting opinions about how effective government programs are. The majority of people were of the opinion that the governmental schemes needed to be changed and updated as they were not as effective in practical settings. A large majority of the populace was found to be uninformed of the new programs, suggesting that there may be a communication gap between the government and the people. These results highlight the value of increasing knowledge and making social welfare programs easily accessible.

Impressions and Expectations from the Society

This survey aimed to gather the perspectives of individuals facing the problems of poverty and homelessness and also how society can contribute to eradicating the problem of homelessness. The survey sought to understand their opinions on the role of empathy, job opportunities, and other societal efforts in addressing the issue.

Out of the 450 participants, respondents shared diverse perspectives on how society can contribute to eradicating the problem of homelessness. About 32.90% of respondents emphasized the importance of empathy and understanding towards their condition. This highlights the need for society to adopt a compassionate approach, fostering a greater sense of empathy and reducing stigma associated with homelessness; 56.70% of respondents suggested offering job opportunities as a crucial step in addressing homelessness. This finding emphasizes the significance of employment in providing stability, empowerment, and a pathway out of homelessness for individuals facing poverty.

Other suggestions provided by respondents included providing financial allowances, urging the government to take action, ensuring access to permanent homes, and providing proper facilities. These suggestions reflect the multifaceted nature of the issue and the diverse ways in which society can contribute to addressing homelessness.

Some respondents also expressed scepticism about the government's ability to fully address the issue of homelessness, indicating a desire for collective action beyond government intervention.

Expectations from the Government

The survey sought to gather insights on their expectations from the government regarding the type of government assistance they would like to receive and to address the issues they face. The respondents expressed diverse opinions on the type of government assistance they would prefer. More than a quarter of the respondents (28.80%) indicated their desire for the government to set up helpful schemes. This suggests a need for targeted and practical initiatives specifically designed to address the challenges faced by the economically disadvantaged population. Another 37.50% of the respondents expressed the preference for the government to provide a certain amount to sustain them in case of joblessness. This highlights the importance of financial support to bridge gaps during periods of unemployment and enhance stability for individuals facing poverty and homelessness. On the other hand, 31.50% of the respondents emphasized the need for the government to focus on education to increase job opportunities. This finding underscores the value placed on education as a means of empowerment and a pathway to secure employment and economic independence. The remaining respondents suggested various measures, such as providing permanent homes and implementing schemes in a manner that actually benefits them. These suggestions reflect the broader aspirations for long-term housing stability and a call for effective and impactful government initiatives.

Key Findings

1. Most of these people are migrants having rural background and have come to the city with the aspiration of improving their socioeconomic conditions.
2. Lack of education is one of the main reasons behind homelessness in Chandigarh. Education empowers people, and it is one of the most effective tools to alleviate poverty. Since, majority of the homeless people were found to be lacking knowledge and skills required for availing the employment opportunities, they ended up either jobless or getting low paid jobs in unorganized sector with no economic security.
3. Overcrowding emerged as a pressing issue within the temporary living spaces of economically disadvantaged individuals. Crowded conditions can exacerbate health risks, strain available resources, and impact overall well-being.
4. Hopelessness and despair seem to have engulfed the homeless in the city. Apart from the daily socioeconomic problems faced by them, many respondents seemed to have lost hope in the government, in themselves and for things to change for them.

Suggestions

Since Chandigarh, when designed originally, was for just half a million people, it is understandable now that with the population expansion, the city would get suffocated. Despite so much thought and action being taken by the government at both the national level and the city, for the urban homeless, there are still many shortcomings we must overcome. These policies and initiatives originate fabulously at the cognitive stage; however, poor execution needs to be taken care of. The setting up of shelters and schemes would do no good if people cannot have access to them. Like, despite the night shelters being set up, the majority of the homeless still prefer to sleep on the streets due to lack of space, the fear of thefts, the unhygienic conditions, or perhaps a dispute between people in the area, etc., so such factors will prevent the homeless to take advantage of the opportunity provided.

In the light of all the findings and analysis, the best strategies to ameliorate the problem of homelessness in the city are as follows:

1. Less complex schemes that are more practical and reality based so that people can actually have a chance to get access to them.

2. Instead of building additional shelters and houses, Government should take charge of better maintaining the ones already in place and rectifying them. Tin shacks cannot be lived in the summer months as it would be unbearable hot, so the temporary shelter stations can instead provide tents, etc., that can at least offer some livability. The issues at the ground level like theft, arguments, and fear of safety, etc. are the ones that really need to be worked upon. A night guard or two could be deployed to help control such instances.
3. Understanding the extent of overcrowding is crucial for policymakers, organizations, and community members to develop effective interventions that address this specific challenge. These results emphasize the urgency of implementing targeted strategies to address overcrowding, ensuring that individuals facing homelessness have access to safe, secure, and adequately sized living environments.
4. The government can start an initiative to assist the homeless people getting employment which will help them afford a decent place of living. Through this initiative, there can be a creation of a database holding records of the homeless including their identification cards, other personal details along with their skill set and job preference such as cook, driver, baby sitter, labourer, mason, household help, etc. This database will help homeless people to find desirable jobs while it will also be of great help to the city residents who usually struggle to find reliable and dependable workers at affordable prices.
5. A practical and viable solution to solve the problem of homelessness in the long run can be initiation of the 'Housing for Work Scheme' which, on one hand, will provide housing as along with employment opportunities for the homeless people and on the other hand, generate revenue for the Chandigarh administration. Under this scheme, vacant lands under the U.T. administration can be converted into landscapes depicting miniature villages of different parts of India such as a village displaying the natural and cultural landscape of North-East India, a Tribal-theme village of Central India or a village showcasing a simplistic lifestyle of Western India. These theme-based landscapes will be owned and maintained by the Chandigarh administration along with providing employment to the homeless people based on their skills and housing on the same premises on a very nominal and affordable rent amount. Such places will become one of the main attraction centres of the city, which is already known for its various tourism sites, and hence, facilitate revenue generation of the U.T. administration.

Within the framework of tackling homelessness in Chandigarh, India, the solidarity dimension highlights the need of solidarity, mutual aid, and shared accountability among individuals and communities. Recognizing the humanity and dignity of the homeless and banding together to solve the underlying causes of homelessness are examples of solidarity. This strategy offers long-term solutions including accessible healthcare, affordable housing, educational chances, and work opportunities, rather than just band-aid repairs. Solidarity strives to establish an inclusive atmosphere where homeless people can rebuild their lives with dignity and respect by encouraging empathy, understanding, and cooperation across all sectors of society, including government and civil society.

Proposals like the 'Housing for Work Scheme' are prime examples of this strategy because they provide both homes and jobs, helping the homeless achieve economic empowerment and self-sufficiency. All things considered, solidarity provides a revolutionary framework that puts human rights, social fairness, and group accountability first when tackling homelessness.

Conclusion

In many urban locations around the world, including Chandigarh, a city in northern India, homelessness is a significant issue. Being homeless describes a situation in which people or families are compelled to live on the streets or in makeshift shelters because they lack a secure place to call home. Homelessness in Chandigarh is a complicated problem that is impacted by many variables, including poverty, unemployment, migration, natural catastrophes, and a shortage of affordable housing. Chandigarh serves as a hub for migrants from nearby states who travel there in pursuit of better employment and living conditions. However, a lot of people have trouble finding dependable employment and inexpensive accommodation, and they frequently end up living on the streets or in squalid neighbourhoods.

Numerous initiatives have been put in place by the Chandigarh Administration to deal with the issue of homelessness in the city. These actions consist of building inexpensive housing units, such as the Chandigarh Small Flat Scheme 2006, and providing shelters and community kitchens as well as basic utilities and services to homeless people and families. However, despite these initiatives, homelessness is still a persistent issue in Chandigarh, necessitating continued efforts to address its underlying causes and offer long-term solutions.

This study highlighted the main challenges faced by the homeless population as being lack of employment opportunities, improper housing conditions, lack of sufficient sleep time along with continuous harassment by the law enforcement body for vacating the land occupied by them. Their main financial priorities revolve around their children's education, owing a proper accommodation and saving money. They also feel dejected with the usage of terms like downtrodden used for them by the mainstream population. While government schemes have had some positive impact, there is a need for better awareness and more comprehensive measures for addressing the socioeconomic difficulties face by this population group. By suggesting some effective measures for addressing homelessness and associated issues, this paper provides important insights for policymakers and stakeholders to create interventions and policies that tackle these issues and promote socioeconomic development in a way that touches the hearts and gains sympathy for the affected population.

References

Chandigarh city population 2023. Literacy and Hindu Muslim population. (n.d.). Retrieved from https://www.census2011.co.in/census/city/22-chandigarh.html

DISD. (n.d.). *First-ever United Nations resolution on homelessness*. Retrieved from https://www.un.org/development/desa/dspd/2020/03/resolution-homelessness/

Fazel, S., Geddes, J. R., & Kushel, M. (2014). The health of homeless people in high-income countries: Descriptive epidemiology, health consequences, and clinical and policy recommendations. *The Lancet, 384*(9953), 1529–1540. https://doi.org/10.1016/s0140-6736(14)61132-6

Filipenco, D. (2023). *Homelessness statistics in the world: Causes and facts*. Retrieved from https://www.developmentaid.org/news-stream/post/157797/homelessness-statistics-in-the-world

IBEF. (n.d.). *Industrial development & economic growth in Chandigarh*. India Brand Equity Foundation. Retrieved from https://www.ibef.org/states/chandigarh#:~:text=Chandigarh's%20Gross%20State%20Domestic%20Product,Product%20(NSDP)%20was%20Rs

OHCHR. (n.d.-a). *International Bill of Human Rights*. Retrieved from https://www.ohchr.org/en/what-are-human-rights/international-bill-human-rights

OHCHR. (n.d.-b). *Homelessness and human rights*. Retrieved from https://www.ohchr.org/en/special-procedures/sr-housing/homelessness-and-human-rights#:~:text=Homelessness%20violates%20the%20principle%20of,Economic%2C%20Social%20and%20Cultural%20Rights

Sheltering the urban homeless. (n.d.). orfonline.org. Retrieved from https://www.orfonline.org/expert-speak/sheltering-the-urban-homeless

Sustaining Solidarity: Dharavi's Socio-Economic Landscape in *Rediscovering Dharavi* by Kalpana Sharma

Kasturi Saikia and Nigamananda Das

Introduction

The urban poor are an inescapable reality of modern cities. The spaces occupied by the urban poor can be seen as a contested terrain that, at times resists monolithic conceptualisations as enclaves of poverty. This is the result of migration from rural to urban areas which is an age-old phenomenon. With varied reasons for migration, the migrated section of people is often seen as the surplus population and, in worst cases, the discarded population subjected to the whims of the powerful section of society. In modern cities, this migrated population is often seen to occupy the peripheral regions of the cities with temporary makeshift shelters subjected to uninformed demolitions and evictions. These areas grow to be slums primarily characterised by poverty and lack of proper means of survival. This urban sprawl continues to extend its peripheries and is a potent rising issue in the Indian subcontinent. The volatility of cities brings changes in the lives of these urban poor. A German Sociologist and Philosopher, Georg Simmel's conceptualisation of poverty is significant in this context as it demonstrates the multidimensional nature of poverty. Poverty, according to him, is a lack of means to provide for themselves and is a 'unique sociological phenomenon, where several individuals who, out of a pure individual fate, occupy a specific organic position within the whole' (Simmel & Jacobson, 1965, 140). The nature of the formation of slums aligns with this sociological understanding of poverty.

Nonetheless, defining a slum in common parlance without a reference to its history becomes a difficult task ethically. Even though a section of the urban poor owns

K. Saikia (✉) · N. Das
Department of English, Nagaland University, Meriema, Nagaland, India
e-mail: kasturi_rs2021@nagalanduniversity.ac.in; nnddas@nagalanduniversity.ac.in

K. Arora, R. R. Anand (eds.), *The Solidarity Approach in Geography*,
https://doi.org/10.1007/978-3-031-86753-8_6

their agency, the paternalistic interventions of the State cannot be overlooked. The sense of wanting to be fulfilled by the privileged section jeopardises their independence. It is crucial to note that the heterogeneous nature of the urban poor category should not be overlooked as poverty and seclusion have many dimensions that extend beyond the lack of material means. This chapter shall delve into significant questions regarding the nature of the urban poor.

It is pertinent to note that the inhabitants of the slums have their individual history of being slum dwellers. They are generally the migrant population who migrates from the village to the city for economic pursuits. The lives of the urban poor are a socially relevant issue that is frequently dictated by the ruling elite. However, the slum as a concept has shifted from being perceived through a series of metaphors of disease and dread to an inseparable part of the city contributing to its growth. The shifts and evolution of its meaning are numerous and subjected to continuous change and accommodation.

Dharavi, generally regarded as Asia's largest slum shall be the base of the study. This case study of Dharavi shall highlight the significance of interconnectivity and interdependence as a crucial factor for establishing itself as the geography of care, responsibility, and a complex social conglomeration. In the context of Dharavi, the study of its geographies of care and responsibility shall enhance the understanding of the internal dynamics of the place. The chapter shall further integrate a humanitarian perspective into the geographical study of the place. The study shall espouse the consideration of this perspective while implementing developmental processes in the slums.

The new critical method of close text reading shall be applied. The chapter will re-examine the available critical perspectives on the selected authors and texts to engage the study in the desired field of research.

Literature and Geography

It is possible to appreciate and reflect on one's surroundings through literature as a means of cognition and expression, which leads to an emotive interaction with the physical environment. Considering this, literature can be an empathetic means of engaging with a geographical environment and can assist in the comprehension of the changing realities of the geographical environment. Application of humanist concepts while analysing the dynamics of geographical territory shall provide substantial insights into a place. The objective geographical reality combined with the human experiences of a place shall enhance one's understanding. This aspect of geography that focuses on the human experiences forms the basis of humanist geography and shall play a crucial role in the entire discussion. This chapter demonstrates the relation between literature and geography with a particular focus on the Dharavi Slum in Mumbai. With non-fictional accounts of Dharavi as the base, the chapter further seeks to offer a novel perspective to the discipline of geography in comprehending the socio-cultural landscape of a place. This shall highlight the universality of literature in describing the human condition as literature can be perceived as a means to create possibilities of alternative worlds to foreground and critique the present conditions. It is crucial to note that like other fields of knowledge, literature too beholds the power of expression of the plight, struggle, and stories of the lives of the urban poor. In a way, literature renders power to the voiceless by documenting their experiences, identity, and their community ties. Noted journalist and writer Kalpana Sharma's non-fictional account of Dharavi in her book *Rediscovering Dharavi: Stories from Asia's largest slum* (2000) shall form the base of the study. Further, Liza Weinstein's *The Durable* Slum (2014) and Marie-Caroline Saglio-Yatzimirsky's *Dharavi: From Mega-Slum to Urban Paradigm* (2013) shall assist in more profound understanding and conceptualisation of the slum selected.

Slums as Texts: Human Geography and Phenomenology

Slums as texts can be read and interpreted for their historical, cultural, and social meanings. This leads us to a relational and diversified understanding of slums. This relational understanding of slums shall guard against preconceived and absolute notions. Instead, an inclusive approach should be adopted that incorporates spatial dynamics and the nature of human relations. Slums, being the focus of the study, spontaneously evoke a sense of empathy combined with a feeling of unexpressed fear and disgust. The chapter seeks to demonstrate and analyse the presence of solidarity among the diversity in the slum. This shall, in turn, seek to delve into alternative possibilities of perceiving the slum.

A noun, verb, and adjective, slum refers to a place, a phenomenon, and a way of life. The word slum was popularised among the urban working class in nineteenth-century London where people lived in proximity with the factories and industrial plants. With its genesis in London, slums were gradually recognised as an international phenomenon. However, given their social and physical constitution, slums are often presented as bodies existing in the peripheries of an urban space, as *terra incognita*. Nonetheless, slums are a defining feature of urban growth and urban spaces and developed in the latter half of the twentieth century. However, this cannot be generalised for all the cities. In various cities, slum developmental measures have been initiated for the well-being of the slum dwellers.

The meaning of slum gradually transformed to embody a wide array of realities. The established definitions of slums based on binaries paint a rather grim and gloomy picture of the slums. Reinforcing negative stereotypes, such definitions are inadequate and improper. The definition of slums suggesting inglorious associations and dark spaces which turned into a generic and widely spread definition was explored by Charles Dickens, an English novelist and social critic. He is regarded as the most famous early chronicler of these impoverished urban living conditions. The Dickensian definition of slums explores the dark underbellies of the cities, thus foregrounding a stereotypical notion of slum.

It is pertinent to initiate an understanding of place and space for the conceptualisation of slums. The concepts of place and space are vital to humanist geography. As Christopher Tilley writes, 'Places constitute space as centers of human meaning' (Tilley, 1994, 14). Unlike places, which refer to specific locations or territories, space holds the significance of the places. Associated with specific places, spaces are intangible. Geographical spaces are loaded with the significance of the lived experiences of the humans inhabiting a particular space. Thus, space, here, refers to the nexus of social relations that constitutes a place.

Humanist geographers consider places as fundamental aspects of human experience. Edward Relph's thoughts align with the idea that place is crucial to the human experience. In his book *Place and Placelessness*, Relph argues for the phenomenological understanding of geography when he states, 'Geography is not to be understood as just another branch of knowledge with geographical reality being primarily an object' (Relph, 1976, 1). Instead, one should recognise that there underlies a complex amalgamation of human experiences that leads to their understanding of the built spaces as pleasant, threatening, or menacing.

Ralph draws reference from another geographer, Lukermann, who posits place as a 'complex integration of nature and culture that have developed and are developing in particular locations and which are linked by the flow of people and goods to other places' (Relph, 1976, 5). This notion of places embodies a broader meaning not limited to restrictive definitions. Instead, it focuses on the continuous becoming and emerging of spaces with historical and cultural changes over time. Places, when understood through this perspective, embody meaning that highlights the beliefs and values people hold of a place instead of mere factual territorial details. This emphasises acknowledging the instrumentality of constructed environments and their connections to various aspects of human life, such as cognition, self-identity, and socio-cultural interactions. It is safe to conclude here that places are integral aspects of human life, security, and identity. These dynamics of place and space are inseparable from power politics. From a Foucauldian perspective, power structures underpin built environments, which in turn largely articulate the relationships that control the inhabitants of those environments.

Slums, when approached through the humanist geography perspective, foreground a subjective understanding of the slum landscape. Humanist geographers highlight the significance of people's lived experiences and the socio-political and cultural powers that shape the environment. When seen through this perspective, the complexity of human experiences in the slums can be captured through an understanding of the subjective experiences of the slum dwellers. Adopting this approach in deciphering the life of the slum dwellers in the novel shall foreground the lived experiences of the slum dwellers.

Humanist geography draws on phenomenological concepts. The engagement of geography with phenomenology begins with human geography, where human experiences and perceptions of a place are taken into account. Man's experiences of the geographical environment are valuable and crucial to the meaning-making process. In the article 'Grasping the Dynamism of the Life World' Anne Buttimer addresses the idea of a life-world to comprehend people's experience of space. Life-world refers to the sharing of the physical and the social world where 'experience is constructed' (Buttimer, 1976, 280). Mundane activities would be crucial here in understanding human engagement with the place. The phenomenological understanding of place here foregrounds an understanding of mundane activities. People's manner of experiences and knowledge of the world are key to any phenomenological approach. This is expressed by Christopher Tilley who states that 'Phenomenology involves the understanding and description of things as they are experienced by a subject' (Tilley, 1994, 11).

With a basic understanding of the notion of place as analysed by various geographers, slums can be studied. When examining various aspects of the urban fabric, Marxist geographers like David Harvey have focused on the special function that constructed space plays in upholding the socio-economic structure of society. Doreen Massey, the famous geographer, highlights the meanings and significance attributed to a place by people varying in age, gender, class, and race and the implications generated by their inhabited spaces (Massey, 1994, 2). The socio-political and cultural properties of a place bring forth a multidimensional perspective.

It is a known fact that the visual form of every place is the landscape and every slum has its landscape. As expressed by Relph, 'the spirit of a place lies in its landscape' (Relph, 1976, 30). This brief conceptualisation of the landscape has been intended to assist in a precise understanding of the slumscape. As a landscape, Dharavi has distinct and prominent features such as narrow alleys, dumping grounds, tin roofs, etc.

Several variables, including economic inequality, urbanisation, rapid population expansion, the need for employment, and the slum's history, have molded the slumscape. The perception of slums as a geographical territory with precise definition would be fallacious and naïve. However, an in-depth understanding of the place and its way of life demands an understanding of the place combined with the phenomenological foundation of its geography. Phenomenological understanding imbibes the individual experiences of the place, highlighting an individual's perception of a place.

It is vital to remember that the slumscapes vary according to their location, socio-cultural history, and the people who inhabit them. The concept of the slumscape highlights the physical manifestations of poverty, inequality, and marginalisation, providing a lens to understand the lived experiences of residents and the urban environment they navigate daily.

Dharavi: From a Fishing Village to an Entrepreneurial Hub

Dharavi, regarded as Asia's largest slum, is a multi-ethnic space bustling with local economies and offering alternative forms of solidarity, mobilisation, and governance. Superficially seen, Dharavi is regarded as synonymous with chaos, poverty, squalor, and unhygienic living conditions. Dharavi is not exempted from such sloppy generalisations as it is perceived to be an area infested with all sorts of crime.

This monolithic reality of Dharavi is challenged by its people and their way of life as a resistance against hasty inadequate generalisations. However, an in-depth analysis of the socio-cultural and economic structure of Dharavi reveals its complex internal organisations. Dharavi through its surprising economic vitality posits itself as an entrepreneurial hub contributing to the city's economy. It is known for its urban informal economy, which encompasses all small trades, establishing it as the most industrious zone of Mumbai. The functioning of the informal economy lies at the core of Dharavi's prosperity and can be considered a significant lens through which the solidarity of the people of Dharavi shall be analysed.

As mentioned in the book, *Dharavi: From Mega-Slum to Urban Paradigm* by Yatzimirsky, a slum is usually considered an 'illegally occupied self-built zone' (Yatzimirsky, 2013, 10). Subject to the violence of nature and bulldozers, the slums are ever-changing in nature and continue to be rebuilt after being destroyed. The status of the people who occupied Dharavi before 1995 has been legalised. However, the legality remains unstable. However, the question arises if Dharavi can be merely considered a slum. Embodying various modes of production, Dharavi, with its size and history, is constantly growing, with its meaning evolving. With Dharavi's interaction with the wider socio-economic conditions of the city, it would be inadequate and incompetent to treat Dharavi only as a marginal space. On the contrary, the marginal status of Dharavi cannot be completely overlooked even if its development fits into the urban framework of the city.

Geographically, Dharavi is located in the heart of Mumbai and stretches between the western and central railway lines along the border between the city and the suburbs. Dharavi, located close to the employment areas, becomes a residential area for the workers as the city's rail, and highway routes are easily accessible from here. Initially, Dharavi was an illegal garbage dump for nearby construction sites. Further, it was also once a human dumping ground as the unwanted and surplus population including the labour migrants and refugees settled here through waves of migration. Dharavi's industrial growth aligns with Bombay's industrial development in the nineteenth century. As new economic activities spurred in Bombay, there was demand for space and hence the most polluting industrial activities including leather work, pottery firing, and recycling were moved to the marshy lands of Dharavi. The industrial force required here were the migrant labourers and refugees from the rural areas. With temporary makeshift houses, Dharavi gradually came to be known as a slum that houses these labourers. Dharavi, as a low-cost housing space proliferated as it addressed the needs of the workers as well as provided spaces for the industrial demands of the city. As years passed, real estate companies built apartments in Dharavi whose value rose monthly. In 2004, the government adopted the Dharavi Rehabilitation Plan. However, it must be taken into account that though some areas of Dharavi have been demarcated as legal by the municipality, there still exists areas like the Naik Nagar, which consists of the huts of the fishermen that have been illegality built along the marshy land. The population of Dharavi is considered to be more than the population of Mumbai taken together with 100,000 inhabitants/sq. km (Yatzimirsky, 2013, 44). Owing to the diversity of people that reside in Dharavi in terms of class, caste, community, and religion, Dharavi embodies the cosmopolitanism of the city. People speaking different dialects and languages are united by Hindi which facilitates day-to-day communication.

The physical architecture of Dharavi further adds to the volatility of the term 'slum'. With the growing economic prosperity and modern infrastructure, Dharavi elevates itself from the distinction as a slum. However, the narrow alley, stinking drains, sewage system, and open garbage disposal system bring forth the characteristics that qualify a slum. Illegal water and electricity connections from the city to Dharavi are an everyday affair. Makeshift shelters with plastic or jute bags characterise the place meant for newcomers. The older inhabitants stay in 'jhonpri that have bamboo, wood, and cob walls' (Sharma, 2000, 48). Roofs with sheet metals, tarpaulin sheets, and bamboo floors add to Dharavi's distinctive features. Living conditions in Dharavi are not homogenous. The conditions in the cleaner areas with regular water and electricity are better than the overcrowded ones.

Setting aside these objective descriptions of Dharavi, an in-depth understanding of the history of the evolution of Dharavi would help in the comprehension of the socio-economic dynamics of the place. Dharavi's growth from a fisherman village to a prosperous leather colony unveils tales of hardship, unity, and solidarity. The history of Dharavi has been curated in the non-fictional accounts of Kalpana Sharma, Liza Weinstein, and Yatzimirsky. Bombay was a group of seven islands, and Dharavi was originally a Koli fisherman's village located on one of these seven islands. As mentioned by Yatzimirsky in her book *Dharavi: From Mega Slum to Urban Paradigm*, the name of Dharavi comes from *Dharevarca Bhag* which means the 'creek's shore' (Yatzimirsky, 2013, 52). The industrial explosion in Bombay led to the influx of leather community migrants from Tamil Nadu. With other activities being gradually developed, the area was organised into districts where small activities were performed initially. The tailors, cobblers, and barbers established themselves in makeshift structures. With the migration from Tamil Nadu and Kerala, the population in Dharavi grew, with Tanning being the most important activity. Other small trades that contributed to leather making developed and Dharavi slowly became a small-scale leather industry.

Dharavi through Solidarity Lens

Diversified reasons function as crucial factors that lead to the ties of solidarity in Dharavi. These factors can be drawn from Dharavi's history of growth, economic development, and socio-cultural amalgamation.

Solidarity, derived from the French term solidarity, is the sharing of common interests and goals that leads to the feeling of social connection and unity. Shared responsibilities that lead to self-sustenance and identity assertion are crucial attributes of solidarity. Solidarity necessitates a particular level of unity, involvement, and cooperation. According to Hollinger, 'solidarity entails a greater degree of conscious commitment, achieved only when parties to an affiliation exercise at least some measure of agency' (Hollinger, 2006, 24). Further, the concept of solidarity has associations with labour and socialist movements.

In other words, solidarity can be regarded as human cohesion to battle uncertainties and injustices. Shared culture, norms, goals, and ideals can be considered as the basic sources of solidarity in society, as Putnam declares that 'shared norms and values are bound up with fostering of shared goals' (Oosterlynck et al., 2017, 4). The social fabric of the community which is the major bearer of shared ideals is regarded as the main source of solidarity. Oosterlynck's sources of solidarity, as discussed in the book *Place, Diversity, and Solidarity* shall be crucial in analysing the nature of solidarity in Dharavi. The pressing challenges for the urban poor include social marginalisation, negotiation of differences, and assertion of identities calls for solidarity among the urban poor which shall contribute towards the common good. The discussion on solidarity shall encompass the thoughts of Emile Durkheim, the French social philosopher. Durkheim, in his work *The Division of Labor* (Durkheim, 1984) in Society, analyses the concepts of mechanical and organic solidarity. Social solidarity, according to Durkheim, is crucial to all societies. Societies governed by collective consciousness practiced mechanical solidarity. These societies were traditional and undifferentiated societies with homogenising tendencies. On the other hand, modern societies which are characterised by division of labour with interdependence practiced organic solidarity. The Durkhemian concept of mechanical solidarity foregrounds the significance of group minds rather than individual minds. Durkheim espoused this with particular reference to primitive societies. The individual minds being merged into the group minds can bear positive and negative connotations. A tendency of forced homogenisation can be traced through the erasure of individual differences. Durkheim's notion of solidarity has been challenged by many researchers based on its functionality. However, the sense of interconnectedness manifested through a shared identity which is core to Durkheim's understanding of solidarity cannot be overlooked. This aspect of his understanding resonates in modern times where the marginalised section of people shares a common history and works collectively to assert their identity and resists succumbing to the whims of the powerful section of society. Durkheim's application of solidarity from historical and cultural perspectives aligns with the solidarity among the social groups in Dharavi. This, according to Durkheim is organic solidarity, where people work together in small groups to fulfill various societal functions.

The vivacity of Dharavi is evident in the stories by Kalpana Sharma in *Rediscovering Dharavi* which narrates stories of the triumph of the human spirit and bravery over adversity. It emphasises the value of a comprehensive understanding of the slum by capturing its social and physical environment. This amalgamation of material and symbolic meanings of a landscape creates the sense of a place and is the primary preoccupation of humanist geographers. Hence, the approaches of humanist geographers have been adopted to comprehend Dharavi's social and cultural landscape. The humanist perspective interprets the landscape as a carrier and repository of symbolic meaning. The humanist geographers, drawing on phenomenological ideas, focus on the 'social construction' of the place through the thoughts and actions of the human agents (Lando, 1966, 6). For instance, the essence of slums lies primarily in the human experiences of the place. When viewed in this light, Dharavi is more about the people inhabiting it. It is about the never giving up spirit of Manish Mane, who runs a flourishing leather business, Waqar, who earns 70 lakhs per annum by selling shirts, and Abdul Baqua, who runs a factory registered as Ideal Trading Company in Dharavi. These individuals moved to Dharavi for a variety of reasons and contributed to Dharavi's evolving reputation. Remarkably, every person's accomplishment is the result of a group of people working together. The connections and collaboration among residents of Dharavi are essential to the development of their unique and collective identities and help people understand Dharavi as a place.

In *Rediscovering Dharavi*, the author introduces a cast of colourful characters, including the stoic potter Ramjibhai Patel, who represents six generations of Saurashtrians who have lived and worked in Mumbai, and women like Khatija and Amina, who helped quell racial tensions during the 1992–1993 riots and maintains the harmony of the social fabric of Dharavi. Some of the earliest migrant communities to have settled there are Kumhars (potters) from Gujarat, leatherworking castes like the Charmakars, Dhors, and Mahars from other parts of Maharashtra and Tamil Nadu, and artisans from Uttar Pradesh who set up industries dealing in ready-made garments, shoes, bags, and metal work. Today, Dharavi is known for its garments, leather industries, recycling business, and food products. This category of people contributed to Dharavi's transformation from a simple marsh into a flourishing land.

Solidarity through Migration

According to Kalpana Sharma, the inhabitants of Dharavi were 'economic and environmental refugees' (Sharma, 2000, 40). There were several migratory waves to Dharavi and thus, the population settled in Dharavi is not homogeneous. Delving into the history of migration, it can be seen that mass exodus occurred during droughts or famine. For instance, as noted by Yatzimirsky, the droughts and floods that struck Maharashtra in the early 1970s led to severe famines that forced people from rural areas to migrate to Bombay. These people either settled in Dharavi or further away. These people from Maharashtra were engaged in the activity of Pottery. The second migratory movement consisted of villagers from Bihar and Tamil Nadu who migrated because of poverty and began working in the leather manufacturing tanneries in Dharavi, and those from Uttar Pradesh started working in garment factories. Thus, both voluntary and involuntary migration can be observed. As described by Kalpana Sharma, Dharavi's history is entwined with the migratory pattern of Mumbai city. Dharavi emerges from different life stories of the people who inhabit them. Haji Shamshuddin, a migrant from Tamil Nadu, began his career as a rice smuggler and then worked in a printing press and finally got engaged in the peanut brittle-making business. Selvaraj from Tamil Nadu began working in Tanneries and then moved on to other businesses. These few instances described by Sharma foreground the dynamic collection of individuals surviving in dire circumstances.

Dharavi, being the salad bowl of people, is often seen as a place of integration of communities, religions, and professions. However, it is crucial to decipher the nature of integration. This migration of people can be seen as strengthening the social ties. The different communities that migrated to Dharavi are closely knitted among themselves as well and they share common tendencies towards work among the rest of the inhabitants. The spaces in Dharavi are organised along with a sense of belonging to the ethnic, linguistic, and religious community. They include craft-related affiliated areas like the Dhorvada (tanners) or the Kumbharvada (potters). These groupings create a sense of belonging, leading to solidarity and unity among themselves.

The varied castes in Dharavi, like Chambars, Dhors have their associations and identities. It is a crucial organisational factor that brings together communities based on linguistics or religious grounds. There are migrant groups that migrated from the same place but belong to different castes. However, these different migrant groups form associations through common link factors such as the homeland, language, or religion. Thus, despite particular distinctions, they identify as a member of the more significant group and share principles, a work ethic, and communal values. These groups express their identity through repeated practices, including religious processions, meetings, or other festivals.

A Collective History

The collective history that the people of Dharavi share is another significant source of solidarity among them. The inhabitants of Dharavi share a collective memory of growing up in a fishing village. A group's shared, created, or inherited memory is referred to as collective memory, according to French philosopher and sociologist Maurice Halbwachs (Halbwachs, 1992, 22). These memories are inherited from generation to generation, which in a way, enable the future generations to construct their past, affiliate themselves to the past, and comprehend their present predicament. The individual remembrances of the past can never escape the influence of the prevailing social discourse. The notion of belongingness, which lies at the heart of Dharavi, comes from a shared history of evolution from a fishing village to an entrepreneurial hub. Their shared collective memory stores images of the Koli fisherman as their pioneers, the slow development of the leather business, and the growth of hundreds of diversified industrial units. This instills in them a sense of shared belongingness with their fellow inhabitants who bind over notions of community and ethnicity. The inter-caste, communities, and professional dealings gradually expand the group borders and span across Dharavi's districts. Through their professional transactions, the community ties strengthened, and hence, a monolithic identification as being a resident of Dharavi overpowers the diverse religious and caste identifications. Their strong social networking lies at the base of the sense of security that prevails in Dharavi. A detailed understanding of Dharavi's history, location, and inner workings brings forth the polysemous nature of the place. The understanding of Dharavi as a 'slum' through the economic and geographical realities seems inadequate in capturing its polysemic nature. Hence, the focus on human realities and their perception of themselves is crucial in understanding the nature of Dharvai. Dharavi can also be regarded as a way of life its people developed for themselves, thus fostering a unique identity and specificity of the place. The objective details of the geography of Dharavi, along with the humanistic aspect of the Geography, present a wholesome understanding of the complex nature of Dharavi.

Dharavi's Culture

Dharavi developed a way of life of its own, unique to itself. With complex historical backgrounds, Dharavi developed its own culture or way of life for its people which stands apart from the culture of the city. The work culture developed in Dharavi remains undisturbed by its diversified inhabitants. Dharavi's unwritten but ingrained marginal status led its people to survive on their own without much assistance from the government. The people of Dharavi transformed this marginal space into a space of resistance through their flourishing activities. The inhabitants of this place asserted their identities through their work and thus resisted succumbing to the power of the ruling class. Notably, the concept of Marginality is linked to the idea of Resistance. Bell Hook's concept of Marginality as a site of resistance can be employed to understand the changing dynamics of Dharavi. The exploited section must comprehend marginality as a position and a probable site of resistance. Bell Hooks, in the article 'Choosing the Margin as a Space of Radical Openness' mentions marginality being a site of resistance. She writes, 'A message from that space in the margin that is a site of creativity and power, that inclusive space where we recover ourselves where we move in solidarity….' (Hooks, 1989, 22). Marginality as a site of resistance gives rise to liberation and unity. Transformed and appropriated through economic practices, Dharavi transformed itself through collective practices. Thus, speaking from the margins, Dharavi speaks in resistance and posits itself as a location of myriad possibilities.

The socio-cultural ties in Dharavi reflect the solidarity among people. Dharavi is home to varied castes, religions, and professional groups. The Dargah Chawl in Dharavi, as mentioned by Sharma, is home to mixed settlements of Hindus and Muslims. In one particular year, Bakri Id and Mahavir Jayanti coincided and Dharavi was filled with an air of festivity. Khatija who resides in the Dargah Chawl regards Dharavi as 'mini-India' (Sharma, 2000, 36). Each group upholds its own identity and yet participates in the larger interests. Rather than a 'chaotic collection of structures', Sharma considers Dharavi as a 'dynamic collection of individuals' who figured out self-sustenance strategies in the face of adversities (Sharma, 2000, 40). However, Sharma also warns against the excessive valorisation of their struggle, which would jeopardise the gravity of their living conditions.

Different community of workers of one profession upholds their distinctive personality. For instance, Kumbharwada, an area where a community of potters resides, upholds their distinct personality with every house accommodating the potter's wheel and a shared kiln. Apart from maintaining solidarity among their community, there are instances where cooperative systems evolved with other communities. As cotton for burning their kiln is bought from other communities, hence developing a system of interdependence among the communities.

Despite the diversity of communities, Dharavi is considered to be more peaceful than many other slum settlements. The work of the Muslim and Hindu tanners from Tamil Nadu unites them. The reason behind this would be challenging to be specified. One reason might be traced in Ramjibhai Pithabhai Patel's statement, a resident of Dharavi. He states, 'We do our work, We are working people. We don't have time for anything else' (Sharma, 2000, 70).

Exclusionary Politics

The residents of Dharavi share a troubled existence. Dharavi's history and growth foreground the problems with urban planning. Initially built on marshy land, Dharavi was seen as a place perfect for hazardous industries to grow. When it slowly turned into a residential area catering to the needs of the poor migrant labourers, the structure of the city was challenged. Dharavi, situated on the fringes of Bombay, became a favourable site for the State to transform it into a dwelling space for the illegal occupants of the city. The people were seen as illicit occupants, and the place gradually was recognised as a slum owing to its illegal status. Selective services were offered in attempts to develop the slum. However, time and again, the people were reminded of their illegal status. The slum dwellers share a precarious existence and make peace with their illegal means of survival. Associating slums with negative notions of poverty brings forth their othering. It presents them as the urban subaltern who seemingly rely on the government or the powerful sections to uplift their living conditions.

The territorial exclusion of the slums can be analysed through Bourdieu's concept of 'effects de lieu' (side effects) (Bourdieu et al., 1993, 123). This concept foregrounds place-generated impacts as the key factors of exclusion. Apart from the lack of access to material facilities and some public services, including quality education, the social stigma attached to the slum life is one of the glaring effects of effect de lieu. The incorporation of slum inhabitants in urban society raises concerns because they are consistently seen with some degree of hostility, mistrust, and fear.

Dharavi's Growing Enterprise

Haji Shamsuddin's story, as narrated by Sharma is reflected in the lives of many people in Dharavi. Dharavi is distinguished from other slums by its commercial success and is renowned for its innovation and entrepreneurship. In every corner of Dharavi lies productivity. Notably, the place endorses illegal activities ranging from child labour, hazardous industries, adulteration, etc. Although much of the production is illegal, people nonetheless engage in it. There lies solidarity in the thoughts of the people that nothing would hamper them despite the mass production of illegal activities because they contribute immensely to the city's economy. Surprisingly, Dharavi's daily turnover, as stated by Sharma is around Rs. 5 crore per day; hence, it rightfully earned the title of a 'goldmine' (Sharma, 2000, 79). The enterprises share the commonality of establishing themselves without the help of the government. With roughly 1700 industrial units, Dharavi is home to around 43 big industries that established themselves without the government's help. The diversity of the manufacturing activities of Dharavi is appalling and has grown and flourished because the state benefits from the illegal enterprises of Dharavi. Dharavi's leather tanning business grew to produce finished leather goods and continues to be the dominant trade of Dharavi. As mentioned by Sharma, the Sion-Mahim link road is the famous leather street known to many. Apart from the leather business, Dharavi's recycling business is the largest in India. This commercial success of Dharavi would not have been possible without solidarity arising out of interconnectedness and responsibility.

Conclusion

According to Kalpana Sharma, the development efforts in Dharavi, rather than being driven by an altruistic mindset, were more heavily influenced by objectives. Reorganising the informal settlements and converting them to formal housing areas were possible only in those areas where the communities cooperated. It must be kept in mind that Dharavi is a composite of its history, development, migration, and industrialisation patterns, coping mechanisms, and survival tactics. However, it would be an ambitious or a herculean endeavour to provide simple answers to the urban poor's questions. The chapter intended to foreground a reality of the cities, often overlooked. The chapter begins by analysing the prevailing as well as probable presumptions on slums and later analyses and conceptualises the nature of life in Dharavi. Apart from the affective appeal, the chapter delves into deeper questions of their existence. Dharavi's development and its people with the never giving up attitude convey the possibilities of the subalterns who speak to be heard and understood as conceptualised by Gayatri Chakravarty Spivak. The labelling of a location as a slum comes with severe implications for its inhabitants as it develops a culture of fear in the minds of outsiders. Without falling into the trap of romanticising poverty and eulogising the survival spirit of the people of Dharavi, the chapter intends to raise deeper questions regarding the status and socio-economic dynamics of the slum.

Nonetheless, the people in Dharavi must sustain the solidarity that exists. Dharavi presents the real challenges of the management of urban poor. This study suggests that Dharavi should be considered as a reminder not of triumph over poverty but as a triumph amidst poverty. The unsanitary living conditions and lack of access to quality public services cannot be overlooked in favour of the commercial success of the place. The solidarity that exists there is crucial for their survival through which the geographical landscape of Dharavi shall be able to maintain the complex conglomeration of caste, culture, and responsibilities. This humanitarian and ethical perspective in the study is intended to create awareness of the lives that inhabit spaces like Dharavi.

References

Bourdieu, P., et al. (1993). *The Weight of the World: Social Suffering in Contemporary Society* (P. P. Ferguson et al., Trans.). Stanford University Press.

Buttimer, A. (1976). Grasping the dynamism of the life world. *Annals of the Association of American Geographers, 66*(2), 277–292. Retrieved September 15, 2023, from https://www.jstor.org/stable/256247

Durkheim, E. (1984). *The Division of Labour in Society* (W. D. Halls, Trans.). The Macmillan Press.

Halbwachs, M. (1992). *On Collective Memory* (L. A. Coser, Trans.). University of Chicago Press.

Hollinger, D. A. (2006). *Cosmopolitanism and Solidarity: Studies in Ethnoracial, Religious*. The University of Wisconsin Press.

Hooks, B. (1989). Choosing the margin as a space of radical openness. *Framework: The Journal of Cinema and Media, 36*, 15–23. Retrieved July 29, 2023, from https://www.jstor.org/stable/44111660

Lando, F. (1966). Fact and fiction: Geography and literature: A bibliographic survey. *Geography and Literature, 38*(1), 3–18. Retrieved September 3, 2023, from https://www.jstor.org/stable/41146699

Massey, D. (1994). Space, Place, and Gender. University of Minnesota Press.

Oosterlynck, S., Schuermans, N., & Loopmans, M. (Eds.). (2017). *Place, diversity and solidarity*. Taylor & Francis.

Relph, E. C. (1976). *Place and Placelessness*. Sage Publication.

Sharma, K. (2000). *Rediscovering Dharavi: Stories from Asia's largest slum*. Penguin Books.

Simmel, G., & Jacobson, C. (1965). The poor. *Social Problems, 13*(2), 118–140. Retrieved September 14, 2023, http://www.jstor.org/stable/798898

Tilley, C. (1994). *A Phenomenology of Landscape Places, Paths and Monuments*. Berg Publishers.

Weinstein, L. (2014). *The Durable Slum*. Orient Blackswan.

Yatzimirsky, C. S. M. (2013). *Dharavi: From Mega-slum to Urban Paradigm*. Routledge.

Part III
Sustainable Development of Rural Societies and Indigenous Knowledge Base

Socio-Economic Development through Ecological and Social Solidarity: A Case of NABARD Springshed-Based Participatory Watershed Development Program in the North East Region

A. K. Mishra and Atosü Hibo

Introduction

The various dimensions of the sustainable development model for the socio-economic group of masses rest on poverty alleviation and the conservation of nature. In an interdependent and multi-plural economic, political, economic, and administrative environment, it is still being determined to build a consensus on key factors influencing global issues like climate change, food security, water security, human health, conservation of biodiversity, etc. Mother Earth, which is home to mankind and other living beings, provides food, shelter, water, and a platform for healthy living from generation to generation. The socio-economic issues that need attention are centred on health and hygiene, poverty alleviation, conservation of nature, rational use of natural resources, and more and more use of renewable energy.

Objectives of the Study

To determine the role of NABARD in fostering Social and Solidarity Economy (SSE) in the Northeast.

To examine the NABARDs Springshed program in the North East Region (NER) and its impact on community empowerment, equitable development, and sustainable economic practices.

A. K. Mishra
Department of Commerce, Nagaland University, Meriema, Nagaland, India
e-mail: akmishra@nagalanduniversity.ac.in

A. Hibo (✉)
Department of Commerce, Patkai Christian College (Autonomous), Chumoukedima, Nagaland, India

K. Arora, R. R. Anand (eds.), *The Solidarity Approach in Geography*,
https://doi.org/10.1007/978-3-031-86753-8_7

Research Methods

The paper explores the intersection of sustainability and livelihood within the framework of SSE. Information has been gathered from secondary sources, including published reports from NITI Aayog and the National Jal Jeevan mission, as well as data from the website of Bhuvannabard, and various books and journals focusing on social sustainability approaches.

To evaluate the socio-economic effects of the Springshed-based Watershed Program in the Northeastern Region (NER), a micro-level case study was conducted. This involved administering a questionnaire to participants from 17 villages across 11 districts of Nagaland, where the program was implemented by NABARD between 2017–2018 and 2018–2019. The collected responses underwent analysis via a *t*-test to ascertain the impact of the intervention.

Dimensions of Socio-Economic Development: Indian Experience

India, home to the world's most populous country with 1.4 billion people, follows a liberal, democratic, social, and economic model and advocates for equality in economic growth with sustainability. The Indian constitution ensures social justice which emphasizes the concept of distributive justice to secure social equality among the different strata of the society.

In the present millennium, development is identified not only with economic growth but also with growth in the social context. Policymakers and researchers have long recognized that economic development cannot happen without social change (Carlson, 1999). The importance of the social dimension in economic development has thus been strongly emphasized. This deviation from the conventional idea garnered more validation when the United Nations strongly emphasized social well-being as a measure of development (Litwiński, 2017). Measures to improve social conditions such as poverty reduction, education, food security, and health are now taken as major determinants for socio and economic development. Sustainable economic development can only be achieved if social development is at par with economic growth. Measuring social performance on equal footing with economic performance forms the subsequent consideration for development. Economic and social issues are related and thereby failure in one area will lead to failure in the other (Carlson, 1999).

Social Solidarity: A Sustainable Development Approach

Formation of a Social and Solidarity Economy (SSE) encompassing, an inbuilt robust participatory as well as collaborative system of management, will facilitate the eradication of poverty as well as accelerate socio-economic growth with sustainable development.

The 110th International Labour Conference adopted a resolution relating to SSE, which reads, 'The SSE encompasses enterprises, organizations and other entities that are engaged in economic, social and environmental activities to serve the collective and general interest which are based on the principles of voluntary cooperation and mutual aid, democratic and participatory governance, autonomy and independence and the primary of people and social purpose over capital in the distribution and of surplus and/or profits as well asset' (International Labour Organization (ILO), 2022).

The inputs that strengthen the formation of a robust SSE would require a multi-layer structural interactive process with sound policy guidelines and proper implementation with accountability. The multi-layer structural interactive forces can be categorized into three basic fundamental premises to strengthen the solidarity economy. It will result in a collaborative, participative, and inclusive economy. The characterization of the entire premises concludes that our economic development model be designed or based on socio-economic aspects with a participatory style of ownership and management with a startup culture as well as more and more use of conventional energy. The formation of social capital through an institutional intervention by way of capital inclusion or investment, as well as through policy formulation and follow-up measures, could lead to sustainable socio-economic development.

The supply and demand dynamics of the economy substantially influence the growth part of the nature of the economy. The Indian economy is agrarian, and sustainable economic growth of the country would depend on the creative functioning of small-scale industries (MSMEs) to strengthen the rural market economy. The creation of a strong, vibrant social and solidarity economy is depicted in Fig. 1, with various influencing variables that act as well as interact with each other to shape and influence sustainable growth.

The social solidarity economy draws strength from social capital and social entrepreneurship. To make economic growth sustainable, the fundamentals of socio-economic development need to be revitalized with the formation of group-based startup units as per the market needs. India, being a multi-cultural, multi-lingual, multi-ethnic country with diversified interest groups and subgroups of regional social entities needs a multidimensional approach for long-term socio-economic growth. Thus, it is pertinent to develop a multidimensional approach to create economic units which is shown in Fig. 2.

Institutional Role in Social Solidarity Economy

Contemporary conservation policies in the social and ecological context require human actions that integrate with sustainable development practices. Ecological and social solidarity work on the principle of collective efforts of individuals united towards a common goal with shared interests and responsibility. This calls for an interaction of human activities from an institutional approach with socio-economic dimensions. A participatory and collaborative model would thus form the structural framework for its successful implementation. This shared responsibility of the local

Fig. 1 Inputs/variables depicting the formation of SSE. (Source: Authors)

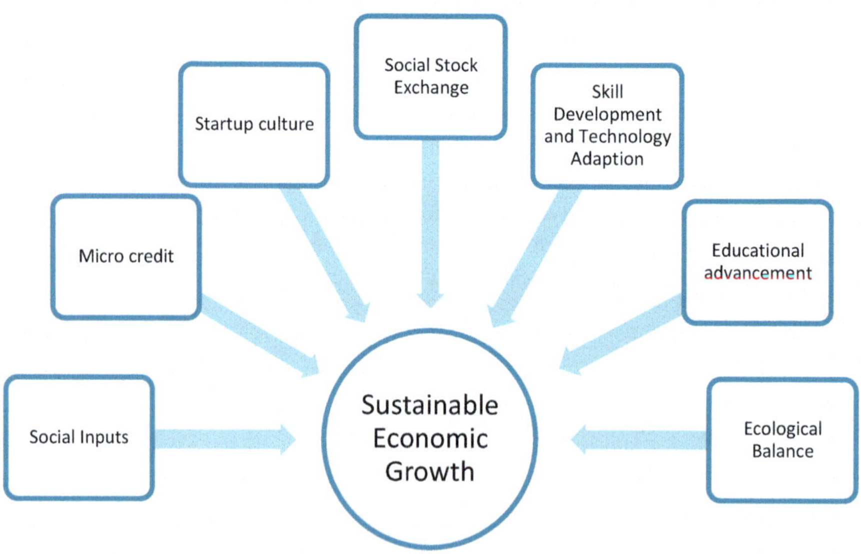

Fig. 2 Multidimensional approach for sustainable economic growth. (Source: Authors)

community with the local government would be a conscious step towards fulfilling the conservation efforts.

North East Region (NER) and Its Hydrogeological Topography

North Eastern Region of India comprising seven states (Fig. 3) comes under the North Eastern Himalayan range. The mountainous region with slopes and valleys constitutes the topography of the area. The NER being a part of the Indian Himalayan Region is a bio-diversity hotspot and is unique in its flora and fauna.

Mountain springs serve as the primary source of water for many rural households in the North East Region (NER). The springs serve as a lifeline to meet their domestic and agricultural needs. The NER receives an abundant amount of rainfall annually. The mountain springs which are natural discharges of ground water can be managed effectively under Springshed management. The socio-economic issues that influence the life of the people of the region are closely associated with the conservation of nature, flora and fauna, protection of wildlife, etc. The tapping of groundwater resources for maintaining the water cycle calls for efficient and effective water resource management. The geological as well as geomorphologic studies for hilly terrains in any region could streamline the effectiveness of Springshed management.

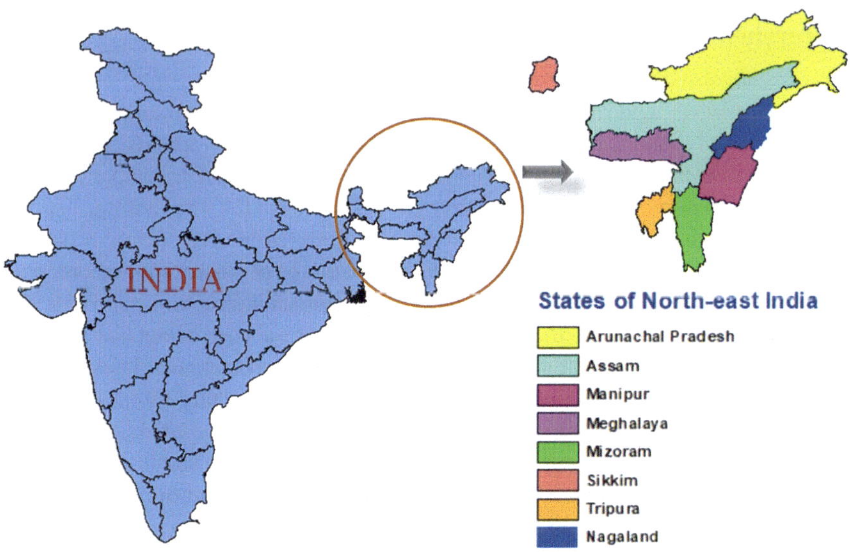

Fig. 3 NER map outline. (Source: Authors)

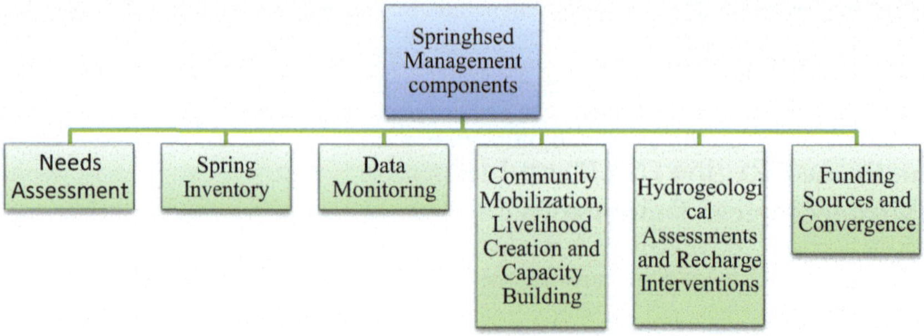

Fig. 4 Springshed management components

The springshed-based participatory watershed development program under NABARD commenced in 2017 in the North Eastern Region including Sikkim on a pilot basis with financial support under the Watershed Development Fund (WDF). The program has now been extended to the hilly areas of other states, viz. Kerala, Maharashtra, Uttarakhand, Himachal Pradesh, and Jammu & Kashmir. A total of 143 springshed development projects have been initiated under NABARD with a cumulative financial outlay of 35.87 crores in the NER as of 31 May 2023.

The springshed management component, as outlined by IWMI, NITI Aayog, and SD (Rathod et al., 2021) is shown in Fig. 4.

A Bird's Eye View on Springshed-Based Watershed Development

The conservation and management of Springshed is multidimensional in its approach, involving a convergence of various stakeholders towards its sustainability. Springshed management must be integrated with community participation, governance, and monitoring through formal institutions for maintaining spring inventory, handholding support, and capacity building from civil society organizations with active community involvement at all stages of planning, implementation, and monitoring. From a scientific approach to the social structure approach, the long-term sustainability of the Springshed is built on a comprehensive strategy. There is also a need for the dissemination of scientific and technical knowledge to the community in their local languages on the Springshed recharge area, sustaining mechanisms through the aquifer system, treatment of recharge areas, and use of equipment for monitoring water quality, spring discharge, etc. The involvement of the local community will facilitate the integration of traditional and local knowledge in the planning and implementation of the Springshed management. The Springshed management module formulated under the Jal Jeevan Mission (2020) is indicated in Table 1.

Table 1 Springshed management module

Steps	Activity	Objective	Equipment	Outcome
Step 1	Spring mapping	To locate spring emergence point tagging/geo-point	GPS device, Android phone	Spring Atlas/ inventory
Step 2	Spring discharge	To measure spring Spring discharge	A bucket whose volume is known, Stopwatch	Spring hydrograph
Step 3	Rainfall	To measure rainfall	Rain gauge	Spring hydrograph
Step 4	Water quality	Water quality monitoring	Tracer, on-spot water quality testing kit	Water quality report
Step 5	Baseline (socio-economic)	Baseline survey	Formats, structured questionnaire	Village water resource map, demand and supply status, vulnerable springs
Step 6	Hydrogeological mapping	To collect hydrological and hydrogeological and from the field survey	GPS, Brunton, Clinometer, Hammer, Google Earth and Sketch (Software)	Hydrogeological conceptual layout and spring site cross-section
Step 7	Designing recharge interventions and management protocols	Physical and biological measures, social fencing Behaviour change	Tools for community mobilization experts such as SARAR kit	Spring recharge and protocols for spring shed management
Step 8	Impact assessment of Springshed works	Measurement of benefits from Springshed works with other socio-economic aspects	A bucket whose volume is known, stopwatch, tracer, on-spot water quality testing kit, formats, structured questionnaire	Improvement in water availability and water quality

Source: National Jal Jeevan Mission, 2020

Community Participation

The community's involvement in the development, monitoring, and maintenance of Springshed is of paramount importance for its effective implementation. Creating awareness and disseminating technical knowledge to the community will help them understand and develop best management practices for springshed. This should also be supplemented by adopting local indigenous knowledge as local inhabitants can provide social-cultural, ecological, economic, and technical local knowledge to ensure consistency between objectives of development and community values and preferences (Bagdi & Kurothe, 2014).

Community participation also brings a sense of ownership and responsibility, which can ensure long-term sustainability through their active involvement. The participation of women and marginalized groups such as Dalits and tribals in decision-making would enable social and gender equity in the program.

Resource Mobilization

The Springshed program is participatory, involving the community members in its planning and implementation. The involvement of the people generates their interest and concern in the conservation of the soil and water structures. This calls for the mobilization of resources from the community members by contributing cash, labour, and materials towards the construction of the structure.

Governance of Springshed Management

The government and non-governmental organizations have a dominant role to play in the success of the Springshed program. In the NER, a few state governments, Civil Society Organizations (CSOs), and Non-Governmental Organizations (NGOs) have been actively contributing towards promoting awareness of the importance and management of springs and providing handholding support for its effective implementation.

NITI Aayog in their report highlights success stories led by a few state governments in the implementation of the program through a scientific management approach. Departmental undertakings such as the Rural Development Department (RDD), Sikkim; Meghalaya Basin Development Authority (MBDA), Govt. of Meghalaya; Meghalaya Community-Led Landscape Management Project (MCLLMP) have trained field staff and village data collectors on hydrogeology, social and governance aspects, and data monitoring (NITI Aayog, 2017). Automatic data collection instruments have been installed to monitor the spring discharge, rainfall, and water quality. The Department of Land Resource (DoLR), Nagaland has displayed 3D models of the Springshed to explain the concepts of Springshed management to the village community. The departmental undertakings have helped to simplify the scientific concepts by explaining in their local languages to get a better outreach from the community.

Civil Society Organizations (CSOs) have also facilitated the Springshed management by acting as Project Facilitating Agencies (PFAs) and assisted in building skills and filling the knowledge gaps for effective implementation of the springsheds. Some notable ones that have been assisting in the North East Region (NER) are the Advanced Center for Water Resources Development and Management (ACWADAM), Pune; North East Initiative Development Agency (NEIDA), Kohima; People's Science Institute (PSI), Dehradun. Table 2 indicates the number of NABARD-assisted programmes in the North East Region. The spring population across the North East Region is indicated in Table 3.

Table 2 Current status of NABARD-supported Springshed programme as of 20 October 2023

Sl no	State	Districts covered	No of springshed	Area (Ha)	Households covered	Total PFAs[a]	Amount sanctioned	Amount disbursed	Amount utilized
1	Arunachal Pradesh	2	4	126.5	37	2	15.9	15.41	15.01
2	Manipur	5	10	540	167	7	57.85	19.95	19.95
3	Mizoram	4	6	2664.1	291	4	93.3	71.36	67.21
4	Nagaland	13	18	10,947.53	4203	6	469.94	249.57	239.23
5	Meghalaya	2	4	1200	1238	2	108	102.83	87.65
6	Sikkim	2	5	0	0	3	0	0	0
Total		28	47	15,478.13	5936	24	744.99	459.12	429.05

Source: NRSC Bhuvan NABARD portal

[a] Project Facilitating Agencies (PFAs)

Table 3 Spring population across NER

NER states	Number of villages with springs	Total number of villages	Percentage of villages that report having springs	Spring channel-based surface flow irrigation	Total number of surface flow irrigation schemes	Percentage of spring channel-based surface flow irrigation systems
Arunachal Pradesh	2086	5589	37.3	15	4953	0.3
Assam	2997	26,395	11.4	265	2815	9.4
Manipur	1405	2581	54.4	0	516	0.0
Meghalaya	3810	6839	55.7	246	8220	3.0
Mizoram	453	830	54.6	1252	5371	23.3
Nagaland	639	1428	44.7	27	20,765	0.1
Sikkim	425	451	94.2	29	1485	2.0
Tripura	141	875	16.1	2	341	0.6

Source: NITI Aayog, 2017

Table 4 Profile of Springshed program in Nagaland

Year	Districts	No of villages	Name of villages	No of households	Amount (Lakh INR)
2017–2018	Wokha	1	Riphyim	74	35.80
	Zunheboto	1	Lukikhe	52	37.80
	Kohima	1	Thizama	96	37.80
	Phek	1	Chesezu	113	37.80
	Mokokchung	1	Aliba	32	37.80
	Mon	1	Lapa	68	37.80
	Dimapur	1	Seithekema	40	37.64
	Tuensang	1	Tuensang	95	38
	Peren	1	Jalukie	64	37.80
		9		**634**	**338.24**
2018–2019	Wokha	1	Chudi	82	37.62
	Longleng	1	Dungkhao	77	37.80
	Peren	1	Pelhang	59	37.73
	Kiphire	5	Vongva, Tikenvong, Luther, Tsutovong, Cedeyevong	275	37.26
		8		**493**	**150.41**

Source: Field Survey (NABARD Regional Office, Dimapur)

Intervention Impact

To assess the program's impact on beneficiaries, a mini-study was conducted across 17 villages spanning 11 districts of Nagaland state. Total number of respondents collected for the study was 224. The intervention study was determined on programs implemented by NABARD during the years 2017–2018 and 2018–2019. The details are provided in Table 4.

Benefits of the Program

Table 5 shows the summary of benefits received from the projects by the rural villagers. The frequency and percent figure of the 11 benefits are shown.

Table 6 presents the results of a *t*-test analysing the intervention impact before and after across various aspects, including total cropped area, livestock production, crop production, access to medical facilities, participation in community decision-making, sanitary habits/toilet facilities, and perception/knowledge on understanding and conservation of springshed. The mean values for both categories (before and after) intervention are 3.118 (before) and 4.628 (after). These results indicate a significant improvement in all areas of social and economic conditions due to the intervention.

Table 5 Benefits of the program

Sl no	Benefits of the project	Frequency	Percentage
1	Additional employment	80	6
2	Increase in income	143	10
3	Increase in fodder	189	14
4	Increase in livestock	41	3
5	Improve in living standards	21	2
6	Improve sanitation habits	102	7
7	Reduction in workload	174	13
8	Reduce debt position	153	11
9	Soil conservation	145	11
10	Availability of drinking water throughout the year	139	10
11	Reduced migration	189	14
Total		1376	100

Source: Developed from Field Survey

Table 6 The t-test: Paired two sample for means of intervention impact (before and after)

Particulars	Intervention	
	Before	After
Mean	3.118	4.628
Variance	0.079	0.040
Pearson correlation	0.517	
df	12	
t Stat	2.214	
$P(T \leq t)$ one-tail	0.001	
t Critical one-tail	1.782	
$P(T \leq t)$ two-tail	0.001	
t Critical two-tail	2.179	

Source: Developed from Field Survey

Conclusion

Contemporary conservation policies in the social and ecological context require human actions that integrate with sustainable development practices. Ecological and social solidarity work on the principle of collective efforts of individuals united towards a common goal with shared interests and responsibility. This calls for an interaction of human activities from an institutional approach with socio-economic dimensions. A participatory and collaborative model would thus form the structural framework for its successful implementation. This shared responsibility of the local community with the local government would be a conscious step towards fulfilling the conservation efforts.

In hilly topography regions where terrace cultivation is predominantly practiced and to keep the area thickly vegetated, the Springshed management plays a significant role. The Springshed programs are participatory and require the combined efforts of various stakeholders for their successful implementation. A community and participatory-based approach form the basis for the successful implementation of the program in the Northeastern states.

The effects of climate change have adversely affected the production and productivity of agricultural produce. Skewed distribution of rainfall and intermittent dry spells compound the farmer's agricultural woes leading to low productivity, low-income generation, poor quality and quantity of produce, and other agriculture-related problems. This negates the collaborative efforts of policymakers to improve the socio-economic life of rural communities. The drying Springshed needs to be rejuvenated with regular recharging treatments. The Springshed management will facilitate the promotion of green entrepreneurship and will generate employment opportunities in rural areas in the long run.

References

Bagdi, G. L., & Kurothe, R. S. (2014). People's participation in watershed management programs: An evaluation study of Vidarbha region of Maharashtra in India. *International Soil and Water Conservation Research, 2*, 57–66.

Carlson, B. A. (1999). *Social dimensions of economic development and productivity: Inequality and social performance*. United Nations Publication.

International Labour Organization (ILO). (2022). *Resolution and conclusions concerning decent work and the social and solidarity economy (SSE)*. Presented at the 110th International Labour Conference, Geneva, ILO.

Litwiński, M. (2017). The evolution of the idea of socio-economic development. *Ekonomia i Prawo, 16*, 449. https://doi.org/10.12775/eip.2017.031

National Jal Jeevan Mission. (2020). *Provision of potable drinking water in mountains through participatory springshed management*. Government of India, Ministry of Jal Shakti Department of Drinking Water and Sanitation National Jal Jeevan Mission.

NITI Aayog. (2017). *Inventory and revival of springs in Himalayas for water security as part of initiatives on Sustainable Development of Mountains of Indian Himalayan Region*. NITI Aayog.

NRSC. https://bhuvan-app1.nrsc.gov.in/nabard/nabard.php?type=citizen

Rathod, R., Kumar, M., Mukherji, A., Sikka, A., Satapathy, K. K., Mishra, A., Goel, S., & Khan, M. (2021). *Resource book on spring shed management in the Indian Himalayan Region: Guidelines for policymakers and development practitioners*. NITI Aayog.

Development under the Aegis of MGNREGA: A Case Study of Riyali Panchayat of District Kangra, Himachal Pradesh

Shweta Rana, Deeksha Rana, and Seema Choudhary

Introduction

Mahatma Gandhi National Rural Employment Guarantee Act was implemented by the Government of India to provide guaranteed employment to every rural household whose adult members are willing to work for a minimum of 100 days on one hand and to build sustainable infrastructure in the villages to develop them on the other (MGNREGA Sameeksha, 2012). This scheme precisely deals with rural areas. Unskilled people who can do unskilled work manual work are given priority under the scheme so that they can work and earn (Ranaware et al., 2015). Development is a continuous process, and it presents growth and welfare for all. Mountainous regions having rough topography are the tedious ones to develop. There are limited economic opportunities present there (Husain, 2021). Hence, MGNREGA is a boon to the people of such hilly regions since it helps not only in income generation but also in infrastructure development.

Riyali panchayat of Fatehpur block of district Kangra is agrarian. The performance of Fatehpur block is commendable under the MGNREGA scheme in Kangra district of Himachal Pradesh. It was found out that Riyali panchayat is the best-performing panchayat under this block based on some selected indicators. Here, people from three distinct religious communities live together. These three communities are Hindu, Sikh, and Muslims. Social cohesion and mutual respect for traditions, and customs become important whenever people from different communities live or work together. A strong solidarity presents good opportunities for the growth and development of the region. This panchayat consists of five villages, and agriculture is their primary occupation. The present study tries not only to examine the impact of MGNREGA on the socio-economic development of the panchayat but

S. Rana (✉) · D. Rana · S. Choudhary
Department of Geography, Himachal Pradesh University, Shimla, Himachal Pradesh, India

K. Arora, R. R. Anand (eds.), *The Solidarity Approach in Geography*,
https://doi.org/10.1007/978-3-031-86753-8_8

also on the status of women as well. Apart from this, it will also try to see the impact of the presence of different communities on the working of the scheme on the one hand and their perception of the status of women working under MGNREGA.

Objectives

1. To analyze the socio-economic development in Riyali panchayat under the impact of MGNREGA.
2. To analyze the impact of MGNREGA on the status of women.
3. To explore the impact of the presence of different communities on the working of MGNREGA and how the scheme is further enhancing the solidarity among these communities.

Materials and Methods

The study is primarily based on primary sources of data. Primary data has been collected through schedules created separately for MGNREGA workers, non-MGNREGA workers, and panchayat heads of Riyali panchayat. A total of 101 samples were collected among which 64, 30, and 7 were collected for MGNREGA workers, non-MGNREGA workers, and panchayat heads, respectively. The data has been tabulated and interpreted using Statistical Package for Social Sciences (SPSS). A simple Likert method using percentage has been used for the analysis of data. The data has been represented using suitable charts, diagrams, and maps. The map of the study area has been created using Arc GIS 10.5. Figure 1 presents the flow chart of the methodology for the present study.

Kangra district lies in the northwestern part of Himachal Pradesh. It is the most populated district of Himachal Pradesh. The area of the district is 5739 sq. km and ranks fourth area-wise among all the districts of Himachal Pradesh. The altitude generally varies from 400 m to above 6000 m. Beas is the main river of the district and covers almost the entire drainage of the district. There are 15 blocks in the Kangra district, namely Baijnath, Bhawarna, Dehra Gopipur, Dharamshala, Fatehpur, Indora, Kangra, Lambagaon, Nagrota Bagwan, Nagrota Surian, Nurpur, Panchrukhi, Pragpur, Rait, and Sulah (Balokhra, 2022). Fig. 2 shows the location map of the study area.

Discussion and Results

Kangra Valley is widely known for its pristine beauty and fertile land. Riyali panchayat is part of Fatehpur block and consists of five villages, namely Belli Thakran, Belli Ladarchian, Belli Jata, Riyali, Belli Rajianala, and Belli Bhurian (DCHB,

METHODOLOGY: FLOW CHART

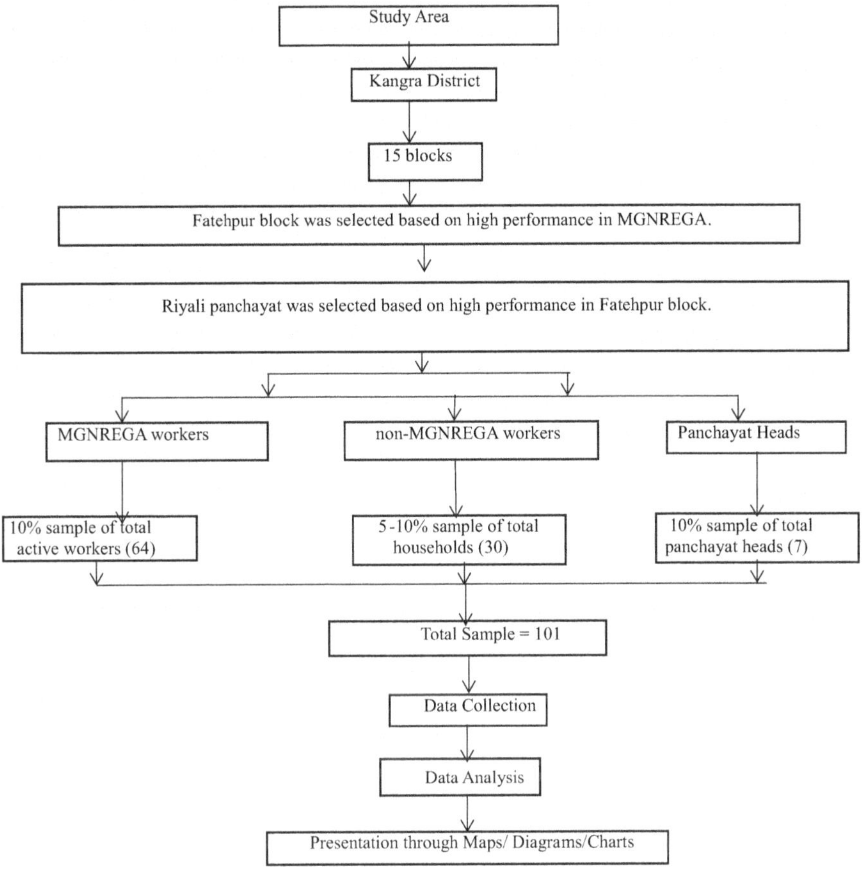

Fig. 1 Flow chart of research methodology. (*Source*: Prepared by the author on the basis of the research methodology used in the present study)

2011). These villages are fertile, and agriculture is practiced in them. Agriculture is the main source of income generation in the Riyali panchayat. Three different communities live in this panchayat, namely Hindus, Muslims, and Sikhs. This panchayat holds great potential to understand the solidarity among these villages along with the impact of MGNREGA in social economic development along with women empowerment. The schedules were the main instrument of the study and dealt separately with MGNREGA workers, non-MGNREGA workers, and panchayat heads such as Pradhan, Secretary, and Ward Panchs. The perception of the people about various aspects related to MGNREGA is discussed below.

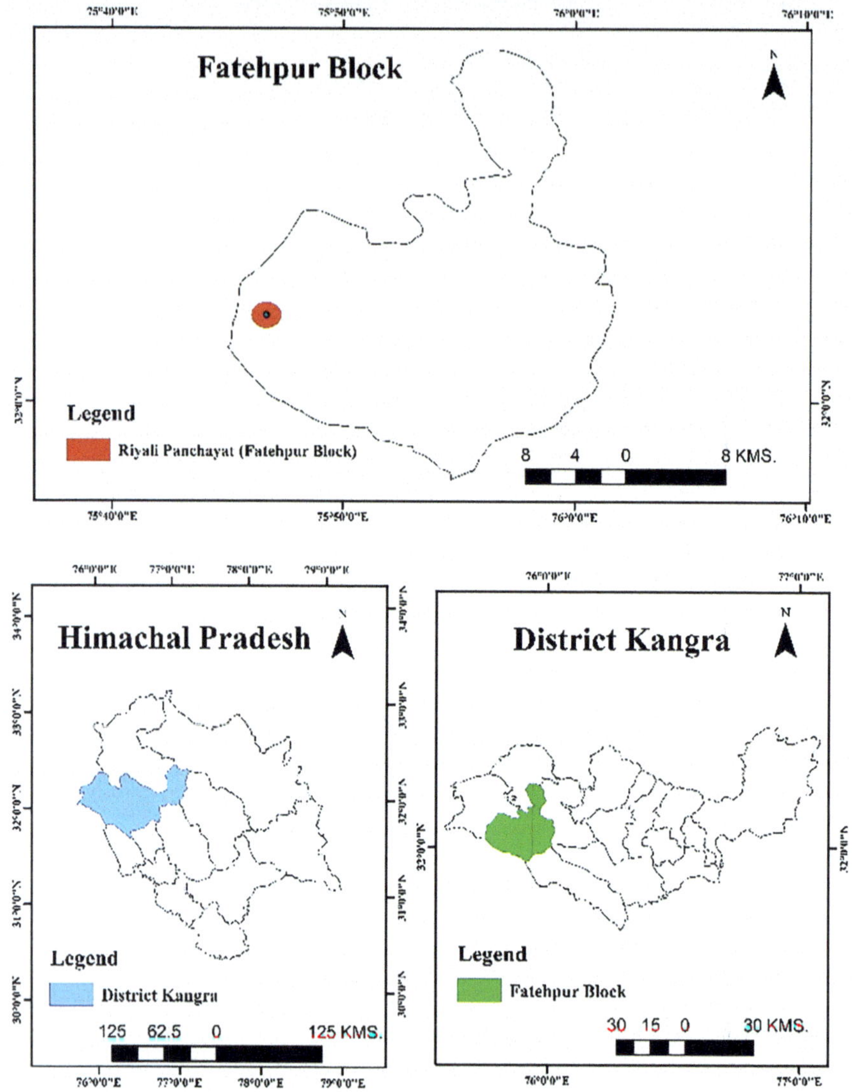

Fig. 2 Location map of study area. (*Source*: Prepared by the author based on secondary data using Arc GIS 10.3)

People's Perception about the Impact of MGNREGA on Socio-Economic Development

The MGNREGA scheme was implemented by the Government of India to provide direct employment to rural people on the one hand and to generate sustainable assets which can directly contribute to the development of villages. For the present study, it was assumed that the three groups of respondents, namely the MGNREGA workers, the non-MGNREGA workers, and the panchayat heads would differ in opinion on the aspects related to MGNREGA. Hence all three were included to gain information about the workings of the scheme and how it is shaping the lives of people in rural areas.

Working Days under MGNREGA and Unemployment Allowance

The mandate of the MGNREGA scheme is to provide a minimum of 100 days of employment to workers who have registered for work in the panchayat. Hence, an attempt was made whether MGNREGA workers get a minimum of 100 days of work or not. It was found that all MGNREGA, non-MGNREGA, and panchayat heads of Riyali panchayat registered positive responses in this regard. It indicates that people do get a minimum of a 100 days of work under the scheme. Also, people from all three communities could be seen working during the field visit.

Basic Facilities Provided under MGNREGA

To work efficiently, the government provides funds for expenditures like labor wages, and material costs along with facilities for laborers. So, an attempt was made to find whether these basic facilities are availed by workers or not.

On enquiring about the issue with the respondents of all categories, it becomes evident from Table 1 that all the respondents of Riyali panchayat have registered the availability of working tools from the MGNREGA funds. However, other facilities have been neglected.

Table 1 Facilities provided to MGNREGA workers

Facilities provided to MGNREGA workers	In Riyali		
	MGNREGA workers	Non-MGNREGA workers	Panchayat heads
Food	–	–	–
Tools	64 (63.4)	30 (29.7)	7 (6.9)
Water	–	–	–
Transport	–	–	–
No facility is provided	–	–	–

Source: Based on Field Survey, 2022
Figures in parentheses show the percentage

People's Perception of Change in Socio-Economic Conditions after the Implementation of MGNREGA

Since the MGNREGA scheme was implemented to generate direct employment, it has the potential to directly impact the socio-economic condition of the people. Henceforth, people were enquired about the change in their socio-economic condition before and after the implementation of the scheme. Table 2 reveals that the majority of MGNREGA workers (98.4%) were enjoying good socio-economic conditions even before the initiation of the MGNREGA scheme. However, it was also noticed that the scheme has benefitted some workers. It is so because, after the implementation of the scheme, 100% of MGNREGA workers have good socio-economic conditions.

Perception of the People about the Most Beneficial Factor of MGNREGA

It is a general perception among the people that MGNREGA has been a very beneficial scheme for the rural population. Thus, the study attempted to enquire about the most beneficial factors of MGNREGA. It was noticed that MGNREGA workers in Riyali are more benefited economically and socially as shown in Table 3. People from all three communities could be seen working together. Also, it has helped in enhancing social cohesion. The major impact has been registered on the economic side as it has generated a feeling of independence among them.

People's Perception of Biasness in the Distribution of Work and Work Selection under MGNREGA

It has been seen that sometimes people feel that the selection of workers as well as works taken up under MGNREGA has been choice or preference-based. Hence, an attempt was made to inspect both these issues in Riyali panchayat. It was noticed that all the MGNREGA workers and non-MGNREGA workers responded that they

Table 2 Status of socio-economic conditions before and after MGNREGA

Socio-economic condition of MGNREGA workers	MGNREGA workers (Riyali)	
	Before	After
Very good	–	–
Good	63 (98.4)	64 (100)
Not so good	1 (1.6)	–
Bad	–	–
Very bad	–	–

Source: Based on Field Survey, 2022
Figures in parenthesis show the percentage

Table 3 Peoples' perception on how MGNREGA has benefited them the most

MGNREGA has helped the most	MGNREGA workers	Non-MGNREGA workers	Panchayat heads
Socially	34 (12.8)	25 (9.5)	7 (2.6)
Economically	64 (25.2)	30 (11.4)	7 (2.6)
Psychologically	34 (12.9)	24 (9.0)	7 (2.6)
Morally	2 (0.8)	24 (9.0)	7 (2.6)

Source: Based on Field Survey, 2022
Figures in parentheses show the percentage

Table 4 Procedure of work selection to be done under MGNREGA

Selection of work to be done under MGNREGA	Non-MGNREGA workers	Panchayat heads
Randomly selected	2 (5.4)	–
By Sarpanch	26 (70.3)	–
Priority based	2 (5.4)	7 (18.9)
Biased decision	–	–

Source: Based on Field Survey, 2022
Figures in parenthesis show the percentage

never faced or noticed any bias in the distribution of work under MGNREGA. Also, Table 4 presents the procedure for the selection of work under MGNREGA. It becomes evident from the table that the majority of the non-MGNREGA workers of Riyali panchayat (70.3%) responded that the selection of work to be done in MGNREGA is decided by Sarpanch. It is also interesting to find that all the panchayat heads of Riyali believe that work selection is priority based.

Work other than MGNREGA

It has been observed that people in rural areas are involved in other economic activities apart from MGNREGA. Table 5 presents the other category of work that MGNREGA workers do besides MGNREGA. The data reveals that 100% of the people working as MGNREGA workers in Riyali are practicing agriculture as a subsidiary work. It is so because the MGNREGA scheme has been launched primarily for rural areas and the study is also rural and farming is the primary activity there.

Perception about the Section in which Most of the Work has been done under MGNREGA

Table 6 presents the perception of people about the section in which most work has been done in Riyali panchayat. MGNREGA schemes offer a variety of works dealing with different arenas. Since agriculture is the primary activity in the Riyali

Table 5 Work other than MGNREGA

Work	MGNREGA workers (Riyali)
Farming	64 (100)
Labor	–
Shop	–
Any other	–

Source: Based on Field Survey, 2022
Figures in parentheses show the percentage

Table 6 Section in which most work has been done

The section in which most work has been done	Panchayat heads (Numbers)
Water conservation and water-related works	7 (100)
Work on lands (land development)	7 (100)
Rural connectivity	7 (100)
Others	–

Source: Based on Field Survey, 2022
Figures in parenthesis show the percentage

panchayat. Hence, much work under MGNREGA is associated with it. About 100% of Panchayat heads of Riyali say that most of the work has been done in water conservation and water-related works, work on lands (land development) along rural connectivity. It was interesting to find during field investigation that a lot of work has been done in Riyali panchayat on internal road connectivity.

Information about Appointments under MGNREGA

Table 7 presents how information about appointments under MGNREGA is provided to MGNREGA workers. The majority of MGNREGA workers of Riyali (36.6%) believe that they get information about appointments through the panchayat secretary. However, the majority of Non-GMNREGA workers (12.9%) believe that the information is provided by Sarpanch. Though, all the panchayat heads say that information is provided by other mediums or more specifically the information is listed on the notice board of panchayats and people come to know about appointment information from the panchayat's notice board.

Worked in the Panchayat Heads' Land

An attempt was made to analyze whether MGNREGA workers have worked in the panchayat heads' land or not. So, MGNREGA workers were asked whether they had ever worked on the officials' land during the working hours of MGNREGA. The data discloses that 100% of the MGNREGA workers have not worked in the official lands during the working hours of MGNREGA.

Table 7 Information about appointments under MGNREGA

Information about appointments under MGNREGA	MGNREGA workers	Non-MGNREGA workers	Panchayat heads
Through Sarpanch	27 (26.7)	13 (12.9)	–
Through Secretary	37 (36.6)	12 (11.9)	–
Through aam Jalas	–	–	–
Others	–	5 (5.0)	7 (6.9)

Source: Based on Field Survey, 2022
Figures in parenthesis show the percentage

Table 8 MGNREGA workers have difficulty getting wages

MGNREGA workers have difficulty getting wages	MGNREGA workers	Panchayat heads
Yes	–	7 (9.9)
No	64 (90.1)	–

Source: Based on Field Survey, 2022
Figures in parenthesis show the percentage

Difference in Wages of Men and Women

The study also attempted to analyze whether both men and women are getting equal wages and working for the same hours or not. It was commendable to note that a combinedly 100% of MGNREGA workers, non-MGNREGA workers, and panchayat heads admitted that men and women get equal wages and work for the same hours in MGNREGA. It is a good indication of the successful working of MGNREGA as there is no disparity of any kind.

Difficulty in Getting Wages

Table 8 discusses whether MGNREGA workers face any difficulty in getting their wages or not. The data reveals that all the MGNREGA workers of Riyali do not face any problems in getting wages. However, opposite to it, all the panchayat heads of Riyali believe that MGNREGA workers do face problems in getting wages. In Riyali, the problem is the lack of banking facilities. The nearest accessible banking facility which is available to them is at a distance of 6–8 km. Some of the villages are located in the very interior part of Riyali panchayat and only one bus goes from Riyali, which takes passengers from Riyali in the morning and comes back to the village in the evening. There is only one bus available as a mode of transportation from the village to the bank.

Preferred Mode of Payment

It was tried to analyze that what is the preferred mode of payment by MGNREGA workers and panchayat heads. It is evident from Table 9 that the majority of MGNREGA workers (84.5%) preferred mode of payment for work is through the bank. However, a minority (5.6%) of MGNREGA workers prefer wages by cash because of the distant banking facilities in their area. However, in Riyali, all the panchayat heads preferred payment by bank.

People's Perception about the Status of Women under MGNREGA

People's perception was also collected to analyze the status of women and other related issues. Women form half of the society and without their advancement, one cannot think about the development of society. The present study tried to analyze the status of women based on the perception of people and is discussed briefly below:

Harassment on the Working Site

Harassment is commonly understood as a behavior that demeans, humiliates, or embarrasses a person, and it is characteristically identified by its unlikelihood in terms of social and moral reasonableness (Wikipedia, 2023). All the respondents were asked whether they had seen, faced, or noticed harassment of the women MGNREGA workers on the working site. It is commendable to note that all MGNREGA workers, non-MGNREGA workers, and panchayat heads have never witnessed any kind of harassment of MGNREGA women workers at the working site.

Table 9 Preferred mode of payment

Preferred mode of payment	MGNREGA workers	Panchayat heads
By bank	60 (84.5)	7 (9.9)
By cash	4 (5.6)	–

Source: Based on Field Survey, 2022
Figures in parentheses show the percentage

Table 10 Women's participation in MGNREGA

Support women's participation in MGNREGA	MGNREGA workers	Non-MGNREGA workers	Panchayat heads
Yes	64 (63.4)	27 (26.7)	7 (6.9)
No	–	3 (3.0)	–

Source: Based on Field Survey, 2022
Figures in parenthesis show the percentage

Justification of Women Working under MGNREGA

A general perception of the people was also registered as to what they think about the participation of women in MGNREGA. Table 10 depicts that all the MGNREGA workers and panchayat heads believe that women should work equally to men in MGNREGA. However, a minority of people (3%) do not support women's participation in Riyali. This restriction was commonly noticed by uneducated and old-aged males coming mostly from the Muslim community. Using the observation method, it was noticed that unawareness, illiteracy, and a strong orientation toward conservative rituals were the main reasons behind this perception. Overall, Table 10 reflects a strong picture of healthy women's status in the study area.

Comparability of the Working Intensity of Women and Men under MGNREGA

It is often believed that the physical strength of males and females are not the same and hence the intensity of working of the two also varies. The present study also tried to obtain people's perception on the issue. Table 11 reveals that all the MGNREGA workers and panchayat heads believe that women work with the same intensity as men in MGNREGA. Only 3% of non-MGNREGA workers don't agree with this. They had the very rigid perception that women could not work as efficiently as men.

Changes in the Financial Status of Women

The study tried to analyze whether MGNREGA has improved the financial condition of women and has made them financially independent or not. Table 12 shows that all the MGNREGA workers and panchayat heads affirm that MGNREGA has made women financially independent.

Likewise, the majority of non-MGNREGA workers (28.7%) also feel that women have become financially independent after the introduction of MGNREGA. This is true as the money earned by the workers is transferred to their bank accounts which has encouraged savings among them and has also aroused confidence to move out of the house and earn their own money as well.

Table 11 Perception regarding intensity of work

The intensity of working for women is equal to men	MGNREGA workers	Non-MGNREGA workers	Panchayat heads
Yes	64 (63.4)	27 (26.7)	7 (6.9)
No	–	3 (3.0)	–

Source: Based on Field Survey, 2022
Figures in parenthesis show the percentage

Table 12 Financial status of women

Financial independence for women	MGNREGA workers	Non-MGNREGA workers	Panchayat heads
Yes	64 (63.4)	29 (28.7)	7 (6.9)
No	–	1 (1)	–

Source: Based on Field Survey, 2022
Figures in parentheses show the percentage

Table 13 Difficulty in managing household chores

Difficulty in managing household chores	MGNREGA workers	Non-MGNREGA workers	Panchayat heads
Yes	20 (19.8)	27 (26.7)	7 (6.9)
No	44 (43.6)	3 (3.0)	–

Source: Based on Field Survey, 2022
Figures in parenthesis show the percentage

Challenges faced by Women in the Management of Household Chores

It has been a general perception of the people that it is difficult for a woman to work outside the house along with household chores. It was this thought that kept women secluded into the fine walls of the houses. However, with the change of time, perceptions have also changed. The present study also tried to find out whether women face difficulty in managing household chores while working in MGNREGA or not. It is evident from Table 13 that the majority of MGNREGA workers do not find it difficult for a woman to manage household chores while working in MGNREGA. The majority of non-MGNREGA workers and 100% of panchayat heads find it difficult for women to work in MGNREGA and manage household chores together. MGNREGA workers of Riyali asserted that their family members help with household chores during the working hours of MGNREGA.

The Problem of Raising Children

One of the most negatively understood aspects of working women is often viewed as a lack of proper attention to children. It is evident from Table 14 that 100% of MGNREGA workers affirmed that they face problems in giving time to their children while working in MGNREGA. Sometimes they bring their little ones to work and sometimes they leave them at home with their grandparents.

Agriculture and MGNREGA

Different perceptions related to agriculture were obtained from people in the study area. These are discussed as follows:

Table 14 Problem of raising children

Face problems in giving time to children	MGNREGA workers
Yes	64 (100)
No	–

Source: Based on Field Survey, 2022
Figures in parenthesis show the percentage

Table 15 Practice agriculture

Practice agriculture	MGNREGA workers	Non-MGNREGA workers	Panchayat heads
Yes	64 (63.4)	30 (29.7)	7 (6.9)
No	–	–	–

Source: Based on Field Survey, 2022
Figures in parentheses show the percentage

Firstly, an attempt was made to know whether people practice agriculture besides MGNREGA. Table 15 discloses that all the MGNREGA workers, non-MGNREGA workers, and panchayat heads practice agriculture. Since agriculture is seasonal hence both activities are easily practiced within a year. In fact, during the field investigation, it was found that in Riyali panchayat people earn more from agriculture as compared to working under MGNREGA.

Secondly, the study attempted to enquire whether people face any problems in practicing agriculture while working in MGNREGA. Table 16 reveals that all MGNREGA workers asserted that they do not face any problems in agriculture while working at MGNREGA. The majority of non-MGNREGA workers also believe that MGNREGA workers do not face any problems in practicing agriculture while working in MGNREGA. All the panchayat heads also feel that MGNREGA workers do not face any problems in agriculture while working in MGNREGA.

Thirdly, an attempt was made to record people's perception as to whether MGNREGA has helped in agriculture in some way or not. The recorded response is presented in Table 17. The data reveals that all the MGNREGA workers, non-MGNREGA workers, and panchayat heads affirm that MGNREGA has helped in agriculture. There are many works like the construction of kuhls or irrigation canals, and check dams which have directly helped in agriculture.

Lastly, an attempt was made to analyze people's perception of whether MGNREGA has helped in the development of their village or not. As per Table 16, all the MGNREGA workers, non-MGNREGA workers, and panchayat heads stated that MGNREGA has helped in the development of their village. This development has been in several forms like repairing and construction of roads, and traditional water bodies.

Table 16 Problems faced by MGNREGA workers in agriculture practice

MGNREGA workers face problems in doing agriculture	MGNREGA workers	Non-MGNREGA workers	Panchayat heads
Yes	–	3 (3.0)	–
No	64 (63.4)	27 (26.7)	7 (6.9)

Source: Based on Field Survey, 2022
Figures in parentheses show the percentage

Table 17 MGNREGA helped in agriculture

MGNREGA helped in agriculture	In Riyali		
	MGNREGA workers	Non-MGNREGA workers	Panchayat heads
Yes	64 (63.4)	30 (29.7)	7 (6.9)
No	–	–	–

Source: Based on Field Survey, 2022
Figures in parentheses show the percentage

Perceived Overall Satisfaction with MGNREGA

Several variables were considered, and people were asked about their perception and level of satisfaction. The respondents' satisfaction has been registered on a five-point scale that is, highly dissatisfied, dissatisfied, neutral, satisfied, and highly satisfied. Table 18 presents the perception of the people on different variables and has been discussed as follows:

Highly Dissatisfied Category: None of the respondents registered any variable in the very highly dissatisfied category.

Dissatisfied Category: It is evident from Table 18 that dissatisfaction has been registered by people from the amount of wages and increase in wage rate, which they consider as negligible and is not at par with other areas.

Neutral Category: The majority of MGNREGA workers (51.5%) are normally satisfied with the opportunity to do activities of the house along with MGNREGA.

Satisfied Category: The majority of people 57% in Riyali are satisfied with the quality of work under MGNREGA; 59.4% of the MGNREGA workers in Riyali believe that they are satisfied with the quality of work under MGNREGA and agree that MGNREGA has changed the lives of women.

Highly Satisfied Category: The majority of people (65.3%) are highly satisfied with the development of their village from MGNREGA as much work has been done in different sections under MGNREGA.

Riyali panchayat has a very large scope for development. Being the interior part of Fatehpur block, many things could be done under the scheme. People are already working together under the MGNREGA scheme, for instance, through the

Table 18 Level of satisfaction with different variables related to MGNREGA

Variables	Highly dissatisfied	Dissatisfied	Neutral	Satisfied	Highly satisfied
Satisfaction with the amount of wages under MGNREGA	–	84 (83.16)	17 (16.83)	–	–
Satisfaction with quality of work under MGNREGA	–	–	4 (3.96)	58 (57.42)	39 (38.61)
Satisfaction with the development of your village	–	–	–	35 (35.65)	66 (65.34)
Satisfaction with an increase in wage rate[a]	–	71 (100)	–	–	–
Satisfaction with the opportunity to do activities of a house along with MGNREGA[b]	–	–	33 (51.56)	31 (48.48)	–
Satisfaction with MGNREGA has changed the lives of women[c]	–	–	–	22 (59.45)	15 (40.54)

Source: Based on Field Survey, 2022
[a] Question pertains to MGNREGA workers and panchayat heads
[b] Question pertains to MGNREGA workers only
[c] Question pertains to Non-MGNREGA workers and panchayat heads
Figures in parenthesis show the percentage

interlinking of villages, construction of kuhls for agricultural enhancement, etc. which is not only contributing to the development of panchayat but also helping them in their socio-economic development. Social cohesion among people apart from belonging to different communities draws a commendable example of solidarity in achieving development.

MGNREGA and Solidarity

MGNREGA has positively impacted the solidarity of the Riyali panchayat. Despite having populations belonging to different religious and cultural backgrounds, people work together in harmony under the scheme. It has been noticed during the survey that they have mutual respect. Moreover, none of the workers reported problems with one other. They were satisfied with the scheme and agreed that the scheme has brought tremendous development for their village in terms of road and Kuhl's construction specifically. Riyali panchayat sets a great example of how the MGNREGA scheme could be utilized for socio-economic development on one hand and increasing solidarity among different communities at the same time. Solidarity in any region is enhanced when people spend time together and show mutual respect toward each other. Understanding each other's ways of living and considering them with esteem are key to maintaining harmony and solidarity.

Conclusions and Recommendations

From the above discussion, it is evident that Riyali panchayat is performing efficiently not only in terms of socio-economic development but also in improving the status of women under the impact of MGNREGA. Moreover, people from different communities are participating in the MGNREGA scheme which presents strong solidarity among them as they are working together in all the works taken up in the panchayat without any mutual problems. Even during the survey, people were seen working together in harmony. MGNREGA has offered them a platform where they communicate and work collectively and hence, it enhances the solidarity among the people belonging to different communities. However, people are dissatisfied with the increase in wage rate under the scheme as they perceive it is very low to carry out their living activities.

Therefore, the amount of wages provided to MGNREGA workers must be increased as they are very low in comparison to other activities. Also, the government should take some steps or measures for the time-to-time checking of works that are being taken up by the panchayats. It has been noticed during the study that the social and economic development made through MGNREGA is declining. It was good in the beginning. Hence, there is a dire need to address the issues due to which the performance under MGNREGA is decreasing. Some of these issues are low wages and low wage rates. MGNREGA schemes are for the people of rural India. It has brought substantial changes in their lives. Therefore, addressing the issues that hamper its beneficial effects is important and should be prioritized.

References

Balokhra, J. (2022). *The wonderland Himachal Pradesh: An encyclopedia on a tiny state of Western Himalaya*. H.G. Publications.

District Census Handbook. (2011). Retrieved from http://www.censusindia.gov.in/2011census/dchb/0202_PART_B_DCHB_KANGRA.pdf

Husain, M. (2021). *Human geography*. Prem Rawat for Rawat Publications.

Ministry of Rural Development. (2012). *MGNREGA Sameeksha: An anthology of research studies on the Mahatma Gandhi National Rural Employment Guarantee Act, 2005: 2006–2012*. Orient Blackswan. mgnrega_sameeksha.pdf. Retreived from indiawaterportal.org

Ranaware, K., et al. (2015). MGNREGA works and their impacts: A study of Maharashtra. *Economic and Political Weekly, 13*, 53–61.

Wikipedia. (2023). *Harassment*. Retrieved from https://en.wikipedia.org/wiki/Harassment Accessed 30 Sep 2023

Enhancing Indigenous Knowledge through Community Solidarity and Sustainable Development in Rural Society

Gopal Sonkar

Introduction

The present study explores enhancing indigenous knowledge through community solidarity for sustainable development in rural society. The study is determined through Santhal and Chakma rural community case studies in the Indian subcontinent context. These two rural community are the plain tribe' community 'Santhal' and the hill tribe community 'Chakma'. The first study defines these three aspects (indigenous knowledge, community solidarity, and sustainable development), and then their co-relationship between them further their case studies.

Indigenous Knowledge: Indigenous knowledge for the present study denotes such knowledge components available in the respective communities and practiced either regularly or from time to time and are not available in the 'Greater society' beyond the community. So, in this perspective, Indigenous knowledge pertains to the inherited wisdom transmitted through generations within rural community, fostering a sense of collective identity and unity over an extended period. Indigenous knowledge relates to the circumstantial information unique to a specific ethnic group or community. In essence, the concept refers to the specific knowledge of Indigenous individuals or the distinctive knowledge specific to a particular culture or community (Slikkervee, 2019).

Indigenous knowledge is 'the knowledge that the people in a given community have developed over time and continue to develop'. Indigenous knowledge is a distinct knowledge specific to each culture or community. It serves as a foundation for making decisions at the local level in areas such as improving quality of life, sustaining livelihood, shaping social systems, preserving biodiversity, and guiding

G. Sonkar (✉)
Department of Geography, Aditi Mahavidyalaya College, University of Delhi, New Delhi, India

K. Arora, R. R. Anand (eds.), *The Solidarity Approach in Geography*,
https://doi.org/10.1007/978-3-031-86753-8_9

religious practices (Warren et al., 1989; Warren, 1990, 1991; Quiroz, 1994; Maundu & Patrick, 1995). Indigenous knowledge is typically collectively held by communities rather than being possessed by individuals. Indigenous knowledge, being of a tacit nature, presents challenges in its documentation, as it is deeply ingrained within community practices and serves as a valuable resource for problem-solving tactics within communities. It is possible to improve the understanding of local conditions and improve efforts to help the local community by learning from Indigenous knowledge by first investigating what local communities know and have on hand. There are at least two main reasons indigenous knowledge is essential for self-sufficiency. Indigenous knowledge is the basis for self-sufficiency for at least following two main reasons (Godbole, 1996–1997; Maundu & Patrick, 1995):

1. Indigenous practices and technologies have stood the test of time and are well adapted to the local environment. They can understand, handle, and maintain them better than those developed by scientific knowledge and technology.
2. The concept of Indigenous knowledge is derived from utilising natural resource within the local environment. The individuals have a reduced reliance on external resources, which may incur substantial expenses and limit their access to resources. The wisdom of indigenous cultures can be a viable alternative to scientific expertise.

Recently, it has been experienced by all concerned that along with the standard package of practices recommended by the scientists, a parallel set of technology schedules generated out of Indigenous people's own experience or knowledge or wisdom about their environmental situation and own resource endowment is really in fashion at field level, these store of Indigenous knowledge which they have earned through generations of experience and which one essentially indigence, not systematic, lacks a scientific explanation of cause and effect relationship but authentic in application and use can be termed as 'Indigenous knowledge system'. Indigenous knowledge is currently recognised as a valid and valuable store of knowledge (Sana et al., 2006).

According to the WIPO Fact-Finding Missions on Intellectual Property and Traditional Knowledge (WFMIPTK) (1998–1999), 'indigenous knowledge' is knowledge held and used by communities, peoples, and nations. 'Indigenous knowledge' refers to the traditional knowledge of indigenous cultures. According to WFMIPTK indigenous knowledge category includes traditional knowledge, but traditional knowledge is not necessarily indigenous. WFMIPTK classified indigenous knowledge as a following criteria: (i) generated, preserved and transmitted in a traditional and intergenerational context; (ii) distinctively associated with a traditional or indigenous community or people which preserves and transmits it between generations (through community solidarity); and (iii) integral to the cultural identity of an indigenous or traditional community or people which is recognised as holding the knowledge through a form of community solidarity, community custodianship, community guardianship, community collective ownership or cultural solidarity responsibility (WIPO, 1998–1999).

Community Solidarity: A central component of this study is the notion of solidarity within the community. Rural indigenous communities seem to have a sense of solidarity in the practice of their way of life patterns concerning all aspects of indigenous knowledge. This solidarity has enabled them to maintain their identity and cultural values. It has also allowed them to develop a strong communication and cooperation system, enabling them to survive and thrive in their environment. Solidarity that developed from the content of commonalities involved maintaining indispensable relationships by enhancing the use of indigenous knowledge and by supporting and assisting them. Consequently, the shared characteristics within the collegial connections were indeterminate and subject to fluctuation between optimistic and relatively positive states (Chaklader, 2018; Chinn, 2007).

The concept of solidarity originates from the Latin word 'solidum', denoting an undivided and unaltered entity (Kortner et al., 1980) that encompasses the entirety (Løgstrup, 1993). Løgstrup's conceptualisation of solidarity is grounded in the fundamental premise of relationality, wherein human beings are inherently interdependent. This understanding forms the basis for his argument on the unmitigated nature of solidarity. Individuals conceptualise relationality as an intrinsic component of solidarity. Solidarity is commonly linked to reciprocal relationships characterised by considering and supporting each other's welfare and all aspects of their lives that rely on the available natural environment. In substitute terms, solidarity can be understood as the foundational element that nurtures the growth and development of a particular society. Nevertheless, it is important to note that relationality and solidarity are distinct concepts (Løgstrup, 1993).

Solidarity encompasses collective identity, fostering a sense of brotherhood and adherence to behavioural norms. By working together, a community can increase Indigenous knowledge accessibility, applicability, and appreciation. By joining in solidarity, a community can open the door to greater understanding and respect for indigenous knowledge. Through solidarity, the community can share traditional ideas and strategies incorporating indigenous knowledge and creating a more inclusive, equitable, and sustainable future. Indigenous knowledge with solidarity combination can lead to the conservation of natural resources, protection of cultural heritage, and development of more sustainable practices. It can also foster stronger relationships between indigenous communities, thus creating more social equity and solidarity. Solidarity is like a potluck dinner, where everyone brings a dish to share. Everyone can benefit from what others are bringing to the table, and the combination of contributions leads to a fuller and more nourishing experience for the whole community. Solidarity also helps build relationships between members, providing protection and support. Working together also allows members to share resources and wisdom, which can help them plan and work together more efficiently. Indigenous knowledge is typically cultivated via community solidarity efforts. It is considered the shared heritage of Indigenous or local communities as well as the groups of persons affiliated with these communities (Cramm et al., 2013; Itzkovich & Heilbrunn, 2016; Prainsack & Buyx, 2018).

Certain countries have implemented national legislation that includes measures recognising 'collective ownership or custodians of Indigenous knowledge, especially for Indigenous and local communities and specifically regarding biodiversity or genetic resources'. This statement indirectly reflected the community solidarity. In the context of Peru country's indigenous community solidarity enhancement reflected from its legislation measurement such as, Article 2(b) of Law No. 27811, which pertains to the establishment of a protection regime for the collective knowledge of Indigenous people's solidarity that is derived from biological resources defines 'collective knowledge' (community solidarity) such as 'the accumulated, trans-generational knowledge evolved by Indigenous peoples and communities concerning the properties, uses and characteristics of biological diversity'. Moreover, it should be noted that Article 10 stipulates that the communal wisdom safeguarded within the framework of solidarity shall be attributed to an indigenous community as a whole rather than to specific individuals within that community. The previously mentioned communal knowledge may pertain to multiple Indigenous communities (Repblica, Poder Legislativo Congreso De La, 2007).

ARIPO's 'Swakopmund Protocol on the Protection of Indigenous Knowledge and Expressions of Folklore provides both solidarity and individual Indigenous knowledge'. Traditional practices and indigenous knowledge of various community's solidarity are rather intricately interwoven with the tenets of sustainable development. These practices benefit humanity in all living sectors, such as culture, customs, water, agriculture, food and health, biodiversity, and infrastructure. Indigenous and local groups' solidarity and long-held knowledge on managing ecosystems and protecting limited natural resource is increasingly seen as an effective tool in sustainable development. (ARIPO, 2015).

Sustainable Development: In consonance with the component of sustainable development, the present study denotes sustainable development as the development which achieves sustainable growth and development through the utilisation of Indigenous knowledge. Therefore, in this perspective, sustainable development is a form of development that effectively addresses the immediate requirements of the current generation. In doing so, we ensure that future generations' ability to meet their own needs is not jeopardised. The concept is frequently cited, especially within the governmental sphere, from the Brundtland Report (World Commission on Environment and Development, 1987). In essence, the concept entails the pursuit of development that adequately addresses the requirements of the current generation. While doing so, the ability of future generations to meet their needs is protected.

The Global Knowledge Conference 97 in Toronto in June 1997 highlighted the pressing necessity to acquire, safeguard, and share Indigenous knowledge embedded in practical community solidarity. The conference brought together scholars, activists, and practitioners worldwide to explore Indigenous knowledge's potential to create sustainable change. It also provided a platform for Indigenous leaders to share their knowledge and experience. It also provided a platform to work together to develop strategies for protecting and preserving their knowledge and culture in solidarity. This would enable its replication in other contexts and use in development processes (The World Bank, 1998). Over the past few years, UNESCO has been engaged in the Man and Biosphere Programme, with the primary aim of establishing a connection between the comprehension of geo-biophysical processes at the ecosystem and landscape levels and the human aspects of the issue in the field of natural resource management, specifically focusing on the sustainable development of rural and traditional societies and solidarity (UNESCO, 1971) There exists a gap between the rural poor with rich traditional wisdom on the one hand and the more significant scientific elites endowed with modern scientific knowledge on the other. Understanding the Indigenous knowledge of rural communities living close to nature and its resources becomes essential to setting the tone towards understanding nature and its resources. Based on a range of issues dealing with ecosystem fragility, such as water and soil as key resources for land management, indigenous versus modern agriculture, and nature-friendly forestry and sustainable management/conservation of these resources.

Consequently, it is essential to preserve indigenous knowledge to raise awareness about the value of indigenous knowledge for sustainable development. Community knowledge will be preserved and documented, making it available to the communities from which it was derived. It will also make it available to development workers (UNESCO, 1971). Indigenous knowledge in rural societies (including tribes and isolated human habitats) is always targeted for productivity, sustainability, suitability, and equitability (Das, 2011).

There is a need to emphasise Indigenous knowledge research from livelihood-oriented practices to rural societies and their communities-based survival. The concept of Indigenous knowledge base development has gained importance in recent years. It is the development of rural societies by using their Indigenous knowledge and innovativeness. It has been emphasised that a comprehensive understanding of the Indigenous knowledge and expertise of rural society is a sine qua non for the sustainable development needs of our country. It is now felt that a breakthrough in Indigenous knowledge research, particularly in the use of bio fertilisers and bio pesticides, will be of immense help to the framers of our country since such inputs can be generated by the rural societies in their situations by utilising the available resources with the help of their Indigenous knowledge. This will also make the rural societies accessible from the development of commercially based essential input for raising the produce. In addition, the ecological impact of such input will be well within the control of the rural society. Indigenous knowledge will thus help correctly identify and develop geographical-specific technologies utilising the local investigator for its fruitful use (Godbole, 1996–1997). The intangible aspect of biological and social resources encompasses indigenous knowledge. The conversion of Indigenous knowledge into commercial benefits is facilitated by its capacity to offer vital insights for advancing practical products and processes. Using Indigenous knowledge offers significant benefits in time, cost, and investment savings within the changing climate change and environment due to pollution. Therefore, producers and custodians of Indigenous knowledge must receive a fair portion of the benefits.

The Co-relationship between Indigenous Knowledge, Community Solidarity, and Sustainable Development

Indigenous knowledge varies geographically—the originally different kinds of indigenous knowledge developed in different forms of geographical similarity (landscape) of community solidarity. Indigenous knowledge is a system of shared meanings and collective practices from distinctive regional conditions and traditions. A regional and community's Indigenous knowledge intersects fundamentally with social structures and livelihood. The study of how a region's Indigenous knowledge geographies and patterns lend a certain distinctiveness to the way of life there and explore how dominant Indigenous knowledge emerged from specific areas. Geographers and anthropologists understand Indigenous knowledge to be a set of distinct meanings and practices shared by a group of people or community solidarity (Anderson, 2003).

Indigenous knowledge and community solidarity concepts can be understood through a simple example: a tree. For members of one culture or community, a copse of trees might be seen as timber or a source of shade. However, those same trees might be the residence or connection to family ancestors in other communities. We are part of a shared indigenous knowledge to the degree that we share our meanings in common with people around us. 'Indigenous Knowledge Practices' refers to how a particular community performs cultural actions. Think about how humans eat. In rural India, no silverware is used; people eat from large plates with their hands. In a restaurant in Paris, however, table settings and appropriate use of silverware are understood. In the earliest human prehistory, herding communities moved their animals across the plains; hunters and gatherers lived in forests; and agricultural peoples settled in irrigated river basins. Each of these separate peoples had their worldview, way of performing actions, and language to communicate and capture their understanding of the world.

A distinctive geography and way of life are connected to systems of meaning and practice that we call indigenous knowledge. 'Indigenous knowledge' refers to understanding indigenous communities and indirectly represents their sense of solidarity. Indigenous knowledge is often passed down through generations and is integral to the communities' spiritual, cultural, and economic survival. It is a source of pride and belonging and strength and resilience in the face of adversity:—Indigenous knowledge is described as a broad field that takes a plural form. The study identifies those knowledge systems that have developed from community solidarity. 'Knowledge described as Indigenous is neither static nor ancient and thus must not be treated as a historical artefact for removal from contemporary life'. Before colonialism, Indigenous communities did not exist in isolation from the global community, facilitating a constant enrichment of their knowledge systems by incorporating many ways of knowing (Chinn, 2007).

Indigenous knowledge systems and community solidarity serve as foundational elements for local decision-making. This process is often facilitated through recognised and informal community solidarity and organisation mechanisms.

Communities engage in the identification of issues and pursuit of resolutions within localised forums. They harness their inherent creativity and create an environment that contributes to experimentation and the emergence of innovative approaches. Indigenous knowledge and community solidarity, although often perceived as primitive remnants of the past, are complex systems embraced and adapted by various geographical communities. These systems respond to challenging and frequently hostile environmental conditions and embody centuries, if not millennia, of adaptive assessment. This evaluation encompasses climate change, land and water availability, and the fundamental requirements of sustenance, shelter, and health for human populations and their livestock. These elements have facilitated societies' survival and advancement in the face of enormous odds (Flavier, 1995). There are varying perspectives about the notion that 'Indigenous or traditional knowledge can be localised through solidarity and constrained by local environmental factors, with community solidarity typically governing its development'. The prevalence of this phenomenon differs across countries, regions, and communities (Maundu & Patrick, 1995; Itzkovich & Heilbrunn, 2016).

The interaction between Indigenous and non-local knowledge is a dynamic process that has a transformative impact on a particular region's socioeconomic solidarity and ecological conditions. Therefore, it is imperative to recognise that Indigenous knowledge should not be perceived as a fixed reservoir. The quantum of Indigenous knowledge content experiences ongoing enhancement, and this dynamic aspect is perhaps of comparable significance to community solidarity (Basant, 1990). Knowledge and wisdom result from experimentation, natural selection, and meticulous observation over centuries. It encompasses various topics from both technical and non-technical domains, including social-religious taboos, beliefs and practices, ecology, negotiation, climate, and more. Indigenous knowledge is characterised by its dynamic nature, rather than static, as the term 'traditional' often suggests, as it encompasses community solidarity and spatial concerns (Warren et al., 1993).

Indigenous knowledge systems include technologies and practices in the past and present by indigenous communities for survival in various environments. Some knowledge may be familiar to all members of communities (this reflects the community solidarity), and as such, may be easier to share with outsiders. In contrast, certain group members only hold other, more speculated (sometimes secret) knowledge (Olaide & Omolere, 2012). 'Indigenous knowledge is place-based or situation-based' (Chinn, 2007) 'although there may be common knowledge amongst the community in different localities, there are features of indigenous knowledge peculiar to individual groups solidarity' through indigenous language (Chaklader, 2018). Knowledge is downed through generations, as evidenced by the indigenous knowledge that young people possess. Indigenous knowledge develops rationally and empirically and has been tested and nurtured through community solidarity for generations (Kamduk, 2016). A process that authenticates and strengthens its relevance and utility. This knowledge is called the survival strategies of the poor people, and these traditional spirits have many authentic aspects for the various kinds of risk management that are environmental, economic, and societal (Warren, 1990).

Rural community solidarity practices have many relevant aspects relevant to sustainable development goals, and by utilising ‘community solidarity-centric knowledge,’ these goals can be achieved. These ‘community solidarity customary practices’ were developed and nurtured by the older generation and transferred from generation to generation through various means, verbally and sometimes overall, perceiving situations (Das, 2011).

At first, one needs to explore the term development from the Indigenous point of view; the word development is contextualised and associated with the totality of life and living beings. Though the term ‘development’ is commonly used in several contexts, determining its exact definition can be challenging. ‘Sustainable development, though it undeniably includes economic components, is best described in terms of enhancing human choices and attention to the range of human needs’ (Ben-Eli, 2018).

However, it is indisputable that a lot of sustainable development work is done among the target rural communities themselves; when it is successful, these local populations can make choices and impact their circumstances for the better. The primary emphasis lies on activities about healthcare, education, and individual capacities to foster sustainable development. The primary emphasis will be on sustainable development, which entails incorporating indigenous knowledge, encompassing its acquisition, preservation, dissemination, and application. In addition, it assesses novel approaches that align with their objectives. This approach is rooted in recognising sustainable development’s efficacy and longevity. However, the primary emphasis will be on sustainable development to promote viable alternatives that yield positive outcomes (Das, 2011).

Indigenous knowledge has so many utilisations for the many points of view; Indigenous knowledge is highly associated with the environment and has many environmental risk management strategies, and it shows how one would proceed towards sustaining birch to a common goal. Indigenous knowledge has much utilisation because accumulation occurs throughout successive generations of people living in a specific environment. Indigenous knowledge refers to the knowledge systems possessed by traditional communities and rooted in their collective solidarity. This knowledge is derived from their experiences and adaptation to the local culture and environment. Indigenous knowledge is used to sustain the community and its culture. Placing value on such knowledge could strengthen cultural identity and the enhanced use of knowledge to achieve appropriate public health and conservation of biodiversity. Areas of application of Indigenous knowledge and how it helps to proceed towards sustainable development: The following are the dominant areas of Indigenous knowledge: agriculture, environmental conservation, food preparation, health care, natural disaster management, natural resource management, poverty alleviation, and transmission of knowledge. Indigenous knowledge also has social and developmental goals, which this knowledge can fulfil the SDGs Goals such as sustainable agriculture, sustainable resource management, and so on (Table 1).

Indigenous knowledge plays a role in conserving nature, food production, forestry development, medicine, sustainable practices, land and resource management

Table 1 Various component of indigenous knowledge which developed in community solidarity helps us to achieve the sustainable development

Sl. no.	Components of indigenous knowledge	Uses for sustainable development
1.	Resource management	House management, water resource, soil erosion, forest, and family management, control environmental hazards handcraft
2.	Health care	Uses tree as a medicine kalmegh, sarpaganda, neem, arjun, baira, hartchandan, and so on
3.	Women empowerment	Domestic product preparation of indigenous food and animal nurturing
4.	Land management	The community has a unique knowledge pattern regarding land protection, and the elderly community members use various methods for land protection, such as tree plantation building a wall
5.	Reduction of infant mortality	Use indigenous medicine, and elders of the family take responsibility for childrearing. Community elders use various techniques for nurturing infants
6.	Agriculture	Cultivate paddy, wheat, and oil
7.	Agriculture	Cultivate paddy, wheat, and oil
8.	Political stability	A corporation is the core value of the indigenous group
9.	Poverty reduction and economic	Use animals for land cultivation and transport
10.	Cultural	Every community member is considered an integral part of the family, so their family structure became vital for cultural transformation

Sources: Comple by authors from various literature

and ecotourism, climate change, and disaster risk reduction (Zakaria et al., 2018). Pandey studied the Application of Indigenous Traditional Knowledge for Sustainable Agricultural Development. The study concluded that traditional knowledge is vital for well-being and sustainable development as it has evolved after thousands of years of observation and experience. By linking the indigenous knowledge systems of the people with farmers, research, and extension, the research output can be made more fertile and usable. The participatory technologies developed through Indigenous Traditional Knowledge integration will provide diversified technological options. Being low in cost will also benefit the national economy, besides sustainable agricultural development (Pandey et al., 2017). Giorgia studied indigenous knowledge and its implications for the sustainable development agenda. How indigenous peoples respond to ecological and development challenges, and how their cultures and knowledge systems can contribute to the sustainable development agenda. Indigenous knowledge characteristics and Indigenous peoples, to comprehend the principles underlying conventional knowledge, it is imperative to grasp the notion of progress (Giorgia, 2016).

Magni examines how indigenous communities address ecological and developmental challenges, and how their cultural and knowledge systems might assist in the pursuit of sustainable development goals, particularly in climate change adaptation and disaster risk reduction. The significance of establishing a comprehensive framework of knowledge, wherein Indigenous communities can contribute their experiences to overcome forthcoming challenges and assume proactive roles in effecting change through participation in decision-making processes (Magni, 2016). Another study explores the indigenous knowledge systems' viewpoint on sustainable development, focusing on two alternative development approaches: integrated local and global knowledge systems. Specifically, it analyses the role of integrated microfinance management (IMM) and Integrated Sustainable Community Development (ISCD) as contributions based on Indigenous Knowledge. The aim is to assess their potential in reducing poverty for the people of Indonesia and other regions worldwide. Incorporating the cultural dimension of development into all sustainable development goals is highly recommended (Slikkervee, 2019).

Santhal and Chakma Communities' Solidarities Centric Indigenous Knowledge and Sustainable Development

The Indigenous knowledge structure of two Indigenous communities' solidarity, the Santhal and the Chakma communities, is essential to understanding the study's objective: both are tribal communities that accord deep knowledge from their ancestors. The Santhal tribe is those tribe communities usually residing in the plateau and plain areas amidst forest. The prominent Santhals tribe resides in the Purulia district of the Chhotnagpur plateau. Usually, the Chakma community or tribes denote such tribe groups who reside in the hills and the hill tracks in Mizoram, Tripura, Chittagong, and surrounding areas. Both communities' knowledge is developed by various methods of Indigenous knowledge, trial, and error, doing by the content, perceiving the nature or long-term observation, imitation, working together, observing by self, and so on. Indigenous knowledge is helpful for better living, livelihood, social system, biodiversity, and 'religious practice'and fosters sustainable development.

The Santhal Community

The Santhals, a prominent proto-Australoid ethnic group, hold the distinction of being the largest tribe in the state of West Bengal, constituting almost 50% of the tribal population in the region. In addition to West Bengal, Jharkhand, Odisha, Tripura, and Assam. Despite being located in various districts of Paschal Manipur District. The predominant profession in contemporary times is agriculture. There is a growing trend among young individuals to pursue higher education and strive for self-sufficiency. However, the precise origins of the Santhal community remain a subject of controversy. The Kherwar tribe, known by this name in ancient times, is believed by some scholars to have moved to the Indian subcontinent from southeast Asia around 1500 BC. This migration occurred before the arrival of the Aryans in the Indian subcontinent, which is considered the indigenous knowledge homeland of the Chhotanagpur plateau. It is widely acknowledged that the regions of Assam and Bengal are characterized by their mountainous terrain and dense forest cover. It is also recognised that the inhabitants of these regions had engaged in shifting cultivation long before migrating to their central territory, the Chhotanagpur plateau. Cultivation predates individuals' migration to their primary settlement area, namely the Chhotanagpur plateau. The primary livelihood of the community under consideration encompassed agricultural practices, supplemented by activities such as hunting, gathering, and fishing. The Santhal community prefers self-identification as 'Hor', a term denoting the man (Boddind, 1925; Bodding, 1927, 1929; Guha et al., 2015).

Santhal Community and their Unique Solidarity

The Santhal tribe is widely recognised as one of India's most populous Indigenous communities. The Santhal belong to a non-Hindu demographic, so they establish a distinct racial distinction between themselves and adherents of Hinduism. The tribe in consideration has been categorised as a 'Pre-Dravidian' group. Santhali is one of the languages spoken in India and belongs to the Munda (or Mundari) subgroup of the Austro-Asiatic language family. Dialectical variants in the Santhali language exhibit regional differences between northern and southern regions.

Nevertheless, Guha has suggested that there exists a comprehensive level of mutual intelligibility among the Santhal people across their whole community (Guha et al., 2015). However, this mutual intelligibility serves as one of the vital bases of their social solidarity. Their social organisation is characterised by solidarity and equality among them. The Santhal society exhibits a notable absence of caste divisions, a patrilineal kinship system, and a relatively small degree of political unification (Culshaw, 1949).

The societal structure appears to be organised into 12 distinct clans, each containing multiple subclans. These subclans, in turn, are further divided into local patrilineal lineages, which serve as the primary entities' kin groups within the society. The primary social unit within a given society is sometimes called the extended patrilocal family. Typically, a village consists of multiple clans. The primary political unit in the region is often represented by distinct villages, although the largest formally organised territorial entity, known as a Pargana, functions as a grouping comprising around 12 villages. This organisation was established to address and resolve diverse legal problems collectively. Bodding discusses the 119 Pargana, which seems to be the unit. The mentioned entity can be characterised as an administrative assemblage of rural settlements overseen by a Pargana. However, the Santhal society promotes the practice of democracy, making various decisions through consensus and respecting every individual in the society (Bodding, 1940).

The Santhal community has unique solidarity in many dimensions, and they have an independent history of life, whether it is the Indian freedom struggle or any means. The Santhals have no recorded history and lack of the historical evidence. It is not possible to know all the cultural background of the Santhal. Consequently, all existing knowledge about them has been transmitted orally from one generation to the next. However, the culture and history of Santhal are not particular, and they have their solidarity, which gives them a unique identity.

Santhal Solidarity-Centric Indigenous Knowledge

The significant finding of the indigenous knowledge has been represented in the following section in a tabular format.

Table 2 presents the various Santhal solidarity-centric indigenous knowledge, including major indigenous knowledge categories and related areas. Most Santhal residents in Assam and West Bengal can be characterised as an established agricultural group. The Santhals, who engage in agricultural practices, integrate religious rituals that reflect a societal emphasis on fertility, fecundity, generation, and other aspects closely tied to their fundamental life. Rice is their primary crop, but they also cultivate millet, maize, and a wide variety of vegetables, some of which they sell. They have adopted many of the common agricultural practices of the majority culture despite their lack of experience with or familiarity with techniques like crop rotation, irrigation, fertilisers, pesticides, and so on. The Santhals rely heavily on rice as a staple crop for their diet and economy. Buffaloes and oxen are commonly used in farming. As a form of fertiliser, most Santhals rely on cow dung. The majority of their food supply comes from crops grown with the help of rainwater. The Santhals have earned a reputation as some of the hardest workers in the history of humankind. Despite the gendered labour divide, both men and women are involved in farming.

Hunting, fishing, and gathering have mostly been rendered irrelevant as economic activities. The male population still participates in the yearly dehiri I disom sendra hunt. They also raise domesticated animals like cats, dogs, oxen, buffaloes, goats, lambs, pigs, and cows. Land is more valuable than human life. The land for Santhals provides economic stability and a strong connection to their ancestors. Their spiritual and material well-being are inextricably intertwined with the land. Traditional subsistence activities, including gathering, hunting, and fishing, have fallen from prominence but still provide some sustenance. Animal husbandry is widespread but contributes little to the stability of the food supply. The Santhals' subsistence relies heavily on hunting and fishing.

Gathering wild edibles, including berries, nuts, roots, sprouts, tubers, flowers, mushrooms, etc., not only supplements food supplies but also protects against seasonal food shortages and natural disaster-induced crop failures. Most collectors are women, though men have been known to join in on the action in the past. Animals and birds can be killed and captured with bows and arrows, swords and clubs, nets

Table 2 Santhal solidarity-centric indigenous knowledge practiced categories

Better living	Livelihood	Social system	Biodiversity	Religious practice
Traditional health care	Occupation depends on their sects or clans (the Santhal has 12 clans)	Santhali songs are the basic foundation of Santhali narration; with songs, they hail the companions, invite guests, and welcome the gathering in the social and cultural events	Santhali accepted natural phenomena as a god	'*Jom-im-Binti*'
Child rearing process	Agriculture	Santhal culture is beautiful	They are expressing their gratitude towards nature	'*Chand o bonga*'
Utilised local resources for livelihood	Food preservation	'Bin-ti'—the transformation of oral tradition/oral description through verbally	Earth, water, fire, trees, flowers, and mountains are ritual objectives	Religion is the existential connection between nature and culture
Physical exercise	Using natural fertiliser technique (bio-fertiliser)	Santhal festivals (*sairai, da nsai*)	Bonga's nature, or the degree of reality, is known by its spiritual presence and power	'*Maran g Buru*' is the spirit who dwelt with the first Santhal
Santhal community member is attempting to revive their culture	To participate in agricultural events with a sense of togetherness and concern	The '*Sarhul*', the 'Feast of Sal Blossoms', also known as the marriage of earth'	Erok or Eroh is an act of worship for sowing seeds	Sima bonga is the boundary spirit
Communicate through the Santhali, their mother tongue	The source of both plant and human life is seed	'Bitlaha' is a punitive mass gathering for significant offences	Sura Sagen Mahamane is a ritual act invigorating a slow-growing corp	Manjhi Haram is the spirit of the founder of the village

(continued)

Table 2 (continued)

Better living	Livelihood	Social system	Biodiversity	Religious practice
For preparing medicine, the Santhal community used the leaves of a tree, the roots of a tree, and the skin of a tree	Santhal have their methods of preparing medicines from animal sources	The village council (Manjhi-Manchor) controls the entire social system of the Santhals	'Jantral' is a worship for the protection of crops from birds	Bir bonga is the forest spirit
For the betterment of the child, the Santhal community mixes up the turmeric and oil-wash child	The Santhal love land more than their life. It provides economic security and is a powerful link with his ancestors	Santhal village nominates its headman (Manjhi), his deputy (Jog Manjhi), and his secretary (Godet)	'Janthar' is a festival of first fruit	Jaherer is the spirit of a sacred grove
Santhal eats vegetables produced by nature spontaneously	The land holds significant importance within the Santhal community as it is intricately intertwined with their spiritual and economic legacy	Santhal language is the medium to establish links between the community's people	'Gundali' is a festival of eating the first millet	Dadhi bonga is the spring spirit
The Santhals are well-versed in the art of hunting	Irrigation, family planning, and health	Women in the Santhal society enjoy liberty equal to men. Manjhi era or Naeke era is traditionally the leader of women folk	Some primary rituals in the Santhal life celebrate the relationship of the rice (seed) and earth (womb)	The community celebrates every religious or social function with drinks, dance, and music

Sources: Compile by author from the following literature: Saren (2013), Mathur (2001), Jamil et al. (2016), Guha et al. (2015), Dey (2015), Culshaw (1949), Chakraborty et al. (2011), Ghosh et al. (2016), Bodding (1925a, b, 1927, 1929, 1940, 2008), Boddind (1925)

and traps. They use rods, lime, poison, fishing nets, and various traps and hands to catch fish in streams, artificial ponds, and water reservoirs. Day labour is the mainstay of the Santhal economy. That is now their primary source of income and employment. Quite a few families survive by filling land for others, while others work as unskilled labour in tea gardens and factories. Santhals sacrifice their bongos at the Jaherthan or in their homes before beginning any farming. The following are examples of significant annual rituals and celebrations that can be linked to efforts towards social and economic solidarity in support of sustainable development:

- **Erok Sim**: In Asar (June–July), people celebrate planting rice seeds, or Erok Sim. The naeke offers the birds as a sacrifice to the Jaherbongas and the Manjhibonga, calling on both to ensure a fruitful harvest.
- **Hariar Sim**: July–August is reserved for celebrating Hariar Sim. Paddy seeds germinate at this time of year. For abundant paddy growth, the naeke offers birds as sacrifices to the bongas of the hamlet, such as Maran Buru Jaher Era, Gosae Era, Moreko-Turuiniko, Pargana bonga, Manjhibonga, and Sima bongas.
- **Baha**: During February and March (Phagun), Baha is celebrated. Wild fruits and flowers, especially the sarjom (shorearobusta) blossom, are being provided with the first matkom (bassialatifolia) harvest. At the sacred grove, the naeke and kudam naeke sacrifice chickens to the gods Maran Buru, Jaher Era, and Moreko-Turuo.
- **Janthar**: Aghar (November–December) is when the first harvest of the winter rice crop is celebrated or Janthar. To ward off stomach ailments, increase paddy yields, and ensure the safety of their harvest, the 'kudam naeke' will often sacrifice a goat or ram and present it to the pargana bonga alongside the first ears of rice.
- **Magh Sim**: While the sauri (thatching grass) is being trimmed in Magh (January–February), the festival of Magh Sim is celebrated. To increase the yield of 'Sauri', the naeke will sacrifice birds to the village's deities.
- **Sohrae**: Sohrae is celebrated between December and January (Pous). The term 'harvest festival' might be applied as well. The local naeke (priest) brings sacrifices to the Jaherthan (holy grove) every day.
- **Iri-Gundli Nawaiis**: The month of Bhador (September–October) is designated as Iri-Gundli Nawaiis. The milletiri (panicummillaceum) and gundli (panicum-frumentaceum) are sacrificed to the bongas as the season's first fruits. Farmers who grow kudam naeke and millet fruit sacrifice goats or rams to the pargana bonga to ward off pests like rats.

The traditional and Indigenous health system is practiced among the Santhal community—medicinal plants used by the Santhal community. One study looked into the medicinal plants that have been catalogued according to the prescriptions of traditional healers (Dey, 2015). Traditional Santhal medicine treats various conditions, including but not limited to abortion, physical weakness, bronchitis, burning feeling, leprosy, and gout (Jamil et al., 2016). Dey studied the Indigenous peoples of the Purulia district and their usage of medicinal plants in paediatric and Material care. The study focused on the ethnic groups of Purulia using several medicinal plants to treat various ailments occurring in children and mothers (Dey, 2013). Mandal conducted a study on Medicinal plants used by the Santhal tribal people of Birbhum district of W.B. and Dumka district of Jharkhand in India. The study was to explore the uses of medicinal plants used by the Santhal tribal (Mandal, 2012).

Guha focused his research on the Santhal of West Bengal to better understand how social and cultural shifts within tribes affect the natural world. Positive findings include a wide range of cultural and dietary practices, religious observances, social-cultural shifts, marital patterns, and levels of awareness among the Santhal people in the Birbhum area. The study shows that the Santhal Community has many unique religious practices, independent language, dress, food habits, settlement patterns, festivals, marriage systems, and arts and crafts (Guha et al., 2015). Ghosh and others studied the health benefits of Santhal tribal foods from Jharkhand; the people of this area were quite knowledgeable about the local food sources. More than half of all native foods were green leafy vegetables, which provided excellent calcium, iron, vitamin A, beta carotene, and folic acid. Dietary diversification through increased use of traditional foods can be an effective and long-term solution to the underlying hunger problem in this indigenous population (Ghosh et al., 2016).

Santhal Solidarity-Centric Indigenous Knowledge and Sustainable Development

The Santhal community's Indigenous knowledge components for sustainable development are essential for achieving sustainability throughout the social, economic, and environmental spheres (Guha et al., 2015). Sustainable development and environmental risk management measures rely heavily on the indigenous knowledge used by the Santhal group, as given in Table 3. The Santhal community's customs are important to sustainable development in society, economy, and the environment. Health care, culture, festivals, rituals, folklore, traditional laws, music, and indigenous resource management are all examples of Santhal indigenous practices (Guha et al., 2015). Veeresh studied indigenous agricultural systems and their impacts on modern agriculture systems. The study was conducted in the Dumka district of Santhal Pargana region in Jharkhand. The results revealed that the farmers of agriculturally fewer progressive villages were using more Indigenous technical knowledge for sustainable development (Veeresh et al., 2015).

Indigenous knowledge and biodiversity are interconnected for sustainable development. Conserving indigenous knowledge, skill sets, and techniques is important to conserve biodiversity. People cannot achieve sustainable development until they conserve indigenous knowledge, as indigenous and place-based knowledge always works to achieve sustainability. Indigenous knowledge and biodiversity are interlinked; if one is affected, the other may disappear, or if biodiversity is less, the survival of Indigenous people is questioned, and if they lose Indigenous knowledge, they will lose culture and language, which are very closely linked with biotic (Parajuli et al., 2013). Saha conducted a study on *indigenous knowledge of dairy farming in the Birbhum district of West Bengal, India.* This study focuses on the art of dairy farming, profiles of the dairy farmer, the rationality of indigenous knowledge, the suitability of indigenous knowledge, and the incorporation of indigenous knowledge in formal research and development. This study shows that Indigenous knowledge practices persist, given that the benefit of centralised, high-tech-driven, and top-down approaches to animal health care have reached only a minority of the world's stock raisers by farmers (Saha, 2006).

Table 3 Santhal solidarity-centric indigenous knowledge for sustainable development

Types of Indigenous knowledge found in social, cultural, and economic patterns of lifestyle	Types of Sustainable Development associated
The Santhals believe that the mountains and the hills are the home of gods and goddesses (bongas)	Environmental sustainability
The sim sala bongs are worshipped at the 'starting' of cultivating a plot of land or harvest time	Environmental and economic sustainability
'Baha' specially to pay homage to the Sal tree or the Sari Sarjom, symbolising the truth	Environmental and social sustainability
Land fertiliser	Environmental and economic sustainability
Santhal medicine	Environmental and economic sustainability
Religion (Sari-Dharam)	Social sustainability
Santhal judicial system and law	Social sustainability
Child rearing	Social, economic, and environmental sustainability
House building	Economic and environmental sustainability
Woodwork	Economic sustainability
Joma nuya (tribal economic)	Economic sustainability
Baha bonga (the spirit of Baha flower festivals)	Environmental sustainability
Mag bonga (the spirit of *mag* festivals)	Environmental sustainability
Doho-Hor means 'traditional mandates handed down from one generation to the next'	Education for sustainable development
Sari-Sarna is the religion of the sacred grove, both spiritually and socially	Social sustainability
Natural painting (using natural colour)	Environmental sustainability
Seed preservation	Economic sustainability
Resource collection and preservation process	Economic and environmental sustainability
Tree identification process	Environmental sustainability
Seasonal knowledge	Social, economic, and environmental sustainability
Hunting	Social and economic sustainability
Cultivation system	Economic sustainability
Soil erosion preventive technique	Environmental sustainability
Art and Craft	Economic environmental sustainability
Animal nurturing process	Economic sustainability
Santhal administrative	Social sustainability
'*Oteili*' is a ceremonial act connected with land	Environmental sustainability
Erok or Eroh is an act of worship for sowing seeds	Environmental sustainability

(continued)

Table 3 (continued)

Types of Indigenous knowledge found in social, cultural, and economic patterns of lifestyle	Types of Sustainable Development associated
'*Sura Saren Mahamane*' is a ritual act of invigorating a slow-growing crop	Environmental sustainability
'*Nim Dak Mandi*' is a cleaning ceremony	Social and environmental sustainability
'*Bitlaha*' is a penetrative mass gathering for significant offenses	Social and environmental sustainability
Marriage system	Social sustainability
Dowry system	Social sustainability
Family structure	Social sustainability
Bhandan geriyo (death of a family member)	Social sustainability
Karam festival	Socio-environmental sustainability
Soharai festival	Socio-environmental sustainability
Mah more festival	Socio-environmental sustainability

Sources: Compile by author from the following literature: Saren (2013), Mathur (2001), Jamil et al. (2016), Guha et al. (2015), Dey (2015), Culshaw (1949), Chakraborty et al. (2011), Ghosh et al. (2016), Bodding (1925a, b, 1927, 1929, 1940, 2008), Boddind (1925)

The Chakma Community

The word '*Chakma*' or '*Chukma*' (as spelled by some British scholars) is a generic term given to a predominantly hill tribe of Chittagong Hill Tracts dwelling in the corridor link of *Arakan* (Burma) by the British administrators and scholars who ruled over and researched on these tracts of hill from mid-eighteenth century to mid-twentieth century. These ethnic tribal people have been grouped under a common name head in the British census records. Thus, it takes the form of Chakma (*TSAK-THEK* in Burma). The Chakmas are one of the largely indigenous communities of the Indo-Bangladesh land strip. The original habitat of the Chakmas had been in the districts of Chittagong and Chittagong Hill of Bangladesh. They settled in this area before the sixteenth century. Some spread over and settled in Mizoram, Tripura, Assam, and Arunachal Pradesh of North East India. More than 1 lakh members of this community, called 'Da', reside in Myanmar's Arakan and Chin Hill regions. These ancient indigenous people are divided into three main groups: *Anokya*, *Tonchongya*, and *Dainnak*. There are as many as 60 sects (*Gozas*) and over a 150 clans (*Gutthis*). Recently, anthropologically, the Chakmas belong to the Mongoloid division, and in religion, they follow Heenayani (*Therovadi*) Buddhism. Among the rural inhabitants' primitive animistic beliefs and rituals can be found alongside the practice of Buddhism. The Chakmas claim their descent from the Sakya dynasty and believe in Gautama Buddha as their ancestors (Ghosh, 1909).

The Chakma Community and their Unique Solidarity

The village society of the Chakmas has been based on close cooperation and collective activities. The social norms demand that everyone in the community contribute during weddings and funerals. The system persists with uninterrupted continuity. The headman of the village is elected unanimously. If the headman fails in his duties or does something detrimental to the village or its inhabitants, a system of recall is there, and a new Headman is elected (Hoque, 2013). The Chakma customary laws are inviolable and binding for all. Even the Headman or the King cannot violate it. In the past, Chakma villages had village-based youth organisations reinforced by the system of common dormitories for young men. These are called '***Zall-tungies***' (Chakma, 2018).

The Chakmas usually prefer riverbanks for their habitat and named the locality after the river's name, for example, Gumetkul (bank of river Gomati in Tripura) and Thegakul (bank of the river Thega in Mizoram). However, the Chakmas generally do not desert their place for disasters, natural calamities, or epidemics. Formerly, it was customary to build huts on stilts, though it is relatively rare now. The central system of production was jhum cultivation, but they also grew crops by irrigating plain land by constructing dykes and by sowing paddy without tilling the land. This was known as '***Gorjyahunda***' (Majumdar, 1997).

Chakma Solidarity-Centric Indigenous Knowledge

The significant finding of the indigenous knowledge has been represented in the following section in a tabular format (Table 4).

The Chakmas have their traditional social pattern, which is unique and exclusive. Chakmas have ways of managing the decisions; traditionally, they follow the practice whereby the older male and the female members are active and responsible for making the family decisions. These customary laws have won wider acceptability from the community's point of view. So, possibilities might unfurl when everybody can get universal and inclusive justice. These customary laws of the Chakma community have an essential role as they celebrate their knowledge system and sustain their community successfully (Chakma, 2018).

The social relationship depends on the interconnectedness among the community members. The community member builds relationships through various means, which are also means of their socialisation. The visitor is treated as a family member, and uniquely cooked food with improvised cuisine is shared when dining with all the family members. As they believe in what they would say in their own words '*aksate khaibo anado lagbo o sabail saman paibo*'. This unique spirit of the Chakmas is undoubtedly a significant aspect of attaining Sustainable Development Goals because social sustainability begins from the society and community habits at the rudimentary family level (Chakma, 2014). The entire family of the Chakma community is used to taking food together at a time when simultaneously they utilise the time by making various decisions on various family problems (Chakma,

Table 4 Chakma solidarity-centric indigenous knowledge practiced categories

Better living	Livelihood	Social system	Biodiversity	Custom practice
Traditional Health care system	Economic grain produced supori, karpas, coconut, til, master oil	Owners towards their culture	The Chakma community member thought that forest fulfil their needs and river fulfils their thrust	According to the Chakma Myth the universe evolved into four stages Andhakara (Dark), dhandakara (Samsai), khwakara (Foggy), nirakara (noshape)
Pure water, natural water, jungle wall	Daily labour	Believe in the Chakmas village council justification	Usually, chakma does not cut the tree	Puja or worship is performed for the appeasement of the spirits
Health and hygiene	Farming/ ploughing	Mass feeding	Banyan tree worship	'Malakha' utsab
Housing	Cultivation land (jhum as well dhap or ahal chas), Grga has	Through collective worship	Bamboo worship	During the jhum cultivation, Chakmas are observed buju utsab
Child care	Bamboo-bath	Public welfare	Community common sentiment	'Hal paloni' utsab
Taking meals on time	Land fertiliser through bio-resource	No sex discrimination among the boys and girls	Worship of Bantha	Aycya, Than mana, Budd ha Purnima
Heard work	Fruit Preservation	To obey the old member's command, exercise the team's mind, and maintain the old societal heritage	Worship of river	Including the 'Sanga' 14 lord devoted by Chakma at the time of jumping
Chakma is highly sensitive towards nature and natural resource, they thought that if all valuable resources, such as medicinal plants, must be preserved if the forest is missing	Making bamboo toys and musical instruments	Respect to the older member	Festivals are associated with nature and natural laws	Kalia: he is the son of chandi. He is believed to be the master of family life

(continued)

Table 4 (continued)

Better living	Livelihood	Social system	Biodiversity	Custom practice
Preferring traditional foods for the batter living with minimum oil	Collecting livelihood from the forest such as vegetables, fish, meat, but most favourite item is bamboo 'koural'	Celebrated the diversity among the community members	Women member has more love for nature	'Niranjane' flourish the universe

Sources: Compile by author from the following literature: Chaklader (2018), Chakma et al. (2012), Chakma (2014, 2018), Dey (2013), Giorgia (2016), Ghosh (1909), Hoque (2013), Kamal (2011), Kamdu (2016), Kamduk (2016), Mandol et al. (2016), Majumdar (1997), Manda et al. (2009), Saha et al. (2017), Sharma et al. (2006), Uddin et al. (2014), Rubal et al. (2015)

2018). The Chakma community has its perception and pattern of worship of mother nature, including trees, plants, hills, mountains, and rivers as gods. They are compassionate towards natural objects. Even today, the Chakma community has an exchange system. They exchange the goods in a local community exchange centre, in their word, 'Hatt', where they sell the material and purchase different goods. The Chakma community has been accustomed to change, and the entire community assembled in one place: the temple of Lord Buddha. The Chakmas primarily collect their livelihoods from the forest areas and use natural fertiliser techniques for production. Cultivating jum was the mainstay of their economy. They change their lives and relocate to new areas to better their financial situation. They performed numerous rituals as they moved from one natural water source to another (Dey, 2013). Some of the given below (Chakma et al., 2012; Dey, 2013; Chakma, 2014):

- **Pakkon ahruk**: If a valley or hilltop is triangular, jum agriculture is not allowed there because evil spirits are thought to reside there and harm jum farmers. This concept is known as 'pakkon ahruk' (triangular shape of the land).
- **Ahja**: Ahja is the name of the flat, narrow foothill, and swampy area below. Any hillock near Ahja used for growing Jum or settling Jum colonists is doomed to failure. To hunt in Ahja is to cause harm. Leprosy, Ringworm, and other fatal skin infections are spread through personal hygiene practices such as urinating or spitting on the Ahja. The majority of Chakma people have this view.
- **Ekmos che videy**: 'Ek mosche videy' is the Chakma term for a narrow gorge formed by a stream surrounding a hill on three or four sides. No one lives or works on that hill because the family patriarch will die if they do.
- **Gonga nijesh pore** (the breath of the river befalls): Chakmas do not construct houses on the river turning or streams turning. Ganga Nijesh will bring disaster upon every household that builds a home on the bend of a river or stream.

- **Pah (oven)**: Pan means the oven. If three hills form a triangle and remain only a short distance apart, in the same parallel plain, side by side, and naturally formed land position oven, it is called Pah. If anybody cultivates on the land, then the family member must die, so nobody cultivates on that land. Before the construction of a house, it was also seen not to form Pah to avoid the delete riots effect.
- **Nelchumo-get** (water passing natural drain): A nelchumo-get is a deep, underground ditch or drain in flat land or on a hill. When it rains, water flows into the hole and collects until it reaches a river or stream below. The Chakmas do not use the flatlands or the nearby hill for farming or colonisation/settlement. If people start farming or settling there, they might as well all perish from starvation and natural disasters, in their minds.
- **Deo-dhulon-(hammockoftheevil spirits)**: 'Deo dhulon' has a spooky connotation in the Chakma language. The 'Deo dhulon' bow was made of twigs, two trees bending and contacting towards one other in a parallel line, and jungle vines hanging from the trees. The authentic deo dhulon is typically situated in a gorge on the side where the trees are offset to form a hammock or cradle. No one is allowed to live on or colonise that hill, and growing jum there is prohibited.
- **Munuschoro-ahruk** (a shape of the human body): The Chakmas do not live or farm on the tops of hills where the terrain looks like a human being lying down. Farming or settling on that land could be fatal for everyone in the family.
- **Leja Samuga bansh**: (Snail bamboo on top) The Chakmas are not grown in areas where people perceive bamboo to be a long snail. Folklore has said that this is a thunderous anthem.
- **Pagol/pharangibola gareye jaga** (mad leprosy buried place): Natural cremation is preferred by the Chakma people after death. However, they exploit the bodies of people with a mental illness or leprosy. Where the insane and diseased dead are buried is an unsettling place for jum farming and colonial settlements.
- **Puri Khola** (fairy place): The fairy land is called Puri Khola in Chakma. In Chakmas's view, the fairies became human when Adam was created. If Jum cultivates and a colony is developed at the fairy site, the fairies will become angry and try to destroy the humans. At night, the fairies use a variety of noises and disruptions to sign and whisper that it is time for them to leave. Those who stay there after sending warnings are also exposed to potentially fatal pathogens. So, farming and colonising are strictly forbidden in the enchanted land.
- **Var-buwh** (misbalanced/weighted): If two older brothers are in the family, the younger brother cannot build his home between theirs. The older sibling is in the same boat. It is believed that incurable sicknesses will always plague a person and even death if they build a home in the space between the homes of two elder or younger brothers, a practice known as 'Var-buwh'. The designated bedroom areas are also tightly enforced.

- **Ar Nijesh pore** (Long breath is fall): Building a home in front of senior members of the same clan is strictly forbidden. A person's Bor nijesh (long breath) falls upon the youngest in the family if he or she builds a house in front of the older members of their clan.

Chakma community member has vast and primary indigenous practices, among which health care is one. The Chakma community members have strong faith in their indigenous healthcare system. They used various trees as medicine for preventing and curing diseases from various Fatha diseases. The Chakma tribe uses 198 plant species belonging to 74 families with their local names, preparation, route of administration and uses (Pandey et al., 2014).

Chakma Solidarity-Centric Indigenous Knowledge and Sustainable Development

The following section describes the components of Indigenous knowledge practiced by the Chakma community for quite a long time for sustainable development (Table 5).

The Chakma community intensely depends on nature; they live in hilly areas. Naturally, in this place, they face various difficulties in their lives. Because they lack access to contemporary conveniences, Indigenous peoples like these nonetheless use age-old practices today (Uddin et al., 2014). This knowledge provides essential inputs for the community's long-term social, economic, and environmental health. Based on available case studies (Kamal, 2011; Kamdu, 2016), community has sought to explore the nature and components of this fundamental knowledge, providing potential ways and insight for achieving sustainable development. A few selected examples are presented in a table: Indigenous agricultural activity/eco-friendly agricultural practice, natural fertiliser for agriculture, land protection, water prevention, purification, Indigenous food collection and preparation, animal husbandry for economic sustainability, etc. The Chakmas can be credited for building alternative ways of knowing and judging things for a better world for a long time (Mandol et al., 2016).

Table 5 Chakma community solidarity-centric indigenous knowledge for sustainable development

Types of indigenous knowledge found in social, cultural, and economic patterns of lifestyle	Types of sustainable development associated
Rather than the individual, the community is the primary reference for natural and cultural property	Social, economic, and environmental sustainability
Creating a sustainable livelihood system	Economic sustainability
Indigenous knowledge is diverse knowledge deeply rooted in their relationship with the environment and cultural cohesion	Social, economic, and environmental sustainability
Suitable for use and managing natural resources	Social, economic, and environmental sustainability
Utilised the local/situational resources	Social, economic, and environmental sustainability
Alternative ways of living and knowing the world	Social, economic, and environmental sustainability
Co-operative living to promote their smooth ways of living	Social, economic sustainability
Unique ways of observing, adapting, and mitigating have enabled them to face new and complex circumstances	Environmental sustainability
Learning through the natural phenomena	Social, economic, and environmental sustainability
Love for nature, awareness for nature, preservation, and conservation of natural resources	Save the environment for the social and economic environment for ecological balance, thus ensuring life span
Cultural heritage	Social sustainability
Biodiversity	Environmental sustainability
Indigenous ways of looking at and judging things	Social and environmental Sustainability
Cultural value	Social sustainability
Evaluating the knowledge by experiential	Social sustainability
Preserved and promoted indigenous languages by their means	Social sustainability
Learning meaningfully	Social sustainability
Customary laws	Social sustainability
Live in harmony with nature	Social and environmental sustainability
Balanced and complementary rtelationship with human beings	Social and environmental sustainability
Balance and complementary relationships all creature	Social and environmental sustainability
Live with harmony: equality among all the community members, particularly women, is the most prestigious in the tribal society	Social sustainability

Sources: Compile by author from the following literature: Chaklader (2018), Chakma et al. (2012), Chakma (2014, 2018), Dey (2013), Giorgia (2016), Ghosh (1909), Hoque (2013), Kamal (2011), Kamdu (2016), Kamduk (2016), Mandol et al. (2016), Majumdar (1997), Manda et al. (2009), Saha et al. (2017), Sharma et al. (2006), Uddin et al. (2014), Rubal et al. (2015)

Conclusion

There is a remarkable normative space created by the Indigenous knowledge and community solidarity created over generations through their attachment to the rituals, customs, cultural, social, administrative, and livelihood. These reflect a symbolic and symbiotic relationship and coordination between indigenous knowledge of nature and community solidarity. The aspect that runs through it has more profound developmental implications at contextual and solidarity levels. That means it is culturally and ethnically embedded. That is why the members of the respective community are likely to inherit tradition and Indigenous, like temperamental composure, which is natural in the community solidarity ecosystem or the social-cultural settings of the community. This creates the basis of healthy stability in their society. They also create psychological well-being in their minds and develop the capability to withstand the challenges of nature.

Indigenous knowledge of community solidarity and its implications for sustainable development have a more significant relationship. Health care, parenting, education, natural resource management, animal husbandry, agriculture, shelter, art, craft, and an ecological perspective are some areas where indigenous knowledge may be constructive., etc. Throughout the study, it is evident that the Santhal and the Chakma communities possess profound dignity, and all dimensions of their lives reflect significant attachment to Indigenous practice through solidarity. Indigenous knowledge flows from the distant past and has been tested over generations by various experiments and introspection regarding nature. Hence, this is the focus of the study, and it is clear how the entire edifice of this unique cognitive exercise could give way to interrelated practices about multiple aspects of sustainable development, particularly in environmental, social, and economic segments. For the rural poor, indigenous knowledge is a form of absolute social capital. Indigenous knowledge is a vital tool as an intellectual resource for them.

Undoubtedly, this knowledge is mainly context-specific and ethnocentric and integrates Indigenous knowledge in enchantment. It could be done in multiple ways, such as capturing, storing, and disseminating information with cost-effective and sustainable strategies in development planning and developing mechanisms to sensitise information seekers with a professional interest in development and development planning. The study has confirmed that Indigenous knowledge practiced by the Santhal and the Chakma community has a high essence for sustainable development. Both communities show that various Indigenous practices, such as Indigenous health care, agriculture, Indigenous culture, festivals, rituals, folklore, and traditional laws, have exceptional value in this fast-changing era of consumption and commercialisation. Conceptually, through both case studies, it could be concluded that the Indigenous knowledge practiced by two indigenous communities, i.e., the Chakma and the Santhal, has a high potential for sustainable development.

References

Anderson, K. (2003). *Handbook of cultural geography*. SAGE.

ARIPO. (2015). *Traditional knowledge* laws. ARIPO. Retrieved from Swakopmund Protocol on the Protection of Traditional Knowledge and Expressions of Folklore (ARIPO): https://www.wipo.int/wipolex/en/treaties/details/294

Basant, R. (1990). Documentation of indigenous knowledge in Gujrat agriculture. In *A note in Proceedings of International Workshop Generation* (pp. 53–67). Indian Society of Agronomy.

Ben-Eli, M. U. (2018). Sustainability: Definition and five core principles, a systems perspective. *Sustainability Science, 13*, 1337–1343.

Boddind, P. O. (1925). *Santhal folk tales*. H. Aschehong Harvard University Press.

Bodding, P. O. (1925a). *Studies in Santal medicine and connected folklore* (Vol. 3). Gyan Publishing House.

Bodding, P. O. (1925b). *Traditions and institutions of Santhals*. Gyan Publishing House.

Bodding, P. O. (1927). *Studies in Santal medicine and connected folklore*. Asiatic Society of Bangal.

Bodding, P. O. (1929). *A Santal dictionary* (Vol. 36). J. Dybwad.

Bodding, P. O. (1940). *Santal riddles and witchcraft among the Santhals*. A.W. Broggers.

Bodding, P. O. (2008). *Tradition and institutions of the Santhals*. Gayan Publishing House.

Chaklader, M. A. (2018). Rethinking medicalization through the lens of cultural hegemony: Traditional healing and societal wellbeing of Chakma ethnic community in Bangladesh. *Singaporean Journal of Social Science, 1*(2), 35–46.

Chakma, L. B. (2014). *The Chakmas: An introduction, The chakma Darpan patrika*. Prantar Printes & Publishing Pvt. Ltd.

Chakma, S. B. (2018). *Chakma customary law*. Gandachara.

Chakma, S., et al. (2012). The contribution of forests to the livelihood of the Chakma community in the Chittagong hill tracts of Bangladesh. *Journal of Forest Research, 17*(6), 23–63.

Chakraborty, et al. (2011). The health maintenance and disease curative ethno-medicinal and religious practices by the Santhal of Keonjhar District, Orissa. *American Journal of BioScience, 1*(2), 1–23.

Chinn, P. W. U. (2007). Decolonizing methodologies and indigenous knowledge: The role of culture, place and personal experience in professional development. *Journal of Research in Science Teaching, 44*(9), 1247–1268.

Cramm, J. M., Strating, M. M., & Niebohr, A. P. (2013). The influence of organizational characteristics on employee solidarity in the long-term care sector. *Journal of Advanced Nursing, 69*(3), 526–534.

Culshaw, W. K. (1949). *Tribal Heritage: A study of the Santhals*. Lutterworth Press.

Das, G. A. (2011). Does Indigenous Knowledge have anything to do with Sustainable Development? *Antrocom Online Journal of Anthropology, 7*(1), 57–64.

Dey, A. (2015). Traditional Vs Transitional: Health care system of the Santhal at Paschim Medinipur, West Bengal, India. *IOSR Journal of Humanities and Social Science, 20*(7), 84–87.

Dey. (2013). Chakma tribe of Mizoram (North East India): A study of their faith and beliefs. In K. A. Jain (Ed.), *Ethnobotanical studies in North East India: A review* (pp. 99–112). Scientific Publishers.

Flavier, J. M. (1995). The regional program for the promotion of indigenous knowledge in Asia. In D. M. Warren, L. J. Slkerveer, & D. Brokensha (Eds.), *The cultural dimension of development: Indigenous knowledge systems*. Intermediate Technology Publications.

Ghosh, S. (1909). *Chakma Jati (Chakma Nation)*. Sona Moni Chakma.

Ghosh, J. S., Singh, A., Magsumbol, M. S., Kamboj, P., & Goldberg, G. (2016). Exploring the potential of indigenous foods to address hidden hunger: Nutritive value of indigenous foods of Santhal tribal community of Jharkhand, India. *Journal of Hunger and Environmental Nutrition, 11*(4), 548–568. https://doi.org/10.1080/19320248.2016.1157545

Giorgia, M. (2016). The Indigenous knowledge and implications for sustainable development agenda. *European Journal of Education, 52*(4), 1–22.

Godbole, A. (1996–1997). *The use of Indigenous Knowledge (IK) in mountain natural resources management: A case study of Wancho Community, Tirap District: Arunachal Pradesh, India.* Final Report of the ICIMOD Research Fellowship Work 1996–1997.
Guha, et al. (2015). Socio-cultural changes of tribes and their impacts on the environment with special reference to Santhal in West Bengal. *Global Journal of Interdisciplinary Social Science, 4*(3), 148–156.
Hoque, J. (2013). Study on cultural history of the Chakma ethnic tribe in Mizoram. *Applied & Social Sciences, 3*, 23–40.
Itzkovich, Y., & Heilbrunn, S. (2016). The role of co-workers' solidarity as an antecedent of incivility and deviant behavior in organizations. *Deviant Behavior, 37*(8), 861–876.
Jamil, et al. (2016). The folk medicine practices for the treatment of Abortion, Body Weakness, Bronchitis, Buring Sensation, Leprosy, and Gout of Santhal tribes' practitioners at Jamtala village under Sadar Upazila of Chapai New Abagani district Bangladesh. *Postgraduate Medical Journal*, 12–28.
Kamal. (2011). The attitudinal preferences for socio-cultural practices in the Chakma tribe. *Journal of Applied Social Psychology, 20*(8), 639–660.
Kamdu, J. (2016). *Rise of Chakma ethnic consciousness in Arunachal Pradesh: An instrumentalist approach*. Academia Education.
Kamduk, J. (2016). Rise of Chakma ethnic consciousness in Arunachal Pradesh: An instrumentalist approach. *IOSR Journal of Humanities & Social Science, 21*(5), 24–29.
Kortner, O., Munthe, P., & Tveteras, E. (1980). *Aschehoug og Gyldendals Store Norske leksikon [Aschehoug and Gyldendal Large Norwegian encyclopedia]*. Kunnskapsforlaget.
Løgstrup, K. E. (1993). *Solidaritet og kærlighed [Solidarity and love]*. Gyldendal.
Magni, G. (2016). The indigenous knowledge and implications for the sustainable development agenda. *European Journal of Education, Research, Development & Policy, 52*, 65–79.
Majumdar, P. (1997). *The Chakmas of Tripura, Tripura State Tribal Cultural Research Institute & Museum*. Government of Tripura.
Manda, I., et al. (2009). Family structure, economic security and educational status of rural Chakma in CHT of Bangladesh. *Journal of Social Science, 19*(3), 219–224.
Mandal. (2012). Medicinal plants are used by the tribal people of Birbhum district of W.B. and Dumka district of Jharkhand in India. *Indian Journal of Traditional Knowledge, 5*(4), 674–679.
Mandol, N., et al. (2016). Family structure, economic security and educational status of rural Chakma in CHT of Bangladesh. *Journal of Social Science, 12*(2), 1–43.
Mathur, N. (2001). *Santhal world view*. Concept Publishing Company.
Maundu, & Patrick, M. (1995). Methodology for collecting and sharing indigenous knowledge: A case study. *Indigenous Knowledge and Development Monitor, 3*(2), 3–5.
Olaide, I. A., & Omolere, O. W. (2012). Management of indigenous knowledge as a catalyst towards improved information accessibility to local communities: A literature review. *Chinese Librarianship, 35*, 87–98.
Pandey, et al. (2014). Ethno-botanical usage of plants by the Chakma. In Thakuria (Ed.), *On traditional knowledge for sustainable development: A geographical analysis, community of Tripura, northeastern, India* (pp. 43–67). Bulletin of Environment, Pharmacology and Life Sciences.
Pandey, V., et al. (2017). Application of indigenous traditional knowledge (ITK) for sustainable agricultural development. *AJAEES*, 43–56. https://doi.org/10.9734/AJAEES/2017/3148
Parajuli, D. R., et al. (2013). *Indigenous knowledge and biodiversity: Interconnectedness for sustainable development*. IJSTR Research Publication.
Prainsack, B., & Buyx, A. (2018). The value of work: Addressing the future of work through the lens of solidarity. *Bioethics, 32*(9), 585–592.
Quiroz, C. (1994). Biodiversity, indigenous knowledge, gender and intellectual property rights. *Indigenous Knowledge & Development Monitor, 2*(3), 12–15.
Repblica, Poder Legislativo Congreso De La. (2007). Retrieved from https://zh.unesco.org/sites/default/files/per_law27811_engtof.pdf

Rubal, et al. (2015). Conducted a study on indigenous medicinal practices: Medicinal plants of Chakma tribal medicinal practitioners in Rangamati district. *American-Eurasian Journal of Sustainable Agriculture, 9*(5), 28–35.

Saha, A. (2006). *Indigenous knowledge of dairy framing in Birbhum district of West Bengal, India: An explorative study*. Ph.D. Thesis, Institute of Agriculture, Sriniketan, Visva-Bharati.

Saha, S., et al. (2017). *Physical growth and nutritional status of Chakma*. Scientific Publishers.

Sana, A., Mukhopadhyay, S. B., & Das, G. D. (2006). Validation of indigenous knowledge in Livestock Health Care System. *Indian Journal of Extension Science*.

Saren. (2013). Conducted a study on the impact of globalization on the Santhals: A study on migrations in West Bengal, India. *International Journal of Human Genetics, 1*(32), 21–45.

Sharma, et al. (2006). Ethno-medico-botany of Chakmas in Arunachal Pradesh, India. *Indian Forester, 132*(4), 474–484.

Slikkervee, J. L. (2019). *The indigenous knowledge system's perspective on sustainable development, by Munyaradzi Mawere*. Langaa RPCIG.

The World Bank. (1998). *The Global Knowledge Conference 97: Conference Evaluation Toronto 97*. The World Bank. Retrieved from https://documents1.worldbank.org/curated/zh/522001468325221853/pdf/488640NWP0Box338929B01PUBLIC10ES98108.pdf

Uddin, et al. (2014). A survey of traditional health remedies of the Chakma indigenous community of Rangamati District, Bangladesh. *Journal of Ethnopharmacology, 144*(3), 627–637.

UNESCO. (1971). *UNESCO's Man and the Biosphere Programme*. UNESCO.

Veeresh, et al. (2015). *Indigenous agricultural systems and in the impacts on the modern agriculture systems: Indigenous knowledge systems and sustainable development* (1st ed.). Relevance for Africa.

Warren, D. (1990). *Indigenous knowledge systems and development*. Background paper for Seminar Series on Sociology and Natural Resource Management. The World Bank.

Warren, D. M. (1991). Using indigenous knowledge in agricultural development. World Bank Discussion Paper No. 127. In *World Bank discussion on indigenous knowledge* (pp. 23–49). The World Bank.

Warren, D. M., Slikkerveer, L. J., & Titilola, S. O. (1989). *Indigenous knowledge systems: Implications for agriculture and international development*. Iowa State University.

Warren, M. D., Liebenstein, G. W., & Slikkerveer, L. J. (1993). *Networking for indigenous knowledge, indigenous knowledge and development monitor*. Bad Boll.

WIPO. (1998–1999). *Intellectual property needs and expectations of traditional knowledge holders* (WIPO Report on Fact-Finding Missions). Retrieved from WIPO: https://www.wipo.int/edocs/pubdocs/en/tk/768/wipo_pub_768.pdf

Zakaria, et al. (2018). Role of indigenous knowledge in sustainable development. *International Journal of Development Research, 8*, 43–56.

Food Insecurity and Emotional Experiences in Tribal Community of Vagad Region, Rajasthan, India

Swati Rajput, Kavita Arora, and Sanjeev Kumar

Introduction

"Food insecurity is a situation that exists when all people, at all times, do not have physical, social, and economic access to sufficient, safe, and nutritious food that meets their dietary needs and food preferences for an active and healthy life" (www.omicsonline.org). The underdeveloped regions of the country especially the forested area and tribal belts have various challenges to meet in day-to-day life. Their story is of "Need". The basic need for food and water looks like a luxury to acquire. Nutrition and health is an afterthought. The energy to survive is minimal and so is immunity to fight against diseases. Lack of sufficient and nutritious food further mounted with a lack of knowledge. In their eating behavior, human beings are very much affected by their emotions: food choices, quantity, and frequency of meals are all dependent on many variables not necessarily related to their physiological needs (Canetti et al., 2002; library.athabascau.ca). Both poverty and child-related undernutrition are remarkably high among tribal communities in India (Thorat, 2010). The economic deterioration of tribal communities began in the British era. The influx of the non-tribal population and their consequent encroachment of forest areas further aggravated the situation (Guha, 1996). The objectives of this chapter are to understand the prevalence of food insecurity of individuals at different levels of severity in the Vagad Region and also to analyze the level of negative emotions linked to it. The seriousness of food insecurity is ascertained through the Food Insecurity Experience Scale (FIES). The emotional analysis is done by ascertaining the level of sadness and motivation by creating an Emotional Scale of Food Insecurity (ESFI). The sample of 200 households has been taken into consideration. The survey has been done during the pandemic times in December 2021.

S. Rajput (✉) · K. Arora · S. Kumar
Department of Geography, Shaheed Bhagat Singh College, University of Delhi, New Delhi, India

K. Arora, R. R. Anand (eds.), *The Solidarity Approach in Geography*,
https://doi.org/10.1007/978-3-031-86753-8_10

Vagad region is a tribal region of Western India (Fig. 1). It consists of two main districts, namely Banswara and Dungarpur. It has more than 76% of the scheduled tribe population. The ethnic group in this region is called Bhils. "Bhil" means bow, i.e., this ethnic group used a bow and arrow as a weapon. This region has an arid climate with a hot season that lasts from April to June, but it is not as hot as the desert regions of Rajasthan to the north and west. In the summer, the highest

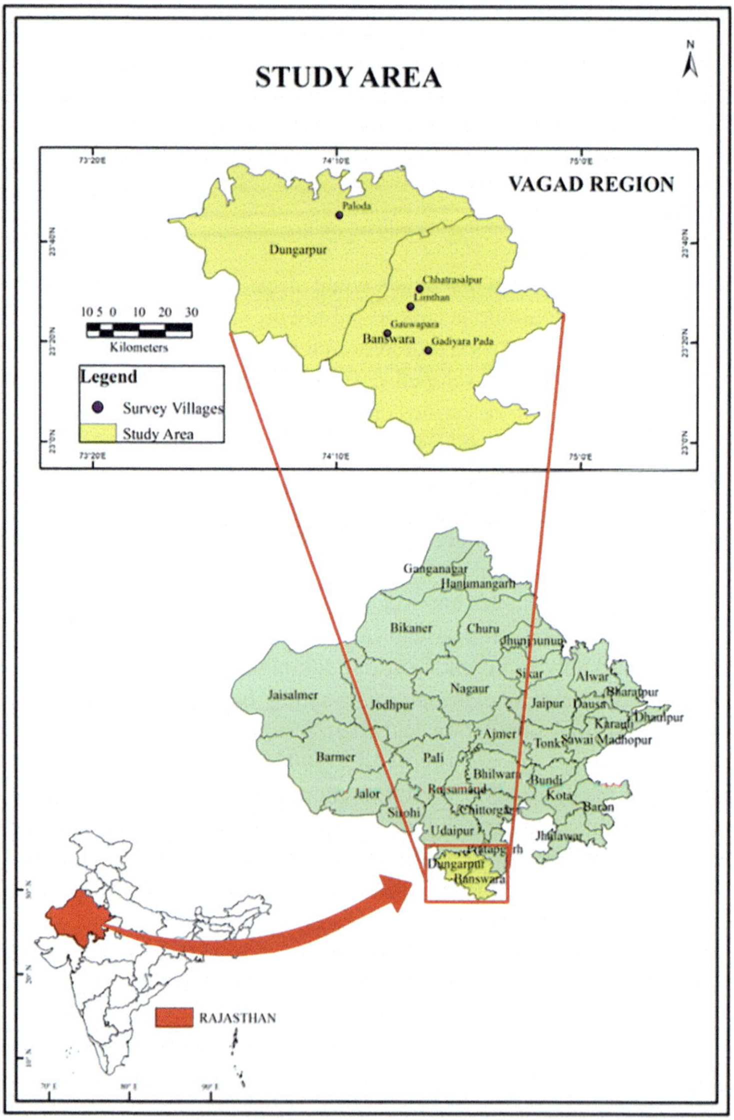

Fig. 1 Study area. (Source: Authors 2023)

temperature is between 40° and 45° C and the lowest temperature in the winter is between 10° and 12° C.

For much of the area, the monsoon season, which lasts from June to September, offers practically the only rainfall. The region is rich in flora and fauna. Around 20% of the geographical area of the region has forest. Dry teak forest is found here. It also has mainly bamboo trees. Animals like Chinkara, Leopard, snakes, squirrels, and four-horned antelope are also found here. Maize is the main crop in the area. The farmers also grow soybeans, rice, and some vegetables. The land size holdings are tiny, i.e., 70% of farmers have less than 1 acre of land. The farmers are dependent upon old traditional methods of agriculture like bullock tilling and family labor are employed to do work on farms.

Material and Methods

This chapter is based on a field survey done in the Vagad region in five villages and 200 households through a random sampling method. A well-structured closed-ended questionnaire has been administered in the area. The respondents gave their responses to questions linked to the Food Insecurity Experience scale. They were asked to share their experience by opting between never, rarely (once or twice), sometimes (three to ten times), and often (more than ten times) in the last 12 months. FIES measures the experience of individual respondents or respondents' household basis food insecurity. Because it can be calibrated against a global reference standard, the Food Insecurity Experience Scale (FIES) methodology is currently the only household or individual food security assessment approach that guarantees global comparability of the measurements (see Cafiero et al., 2018).

Item Response Theory (IRT), a subfield of statistics that allows for the evaluation of unobservable qualities through the examination of test and survey responses, provides the methodology for analyzing FIES data. Food security can only be quantified by looking at its observable manifestations because, like attitude or intelligence, it is an intrinsically unobservable trait. The Rasch model, which is frequently used in psychology, education, and health, is the specific IRT model that was applied to the FIES data. To (1) evaluate whether a set of survey questions (or "items") is appropriate for creating a measuring scale and (2) compare a scale's performance across various populations and survey situations, the Rasch model offers a theoretical foundation and a set of statistical techniques. The analysis of FIES data involves the parameter estimation and the calculation of the severity of food insecurity associated with each survey item and each respondent.

FIES measures the experience of individual respondents or their household basis food insecurity. The scale is currently the only household or individual food security assessment method that ensures global comparability of the measures.

- The typical FIES survey module contains **eight questions** that individuals can easily self-report. These questions cover a range of severity related to food insecurity.
- The FIES analytic protocol transforms qualitative information (e.g., "yes/no" answers) into quantitative, rigorous measures of food insecurity severity.
- Respondents are categorized into classes of moderate and severe food insecurity.

The respondents who did not give the response "never" were food insecure and were further posed 13 questions linked to the Emotional Scale of Food Insecurity. The data thus acquired has been analyzed and illustrated through tables and statistical diagrams.

Theory and analysis: Food insecurity is associated with a 257% higher risk of anxiety and a 253% higher risk of depression. Losing a job during the pandemic is associated with a 32% increase in risk for anxiety and a 27% increase in risk for depression (Fang et al., 2021). Among studies published in the previous 5 years, food insecurity was significantly and positively associated with multiple indicators of psychological distress (Myers, 2020). Ballard et al. in a report entitled Development of a Global Standard for Monitoring Hunger Worldwide highlighted that the Food Insecurity Experience Scale (FIES) is expected to make an important contribution to any suite of food and nutrition security indicators. It is an experience-based measure of food insecurity severity based on respondents' candid answers to a series of inquiries about their ability to obtain enough food. The FAO project Voices of the Hungry (VOH) sought to investigate if the FIES may serve as a new worldwide benchmark for gauging food insecurity. The FIES measure of the severity of food insecurity can be used in surveys that allow disaggregation at subnational levels and across different population groups, making it possible to identify more precisely who the food insecure are and their geographic distribution. This is in contrast to aggregate measures like the FAO Prevalence of Undernourishment or the Global Hunger Index (Ballard et al., 2013).

The previously mentioned ethnographic research, carried out in the USA by Radimer et al. (1990, 1992), revealed that the experience of food insecurity is characterized by changes in the quality of the diet as the situation worsens, such as a less balanced, more monotonous diet. The experience of food insecurity is first characterized by uncertainty and anxiety regarding food access.

With increasing severity, the quantity of food consumed decreases as portion sizes are reduced or meals are skipped (Radimer et al., 1990, 1992). Additional ethnographic studies aimed at understanding the experience of hunger from the perspective of the elderly in the USA and low-income families in Quebec, Canada, revealed similar patterns (Wolfe et al., 1998, 2003; Hamelin et al., 2002).

Results

Food Insecurity Experience Scale: FIES measures the experience of individual respondents or respondents' household basis food insecurity. FIES is an official instrument by the Food and Agriculture Organization.

The FIES highlights and brings out the emotions of people by targeting eight questions. These questions reflect their state of being worried, anxious, apprehensive, afraid, or concerned. The scale also pinpoints respondent's opinions on the quantity and quality of the diet they consume.

The typical FIES survey module contains eight questions focused on food-related behaviors and experiences associated with difficulties in accessing food due to resource constraints. These experiences or conditions are self-reported by individuals reached in the survey. The approach is based on the Rasch Model which is widely used in health, education, and psychology.

Table 1 clearly indicates that the Food Insecurity Experience Scale shows that 33% of households are worried that sometimes they will not have enough food to eat while nearly 26% of households are often worried about that. Around 25% of households are sometimes unable to eat healthy and nutritious food and 37% often experience this. Around 20% of households sometimes ate only a few kinds of food while 23% often. Skipping a meal is very common in this region. 31% of households often skip a meal while 35% of households sometimes skip a meal. 32% of households often ate less than they thought you should while 25% sometimes. 38% of households often ran out of food and 20% sometimes ran out of food. The

Table 1 Food insecurity experience scale

	Food insecurity experience scale (FIES)	Percentage of households			
S.no	Questions	Never	Rarely (once or twice)	Sometimes yes (three to ten times)	Often (more than ten times
1	You were worried you would not have enough food to eat	14	27	33	26
2	You were unable to eat healthy and nutritious food?	15	23	25	37
3	You ate only a few kinds of foods?	35	22	20	23
4	You had to skip a meal?	14	20	35	31
5	You ate less than you thought you should.	28	15	25	32
6	Your household ran out of food?	10	32	20	38
7	You were hungry but did not eat?	15	28	25	32
8	You went without eating for a whole day?	30	27	28	15

Source: Primary Survey

experience is worse with around 32% of households who were often hungry but did not eat while 20% sometimes experience this. 15% of households went without eating for a whole day while 28% sometimes went without eating for a whole day. The households who did not respond "Never" are considered to be moderately or severely food insecure. Furthermore, a total of 140 households have been administered the questionnaire for emotional analysis.

Emotional Scale of Food Insecurity (ESFI): According to a 2002 study by Hamelin et al., the experience of food insecurity is dynamic, highlights the monotony of the diet, characterizes the feeling of estrangement, and distinguishes between the reactions that come with not eating and the actual absence of food. The scale was developed based on the several aspects of food insecurity that the FAO listed in its definition. Respondents have been questioned about their feelings regarding food insecurity.

The scale is made to understand the level of sadness and motivation in food insecure households. The scale consists of 13 queations and all questions have been asked to quantify the emotions of food insecure households.

The villages surveyed have close small communities. People believe in exchange for food and spend a lot of time with each other. Therefore the above questions also analyze the emotions of respondent for their households and their community. Table 2 presents that people are emotional not only for their own families but also for their neighbors and known people. 57% of respondents feel sad when their families do not get a meal to eat. 60% of respondents feel sad when they do not have sufficient food to eat while only 35% feel sad when they do not get nutritious food to eat. This is because of their dietary habits and lack of sufficient knowledge about nutritious food. 45% of respondents feel sad when they consume unhygienic food or unsafe water. The people of the Vagad region have a natural source of water but are not very aware of safety concerns. The food is usually home-cooked and grown on their farms, so not many of them consume unhygienic food. 60% of the respondents feel sad when the family members fall sick while 35% sometimes feel bad when they are severely sick and not always. Lack of food affects the motivation of people in either way, i.e., either people get motivated or they get unmotivated to earn more. The region witnesses the old traditional form of agriculture and does not have enough job opportunities to explore. 45% of people feel motivated to earn more. Many people send their male members to nearby towns to earn more or for seasonal jobs. 25% of people mentioned that they feel unmotivated and helpless. Many such households have old people in their house and due to the rigidity of thought, and attachment to the land and work they do not feel that family members should move out.

The emotional analysis for the neighbor or known shows that only 10% of respondents feel sad when their neighbors skip a meal and 20% feel sad when they do not get sufficient food to consume. 5% of them feel sad when their neighbors do not get safe food and 35% feel sad when they fall sick due to consumption of unsafe food and water.

The final question for the respondents was, "What are the other emotions besides sadness you feel when you do not get sufficient and safe food to eat every day?"

Figure 2 shows that approximately 75% of people experience anxiety and insecurity. Anguish is the most common emotion, with hatred coming in second. About 25% of people experience demoralization, while 22% report feeling sad, or unhappy.

Table 2 Emotional scale of food insecurity in food insecure households

S. no	Question	Responses (Percentage to total food insecure household)		
		Yes	No	Sometimes
1.	I feel sad when my family doesn't have a meal	57	8	35
2.	I feel sad when my neighbor/known doesn't take a meal	10	25	65
3.	I feel sad when our food is not enough	60	5	35
4.	I feel sad when my neighbor or known doesn't have enough food	20	45	35
5.	I feel sad when the food my family consumes is not nutritious	35	20	45
6.	I feel sad when food my neighbor/known consume is not nutritious	15	55	30
7.	I feel sad when we consume unhygienic food or unsafe water	45	10	45
8.	I feel sad when my neighbor/known consumes unhygienic food or unsafe water	25	45	20
9.	I feel sad when we consume food other than our dietary habit	55	20	25
10.	I feel sad when my neighbor/known consume food other than their dietary habit	0	45	55
11.	I feel sad when our family member falls sick due to unsuitable food or water consumption	60	5	35
12.	I feel sad when our neighbor/known falls sick due to unsuitable food or water consumption	35	45	25
13.	I feel motivated to earn more when my family doesn't have sufficient and safe food to consume	45	25	30

Source: Primary Survey

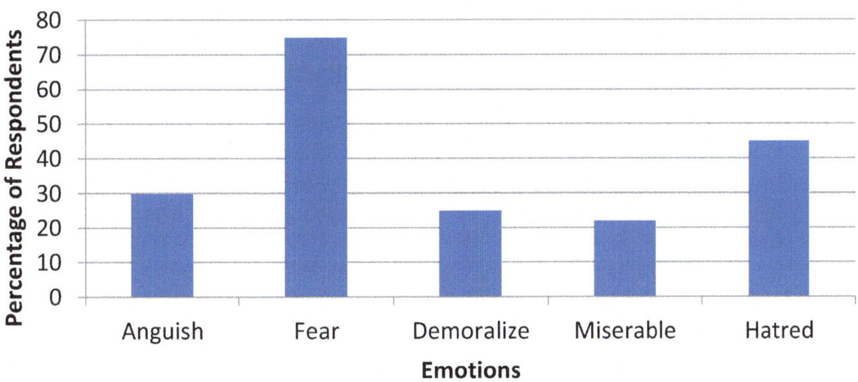

Fig. 2 Emotions linked to Food insecurity. (Source: Primary Survey)

There are numerous challenges faced by the people of this region. According to Krishi Vigyan Kendra, Banswara (kvk2.in), the people in this region face various challenges The majority of farmers are marginal (61.77%) and small (20.56%); most farms are rainfed; low-value, and low-demand crops predominate; there is little market support and no credit availability; livestock productivity is low; there is a dearth of oilseed and pulse productivity; there are few active cooperative milk societies; there is a lack of high-quality fish seed; and there are few opportunities for off-farm employment; Although there are many products grown in the district, including maize, soybeans, cotton, chiles, tomatoes, mangoes, etc., processing businesses have not yet developed. Middlemen continue to receive a substantial share of the benefits. (banswara.kvk2.in)

The district is inhabited predominantly by scheduled tribes and has a very poor and miserable socio-economic status owing to age-old cultivation practices. Dungarpur has been identified as the most backward district by the National Institution for Transforming India, Govt. of India looking.

There is a dominance of the tribal population and its poor socio-economic conditions in the district. The nutrition profile and body mass index show that people are mild to severely underweight. In such conditions, they are easily susceptible to various deficiency diseases. The body mass index determines the poor health condition of people living in the region. Around 59% of randomly selected people have below-normal BMI. Underweight is linked to various health problems like osteoporosis; skin, hair, or teeth problems; feeling tired all the time, anemia, slow and impaired growth, etc.

According to a report by IFPRI, Poshan, there are 33.3% of women below normal weight, and around 76% of women of reproductive age have anemia.

A person can be roughly classified as underweight, normal weight, overweight, or obese using the BMI, which is a handy rule of thumb based on height and tissue mass (muscle, fat, and bone). Underweight (less than 18.5 kg/m^2), normal weight (18.5–24.9), overweight (25–29.9), and obese (30 or more) are the four main adult BMI classifications.

(Surf Report 2005). A study published by the *Journal of the American Medical Association* (*JAMA*) in 2005 showed that overweight people had a death rate similar to *normal-weight* people as defined by BMI, while underweight and obese people had a higher death rate. World Health Organization considers an adult BMI of less than 18.5 as underweight and may indicate malnutrition, an eating disorder, or other health problems, while a BMI equal to or greater than 25 is considered overweight and 30 or more is considered obese. The average height in the region is 153 cm and the average weight is 45 kg.

Fig 3 shows that 44% of the respondents had normal BMI. Around 25% have mild underweight and 24% of respondents have moderate BMI. Around 7% of respondents have a severely underweight body mass Index.

The respondents in these regions face some major socioeconomic challenges. In the Vagad region around 92% of people feel that due to lack of facilities and job opportunities, people are now migrating to nearby towns. Students also go for higher education in other places and then take up jobs. Employment is a major concern in both regions. Almost more than 80% of respondents feel that lack of employment opportunities makes them deprived of having a better lifestyle. Land use change has also become more prominent in the Vagad region as people are shifting to other crops. People are growing safer crops rather than focusing on nutrition. Moreover, the agricultural land is getting saline so has been converted

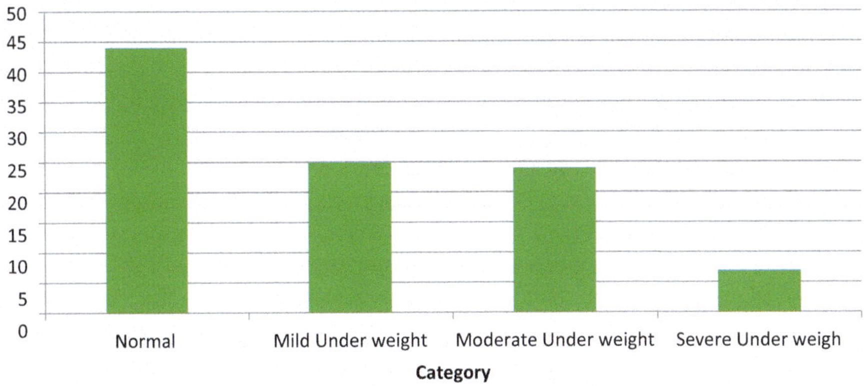

Fig. 3 Body mass index. (Source: Primary Survey)

into other activities. Beggary has been mentioned as a major issue in the Vagad region. Around 54% of respondents feel that there has been an increase in beggary in the region since the past decade. In the Vagad region, alcohol consumption is another major problem. People get alcohol at a reasonable rate and are ready to spend on alcohol rather than nutritious food. The crime rate also has increased in both regions. In the Bharatpur region, people specifically mentioned that poor irrigation facilities, water storage facilities, and government policies are responsible for the backwardness of the region. They feel isolated and ignored. There is a need for a comprehensive, holistic, and inclusive policy for the overall development of the region. Models for dryland farming and water resource management can be built up by micro-level data analysis. The deficiency of micronutrients needs to be studied in great detail, this would rather help to focus on adopting a wise cropping pattern. Agriculture has to give sufficient returns to farmers then they can focus on being innovative and using smart agriculture technology. Moreover Vagad region is predominantly a tribal area, it has more than 75% tribal population. The traditional Indigenous people living here are completely dependent on natural things like forests, traditional food habits, and naturopathy techniques. These people lead a very simple life. People of Vagad don't want to interfere in the natural settings of their area. According to many respondents, they feel more protective and comfortable in the way they are living. They want development but only sustainable development and essential things like electricity, food, housing, and some jobs. They do not want government or other authorities to interfere in their region.

Conclusion

Emotions and food are linked to each other. Food insecurity increases the risk of poor emotional health, and poor emotional health increases the risk of food insecurity (Bruening et al., 2017). In rural areas, dietary habits play a very important role as they are not easy to change or modify. Vagad region has a uniqueness of indigenous food. The indigenous vegetables like *chakod*, *chaulai*, and *luniya* have great nutritive value but all these crops are losing ground to new vegetables. According to a report published by Welt Hunger Hilfe, 2017, over the years farmers have switched to crop planning based on the market demands rather than on the practices and preferences of the local communities. Poor availability of grains and vegetables from their farms and have increased their dependence on markets and the Public Distribution System (food rations). The need for cash (money) has compelled the youth to migrate to adjoining states to work as wage laborers. All these have impacted the dietary diversity and consequently the nutritional status of the community, especially women and children (Banerjee, 2017). Psychological stress is often used to explain the effects of food insecurity on health risks, though no study has been able to tease apart the psychological stress associated with food insecurity, independent of poverty or other socioeconomic stressors (Weinreb et al., 2002; Whitaker et al., 2006). Well-being and health are interlocked; therefore it's important to research the link between the two. Food is a basic need for humans and is linked to physiological and psychological health. The traditional people should be encouraged to grow their indigenous variety of food on the one side and new scientific varieties on the other side. Regular health checkups are important for social well-being. Public health practitioners addressing mental health may consider screening for food insecurity for at-risk populations and vice versa. Better-constructed studies are needed to follow cohorts at risk for both food insecurity and poor emotional health (Bruening et al., 2017). Solidarity efforts in this region can:

1. Advocate for policy changes addressing historical underpinnings.
2. Support eco-cultural restoration activities to enhance cultural foodways.
3. Promote increased access to and consumption of native foods.

Solidarity requires empathy. Understanding the emotional impact of food insecurity helps build compassionate responses. Community resource mobilization plays a crucial role in promoting food access. By providing opportunities for those impacted by food inequities to develop and lead programming, we can mitigate food insecurity. Social solidarity—the sense of collective responsibility and mutual support—plays a central role in community efforts to improve food access. Activating social solidarity within community programs can enhance their effectiveness.

References

Ballard, T. J., Kepple, A. W., & Cafiero, C. (2013). *The food insecurity experience scale: Development of a global standard for monitoring hunger worldwide. Technical Paper*. FAO. Retrieved from http://www.fao.org/economic/ess/ess-fs/voices/en/

Banerjee, S. (2017). Farmers' campaign bears fruit in Banswara, Welt Hunger Hilfe. Retrieved from Farmers' campaign bears fruit in Banswara—Welthungerhilfe (welthungerhilfeindia.org)

Bruening, M., Dinour, L., & Chavez, J. (2017). Food insecurity and emotional health in the USA: A systematic narrative review of longitudinal research. *Public Health Nutrition, 20*(17), 3200–3208. https://doi.org/10.1017/S1368980017002221

Cafiero, C., Viviani, S., & Nord, M. (2018). Food security measurement in a global context: The food insecurity experience scale. *Measurement, 2018*(116), 146–152. Retrieved from https://www.sciencedirect.com/science/article/pii/S0263224117307005

Canetti, L., et al. (2002). Food and emotion. *Behavioural Processes, 60*, 157–164.

Fang, D., Thomsen, M. R., & Nayga, R. M. (2021). The association between food insecurity and mental health during the COVID-19 pandemic. *BMC Public Health, 21*, 607. https://doi.org/10.1186/s12889-021-10631-0

Guha, R. (1996). Savaging the civilised: Verrier Elwin and the tribal question in late colonial India. *Economic and Political Weekly, 31*(35/37), 2375–2389.

Myers, C. A. (2020). Food insecurity and psychological distress: A review of the recent literature. *Current Nutrition Reports, 9*(2), 107–118. https://doi.org/10.1007/s13668-020-00309-1. Retrieved from www.omicsonline.org

Radimer, K. L., Olson, C. M., & Campbell, C. C. (1990). Development of indicators to assess hunger. *Journal of Nutrition, 120 Suppl*(11), 1544–1548.

Radimer, K. L., Olson, C. M., Greene, J. C., Campbell, C. C., & Habicht, J. P. (1992). Understanding hunger and developing indicators to assess it in women and children. *Journal of Nutrition Education, 24*(Suppl 1), 36–45. repository.uwc.ac.za.

Thorat, A. (2010). Ethnicity, caste, and religion: Implications for poverty outcomes. *Economic and Political Weekly, 14*(51), 47–53. Retrieved from www.fao.org

Wolfe, W. S., Olson, C. M., Kendal, A., & Frongillo, E. A. (1998). Hunger and food insecurity in the elderly: Its nature and measurement. *Journal of Aging and Health, 10*(3), 327–335.

Wolfe, W. S., Frongillo, E. A., & Valois, P. (2003). Understanding the experience of food insecurity by elders suggests ways to improve its measurement. *Journal of Nutrition, 133*, 2762–2769.

Hamelin, A. M., Beaudry, M., & Habicht, J. P. (2002). Characterization of household food insecurity in Québec: Food and feelings. *Social Science & Medicine, 54*, 119–132.

Weinreb, L., Wehler, C., Perloff, J., et al. (2002). Hunger: Its impact on children's health and mental health. *Pediatrics*, 110, e41.

Whitaker, R. C., Phillips, S. M., & Orzol, S. M. (2006). Food insecurity and the risks of depression and anxiety in mothers and behavior problems in their preschool-aged children. *Pediatrics, 118*, e859–e868.

Corresponding Mental Health and *Prakriti* in Indian *Ayurvedic* Literature

Jyoti Yadav and K. N. Singh

Introduction

Ayurveda, the traditional Indian medicine system ponders *Panchmahabhuta* (i.e., *Jal, Prithvi, Agni, Akash, Vayu*) to be the cardinal charter of the physical universe comprehending the human body (Agnivesha., 2011). These *Mahabhuta* are evident in *Tridosha* (i.e., *Vata*, *Pitta*, and *Kapha*). Though all three *Dosha* endure in each human being, the pre-eminence of one determines the *Prakriti* of a solitary. *Prakriti* elucidates the peculiar traits of a person. At the time of birth, *Prakriti* is decided by the configuration of *Doshas*, so that every single person has different anatomical, physiological, and psychological characteristics. The term *Prakriti* is made up of two words "*Pra*" and "*Kri*" which means first creation it also means the position of an individual in its natural form. *Prakriti* of an individual marked by genetic and acquired factors. In the genetic factors, it is formed on *Shukra* (Sperm) and *Shonita* (Ovum), while acquired factors append climate, region, age, season, race, and heredity (Kalavade et al., 2023). *Prakriti* is the principal concept of *Ayurveda* which elucidates individuality. There are seven classes of different amalgamation of *Doshas* and the ascendancy of *Doshas* forms the various types of *Prakriti* which are *Vata, Pitta, Kapha* (dominancy of Single *Doshas*), *Vata–Pitta, Vata–Kapha, Pitta–Kapha* (*Dwandaj Prakriti-* dominancy of two *Doshas*), and the *Sama Prakriti* (dominancy of all the three *Doshas*). According to *Ayurveda*, the explanation of absolute health is, when all the vital elements of *Dosha* counting biochemical secretions and hormones in the human body, *Agni*—capacity of an individual to make the

J. Yadav (✉)
Centre for the Study of Regional Development, Jawaharlal Nehru University, New Delhi, India

K. N. Singh
Department of Rachana Sharir, Faculty of Ayurveda, Institute of Medical Sciences, Banaras Hindu University, Varanasi, India

K. Arora, R. R. Anand (eds.), *The Solidarity Approach in Geography*,
https://doi.org/10.1007/978-3-031-86753-8_11

devoured food obtainable for all the tissues, *Dhatu*—All the tissues in the human body and *Mala*—Squander product in the human body in an individual are in equipoise along with an exuberant soul and mind, he is said to be in the state of faultless health (Byadgi, 2009). An educated mind is uniformly important for the healthy condition of the individual. Body, soul, and mind are the three basic obelisks of a person. *Tridosha* is the basic notion of *Ayurveda*, *Vata, Pitta*, and *Kapha Doshas* are factors accountable for the healthy position of an individual. *Tridosha* always plays an important role in the physical constitution of an individual. Seven types of constitutions are formed with various amalgams of *Doshas*. These amalgams are known as *Prakriti* (Yadav & Singh, 2022).

Traits of *Vata Prakriti*: A person of the *Vataja* constitution is wakeful, they are violent and have a habit of biting their nails and grinding their teeth. They are the person with less patience and are capricious and heedless in their friendship, lanky, and lout. They possess garbled habits and irresolute anger. Their minds are stable, talk incoherently. In *Asthanga Hridya,* the features underlined are they possess a short memory, average or below average intelligence, short life span, and sleep little. They are not steadfast, cannot control their senses, give quick reactions, are short-tempered, better grasping power but short memory.

Traits of *Pitta Prakriti*: Inability to face a difficult situation. Moderate strength, moderate life span, moderate spiritual knowledge. According to *Shusruta,* they are averse to warmth and irascible temper, though lower down easily. They possess standard power and love to dominate the conversation. They are robust and alluring in battle. They are brave, are of good behavior, and have the mental power to face fear and enemy.

Traits of *Kapha Prakriti*: The Kapha individual has a pleasing appearance, slow in intake of food, slow in movement, slowness in initiating actions, happiness and softness in voice, good strength, good knowledge, peace, and long life span. They love the sweet taste, are appreciative, courageous, patient, non-greedy, and powerful, and do not form any conviction briskly and quickly in antagonism. Enough strength to bear the pain, stress, and reverent toward their senior, believe in religious text dogged in their friendship, veracious toward their speech, and always acquiescent on the way to their instructors.

Dvandaja Prakriti: Amalgamation of two different *Doshas* in an approximately equal manner is known as *Dvandaja Prakriti*. Like the combination of *Vata–Kapha, Vata–Pitta*, and *Pitta–Kapha*.

Sannipata/Sama Prakriti: When a person is characterized with all three *Doshas* together known as *Sannipata*. Who have the characteristics of *Vata–Pitta–Kapha Prakriti* (Gadre, 1963; Atridev, 2018; Ghanekar, 2017).

Ayurveda, the system of Indian medicine is not only an herbal tradition, but also a science with a holistic approach that focuses on overall development which includes physical and mental health. It has been seen that Mental Health issues are the most blazing problem nowadays. Diminishing Mental Health status leads to insecurity and reduced future potentialities. So, it is important to keep a bird's eye on mental health. According to WHO, mental health is the complete physical, social psychological, and intellectual well-being of the individual not just the dearth of disease or infirmity. It is a state of well-being in which one can realize their potentiality, handle the normal stress of daily life, work productively, and able to contribute to their community (World Health Organization, 2004). For the current study, six mental health determinants are chosen to know the mental health status of an individual which include-

- **Positive Self-Evaluation (PSE)**: It incorporates the attitude about your work and skills, a sense of acceptance, trust, and control in your personal life, to accept yourself and all your personality characteristics as they are whether it is positive or negative, self-identity, the internal sense of being worthy, and self-actualization of own potential.
- **Perception of Reality (PR)**: It includes how an individual sees the world around them, discernment free from perversion, omission of exorbitant imagination, and the comprehensive look toward the world.
- **Integration of Personality (IP)**: It represents the equilibrium of the psychic forces within the person and encompasses many abilities such as empathy, focus while working, and a variety of interests.
- **Autonomy (AUTNY)**: It entails having a set of consistent internal guidelines for behavior and relying on one's ability for growth rather than that of other people.
- **Group Oriented Attitudes (GOA)**: It is related to the attitude the person holds toward teamwork, cooperation, recreation, and aptness to get along with different people.
- **Environmental Mastery (EM)**: It includes how the individual deals the situational needs, one's leadership and responsibilities potential, and the ability to adjust (Jagdish & Srivastava, 1995).

Life experiences and individual factors govern the mental health of every single person differently. In continuation of individuality, *Ayurveda* has a basic prime notion of *Prakriti* which illustrates individuality. At the time of birth, *Prakriti* is decided by the compilation of *Doshas*. Therefore, every single person is different anatomically, psychologically, and physiologically (Agnivesh & Bala, 2017).

Materials and Methods

Mental Health Status is the chief etiological determinant for numerous diseases. If we could be in a position to predict the Mental Health Status of an individual, we would be capable of averting mental health-induced problems. The objective of the current manuscript is to show the correspondence between Mental Health Status and *Prakriti* in Indian Ayurvedic literature.

Population: For the present study, the educated youth of Varanasi City were administered between the age group 21 and 35 with 380 sample size.

Sampling Method: In the current study, cluster sampling techniques were used.

Assessment of *Prakriti*: For the *Prakriti* assessment of the sample, we have taken the *Prakriti* questionnaire which was developed by Prof. Kishor Patwardhan from the Department of Kriya Sharir, Faculty of Ayurveda, Institute of Medical Sciences, BHU, and later on modified by P.K. Tripathi, Prof. Kishor Patwardhan, and Girish Singh from the Department of Community Medicine, IMS, BHU, Varanasi.

Assessment of Mental Health Status: The Mental Health Inventory was used to assess the Mental Health Status of the selected sample which was developed by Dr. Jagdish, Dept. of Psychology, RBS PG College, Agra, and Dr. A.K. Srivastava, Dept. of Psychology, BHU, Varanasi. The Mental Health status was divided into five categories, i.e., Very Good, Good, Average, Poor, and Very Poor.

Statistical Techniques: A total of 380 respondents were taken from the study area, i.e., Varanasi City and ward-wise data was collected. Mental Health and *Prakriti* data was collected by using questionnaires and after that, it was evaluated by applying One Way ANOVA to see the correlation and mean differences.

Results

The total sample size was 380 including 204 Male and 176 Female respondents. The *Vata Prakriti* volunteers were 220, *Pitta Prakriti* had 34 volunteers and *Kapha* had 46 respondents. When we see *Dwandaj* (duel) *Prakriti* individuals, we have 32 respondents from *Pitta–Kapha Prakriti*, *Vata–Kapha Prakriti* had 09 volunteers, and *Vata–Pitta* had 22 respondents and in the category of *Sama Prakriti,* we had 12 volunteers.

A total of 380 sample sizes were taken for the current study (Table 1 and Fig. 1), we have a maximum number of volunteers from *Vata Prakriti* followed by *Kapha* and *Pitta Prakriti.* Moving toward *Dwandaj Prakriti*, we have the maximum candidates from *Pitta–Kapha* followed by *Vata–Pitta* and *Vata–Kapha*, and in *Sama Prakriti*, we have 12 number of respondents.

From the above figure (Fig. 2), it is limpid that we have the maximum number of respondents who had Poor Mental Health Status followed by very poor and average

Table 1 Type of *Prakriti* and no. of respondents

Prakriti		
Category	Numbers	Percentage
Vata	220	57.9
Pitta	39	10.3
Kapha	46	12.1
Vata–pitta	22	5.8
Pitta–Kapha	32	8.4
Vata–Kapha	9	2.4
Vata–pitta–Kapha	12	3.2

Source: Author's own

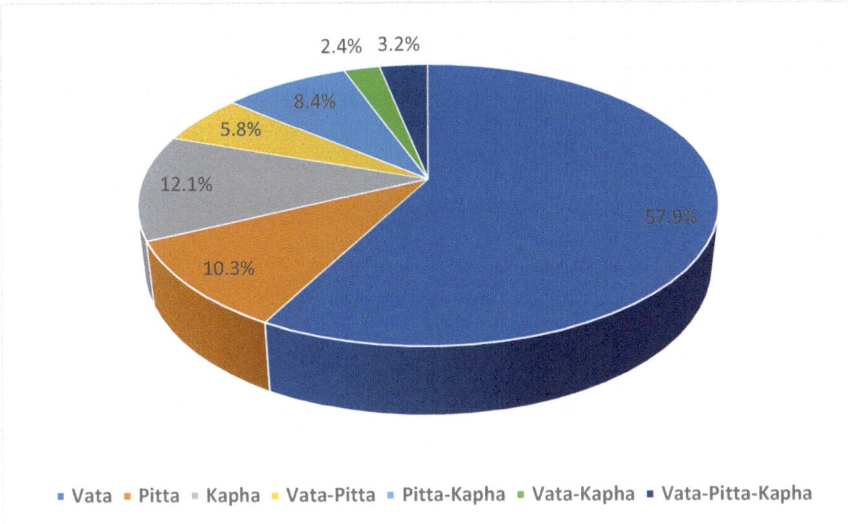

Fig. 1 Percentage distribution of Prakriti in the sample. (Source: Author's own)

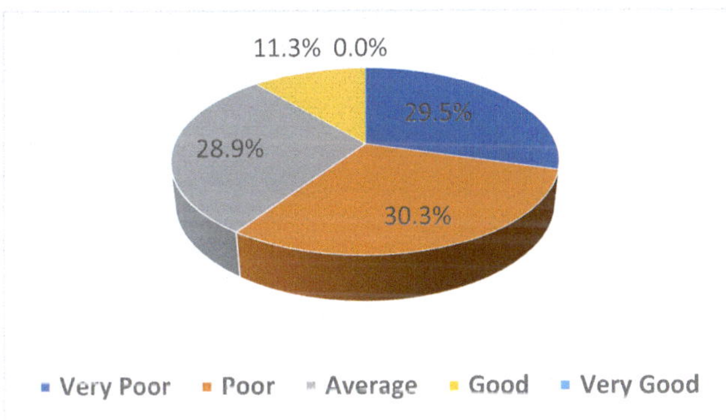

Fig. 2 Percentage distribution of type of mental health status in total sample. Source: Author's own

and it is important to underline here that we don't have any single respondent who had Very Good Mental Health Status.

From Table 2 and Fig. 3, it is limpid that *Kapha Prakriti* candidates have Good Mental Health Status followed by *Vata–Kapha* and *Sama Prakriti* while there is zero number of respondents falling in *Vata, Pitta,* and *Vata–Pitta Prakriti.*

Table 3 and Fig. 4 portray that, out of 380 candidates, 110 candidates had average Mental Health Status. *Pitta Prakriti* had a maximum number of candidates having average MHS followed by *Pitta–Kapha*.

From Table 4 and Fig. 5, it is pellucid that out of 380 respondents, *Vata Prakriti* has the maximum number of candidates having Poor Mental Health Status followed

Table 2 *Prakriti* wise distribution of sample having good mental health status

Prakriti	Number	Percentage
Vata	0	0
Pitta	0	0
Kapha	37	80.40
Vata–pitta	0	0
Pitta–Kapha	3	9.40
Vata–Kapha	1	11.10
Sama (VPK)	2	16.70
Total	43	100

Source: Author's own

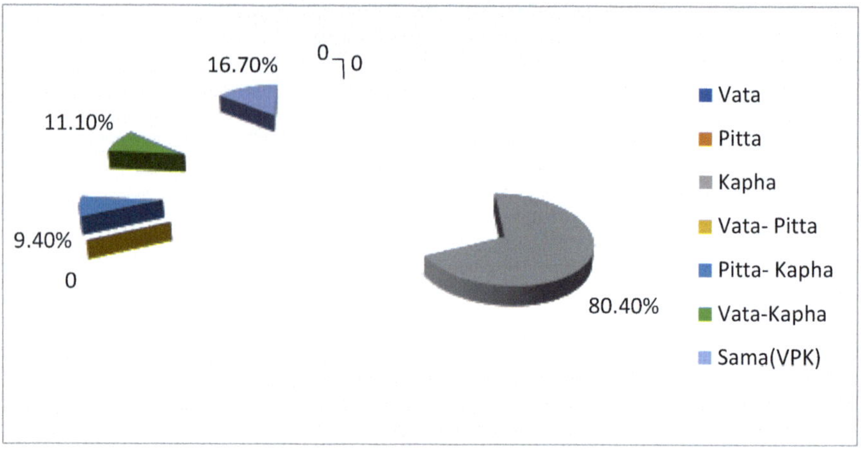

Fig. 3 Percentage distribution of *Prakriti* having good mental health status. (Source: Author's own)

Table 3 Prakriti wise distribution of sample having average mental health status

Prakriti	Number	Percentage
Vata	14	6.40
Pitta	38	97.40
Kapha	7	15.20
Vata–pitta	8	36.40
Pitta–Kapha	29	90.60
Vata–Kapha	6	66.70
Sama (VPK)	8	66.70
Total	110	100

Source: Author's own

by *Vata–Pitta* and we have a minimum of candidates from *Pitta, Kapha,* and *Pitta–Kapha Prakriti*.

Table 5 and Fig. 6 depict that *Vata Prakriti* had Very Poor Mental Health Status as we have a maximum number of candidates falling in this category followed by *Vata–Pitta*, whereas we have minimal respondents from *Pitta, Kapha,* and *Sama Prakriti* who had Very Poor Mental Health Status.

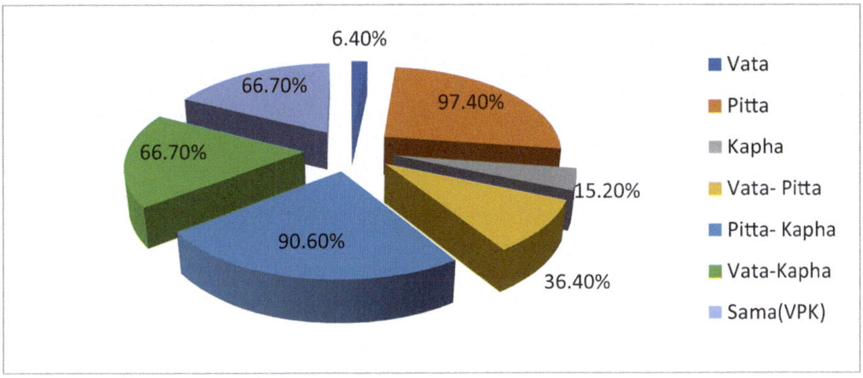

Fig. 4 Percentage distribution of *Prakriti* having average mental health status. (Source: Author's own)

Table 4 *Prakriti* wise distribution of sample having poor mental health status

Prakriti	Number	Percentage
Vata	101	45.90
Pitta	1	2.60
Kapha	1	2.20
Vata–pitta	9	40.90
Pitta–Kapha	0	0.00
Vata–Kapha	10	11.10
Sama (VPK)	2	16.70
Total	115	100

Source: Author's own

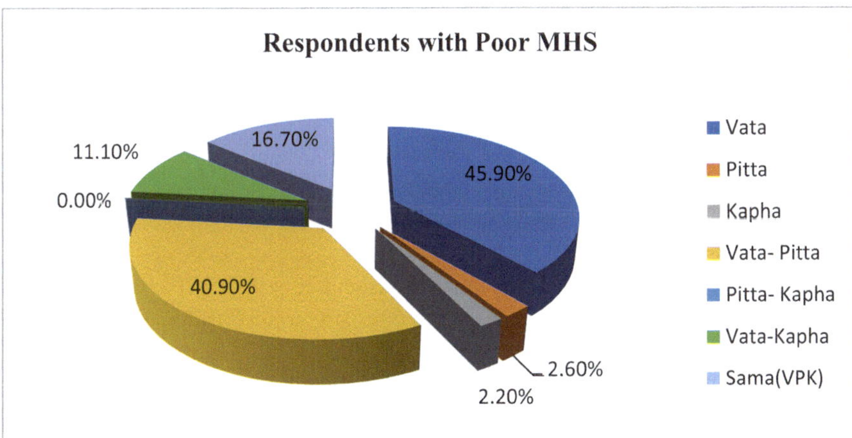

Fig. 5 Percentage distribution of *Prakriti* having poor mental health status. (Source: Author's own)

Table 5 Prakriti wise distribution of sample having very poor mental health status

Prakriti	Number	Percentage
Vata	105	47.70
Pitta	0	0.00
Kapha	1	2.20
Vata–pitta	5	22.70
Pitta–Kapha	0	0.00
Vata–Kapha	1	11.10
Sama (VPK)	0	0.00
Total	112	100

Source: Author's own

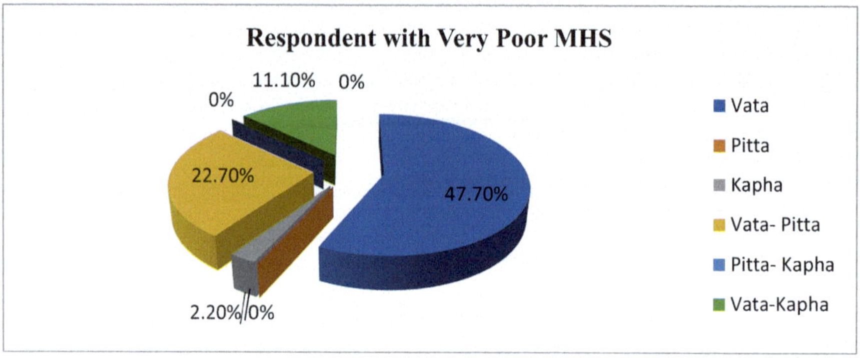

Fig. 6 Percentage distribution of *Prakriti* having very poor mental health status. Source: Author's own

Table 6 Correlation and mean between mental health and *Prakriti* of the respondents

Mental health			
Prakriti	N	Mean	Std. deviation
Vata	220	135.24	13.346
Pitta	39	167.59	6.332
Kapha	46	184.37	14.196
Vata–pitta	22	147.55	13.814
Pitta–Kapha	32	169.16	5.837
Vata–Kapha	9	161.00	23.885
Sama (VPK)	12	167.17	11.304
Total	380	149.69	22.474

Source: Author's own

Table 6 and Fig. 7 depict that the Mean of *Vata Prakriti* for Mental Health status is poor (135.24 ± 13.34) which means *Vata Prakriti* individuals had higher mental distress followed by *Vata–Pitta* (147.55 ± 13.81) and MHS is very good in *Kapha Prakriti* with Mean (184.37 ± 14.19) followed by *Pitta–Kapha Prakriti*.

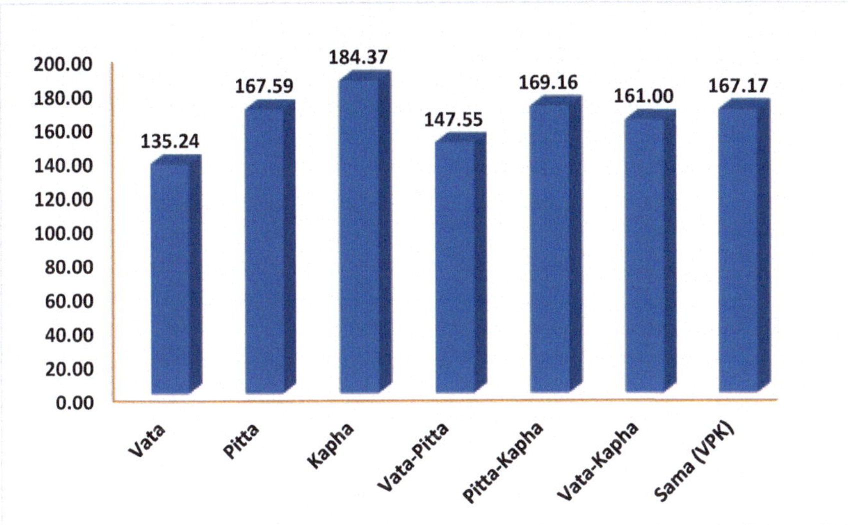

Fig. 7 Bar graph showing mean of mental health status. Source: Author's own

Discussion

Mental Health issues are an ineluctable factor in life. One must know the handling of it. When the mental health status is pestered homeostasis of hormones and supplemental important components of the body (*Dosha*) is also pestered. Physical or mental health problems such as heart disease, diabetes, depression, anxiety, migraine, and hypertension are stress and mental health problems. The psychosomatic amalgam of an individual describes his quintessential physical and psychological attributes (Yaribeygi et al., 2017). The psychological attributes of *Kapha Prakriti* persons are that they are very calm and steady, slow in their actions, they never get very much irritated, and they never get depressed easily. These kinds of individuals are diligent, forbearance, and never get vexed. They readily forgive everyone and are very squashy in voice and nature. They are mature, gentle, and passable people. Generally characterized by a gratifying appearance, contentment on the face, stability of the body and they have also good sleeping pattern (Tiwari & Tiwari, 2019). The psychological manifestation of *Pitta Prakriti* individual is brilliant, brave, and has sharp physical strength. It is not easy to conquer them easily. They are softened to bad and negative people while soft-hearted toward good people. These types of persons are the man of middling strength, mighty, and short-tempered. *Vata Prakriti* individuals are unstable, jealous, violent, impatient, and hot-headed. These individuals experience emotions such as irritability, anger, and fear. Intolerance towards cold things, they have less strength, their sleeping pattern is disrupted and their mind is never undeviating. The persons having *Dwandaj* (duel) *Dosha Prakriti* have psychosomatic features of both the *Dosha*. The supremacy of *Dosha* in dual *Dosha Prakriti* has more impact on the physical and psychological attributes of the person, whereas the individual having *Sama Prakriti* had manifestation of all the doshas. Because of their feature, the *Sama* and *Kapha Prakriti* are considered good *Prakriti* in all the different kinds of *Prakriti* as they are competent in coping the tough situations and psychological distress easily in juxtaposition to other *Prakriti* (Hetalben et al., 2019; Wani et al., 2017). The person who has ascendancy of *Kapha Dosha* in their combination is psychologically balanced and strong. Their mature, forbearance, soft, diligent nature makes them powerful enough to face arduous situations so they can handle mental health problems very well. Also, their quality of good sleeping patterns aids them in perpetuating their hormone level so they keep calm, are less stressed out, and have fewer mental health issues. The individuals having ascendancy of *Pitta Dosha* are mighty, have higher IQs, and are brave. All these aspects in the person make them sanguine enough for any unfavorable situation and challenges. Whereas the individual having dominancy of *Vata Dosha* are mentally unsteady, impatient, and intolerant due to the presence of all these aspects in an individual of *Vata Dosha Prakriti* the person is bad at handling difficult situations and stress that leads them to poor Mental Health Status (MHS). But if the person has *Kapha* dominant *dosha* like *Kapha Prakriti*, *Vata–Kapha Prakriti,* and *Pitta–Kapha Prakriti*, the individual has the advantage of *Kapha Dosha* quality in Mental Health Status (Amin & Vyas, 2016; Joshi, 2004; Shilpa & Murthy, 2011).

The volunteers for the current study are educated youth, being educated youth, they have multiple lifestyle pressures such as career, job, performance, etc., so nobody gauged themselves in Very Good Mental Health Status.

Aggregate 59.80% of individuals had Poor Mental health Status falling for *Vata* and *Vata–Pitta Prakriti*. Talking about those individuals having *Kapha* dominated (*Dosha*), *Prakriti* had Good Mental health Status. 28.90% of individuals had *Pitta* and *Pitta–Kapha Prakriti* falling for Average Mental health Status. The dominancy of *Pitta* and *Pitta–Kapha Dosha* individuals is maximum in Average MHS. The respondents who had dominancy of *Vata Dosha* are minimal in the group of *Dwandaj* (*duel*) *Prakriti* with Average MHS. A total of 11.30% of respondents fall for Good Mental Health Status and these respondents are from *Kapha*-dominated *Prakriti*. In Good MHS, there is dominancy of *Kapha Prakriti* followed by *Sama Prakriti* group. There is not a single respondent of *Vata Prakriti* who had a Good Mental Health Status.

Remarkably, the respondents having Good Mental Health Status had ascendancy of *Kapha*, the Average Mental Health Status had ascendancy of *Pitta* and the Poor Mental Health Status had Ascendancy of *Vata Prakriti*. The psychosomatic features of *Vata Dosha* specify with lamentable MHS. Certain fastidious conditions need prudent attention like hypertension, diabetes, anxiety, renal, respiratory disorder, and heart diseases. Many more related articles are at one's fingertips.

The finding of the present study shows that the *Kapha Prakriti* individuals had Good Mental Health Status and this result was strongly supported by the statement of *Acharya Sushruta* as described in chapter 4 *Sharir Sthana, Sushruta Samhita*. He quotes that *Kapha Prakriti* individuals are self-controlled, have a stable set of minds, do not hastily form any opinion, are true to their words, and are obedient to their preceptors. All these factors make *Kapha Prakriti* stay calm in day-to-day life or in any critical situation that's why they cope easily, this may be the reason why *Kapha Prakriti* individuals had Good MHS. When we observed *Pitta Prakriti,* the result showed that they have Average Mental Health Status and the result was supported by *Acharya Sushruta* as described in chapter "Application of Solidarity Approach to Space Identity Crisis: Issues of Muslim Pockets in Kolkata," *Sharir Sthana, Sushruta Samhita*, said that *Pitta Prakriti* individuals are averse to warmth, having irascible temper, though they cool down easily, they are the person of standard power and lives up to moderate age. All these factors make *Pitta Prakriti* strong enough to keep calm in a complex situation and maintain their mental health up to an average level (Yadav & Verma, 2018; Charak., n.d.).

Taking the *Vata Prakriti* into account the findings of the study show that *Vata Prakriti* individuals have Poor Mental Health Status, the *Ayurvedic* classic literature of chapter "Application of Solidarity Approach to Space Identity Crisis: Issues of Muslim Pockets in Kolkata," *Sharir Sthana, Sushruta Samhita*, he quotes that *Vata Prakriti* individual was impatient, unsteady, ungrateful, lean, and rough. They are muddled in their habit and have an irresolute temper, fast walker, talks incoherently, their minds are never steady hence the presence of all these features in *Vata Prakriti* the individuals are unable to cope-up with the tough situation easily and this may be the reason why this individual has Poor Mental Health Status (Bhishagratna, 2002). Regarding the above findings and statements, we have other modern several research articles (Bagali et al., 2016, Yadav & Singh, 2022) and so on supported this result efficiently.

Recommendations

From the result, it is clear that *Vata Dosha* individuals had a maximum occurrence of Poor MHS, whereas *Pitta* had average mental health status. As we know *Ayurveda* is a holistic approach to keep people fit in every manner so, it emboldens an individual to take food and satiate in those activities which are juxtaposed to their Prakriti to live a hale and hearty lifestyle. The revitalize therapies like *Vajikarana* and *Rasayana* help most when they are considered according to one's *Prakriti* (Laxman et al., 2020).

In *Ayurveda, Dashvidhapariksha* has been unraveled by *Acharya Charak* while diagnosing the solidity of the diseased individual. The first thing to be done is the *Prakriti* analysis while probing a patient. So that the timely prediction may be averted or respited by pertinent diet, regimen, and drugs. *Ayurveda* Diurnal regime (*Dincharya*) and Seasonal regime (*Ritucharya*) have been counseled based on individual *Prakriti*. The individual can espouse this regime as per the need of *Prakriti* to retain health (Khatri et al., 2022). As in the results we found that *Vata Prakriti* individuals had more mental distress, it is suggested that they do less exercise and consume nutritious and heavy food such as in the category of grains *Vata* individuals should take wheat, rice, lemongrass in cereals they could include black grams, green grams, sesame, horse gram, soybean. In vegetables, drumstick, radish, radish leaf, brinjal, tomato, onion, garlic, carrot, Indian bean, tritos cucunerinnas, ash gourd, amaranth, karatotaki, leptadenia reticulata, Cambs, quarters, elephant foot, asparagus, racemosus, nightshade, Indian spinach, Indian sorrel and in the category of spices *Vata* individual should prefer ginger, mustard, fenugreek, camphor, coriander, cumin seed, asafoetida, black pepper, dill seed, turmeric, celery leaves, fennel seeds, roots of piper longum, mint, cinnamon, nutmeg, poppy seeds, saffron, sacred basil, and in fruits Indian gooseberry, ripe mango, papaya, orange, pineapple, apple, banana, grapes, coconut, strawberry, coconut, cherry plum, lemon, pomegranate, jack fruits. Citrus media, sweet lime, pear, sugarcane, ripe mulberry, musk melon, custard apple, gromia, ripe wood apple, dates, cashew nuts, chironji sapida, walnut, pistachio nut, almond, and castor oil, seed oil, sesame oil, rape seed, coconut oil, cotton seed oil, sunflower oil, safflower oil, white mustard oil, linseed oil should be more beneficial for *Vata* individual for being safe and sound.

Whereas *Pitta Prakriti's* diet includes wheat, rice, Barley, corn, jowar, kangu, and Japanese barnyard millet in the group of grain. When we talk about pubes one should have green gram, pigeon peas, horse gram, lentil, sesame, cows peas, aconite leaved kidney, peas, pink beans, buckwheat, soybean, and in the category of vegetable drum stick, radish leaf, bitter gourd, cabbage, potato, sweet potato, lady's finger, beetroot, Indian bean, cucumber, spinach, cauliflower, narrow-leaved sepistan, pointed gourd, vetches, round gourd, pumpkin, white gourd, ribbed gourd, mooth gourd, amaranth, karakotaki, leptadenia reticulata, yam, lambs quarters, asparagus, racemosus, red malabar nightshade, andropogon squarrosus while cloves, coriander, turmeric, murraya koenigii, poppy seeds, fennel seeds. Moving toward the list of fruits for Pitta Prakriti are pomegranate, jackfruit, Indian gooseberry, sweet lime, ripe mango, grapes, apple, sweet melon, banana, coconut, bael, rajadana, sugarcane, ripe jujube, plum, ripe mulberry, custard apple, gromia, apricot, cuddapah almond, water chestnut, walnut, dates, almond, betel nut and in the group of oil one should preferably consume sesame oil, coconut oil, groundnut oil, and safflower oil.

Kapha individuals should do more exercise and intake medium amount of food such as they preferably intake rice, millet, garlic, carrot, cauliflower pointed gourd, pumpkin, amaranth, karkotaki, lambs quarters, elephant's foot, nightshade, cissampelos Pereira, Indian sorrel, and in the category of spices cloves, cumin seed, green chilies, fenugreek, cardamon, asafoetida, red chilies, ginger, mustard, camphor, coriander, fennel seeds, black pepper, turmeric, murraya koenigii, dill seed, celery seeds, roots of piper longum, mint, cinnamon, sacred basil, nutmeg, saffron, soda bi carbonate while when it comes to consuming of fruits we should ideally go for Indian gooseberry, raw mango, pear, rose apple, grapes, pomegranate, citrus medica, walnut, pistachio nut, lemon, fig, betel nut and in the list of oil are castor, sesame, sunflower, safflower, rape seed, and linseed oil to keep themselves fit and healthy (Amin et al., 2014; Pimparkar et al., 2019; Bhat & Lavekar, 2005).

Besides the diurnal and seasonal regime, it is also proposed that the solidarity approach in the environment and society plays the foremost part in nurturing our mental health and physical well-being balance. Nowadays people focusing more on drug treatment which may not be the way to permanent recovery. Currently, the focal point of mental health professionals is on fortifying individual responsibility, helping their habitue to make sound behavioral preferences rather than modifying the surrounding environment that confines choices and restricts opportunities for espousing wholesome behaviors. However, corroboration suggests that in comparison to individual behavior health is shaped maximum by social and economic structures. One of the survey studies conducted by Denton and Walters in the year 1999 included more than 15,000 adults for their health status and set side by side with behavioral factors (like smoking, physical activity, diet, and drinking) and with structural factors (including—occupation, social support, education, and income). From the result, they came to know that health status is largely determined by structural determinants rather than lifestyle determinants.

If the unfavorable living conditions that cause poor health persist, changing behavior alone will not lead to health. An increasing and encouraging body of research shows that, in addition to improving behavior from the inside, organized social interventions and changing the environment's organizational requirements that boost individual performance and interpersonal abilities might benefit mental health.

The Social Basis of Health—One of the psychotherapy research results that spanned many decades shows that 40% of improvement in habitue or patient resulted from extra therapeutic determinants—healthy and sound living conditions, encouraging structure in the surrounding environment, active involvement in the community, and gratifying work. Another 30% was attributed to therapist idiosyncrasy regardless of theoretical intention-invigorate, compassion, empathy, unwavering acceptance, positive look, and mutual affirmation. The rest of the factors were hope, anticipation, and the placebo effects.

Significant human interconnection and close, gratifying social relationships are important for mental health. The behavior of an individual does not occur in a void but it is administered by the casual consequences of environmental conditions, for example, even intelligence can be described, in large part, by environmental and social determinants. After World War II, the children born to White and Black soldiers in Germany had equivalent IQs, whereas the circumstances in the USA, where the IQ test scores of blacks tend to be lower than the scores of whites. It is proposed that similar opportunities, unbiased attitudes, and integration into the community could be possible determinants describing this contrast.

The Role of Environmental Cohesion—Social amalgamation and closeness to the community decreases the feeling of estrangement, agitation, wariness, and victimization that result from degenerating living conditions. For instance, Ross and Jang in 2000 take 2842 samples in Illinois for their study (telephonic interviews). Adjusting for sociodemographic factors, they came to know that understanding and encouraging communities, informal networks and entente among neighbors gives mutual aid and support that buffered the outcome of perceived neighborhood disorder—alcohol, crime, drug, and vandalism.

The physical features of the society vigorously affect neighborhood ties, which in turn, have a productive result on the mental health of residents. A public housing study conducted in Chicago manifests that the residents having surroundings with grass and trees have more informal neighbor ties. Mutual affirmation, in comparison to those who are living in barren. The amount of greenery and vegetation around the residences and public areas was related to a lower rate of stress, mental weariness, and negative mood with more feelings of safety, possessions, adaptation, and satisfaction. These characteristics are representative of cohesive communities; they unnerve antisocial behavior and reinforce control over activities detrimental to the person.

Empirical research advocates that the power of environmental and social determinants in generating and controlling physical and mental illness has been underrated. In lieu, the drug treatment and biomedical explanation have carried the day. There is significant evidence that the society and environment play a remarkable role in the outset and enhancement of mental health and illness. However, the significance of structural changes is not fully appreciated by those who command the economy, the medical profession, and the government. The medication for the treatment of the mentally ill carries the illusion that the reason for illness stems from the person. In the world of medical terms human relationships, affirmation, and conflicts may become lost. The way an individual behaves and thinks is affected, to a greater extent, by interlocking structural factors: technological development, economic situation, political organization, social changes, family patterns, and social compartmentalization. These situations can perpetuate, provoke, and nurture the course of illness. More efficacious than personal intervention is the team-up efforts of all sections of society to annihilate illiteracy, social alienation, unemployment, economic inequality, and poverty and to nourish social solidarity and the pledge of a safe environment. Thus, society needs to support much-required prevention and community-based primary care services and aid structural changes that have exhibited effectiveness (Thakkar et al., 2011; Nikelly, 2011).

Conclusion

Anxiety tests, IQ tests, and many psychological tests have been usual appraisal tools in our society for numerous years. If we gauge the *Prakriti* of the individual at initial age, we can get a hint concerning their mental health status. Management of mental health issues is very indispensable in this modern developing world. The youth is under prodigious mental health-related issues due to their career, education, and jobs. With the aid of unerring *Prakriti* assessment, we can circumvent negative mental health-induced problems. By using the solidarity approach society should emphasize achieving social and environmental solidarity. A positive and friendly environment with social warmth helps in reducing mental health distress.

Acknowledgments We thank Bajrang B. Yadav (Biostatistician) and Divya Kumar Jaiswal (Counsellor) for their undaunted support.

References

Agnivesh, & Bala, D. (2017). Charak Samhita. In B. Tripathi (Ed.), *Vimansthan* (p. 759). Chaukambha Surbharti Prakashan.

Agnivesha. (2011). Charaka Samhita. In D. C. Dutta & D. Bala (Eds.), *Chikitsasthana. 9th chapter, verse 42* (p. 473). Chaukhamba Surabharathi Prakashana.

Amin, H., & Vyas, M. (2016). Evaluation of Pradhana Sharira and Manas Prakriti (bodily and mental constitution) on disease manifestation in generalized anxiety disorder. *International Journal of Clinical and Experimental Physiology, 3*(4), 171–179. https://doi.org/10.4103/2348-8832.196895

Amin, H., Sharma, R., et al. (2014). Recommendation of Ayurvedic diet and lifestyle guidelines for different Prakriti (constitution) people. *Rasamutra, 6*(23), 1–13.

Atridev, G. (2018). Ashtang Hridya. In Y. Upadhyay (Ed.), *Sharirsthan* (pp. 103–262). Chaukhamba Surbharati Prakashan.

Bhat, S., & Lavekar, G. S. (2005). Ayurvedic approach to Pathya (ideal diet planning)—An appraisal. *Bulletin of the Indian Institute of History of Medicine, Hyderabad, 35*(2), 147–156. https://pubmed.ncbi.nlm.nih.gov/17333670/

Bhishagratna, K. (2002). *Susruta Samhita*. In L. Dwivedi (Ed.), *Sharir Sthana* (pp. 180–182). Chowkhamba Sanskrit Series Office.

Byadgi, P. S. (2009). *Ayurvidya Vikriti Vigyan & Roga Vigyan II* (1st ed., pp. 1–39). Chaukhambha Publications.

Charak. (n.d.). *Charak-Samhita (with Chakrapanidatta commentary, chapter 8, verse 16)*. Chaukhambha Sanskrit Sansthan.

Gadre, R. K. (1963). *Ashtanga Hridaya of Vagbhata; Sharir sthanam: Anga Vibhagam (Chapter 3, verse 82)*. Aryabhushan Mudranalaya.

Ghanekar, B. G. (2017). Sushrut Samhita. In *Sharirsthan. Garbha vyakarana Sharir* (p. 141). Meharchand Lachhmandas Publications.

Hetalben, A., Vyas, H., et al. (2019). Role of Pradhana Sharira and Manas Prakriti on the manifestation of hypertension: A cross-sectional survey study. *Journal of Ayurveda Medical Sciences, 4*, 460–466. https://doi.org/10.5530/jams.2019.4.3

Jagdish, & Srivastava, A. K. (1995). *Manual for mental health inventory* (pp. 1–9). Manovaigyanik Parikchhan Sansthan, Chowkaghat.

Joshi, R. R. (2004). A biostatistical approach to Ayurveda: Quantifying the Tridosha. *Journal of Alternative and Complementary Medicine, 10*(5), 879–889. https://doi.org/10.1089/acm.2004.10.879

Kalavade, V. A., Mane, P., et al. (2023). A clinical observational study on Prakriti analysis and its application in Ayurveda. *Journal of Pharmaceutical Negative Results, 31*, 2215–2222. https://doi.org/10.47750/pnr.2023.14.02.272

Khatri, K., Kumar, B., Gautam, M., Bhatnagar, N., Vikash, et al. (2022). Dincharya and Ritucharya concerning circadian rhythm and sleep. *World Journal of Pharmaceutical Research, 8*, 53–56.

Laxman, S., Krishnarao, K., et al. (2020). Critical analysis of Rasayan activity of Dravyas in Ayurveda. *European Journal of Biomedical and Pharmaceutical Sciences, 7*(12), 189–199.

Nikelly, A. G. (2011). The role of environment in mental health: Individual empowerment through social restructuring. *The Journal of Applied Behavioral Science, 37*(3), 305–323. https://doi.org/10.1177/0021886301373004

Pimparkar, P., Chondikar, P., et al. (2019). Conceptual study of diet for different Prakriti. *Journal of Ayurveda and Integrative Medical Sciences, 6*, 147–153. https://doi.org/10.21760/jaims.v4i06.777

Sachin S. Bagali, & Umapati C. Baragi. (2016). Concept of Manasa Prakruti as described in Charaka Samhita. Journal of Ayurveda and Integrated Medical Sciences, 1(03), 81–86. https://doi.org/10.21760/jaims.v1i3.4423.

Shilpa, S., & Murthy, V. (2011). Understanding personality from Ayurvedic perspective for psychological assessment: A case. *Ayu, 32*(1), 12–19. https://doi.org/10.4103/0974-8520.85716

Thakkar, J., Chaudhari, S., et al. (2011). Ritucharya: Answer to the lifestyle disorder. *Ayu, 32*(4), 466–471. https://doi.org/10.4103/0974-8520.96117

Tiwari, L., & Tiwari, S. K. (2019). Scientific assessment of personality based on Doshaja and Manas Prakriti. *International Journal of Research in Medical Sciences and Technology, 8*, 123–126.

Wani, B. A., Andal, S. K., et al. (2017). Prakriti analysis and its clinical significance. *International Journal of Ayurveda and Pharma Research, 8*, 86–90. https://www.ijaprs.com/index.php/ijapr/article/view/775

World Health Organization. (2004). *Promoting mental health: Concepts, emerging evidence, practice (summary report)*. World Health Organization. Retrieved June 31, 2023, from https://iris.who.int/bitstream/handle/10665/42940/9241591595.pdf

Yadav, J., & Singh, K. (2022). Relation between Prakriti and mental health: An observational study. *International Journal of Current Research and Review, 14*(1), 1–7. https://doi.org/10.31782/IJCRR.2022.142101

Yadav, J., & Verma, S. (2018). *Sharir-Kriya Vigyan* (pp. 57–116). Chowkhambha Orientalia.

Yaribeygi, H., Panahi, Y., Sahraei, H., et al. (2017). The impact of stress on body function: A review. *EXCLI Journal, 16*, 1057–1072. https://doi.org/10.17179/excli2017-480

Toward Rural Social Sustainability: An Index to Assess the Sustenance Potentialities

Vipasa Gupta and Soma Sarkar

Introduction

In recent decades, sustainable development has gained prominence in policy formulation and practice, particularly in the context of urban areas. For countries such as India, where 69% of the population still resides in rural areas (Census of India, 2011), addressing rural distress and agrarian crises has become imperative. Unsustainability and underdevelopment in rural areas have consistently hindered a nation's overall progress and dynamism. Notably, sociologist Gupta (2004:11) once asserted that "the village is no longer a site where futures can be planned," attributing this shift not to the diminishing significance of villages but rather to the predominance of urban-focused political debates (Thakur, 2014). The COVID-19 pandemic further revealed that urbanization alone does not guarantee prosperity. The sight of circular migrants walking back to their distant villages during lockdowns demonstrated that not every urban dweller finds a secure foothold. These individuals are typically forced to migrate to metropolitan areas when they cannot sustain themselves in rural spaces. The pandemic laid bare the uneven and unplanned development of urban areas and the lack of viable rural alternatives to support millions.

Strengthening rural sustainability can make rural areas more appealing to their inhabitants, thus, reducing the influx of rural-to-urban migration. Mahatma Gandhi, a prominent social reformer, envisioned rural sustainability as the solution to India's overcrowded urban centers through his concept of *"village swaraj"* (Gandhi, 1963). Gandhi envisioned self-reliant and locally self-governed villages characterized by

V. Gupta
Centre for the Study of Regional Development, Jawaharlal Nehru University, New Delhi, India

S. Sarkar (✉)
SGA, Dr. B.R. Ambedkar University Delhi, Delhi, India

K. Arora, R. R. Anand (eds.), *The Solidarity Approach in Geography*,
https://doi.org/10.1007/978-3-031-86753-8_12

truly sustainable economies. A village is truly sustainable when it possesses all the essential conditions for community living, respects the values of its people, values its environment, provides economic opportunities for all, and actively engages in community improvement. A resilient rural community is the fundamental component of a sustainable village, as articulated by Bryden (1994), who identified it as a sustainable social system with the long-term capacity to regenerate itself economically, socially, culturally, and ecologically. Taking a broader perspective, Bryant (1995) defined sustainability and community development as,

> a process by which the community attempts to influence the processes affecting various activities in the community to improve the quality of life of its residents in an enduring way, compatible with the natural environment, socially and culturally acceptable, and economically feasible.

Since the release of the Brundtland Report in 1987 (WCED, 1987), sustainable development, often framed within the "*triple bottom line approach*", has sought to pursue economic growth while balancing environmental protection and social equity. Despite its inclusion as a crucial dimension in the triple bottom line, academic literature and research have frequently neglected the social dimension compared to economic and environmental aspects (Elkington, 1999; GRI, 2000; Barron & Gauntlett, 2002; Åhman, 2013). Rural sustainability has typically been assessed primarily through economic parameters while overlooking social aspects. However, some researchers have cautioned that a rural system cannot be considered sustainable if its social structure is undermined (Caldwell, 1984; Smit & Brklacich, 1989). Research on the social dimension of sustainability has addressed various issues, from housing problems (Chiu, 2003) to community resilience (Magis, 2010), rural and urban studies (Scott et al., 2000; Dempsey et al., 2011; McManus et al., 2012), and globalization (Koning, 2002).

To understand rural sustainability comprehensively, it is essential to embrace the concept of social sustainability, which encompasses the well-being and social cohesion of rural communities (Scott et al., 2000). Social sustainability recognizes that rural communities are intricately connected, relying on shared norms, values, and mutual aid. In this context, solidarity plays a crucial role. Solidarity among members of a rural community fosters resilience and promotes a sense of collective responsibility (Miller, 2017). Social sustainability is not limited to economic factors but also emphasizes social aspects (Scott et al., 2000), making rural areas more appealing to their inhabitants and reducing migration to urban centers.

Solidarity, as outlined by Spicker (2014), underpins societal cohesion and inclusion, making them interconnected elements rather than distinct entities. The underlying social construct binds individuals together, fostering mutual concern and mutual aid within the community (Miller, 2017). The notion of solidarity, as highlighted by Komter (2010), extends beyond mere ideology or normative ideals. It finds expression in various aspects of society, including humanitarian organizations, political debates, and religious discourses. It has been a core concept in socialist and communist ideologies, representing a fundamental ideal to strive for (Komter, 2010). Durkheim's perspective (Durkheim & Simpson, 1934) on social solidarity underscores its role in maintaining social order and stability by emphasizing the interdependence between individuals in society. It creates a sense that individuals can improve the lives of others, reinforcing the notion of collective action (Hechter, 2001).

In developing countries like India, villages are characterized by strong social systems, such as caste, class, culture, and religion, which even influence the morphology of rural settlements. The basic structure of Indian villages has long been determined by a core-periphery approach based on caste and class hierarchies, with higher castes or dominant classes residing in the core and subordinates in the periphery of the village (Mandal, 1979; Srinivas, 2015). Numerous studies have emphasized that the caste system cannot be ignored in Indian village studies; it is what defines the social reality of a village (Bailey, 1957; Beteille, 1965; Kessinger, 1971; Chakravarty, 1975; Harriss-White & Heyer, 2009). Historically, the role of caste or class has been studied concerning an individual's position in the village social structure (Srinivas, 1954), often neglecting its influence on crucial socio-economic factors such as occupational patterns, land use changes, and access to basic amenities like drinking water and sanitation. Village affairs are governed by ties of mutual obligations, established practices, and social ethics. Therefore, for a village to be truly sustainable, the core social bonds of the community cannot be disregarded and should be treated on par with its economic and biophysical dimensions.

The Government of India launched a scheme called "PURA" (Provision of Urban Amenities in Rural Areas) during the tenth 5-year plan. This plan identifies certain village clusters and aims to develop them as growth centers, providing employment opportunities and urban amenities through public–private partnerships (PPP) between Gram Panchayats (local government bodies responsible for resource disbursement at the village level) and private sectors. Similarly, the Pradhan Mantri Adarsh Gram Yojana (PMAGY), initiated on a pilot basis in 2009–2010, focuses on the integrated development of villages with a population of Scheduled Castes exceeding 50%. In 2014, the Saansad Adarsh Gram Yojana (SAGY) was launched, aiming to develop villages adopted by Members of Parliament (MPs) into "Adarsh Grams," with Gram Panchayats (village councils) as the basic unit. An "Adarsh Gram" is one where people have access to various basic services and is characterized by the upliftment of the poor, gender equality, social justice, ecological balance, peace and harmony, cooperation, self-reliance, and local self-government, among other attributes. Adarsh Grams are intended to act as central villages or growth centers for surrounding areas. Challenges in this regard include the selection of adopted villages or Gram Panchayats, where MPs are free to choose from their constituency or any district in the state from which they are elected. Additionally, identifying key indicators for situation analysis of the adopted villages poses a challenge, with village census data sufficient for the first challenge but requiring more grounded data for the second.

Previous efforts to address social sustainability have primarily focused on theoretical interpretations rather than quantitative assessments, as the latter are considered more rigorous. To comprehensively understand the social and biophysical dimensions of a village, the present study proposes a scoring system called the social sustainability index (SSI). The indicators of this index are selected based on the social sustainability framework of village systems. The index was applied to two villages in the Rajasthan state of India. This study strives to offer insights into micro-level perspectives on social sustainability while providing a measurement mechanism to address development gaps.

Objectives

To address the challenges of rural development in India and ensure the sustainability of rural communities, this study aims:

- To develop and validate the SSI for Indian villages.
- To explore the role of the solidarity approach in enhancing social sustainability in rural communities.

Materials and Methods

Selection of Villages

In this study, two criteria were considered for the selection of villages:

- Medium-sized village (population range of 2000–4999 as per the Census of India definition).
- At least 1/3rd of the population should belong to Scheduled Communities (Scheduled Castes—SC and Scheduled Tribes—ST).

In India, 34.64% of the population lives in villages of medium size, which is the highest in any category. Dausa and Tonk districts in Rajasthan, India, have nearly 10% of the villages classified as medium-sized, while 88% fall under the category of small villages and hamlets. Population size is always an important consideration for providing basic facilities, such as public healthcare services (Ministry of Health and Family Welfare-Government of India, n.d.), as these services are established in India based on population.

Moreover, the reason for considering scheduled communities is that these communities often face major inequalities in rural areas (Gorringe et al., 2017; Saxena & Bhattacharya, 2018). Selecting villages with a substantial Scheduled Community population allows us to explore the dynamics of social sustainability in contexts where inequalities are prevalent.

The population threshold in the study resembles the criteria set by the Saansad Adarsh Gram Yojana (SAGY) for plain regions. Owing to their population size and more equitable distribution in a region compared to larger villages, medium-sized villages have more opportunities to enable the concentration of human and institutional capacity support initiatives. They can take advantage of rural agglomeration economies while allowing sustainable benefits to spread out into their area of influence.

Based on these selection criteria, two villages (one from each district, randomly selected) were chosen: "Balaheri" from the Dausa district and "Soothra" from the Tonk district (Fig. 1).

Study Area

Balaheri Village (27.04° N and 76.82° E) falls under the Mahwa tehsil of Dausa district under Jaipur division, Rajasthan, India. The built-up or habitat area of the village covers approximately 4.2674 sq. km, surrounded by the Aravalli Mountain range to the west.

Soothra Village (26.01° N and 75.99° E) is situated in Uniara tehsil of Tonk district, under the Ajmer division, Rajasthan. The built-up area of the village spans about 14.6643 sq. km and is located near River Banas. Both villages have Gram Panchayats. Demographic composition and characteristics are presented in Tables 1 and 2.

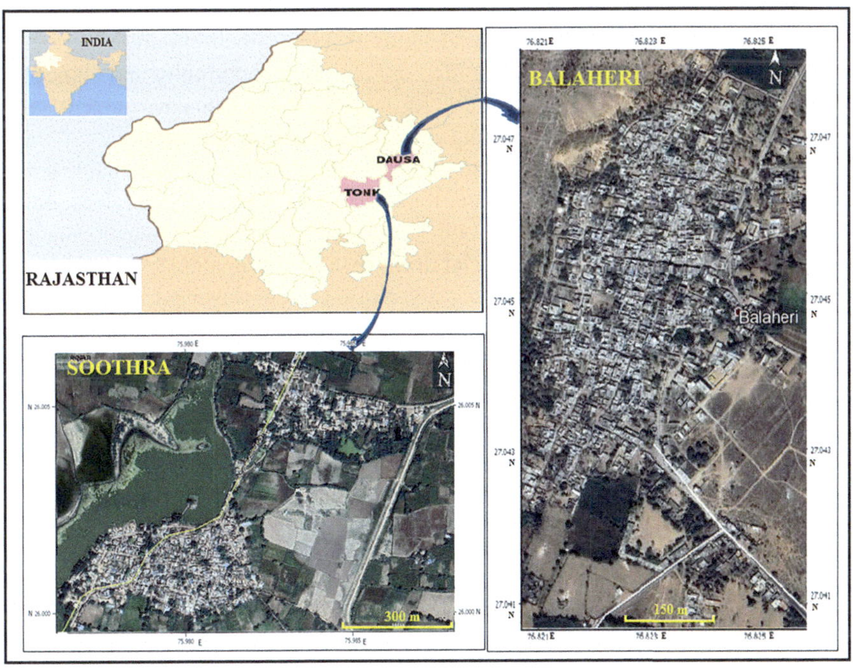

Fig. 1 Location map of village Balaheri and village Soothra. (Source: Satellite imagery taken from Google Earth)

Table 1 Demographic and social indicators

Indicators	Balaheri	Soothra
Population (in number)	3771	3021
No. of households	648	612
Sex ratio	892	963
Literacy rate	61	46
Scheduled caste (%)	38.42	22.57
Scheduled Tribe (%)	21.43	13.7

Source: District Census Handbook (2011)

Table 2 Community division

Village	OBC	SC	ST	Others
Balaheri	Muslim; Yogi; Sanjogi; Dakot; Banjara; Soni; Thatera; Luhar; Kandera; Kumhar	Khateek; Koli; Raigar; Bairwa; Harijan	Meena	Baniya; Brahman; Rajput
Soothra	Jat; Sain; Dholi; Prajapat; Nama; Bagariya; Saini; Jangid; Nath; Suwalka; Sahu; Gujjar; Chapola	Harijan; Bairwa; Khateek	Bhil; Meena	Brahman; Jain

Source: Primary data from village survey, 2018-19

Social Sustainability Framework for Rural Communities

The Egan Review (ODPM, 2004) defines a sustainable community as one that "meets the diverse needs of existing and future residents, their children, and other users, contributes to a high quality of life, and provides opportunity and choice. It also aims to make effective use of natural resources, enhance the environment, promote social cohesion and inclusion, and strengthen economic prosperity."

The Egan wheel further guides us to the eight components of a sustainable community: social and culture, governance, transport and connectivity, services, environment, economy, housing, and the built environment. These components place a strong emphasis on measuring the social dimension of sustainable development. In rural communities, particular attention is given to social cohesiveness, which fosters trust and solidarity among residents. This empowers people, instills a sense of belonging to a group, and ultimately improves their quality of life.

Many approaches have been adopted in recent years to assess the social dimension of sustainability, with most of them taking a theoretical approach (Missimer et al., 2017; Vallance et al., 2011; Åhman, 2013). Quantitative-based approaches have been explored to a limited extent (Magee et al., 2012; Wolski et al., 2016). However, for sustainability assessments, integrating random anthropocentric indicators may not be sufficient to capture the essence of the social dimension of sustainability. This dimension must be analytically addressed in terms of its content, which encompasses aspects like social cohesion, justice, individual needs, and cultural or traditional practices. Indicators related to basic infrastructure and essential services also play a crucial role within the realm of social sustainability.

In the present study, indicators were identified based on three dimensions of social sustainability (Table 3):

This framework draws inspiration from the Sutton Social Sustainability report by Bacon and Arendar (2014), even though the report primarily focused on urban centers. This study adapts the framework to the context of rural areas, where social sustainability takes on unique characteristics and significance.

From a methodological standpoint, the study encompasses three levels:

- Collection of Data: Data collection is conducted by using both primary and secondary sources.
- Data Analysis and Map Preparation: After data collection, analysis is performed, and maps are prepared.
- Preparation of the Social Sustainability Index (SSI): The study involves the development of an SSI to evaluate the social sustainability of the selected villages.

Data Sources

For the present study, primary data sources comprise structured and open-ended household questionnaire surveys, focused group interviews, direct observations, mental maps created by villagers, and Global Positioning System (GPS)-based mapping (Fig. 2). The household survey covered approximately 10% of the total village households. Employing random stratified sampling, roughly 60 households from each village were surveyed and categorized under SC, ST, and others. Focused group interviews were conducted with representatives from various communities. Villagers themselves prepared mental maps of the village settlement morphology. GPS points were taken for community households to validate the accuracy of these maps. Furthermore, GPS was employed to map the existing infrastructural facilities within the village.

Table 3 Dimensions of Social Sustainability Index (SSI)

	Dimensions	Description	Source
1	Amenities and social infrastructure	It represents the public utilities, infrastructures, and vital services to improve the quality of life of a place, this may include public space that can be used by local groups irrespective of age and gender, shared spaces that enable neighbors to meet, healthcare facilities, transport links, and modes	Bacon & Arendar (2014)
2	Social and cultural life	It represents the social and cultural aspects of communities and how they live	Bacon & Arendar (2014)
3	Adaptability, voice, and influence	It represents whether the communities have their say in the decision-making and are their voices heard and implemented	Bacon & Arendar (2014)

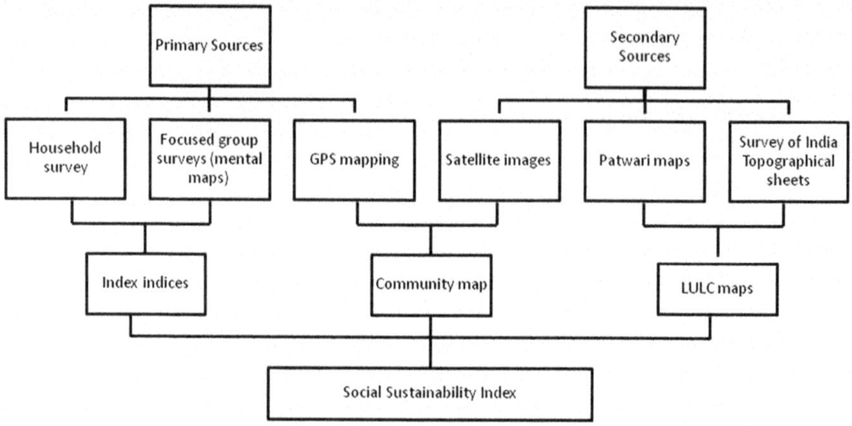

Fig. 2 Conceptual framework of the study. (Source: Compiled by authors)

Secondary data sources encompass data from the Census of India, satellite imagery, Survey of India topographical sheets, cadastral maps from the Patwari (land record or administrative officer at the sub-division or block level), relevant documents from the Gram Panchayat, as well as literature from journals, research papers, and books.

Methods

Neighborhood Mapping

The built environment is recognized to play a significant role in the social sustainability of urban spaces (Dempsey et al., 2011; Arnett, 2017; Shirazi & Keivani, 2017). The urban built environment reflects social interaction and local connectivity within and between neighborhoods (Somanath et al., 2021). Therefore, in rural areas, mapping of settlement morphology is highly essential, as it is guided by a strong social structure. In this study, land use and land cover (LULC) maps, community maps, and village services allocation maps for the selected villages were created by using mental mapping techniques, GPS, and satellite imageries (high-resolution Google Earth images) within a GIS environment. For Balaheri, LULC maps were prepared for the years 2006 and 2018, while for Soothra, the years 2009 and 2018 were selected. The choice of mapping years was limited by the availability of cadastral maps from the Patwari and freely available satellite images. Neighborhood mapping was conducted to assess the availability of services, detect changes in land use and land cover, evaluate the area's capacity to accommodate future village expansion, and gauge the community's resistance to settlement segregation on the village periphery—all of which are indicators necessary for the SSI.

Social Sustainability Index (SSI)

The SSI of the village is a composite index that comprises 18 indicators based on 30 variables, derived from neighborhood mapping, questionnaire surveys, focus group interviews, observations, and the Census of India (Fig. 3). The SSI aims to quantify and assess the social sustainability of both villages. Each dimension includes six indicators, framed with consideration of their desirability in rural areas of India. The Index scores were assigned as 3, 2, and 1, representing good, satisfactory, and poor performance of the indicators. The scoring system and data are available at https://doi.org/10.17632/kvk7bvtrcx.1.

To calculate the scores for community coexistence, we computed the coefficient of variation (CV). The initial step involved determining the standardized scores for each of the four communities (Others, SC, ST, and OBC) across the five sub-indicators (Table 4). Subsequently, we computed the average score for each of these four communities, and from those average scores, the CV was derived. A low CV suggests a higher degree of coexistence among community members, while a higher CV indicates a lower level of coexistence among these communities.

In the final step of the SSI analysis, all the data was scored, and then the indicator scores were combined to generate overall scores for each of the three dimensions. These dimension scores were then used to assess each dimension's level of social sustainability. Each dimension was categorized as unsustainable, less sustainable, potentially sustainable, or sustainable based on the score key developed for the SSI (Table 5). These final scores for each dimension were visualized using a circular bar chart (Fig. 4).

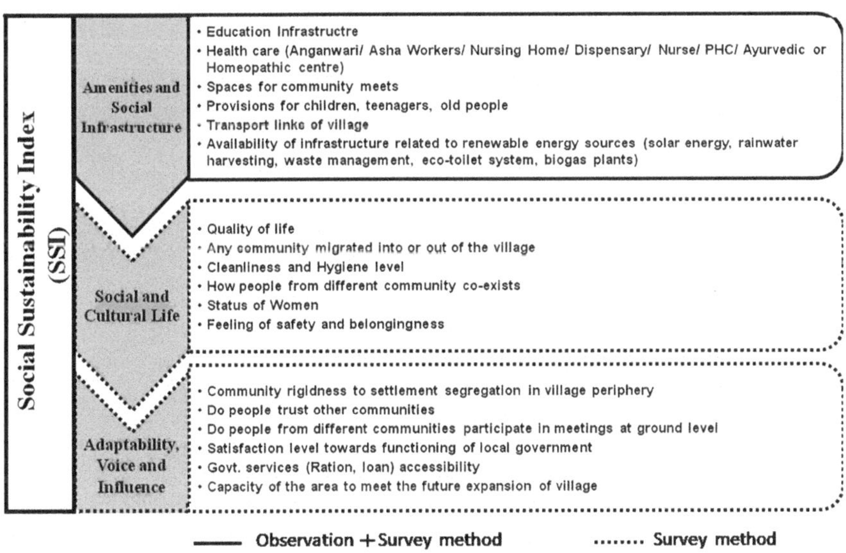

Fig. 3 Indicators of the SSI. (Source: Compiled by authors)

Table 4 Survey questions to understand the level of coexistence among the communities

	How people from different communities co-exist
A	Do individuals from different communities participate in your activities?
B	Do others offer help during challenging times?
C	Are you happy with your current life status?
D	What is the response of the local government?
E	Can people visit other community areas, and do they feel safe doing so?

Source: Compiled by authors

Table 5 Standard scores of SSI

Category	Score
Unsustainable	6–9
Less sustainable	10–12
Potentially sustainable	13–15
Sustainable	16–18

Source: Compiled by authors

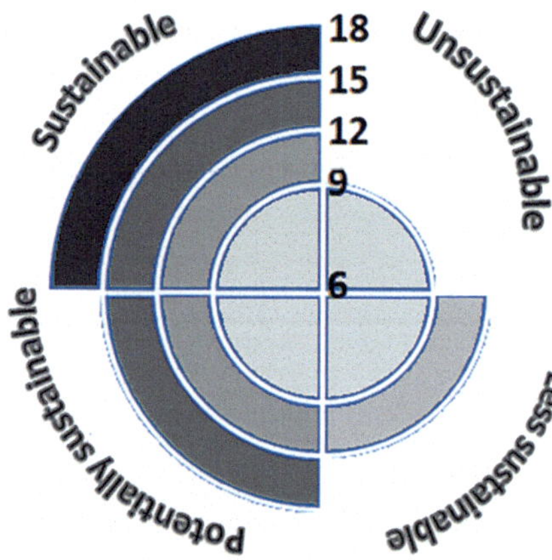

Fig. 4 SSI score key. (Source: Compiled by authors)

Result and Discussion

Neighborhood and Settlement Morphology of the Villages

To understand the settlement morphology and its expansion pattern over the years and assess the future expansion prospects, village LULC maps were prepared on the GIS platform (Fig. 5). The expansion of human settlements and road networks visibly changed between the study periods in both villages. In 2006, people used to live in the plain area near the foothills of Balaheri. The roads within the settlement area were unmetalled. Many households in the periphery consisted of kutcha houses (dwellings made up of mud and wood with thatched roofs). People tended to live at

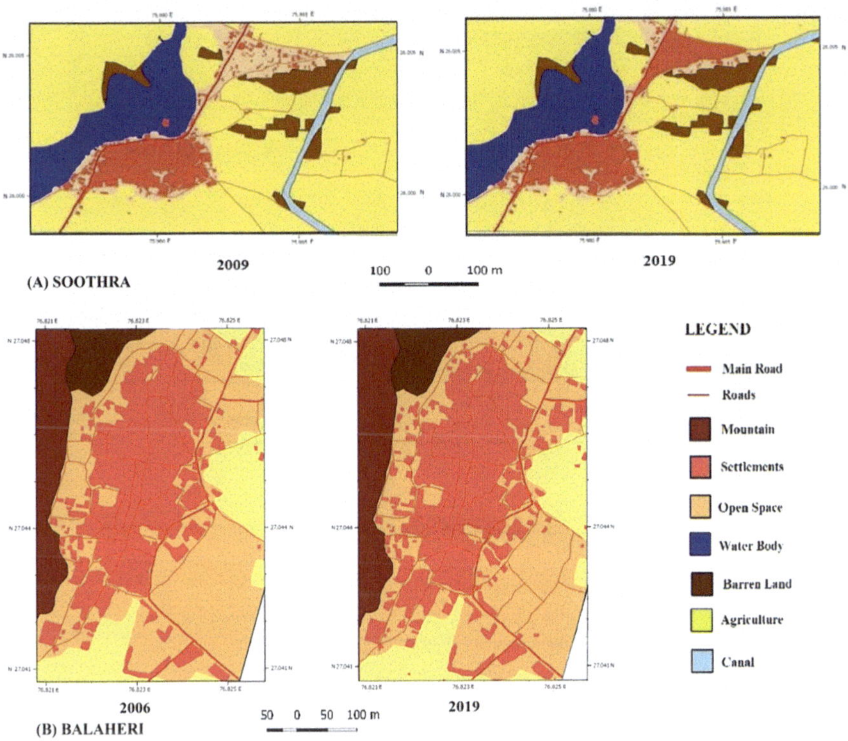

Fig. 5 Land use land cover dynamics of (**a**) Soothra and (**b**) Balaheri villages. (Source: Maps prepared by digitizing high-resolution Google Earth Satellite Images)

their "*dih*" (ancestral site), which led to congestion within the center. In 2018, with an increase in population, people started settling at higher elevations above the foothill plains. During the household survey, it was found that people with low economic status preferred this area as the price of land was lower than in plain areas. These are also the regions where village household wastes are dumped. In recent past, most of the roads have become metalled.

In 2009, Soothra village was a single compact settlement, with extended upper stretches of barren land and open space. The upper stretch space was used for keeping the field products by the villagers. Agricultural fields surrounded the settlement area along with a huge water body on the left of the village, and a canal was under construction on the right. In the 2018 map, however, the canal was operational under the Bisalpur Yojana. The upper stretches of barren land had been converted into a new settlement area by the Meena and Gurjar communities. This stretch had become well-populated due to the increase in family size and people's preference to live close to their agricultural fields. New settlements had also developed near the periphery of the old settlement area, showing the horizontal expansion of the village.

Village community maps for 2018 were prepared using the mental map technique by the villagers and were cross-validated with a GPS survey (Fig. 6). The

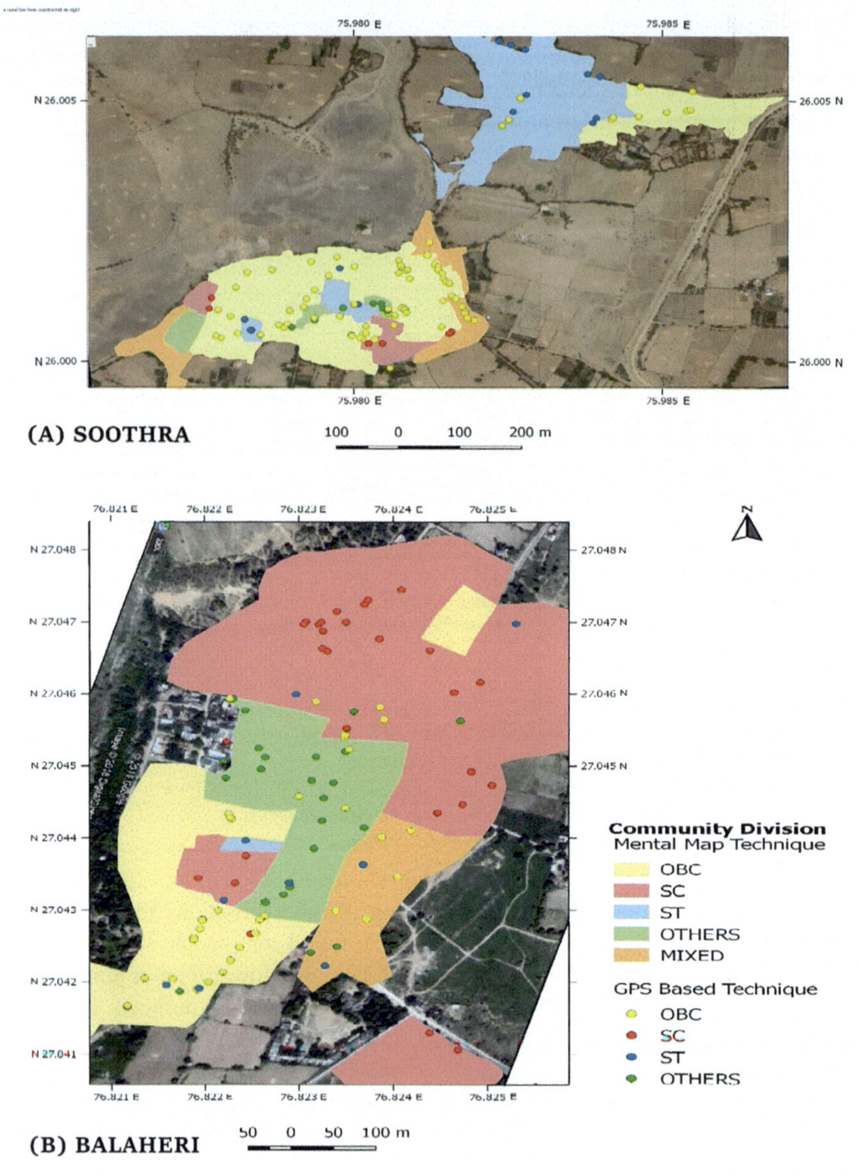

Fig. 6 Community maps, 2018–2019. (**a**) Soothra. (**b**) Balaheri. (Source: Primary data from village survey, 2018–19)

communities residing in these villages were broadly divided into four categories: OBC, SC, ST, and Others. In both Balaheri and Soothra, many communities had coexisted for a long time.

In Balaheri, high-caste people had resided in the center of the village from earlier times. Toward their right was the SC community, and to their left was the OBC community. The proportion of the ST population was low and found in small scattered pockets. According to the community map, the village's morphology bore the imprint of caste prejudices. Still, the center place or the area near the main market lane was inhabited by higher-caste people, mainly the Brahmins and Baniyas. Due to the increase in population, marginalized communities had settled in areas other than their mohalla too. Such areas were now accessible to them as higher-caste people were selling their land due to out-migration for better opportunities. With the passage of time, the stringent practices of different castes seemed to be fading away as well. There was a new community, the "Yogi," who migrated to this village in past years. At the same time, on the extreme outskirts of the village, the "Banjara," a nomadic community, now had permanent settlements. Soothra village was no different from Balaheri, as the morphology of the village still bore the imprint of caste segregation. For a long time, this village was mainly dominated by the "Gujjar" community, followed by the "Meena." Thus, the areas near the main road and pond were inhabited by OBC communities and the fringes by the rest. This segregation was present in recent times, with one exception: the northward stretch of the village, which was earlier uninhabited, is now dominated by the ST community.

The above discussion highlights the social fabric of the villages which could not be ignored while measuring the village social sustainability.

Performance of Villages in SSI

The overall social sustainability of the village was assessed through three dimensions: Amenities and Social Infrastructure, Social and Cultural Life, and Adaptability, Voice, and Influence.

Regarding amenities and social infrastructure, both villages boasted well-developed education facilities, including government primary schools, senior secondary schools, girls' schools, and private English medium schools. The key difference was the presence of a Madarsa in Balaheri catering to the Muslim population. Healthcare facilities were more extensive in Balaheri, offering access to hospitals, dispensaries, nursing homes, and veterinary services. In Soothra, available facilities were limited to Anganwadi centers, an ayurvedic center, Asha workers, and a non-functioning community health center.

Community meetings in both villages lacked dedicated spaces, forcing discussions to open areas or the Bharat Nirman Rajiv Gandhi Seva Kendra (formerly known as Atal Seva Sehkari Samiti), which also served as the Panchayat office compound. Additionally, school playgrounds served as the sole venue for leisure activities, as there were no parks, green spaces, or clubs for residents of all ages.

Transport linkages included metalled roads connecting the main bus stops on the highway to the village interiors in both locations. However, public transport options were notably absent, and both villages suffered from their distance (approximately 2 km) from the highways, necessitating walking or personal vehicles for commutes.

Remarkably, neither village had access to renewable energy sources despite enduring long hours of power cuts. Surprisingly, no efforts were made by villagers or local authorities to establish renewable energy generation structures (Fig. 7).

Under the "Social and Cultural Life" dimension, the first indicator was "Quality of Life" (Fig. 8a). It analyzed various aspects of life quality. Both villages faced challenges in terms of drinking water accessibility. Balaheri experienced irregular water supply, relying on tankers every 3–4 days. In Soothra, people extracted water from community wells using pipes and motors. Both villages suffered from frequent power cuts, with Balaheri enduring 6–12 h of daily cuts and Soothra facing over 12-h cuts. Thanks to the Swachh Bharat Abhiyan, toilet accessibility improved in recent years. The "Kisan Credit Card" (KCC), aiding crop production and agricultural expenses, was prevalent in Soothra but not in Balaheri. As the economic structure improved, residents of both villages transitioned from kutcha to pucca houses.

Women's status played a vital role in "Social and Cultural Life," given its susceptibility to inequalities (Fig. 8b). Female education was advancing in both villages,

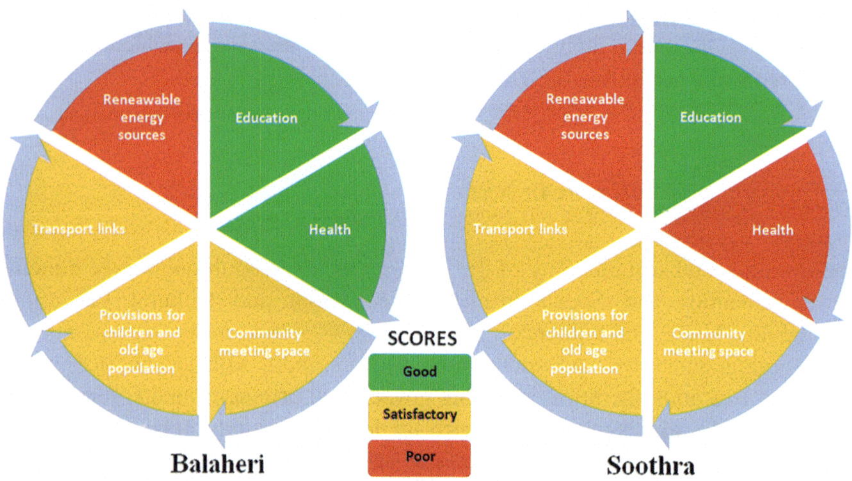

Fig. 7 Scores of amenities and social infrastructure. (Source: Compiled by authors)

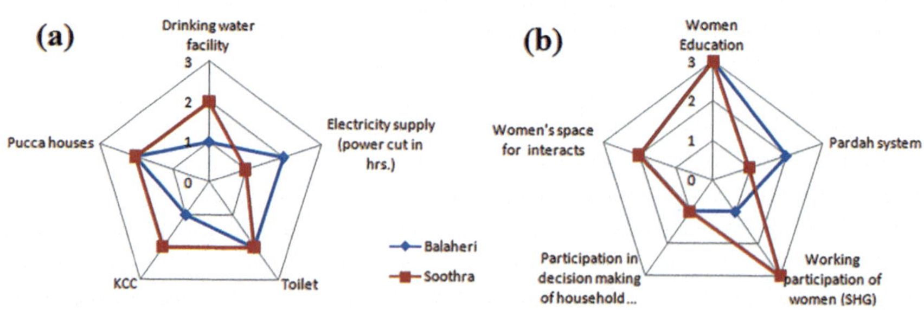

Fig. 8 SSI score: (**a**) Quality of Life; and (**b**) Status of Women. (Source: rimary data from village survey, 2018–19)

with the purdah system less strictly enforced due to improved education and awareness. Balaheri left the choice to follow purdah to its female members, whereas Soothra mandated it for women. While Soothra had a dedicated Self-Help Group (SHG) for working women, Balaheri lacked such an initiative. In both villages, women had limited influence on major household decisions and no dedicated spaces for interaction. They usually conversed near wells or within their households.

Historically, different communities coexisted in this region. In Balaheri and Soothra, SC, and ST respondents reported discrimination by upper-caste or dominant communities in social and cultural activities. However, during emergencies, these marginalized communities received support from others, providing a sense of satisfaction and security in Balaheri. In Soothra, the SC community expressed dissatisfaction due to a lack of cooperation during emergencies and unemployment. Marginalized caste members felt unsafe traveling in upper-caste areas, fearing subordination and humiliation, especially for women. The aggregated scores for the four communities differed, illustrating disparities. The coefficient of variance was calculated to assess community coexistence, with Balaheri scoring 22.26%, indicating low variation, and Soothra scoring 56.34%, illustrating variability among communities and low cohesiveness levels.

Both areas were sustainable in terms of immigration, accommodating different communities within their territory. In Balaheri and Soothra, the "Yogi" community and the "Nama" community, respectively, had migrated in the past years from nearby villages. People from rural areas were not fully aware of environmental issues. They considered their villages clean despite open, unmetalled drains with no sewage treatment and waste management system. In both study areas, more than 90% of the people felt safe in the village territory, having a strong sense of belonging. The overall score of this dimension is shown in Fig. 9.

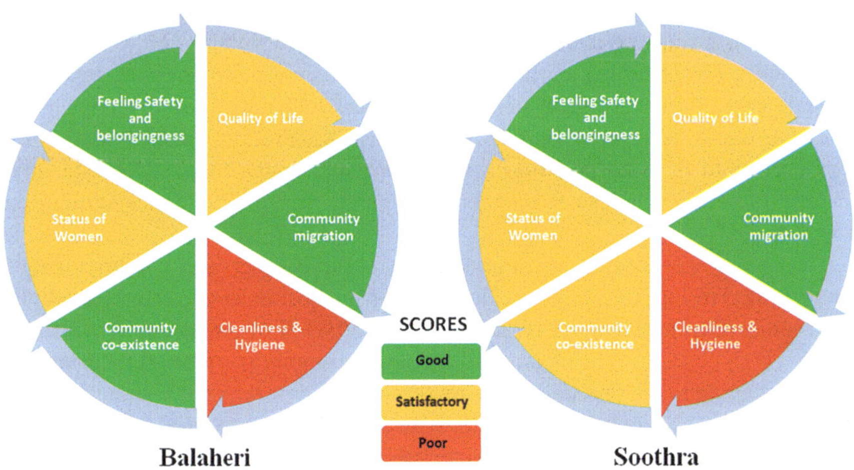

Fig. 9 Scores of social and cultural life. (Source: Compiled by authors)

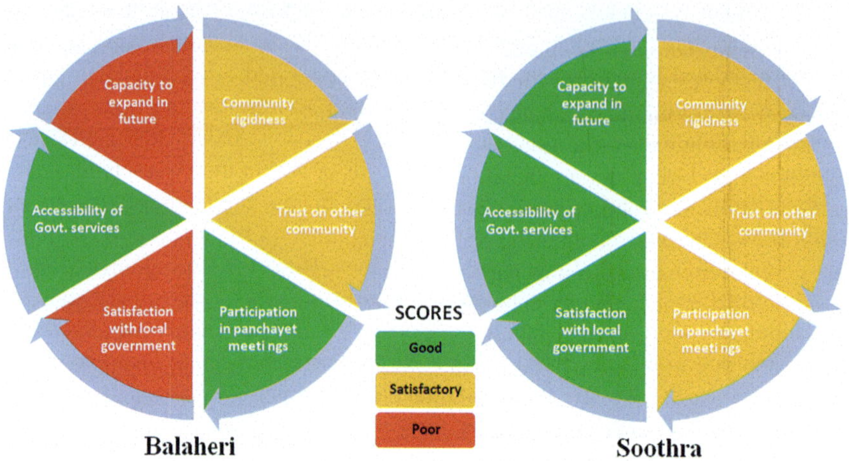

Fig. 10 Scores of adaptability, voice, and influence. (Source: Compiled by authors)

The "Adaptability, Voice, and Influence" dimension of SSI included indicators related to community trust, people's participation in village development activities, satisfaction with the functioning of the local body, accessibility of government services, and the village's capacity for future expansion. Trust among communities had a long history in both villages, both headed by a sarpanch (elected village headman) from the SC community. In Balaheri, the majority of the people participated in panchayat meetings, but they were dissatisfied with the village governance. Participation in panchayat meetings in Soothra was moderate, as residents trusted the local body's decisions and found the government approachable and responsible. In Soothra, the elected body had undertaken various initiatives, such as road development, distribution of KCC loans, installation of LED lights, and school building renovation. Government services, including ration distribution and funds for toilet construction, were accessible to most residents in both villages. However, some respondents believed that certain communities received preferential treatment when it came to government schemes. Balaheri had limited spatial expansion potential, primarily toward the hills. In contrast, Soothra possessed some barren land that could be converted into a settlement area in the future. Consequently, Balaheri was deemed less sustainable compared to Soothra in this dimension (Fig. 10).

The overall scores for all three dimensions in the Balaheri and Soothra villages are presented in Fig. 11. It is clear that none of the dimensions in the selected villages achieved a "sustainable" score. The SSI methodology employed in this study effectively exposed the developmental disparities among the three dimensions of social sustainability. Consequently, a more targeted developmental strategy could have been devised to enhance the performance of each dimension of SSI. However, these developments should have been carried out in a manner that strived to address the needs of all communities and dissolve intra-regional inequalities.

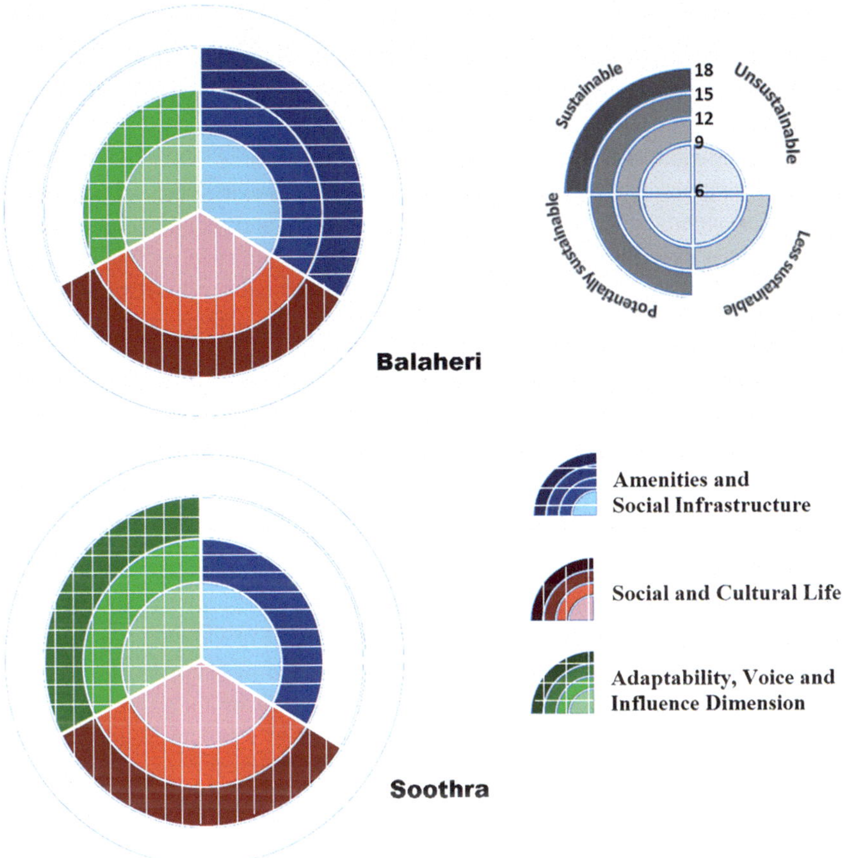

Fig. 11 Social sustainability scores achieved from SSI. (Source: Compiled by authors)

The Forward Looking

This study introduces an index for evaluating the sustainability of regions through a social lens, known as the SSI. The SSI serves as a valuable tool for identifying and comprehending the myriad challenges faced by rural areas, providing a framework to devise effective solutions. Policies and decisions must be formulated with inclusivity at their core, striving to unite all communities under a common purpose. The development process should be a collaborative endeavor, involving both local governing bodies and active community participation. Given the uniqueness of each village, solutions to their respective challenges should ideally emerge from within the communities themselves. In numerous parts of India, many villages have transformed model communities through active participation by local governing bodies and communities.

In the "Amenities and Social Infrastructure" dimension of SSI, the analysis underscores that both of the selected villages currently fall short of achieving sustainability. Notably, Soothra faces more pronounced challenges compared to Balaheri. To catalyze the development of these villages, it is imperative to focus on locally driven solutions. For instance, the absence of palliative care services necessitates that individuals seek medical attention from urban areas. Establishing palliative care institutes, following the successful Kerala model (Kochuvilayil et al., 2023), can significantly benefit these communities and neighboring regions. Furthermore, the implementation of renewable resources such as biogas plants, rainwater harvesting systems, waste management practices, eco-toilet systems, and solar energy solutions is crucial for reducing the dependency on government-provided resources and transforming these villages into eco-friendly communities that contribute to climate solutions (INFORSE—South Asia, 2017). Addressing the issue of low literacy rates among economically disadvantaged segments of the population calls for intervention by local governments, particularly in the development of community libraries (Pyati & Kamal, 2012).

Within the "Social and Cultural Life" dimension, several problem areas are identified, affecting the quality of life for individuals across all social strata. For instance, Balaheri grapples with drinking water scarcity, emphasizing the need for collaborative efforts between the community and local governing bodies to develop rainwater conservation and harvesting infrastructure. The utilization of MGNREGA funds for constructing water conservation structures, as permitted by the Mahatma Gandhi National Rural Employment Guarantee Act 2005, can be instrumental in this regard (Anantha et al., 2021). Additionally, initiatives such as Pradhan Mantri Gramin Awas Yojana can be leveraged to build permanent dwellings for economically disadvantaged individuals. The formation of self-help groups (SHGs) plays a pivotal role in fostering self-development, as exemplified by the transformation of Teliya village in Gumla district, Jharkhand, through initiatives like toilet construction, irrigation facilities, and road development (Walker, 2023). SHGs also raise awareness about government policies and legal rights among individuals from diverse communities.

Within the "Adaptability, Voice, and Influence" dimension, both villages fall below the sustainability threshold, with Balaheri exhibiting a more challenging situation than Soothra. To enhance the effective functioning of local governance, panchayats should adopt an all-inclusive approach that involves people from every stratum of society. Promoting participation in public meetings can facilitate consensus on government policies for rural development (Rasila & Mudau, 2012). Building trust can be achieved by engaging all communities in various activities. Equitable distribution of government services can be realized through the establishment of a well-developed Public Distribution System (PDS).

For the economic progress of villages, social solidarity among communities is very essential. The Asian Solidarity Economy Council (ASEC, 2O19) highlighted the need for revitalizing the rural economy through a social solidarity economy. Drawing lessons from case studies from different Asian countries, they have proposed that community participation in decision-making and policy implementation toward social and economic welfare, will not only foster cohesion among communities but also increase their desire to reside in their place of origin. This forward-looking approach offers a roadmap for sustainable development that is both community-centric and inclusive.

Conclusion

This study sheds light on the often-overlooked yet critically important dimension of sustainability in rural areas. While global research has predominantly focused on the social sustainability of urban neighborhoods, it is essential to recognize the significance of rural sustainability, especially in countries like India with a substantial rural population. The challenges posed by the global COVID-19 pandemic underscored the need for sustainable rural development.

The introduction of the SSI provides a valuable tool for assessing social sustainability in India's rural regions. By quantifying qualitative social indicators and fostering synergy among various dimensions of social sustainability, SSI offers a framework for understanding the potential of villages. It's important to emphasize that SSI is not a one-size-fits-all index; rather, it acknowledges the diversity of Indian villages and allows for customization based on local socio-cultural norms.

The proposed index serves as a schema for measuring social sustainability, addressing village-specific core issues, and enhancing stakeholders' understanding of existing levels of social sustainability. By guiding targeted decision-making processes, SSI can help identify and rectify social problems in rural areas that may have been overlooked during policy formulation.

In a world that seeks comprehensive and inclusive sustainable development, it is essential to acknowledge that social well-being is a foundation of prosperity. This study underscores the social sustainability and solidarity among rural communities, which promotes resilience and collective responsibility. As we move forward, let us prioritize social sustainability and solidarity as essential components of rural development, guaranteeing a brighter and more equitable future for all.

Acknowledgments This research is supported by Indraprastha College for Women, University of Delhi under the Centenary Undergraduate Research Grant. We wish to extend special thanks to the villagers and *panchayat* members for sharing their pearls of wisdom during this research.

Competing InterestsThe authors declared no potential conflicts of interest concerning the research, authorship, and/or publication of this chapter.

Ethics Approval Consent from the villagers was taken for household surveys and individual and focus group interviews. Ethical clearance for the study was obtained from the Research Ethics Committee, Indraprastha College for Women, University of Delhi (IPC/44/10, dated: 12.10.2018).

Appendix

Data availability https://doi.org/10.17632/kvk7bvtrcx.1

References

Åhman, H. (2013). Social sustainability—Society at the intersection of development and maintenance. *Local Environment, 18*, 1153–1166. https://doi.org/10.1080/13549839.2013.788480

Anantha, K. H., Garg, K. K., Moses, D., Patil, M. D., Sawargaonkar, G. L., Kamdi, P. J., Malve, S. H., Sudi, R., Raju, K. V., & Wani, S. P. (2021). Impact of natural resource management interventions on water resources and environmental services in different agroecological regions of India. *Groundwater for Sustainable Development, 13*, 100574. https://doi.org/10.1016/j.gsd.2021.100574

Arnett, H. (2017). The challenges in quantifying social sustainability at a neighborhood level. *Cities & Health, 1*, 139–140. https://doi.org/10.1080/23748834.2018.1460104

ASEC. (2019). Revitalising rural economy, through & to social solidarity economy. Retrieved from Socio Eco. http://www.socioeco.org/bdf_colloque22_en.html

Bacon, N., & Arendar, L. C. (2014). *Measuring social sustainability in Sutton*. London Borough of Sutton.

Bailey, F. G. (1957). *Caste and economic frontier*. Manchester University Press.

Barron, L., Gauntlett, E. (2002). Housing and sustainable communities indicators project: Stage 1 report–a model of social sustainability. WACOSS housing and sustainable communities indicators project. Retrieved from http://www.adelaide.sa.gov.au/soc/pdf/barron_gauntlett.pdf

Beteille, A. (1965). *Caste, class and Power: Changing patterns of stratification in a Tanjore village*. University of California Press.

Bryant, C. (1995). The role of local actors in transforming the urban fringe. *Journal of Rural Studies, 11*, 255–267. https://doi.org/10.1016/0743-0167(95)00020-n

Bryden, J. (1994). Some preliminary perspectives on sustainable rural communities. In J. Bryden (Ed.), *Towards sustainable rural communities: The Guelph seminar series* (pp. 40–50). University School of Rural Planning and Development.

Caldwell, L. K. (1984). Political aspects of ecologically sustainable development. *Environmental Conservation, 11*, 299–308.

Census of India (2011) Directorate of Census Operations, *Ministry of Home Affairs*, Government of India.

Chakravarty, A. (1975). *Contradiction and change: Emerging patterns of authority in a Rajasthan Village*. Oxford University Press.

Chiu, R. L. H. (2003). Social sustainability, sustainable development, and housing development. The experience of Hong Kong. In R. Forrest & J. Lee (Eds.), *Housing and social change. East-west perspectives* (pp. 221–239). Routledge.

Dempsey, N., Bramley, G., Power, S., & Brown, C. (2011). The social dimension of sustainable development: Defining urban social sustainability. Sustainable Development, 19, 289–300. https://doi.org/10.1002/sd.417.

District Census Handbook. (2011). Census of India. Directorate of Census Operations—Rajasthan, Ministry of Home Affairs, Government of India

Durkheim, E., & Simpson, G. (1934). Emile Durkheim on the division of labor in society. *The Journal of Educational Sociology, 7*, 528. https://doi.org/10.2307/2961541

Elkington, J. (1999). Triple bottom line revolution: Reporting for the third millennium. *Australian CPA, 69*(11), 75–76.

Gandhi, M. K. (1963). *Village swaraj*. Navajivan Publishing House.

Gorringe, H., Jodhka, S. S., & Takhar, O. (2017). Caste: Experiences in South Asia and beyond. *Contemporary South Asia, 25*, 230–237. https://doi.org/10.1080/09584935.2017.1360246

GRI. (2000). *Sustainability reporting guidelines*. Global Reporting Initiative.

Gupta, D. (2004). *Whither the Indian Village? Culture and agriculture in "rural India" (Malcolm Adiseshiah Memorial Lecture)*. Madras Institute of Development Studies.

Harriss-White, B., & Heyer, J. (2009). *The comparative political economy of development*. https://doi.org/10.4324/9780203861332

Hechter, M. (2001). *Solidarity, Sociology of* (pp. 14588–14591). Elsevier eBooks. https://doi.org/10.1016/b0-08-043076-7/01969-0

INFORSE—South Asia. (2017). Eco Village Development as Climate Solution. Retrieved from http://www.inforse.org/asia/pdf/Pub_EVD-SouthAsia.pdf

Kessinger, T. G. (1971). *Vilyatpur 1848–1968: Social and economic change in a North Indian Village*. University of California Press.

Kochuvilayil, A., Rajalakshmi, S., Krishnan, A., Vijayanand, S. M., Kutty, V. R., Type, T., & Varma, R. P. (2023). Palliative care management committees: A model of collaborative governance for primary health care. *Public Health Action, 13*, 12–18. https://doi.org/10.5588/pha.22.0028

Komter, A. (2010). Solidarity. In H. K. Anheier & S. Toepler (Eds.), *International encyclopedia of civil society*. Springer. https://doi.org/10.1007/978-0-387-93996-4_81

Koning, J. (2002). Social sustainability in a globalizing world: Context, theory, and methodology explored. ResearchGate.

Magee, L., Scerri, A., James, P., Thom, J. A., Padgham, L., Hickmott, S., Deng, H., & Cahill, F. (2012). Reframing social sustainability reporting: Towards an engaged approach. *Environment, Development and Sustainability, 15*, 225–243. https://doi.org/10.1007/s10668-012-9384-2

Magis K. (2010) Community resilience: an indicator of social sustainability. *Society & Natural Resources 23*. Taylor & Francis: 401–416. https://doi.org/10.1080/08941920903305674

Mandal, R. B. (1979). *Introduction to rural settlements*. Concept Publishing Company.

McManus P, Walmsley J, Argent N, Baum S, Bourke L, Martin J, Pritchard B, Sorensen T. (2012) Rural Community and Rural Resilience: What is important to farmers in keeping their country towns alive? *Journal of Rural Studies 28*. Elsevier BV: 20–29. https://doi.org/10.1016/j.jrurstud.2011.09.003

Miller D. (2017) Solidarity and its sources. In Oxford University Press eBooks, 61–79. https://doi.org/10.1093/acprof:oso/9780198795452.003.0002

Ministry of Health & Family Welfare-Government of India. (n.d.). Infrastructure: National Health Mission. Retrieved from https://nhm.gov.in/index1.php?lang=1&level=2&sublinkid=1220&lid=190

Missimer M, Robèrt KH, Broman G. (2017) A strategic approach to social sustainability – Part 1: exploring the social system. Journal of Cleaner Production 140. Elsevier BV: 32–41. https://doi.org/10.1016/j.jclepro.2016.03.170

ODPM. (2004). The Egan review: Skills for sustainable communities—Digital education resource archive (DERA).

Pyati, A., & Kamal, A. (2012). NGO-developed libraries in India: Impacts, models, and new possibilities. *Libri, 62*, 291–303. https://doi.org/10.1515/libri-2012-0023

Rasila, B. N., & Mudau, M. J. (2012). Effective communication as a strategic tool for rural development: A model to take South African Government beyond mobilization and consultation through public participation. *Journal of Media and Communication Studies, 4*, 134–141. https://doi.org/10.5897/jmcs12.035

Saxena, V., & Bhattacharya, P. C. (2018). Inequalities in LPG and electricity consumption in India: The role of caste, tribe, and religion. *Energy for Sustainable Development, 42*, 44–53. https://doi.org/10.1016/j.esd.2017.09.009

Scott, K., Park, J., & Cocklin, C. (2000). From 'sustainable rural communities' to 'social sustainability': Giving voice to diversity in Mangakahia Valley, New Zealand. *Journal of Rural Studies, 16*, 433–446. https://doi.org/10.1016/s0743-0167(00)00018-8

Shirazi, M. R., & Keivani, R. (2017). Critical reflections on the theory and practice of social sustainability in the built environment—A meta-analysis. *Local Environment, 22*, 1526–1545. https://doi.org/10.1080/13549839.2017.1379476
Smit, B., & Brklacich, M. (1989). Sustainable development and the analysis of rural systems. *Journal of Rural Studies, 5*, 405–414. https://doi.org/10.1016/0743-0167(89)90066-1
Somanath, S., Hollberg, A., & Thuvander, L. (2021). Towards digitalization of socially sustainable neighborhood design. *Local Environment, 26*, 770–789. https://doi.org/10.1080/13549839.2021.1923002
Spicker, P. (2014). Cohesion, exclusion, and social quality. *International Journal of Social Quality, 4*(1), 95–107. https://doi.org/10.3167/ijsq.2014.040107
Srinivas, M. N. (1954). Village studies. The Economic Weekly. Retrieved from https://www.epw.in/system/files/pdf/1954_6/22/village__studies.pdf
Srinivas, D. (2015). Impact of culture on rural built environment. In International Conference on IT, Architecture and Mechanical Engineering (ICITAME'2015) May. 22–23
Thakur, M. (2014). *Understanding ruralities: Contemporary debates. CAS working paper series, centre for the study of social systems*. Jawaharlal Nehru University. Retrieved from https://jnu.ac.in/sites/default/files/u63/manish-thakur-paper.pdf
Vallance, S. A., Perkins, H. C., & Dixon, J. (2011). What is social sustainability? A clarification of concepts. *Geoforum, 42*, 342–348. https://doi.org/10.1016/j.geoforum.2011.01.002
Walker, D. (2023). *How women's self-help groups are transforming rural India*. Ford Foundation.
WCED. (1987). Report of the world commission on environment and development: Our common future.
Wolski, C., Marino, E., Doherty, T. J., Fisher, S., & Green, A. S. (2016). Systems of access: A multidisciplinary strategy for assessing the social dimensions of sustainability. *Sustainability: Science, Practice and Policy, 12*(1), 88–100.

Part IV
Geopolitics, International Diplomacy, Political Movement Within Nation-States

India's Quest for Geopolitical Gains through Metamorphosing NAM Solidarity into Global South Solidarity

Nilotpal Kant

Introduction

"The manifestation of human solidarity which has come to expression on this occasion in the great family of Nations": Yugoslavian president Tito (lauded the importance of international solidarity after the Skopje earthquake in 1963).

In International Relations (IR) theory, the term "solidarity" has been used with social theoretical implications that have often been unclear and limited to academic discussions. Within the discipline of IR, the conceptions of solidarity have played implicit roles. Realist perspectives in international politics, for instance, rely on the concept of the national interest, which is broadly understood as representing the "best interests" of those deemed significant within the society governed by the territorial state in question. Similarly, liberal approaches also operate with a notion of "expanded self-interest" guided by ideas of representation.

However, few international events can be explained through one particular theoretical approach; therefore, this chapter draws over multiple theoretical approaches suitable to explain the historical events, the inbuilt alliances, and evolving identities concerning the Global South. Identity and alliances in IR are not often explained under the Realist-Liberal duet. This chapter adheres to India's navigation to position itself in the global landscape in general and the Global South in particular. In the dynamic landscape of global diplomacy, India's strategic adjustment serves as a significant illustration of geopolitical evolution. Rooted historically in the principles of the non-aligned movement (NAM), which advocated for neutrality amid the Cold War's divisions, India's foreign policy is evolving toward a more intricate and diversified paradigm. This chapter examines India's efforts to leverage geopolitical

N. Kant (✉)
Latin American Studies Program at the School of International Studies, Jawaharlal Nehru University, New Delhi, India
e-mail: nilotp61_isa@jnu.ac.in

K. Arora, R. R. Anand (eds.), *The Solidarity Approach in Geography*,
https://doi.org/10.1007/978-3-031-86753-8_13

benefits by transforming the solidarity of NAM into the inclusivity of the Global South Movement (GSM). It analyzes the shift from political neutrality to a platform focused on collective negotiation and shared advancement.

This chapter explores how India's responses to international developments reflect its ambition to play a responsible and major role in global affairs. Employing interdisciplinary qualitative analysis, this chapter examines the changes in India's foreign policy, highlighting its utilization of the Non-Aligned Movement's legacy to cultivate a broad Global South Movement alliance. This transformation goes beyond mere changes in the terminology, representing a deliberate strategic reorientation of India's engagements with emerging economies and the developing world, aimed at fostering a more equitable international framework. The study aims to decipher the complexities of this shift, analyzing both its theoretical underpinnings and its practical impacts on India's collaborative initiatives within the Global South.

Objectives and Research Methods

This chapter undertakes a conceptual examination of the normative implications of "solidarity." An alternate approach could be to explore the significance of solidarity by conducting a historical-sociological investigation into the development and transformation of solidarity movements, highlighting both progressive and regressive trends through a chronological examination. However, within the realm of IR, solidarity has historically been approached through two theoretical frameworks: the English School of thought and the Neo-Marxist and Neo-Gramscian schools of thought. These approaches utilize solidarity to emphasize the social perspective within analyses of IR.

The objective of this chapter is to present an alternative approach for addressing issues related to the Global South using a solidarity-based framework. The aim is to propose an approach centered on cooperation, knowledge sharing, and socio-economic integration as avenues for addressing challenges faced by the Global South. This study aims to explore the dynamic that is reshaping how movement is perceived and India's role within the Global South. The analysis unfolds in two segments: initially, it examines the historical progression of the solidarity model via the NAM, and subsequently, it investigates contemporary forms of unity within the Global South Movement (GSM). This chapter's goal is to draw connections between NAM and GSM, highlighting their continuity. The fundamental purpose is to delineate the ongoing evolution of solidarity between nations through NAM and GSM, offering evidence of this transformation.

This research employs a descriptive study approach to investigate India's shift from NAM solidarity toward a more expansive solidarity with the Global South. The objective is to comprehend the geopolitical advantages India pursues through this transformation and how it corresponds with the nation's foreign policy objectives and international positioning. It will adopt an interdisciplinary approach in its analysis. The study acknowledges certain constraints, such as the accessibility of primary sources and the subjective nature of interpreting historical and contemporary events in international affairs. To address these challenges, the research will utilize a cross-referencing of multiple sources and incorporate diverse perspectives on the topic. The study anticipates offering a detailed insight into India's diplomatic approaches and the consequences of its heightened emphasis on solidarity with the Global South. Its objective is to enrich discussions on regional collaboration and the shifting dynamics of IR in the post-Cold War period.

Solidarity in IR: Understanding Its Evolution from Philosophy to Practice

Solidarity emerges as a result of circumstances influenced by self-interest, harmonizing with principles such as justice, freedom, and liberty. Consequently, it becomes a crucial component of the social contract (Rawls, 1989). John Rawls' theory of justice, which emphasizes fairness and equality, extends beyond national borders to address global inequalities and IR. This broad application encourages a deep dive into international solidarity through the lens of social justice. In contemporary times, social justice has transitioned from a theoretical concept to a tangible action, prompting a reflective analysis of how research methodologies themselves contribute to the pursuit of a just society. The concept of solidarity became significant in social sciences after Emile Durkheim introduced it in his seminal work, "The Division of Labour in Society," in 1893. Since then, it has periodically attracted the attention of sociologists, as noted by Crow in 2002.

The diversity of meanings associated with the "Solidarity" movement created an opportunity for extensive mobilization by appealing to various reasons for participation. In the context of social movement theory, this can be understood as saturating the movement with different interpretations (Latoszek, 2006). The "Solidarity" movement can be viewed through multiple lenses: it may represent a revolution of the working class, a civic uprising, a Taht for national independence, an effort against religious marginalization, or even a crusade driven by ethical principles. All of these perspectives coexisted within "Solidarity," leading to conflicts in interpreting its essence (Kurczewski, 2006).

This approach obscures the importance of solidarity in constructing relations between places, activities, and diverse social groups. Solidarity in practice can be forged "from below" or through "pressure from without" and such solidarity is being shaped by marginalized groups of states in the international system which ultimately transforms into a new internationalism. The mobilization of transnational solidarities has been extended through working class and anti-slavery politics (Featherstone, 2012).

The concept of solidarity, despite its significance, has not been a central theme in Political Science, often relegated to rhetorical mentions. The discipline has instead delved deeper into various aspects of political association like community, nationalism, democracy, multiculturalism, and human rights. Solidarity at its core embodies a sense of collective empathy and responsibility among members of a group, promoting mutual support (Baker et al. 2006). This concept possesses subjective and emotional components, which, in part, contribute to its relative lack of attention in theoretical discourse.

Solidarity is a practice that is central to the anti-imperialism movement which is often practiced through social, economic, and political instruments. Solidarity is often described as a relation formed through political struggle, aimed at confronting various types of injustice (Featherstone, 2012). In the post-colonial era, solidarity remained the strongest unifying force among decolonized nations that were politically, socially, and economically marginalized in the international political system. Scholars have tried to formulate a general theory of solidarity under the framework of political history but framing the concept of "Solidarity" within the context of political history doesn't provide a complete understanding of its significance as a social and cultural phenomenon. The ambiguity surrounding "Solidarity" arises not solely from conflicting interpretations of political change but primarily from the challenge of comprehending "Solidarity" as a phenomenon in itself. To gain insight into "Solidarity," one can adopt a constructionist viewpoint that presents it as a symbolic movement (Halas, 2010). Solidarity might be understood in terms of likeness toward individuals, communities, or Nations. Solidarity as a movement is a projection of democratic political manifestation and international coalition.

Solidarity in the international sphere is a product of cognitive political cooperation among like-minded nations that recognizes the needs of affected nations in times of catastrophe either natural, social, political, or economic which is not limited to charity, aid, or humanitarian assistance but adherence to the principle of justice and stainability by standardizing equitable share of benefits, equal partnership, and greater well-being of the communities (International Council for Human Rights Policy report 2010).[1]

Commentaries on IR are largely dominated by terms like international solidarity, which is a psychological reconstruction of the morale of the affected community which is followed by a practical solution to the problem for which solidarity is professed (Heyward, 1963). Current international cooperation and flow of aid from the developed world to the developing world is directed from the cognition of a solidarity-based approach upon which the global body United Nations' principle is based. Solidarity is one of the most important aspects of traditional and

[1] Citing the "Report of the independent expert on human rights and international solidarity, Rudi Muhammad Rizki" (A/HRC/15/32 and Corr.1), paragraph 58, the consideration was given to the possibility of establishing a new approach to accountability within the global economy grounded in international solidarity and shared responsibility. Reference to this approach can also be found in the publication titled "Human Rights in the Global Economy" by the International Council for Human Rights Policy (Geneva, 2010), page 11. This section relies heavily on the reports presented to the Human Rights Council by the Independent Expert on human rights and international solidarity in 2009 (A/HRC/12/27 and Corr.1) and 2010 (A/HRC/15/32).

non-traditional security which is often observed among vulnerable societies (Modéer & Eziakonw, 2019).

Most of the international organizations facilitating multilateralism in developing countries are grounded in solidarity. One of the important facets of solidarity is "solidarity in response to what?" Marginalized sections of world communities or nations always collectivize themselves in response to some physical, psychological, natural, and manmade crisis that can transform into a security crisis, food crisis, health emergency, and so on (it can encompass natural crisis like an earthquake, flood, drought, etc. and manmade crisis war, economic sanctions, basic necessity embargo, etc.). NAM and Global South are collective responses to such a crisis in the form of solidarity. However, solidarity has a limited impact on bringing all concerned parties within an organization unless there is concrete interest of participating nationalities. Therefore, NAM and Global South must be seen as normative and cornerstone institutions to locate solidarity in the realm of IR.

Solidarity and NAM During the Cold War

The beginnings of the non-alignment movement (NAM) are often linked to an event that occurred during Jawaharlal Nehru's official trip to the USA in 1949. During his visit, Nehru was questioned by American reporters about India's stance in the developing bipolar global landscape dominated by the USSR and the USA, inquiring if India was in alliance with either of the two powers. Nehru responded by saying, "Why should I align with either of you? Can't I remain unaligned?" Later, when he spoke at Columbia University, he elaborated on his statement. He explained that a small and newly liberated nation like India could not afford the luxury of taking sides in the Cold War between superpowers. Instead, they had to focus their efforts on addressing the significant economic and social challenges within their own country (Mankekar, 1980). Nehru's stance wasn't solely representative of India; rather, it resonated with many decolonized nations. The approach India adopted in its dealings with superpowers served as a model for other countries. This marked a turning point, as it encouraged a collaborative approach to dealing with the superpowers. The idea was that a single nation on its own couldn't effectively maintain a non-aligned stance. Consequently, this approach evolved into a movement based on solidarity among newly independent countries that had recently emerged from the aftermath of the war. Thus, a new international collaboration emerged giving rise to non-aligned internationalism.

In the context of the non-aligned movement, solidarity underwent a gradual transformation, spreading from a regional and domestic context to become an international phenomenon, much like how liquid moves through small tubes due to capillary action. The NAM's doctrine is primarily focused on decolonization and anti-racial movements, which have bolstered the political resolve of developing nations (Lüthi, 2016). In the aftermath of conflict, these nations have united in cooperation to tackle the intensified issues of the post-war period, especially those stemming from the disproportionate sharing of gains yielded by the industrial and technological revolution (Chand, 1983).

Solidarity serves as a potent means of demonstrating allegiance to a group. The initial manifestation of solidarity at the NAM summit was demonstrated by Yugoslav leaders when they drew parallels between the historical experiences of decolonized African and Asian nations and their history of colonialism under the Habsburg monarchy and the Ottoman Empire (Sladojevic, 2015). Communist theories laid the foundation for such solidarity. For instance, they examined the conditions that led India, Egypt, and Yugoslavia to adopt a path of bloc neutrality and non-alignment, asserting commonalities among these three nations in the following ways: These nations shared a common past of being under foreign control, recognized themselves as members of the developing world, accepted their position as third-world countries, aimed to protect their independence and sovereignty from the influence of major world powers, and collectively opposed the dominant international system that emerged following World War II (Tadic, 1976; Spaskovska, 2017).

One of the fundamental elements of NAM solidarity revolved around the commitment to the principles of multiculturalism and multiethnicity. This commitment materialized in the form of a policy promoting "brotherhood and unity" among NAM member nations, signifying its significance (Drapac, 2010). The solidarity of the NAM is deeply rooted in the commitment to equality, human rights, and justice, echoing the Buddhist ethos of harmonious coexistence. This commitment was galvanized by the Bandung Conference and reinforced by the principles of the United Nations. The essence of Bandung fostered two significant global stances: the pursuit of non-alignment, which emphasizes independence from major power blocs, and the adoption of neutrality, particularly during the Cold War era, to avoid entanglement in the superpowers' ideological conflicts. However, a third matter that received insufficient attention during the Bandung conference was the institutionalized concept of Western racial superiority, which was subsequently addressed during discussions in Belgrade. While the issue of race was mentioned during the Bandung conference, it was not thoroughly examined or seriously addressed in the conference's declaration.

In September 1961, the inaugural NAM conference convened in Belgrade with the participation of 25 state delegations, predominantly from Africa and Asia. This gathering aimed to establish a legitimate presence in what we now recognize as global security governance. Notably, it marked the first occasion when this collective identity represented the interests of the third world and articulated its objective to manifest a new identity, known today as the Global South (Ivana, 2017).

Nonetheless, the character of solidarity within NAM during the Cold War was influenced by the national interests of its member states for several reasons: Firstly, the absence of a formal institutional framework in NAM plays a role. Secondly, the limited capacity of NAM members to provide mutual defense in the event of military aggression is a factor. Thirdly, economically and technologically advanced Western nations are reliant on support. Fourthly, the non-compulsory nature of NAM resolutions contributes to this dynamic (Keethaponcalan, 2016). Finally, NAM's dependency on the United Nations for institutional and structural backing is also significant. Following the Cold War, several additional factors contributed to the declining effectiveness of NAM: (1) The liberalization of third-world countries resulted in Western capital, financial resources, and technology coming in the form of foreign direct investment (FDI) instead of traditional aid. (2) The collaboration prioritized economic development, redirecting the focus to South-North cooperation rather than the South's reliance on the North, fostering economic partnerships across political and ideological divides. For instance, the economic cooperation between China and the USA is the key illustration of this shift. (3) Academic discussions regarding NAM's continued relevance after the collapse of the USSR raised questions about NAM's purpose and ideals. This existential debate posed a significant threat to NAM's status as an effective group.

The Non-Aligned Movement remained a significant body representing third-world nations until the end of the Cold War. One of its main goals was for member states to remain unaffiliated with the military alliances of the superpowers during the Cold War period. The end of the Cold War has raised questions about the relevance of the NAM. With the resolution of tensions between the previously conflicting sides, the importance of NAM seems to have lessened (Rauch, 2008). The conclusion of the Cold War was among various international developments that called for a reevaluation of the Non-Aligned Movement's objectives. As key issues were resolved, many of the movement's aims appeared to become less critical. Consequently, a thorough reevaluation had been overdue for quite some time. (Keethaponcalan, 2016). The main aim of NAM has become obsolete, shifting the focus to fostering unity solely among third-world nations within the group. Thus, solidarity among now so-called third-world countries (The end of the Cold War ended the bipolar world) became the sole reproaching means to justify the existence of NAM.

Keethaponcalan (2016) contends that the NAM requires transformation and puts forth three recommendations. Firstly, he suggests a "symbolic change" by altering the organization's name, which he considers to be misleading. He proposes rebranding the organization with a more inclusive and comprehensive name that doesn't solely focus on one of the organization's objectives. Instead of being tied to a specific goal, the organization's name should encompass a broader, integrative concept. Regarding the aspect of solidarity, Keethaponcalan (2016) argues that it shouldn't be rooted in opposition to any particular group but should have an independent identity. For instance, since a majority of NAM member countries come from the Global South, he proposes naming it the "Southern-Solidarity Movement" or "Global South Association." These names not only indicate a geographical region but also signify a shared socioeconomic condition among the member states. Incorporating "solidarity" into the organization's title reflects its fundamental aim to promote unity and strengthen collaboration among nations of the Global South. The essence of the movement's integration within the organization stems from the common experience of decolonization and a shared legacy of emancipation among its member states, fostering a sense of camaraderie.

Secondly, adapting to changing times necessitates structural transformation. The NAM, which comprises nearly 120 countries but lacks a central headquarters and secretariat, has historically encountered opposition to the idea of establishing a permanent secretariat from its founding members. Their concerns primarily revolved around the financial burden associated with maintaining such institutions and the apprehension of a smaller association of countries potentially taking control of the organization. However, with the growing clout of developing countries the Global South has the potential to emerge as a powerful bloc and significant hub of economic expansion (Morphet, 2004). Consequently, the organization should possess the capacity to maintain a well-structured permanent secretariat. The South Commission's, 1990 report also recommended the establishment of a permanent secretariat for the Southern secretariat, arguing that the current ad hoc approach is inadequate to address the needs of smaller developing nations. Currently, numerous significant economies in the Global South possess the capability to facilitate the creation of a dedicated secretariat (The South Commission, 1990).

The continued presence of the NAM with metamorphosed ideals and evolved principles serves as a valid rationale for its existence beyond the Cold War era. Three overarching factors substantiate the relevance of the NAM even after the demise of bipolarity: (1) The emergence of nonconventional challenges such as the rise of non-state actors leading to extremism, terrorism, cyber threat; environmental crisis leading to affect food security, migration, and other crises in the third world necessitates collaborative efforts, (2) Member countries can advance their national interests by leveraging established institutions that bring countries together from the global south, and (3) The evolving nature of the international system presents additional challenges to economies in the global south, which can be collectively addressed by countries from this region.

India's Perspective: Transformation of NAM into Global South After the End of the Cold War

After the dissolution of the USSR, India's leadership transitioned away from its longstanding distance from superpowers and outspoken criticism of the USA, addressing a variety of global issues including disarmament, climate change, and international trade negotiations. New Delhi has embraced a more collaborative approach within international entities such as the World Trade Organization. Furthermore, they stopped championing the causes of the Third World within the United Nations and other global platforms (Ganguly, 2003).

India's interpretation of the NAM continues to align with its foreign policy, but this alignment is now more firmly established and less subject to reversal compared to the Cold War era. On a philosophical level, the essence of NAM has persisted despite the dissolution of one superpower, and this can be attributed to several fundamental factors: one, the shared historical experiences of decolonized nations are unalterable and continue to shape their collective identity (Yadav, 1993); two, by retaining a policy of non-alignment with any specific superpower while accommodating various other forms of cooperation within its organizational structure to achieve political positivity, NAM has managed to maintain its relevance and vitality (Frangonikolopoulos, 1995); and lastly. In a unipolar world, the potential consequences could be even more severe than those in a bipolar world (Ikenberry et al., 2011). Consequently, the presence of a resisting force like NAM has remained essential to counterbalance this unipolar dominance. Multilateralism represents a nuanced legacy inherited from the Cold War-era NAM, upon which India's new foreign policy was poised to be constructed. India's approach to multilateralism began to progress from its early stages toward a more mature form after the end of the Cold War. The introduction of liberalization, privatization, and globalization (LPG) to the Indian economy in 1991 transformed into a mechanism for realizing India's foreign policy objectives, with multilateralism emerging as a crucial component. India's inclusion in international multilateral organizations such as the SCO, BRICS, G20, BIMSTEC, MTCR, WTO, and other significant multilateral institutions after the Cold War period had a pivotal influence in forming India's policy of alignment (Pant, 2021).

Defense and regional security became pivotal to India's foreign policy thus, economic liberalization opened doors to increased foreign investment and a deliberate move to broaden defense partnerships. By the end of the 1990s, India had effectively negotiated defense treaties with Italy, South Africa, the UK, and the USA. From 2000 to 2008, New Delhi managed to conclude 19 more defense agreements, signifying a significant shift from the only seven pacts finalized in the first 53 years following independence (Hedrick, 2009).

Events related to the nuclear regime between 1996 and 2008 changed the discourse of India's foreign policy and its multilateral quest. India's refusal to endorse the US-led CTBT (Comprehensive Test Ban Treaty) in 1996 and its subsequent nuclear weapon test in 1998 led to strained relations with major world powers like the USA, Japan, Germany, France, and even Russia (Ghose, 1997). In a surprising turn of events, it was Pakistan's incursions into Kargil in 1999 that prompted a nuclear war which was evaded after reconciliation between India and the United States. This was the initial occasion when Washington favored India over Pakistan, which was succeeded by an important trip to India by President Clinton in 2000, the first of its kind by a U.S. president in over two decades. Despite some reservations regarding India's possession of nuclear weapons, it appeared that India was drawing nearer to the Western world during this period.

India promptly transformed its multilateral alignment into bilateral collaboration, which gained traction during the 2000s as common interests came into alignment (Raghavan, 2017). The September 2001 terrorist attacks in the USA and the subsequent attack on India's parliament by Pakistani militants that year spurred collaborative initiatives between India and the USA to strengthen counter-terrorism strategies. This partnership led to dialogues on regional security and an escalation of US constraints on Pakistan to curtail cross-border attacks.

The year 2005 was a defining turning point in India's relations with the Global North. During that year, India and the USA reached a significant milestone by working out on the Next Step i.e. Strategic Partnership initiative (Ministry of External

Affairs, Government of India Brief of India-US relations, 2017). This endeavor expanded bilateral cooperation and trade in sectors like space exploration, civilian nuclear programs, and advanced technology trade. Moreover, a significant defense pact was established, mandating combined military drills, a boost in reciprocal defense commerce, and enhanced intelligence exchange. Yet, the most momentous change was the initiation of a novel nuclear collaboration framework in 2005. This culminated in the 2008 US-India Civil Nuclear Accord, which fundamentally altered India's position as a nuclear pariah (Pant, 2015).

Reflecting on the changes in India's foreign policy and its developing alliances, the term "NON" in NAM lost its relevance as India increasingly engaged with various Western agendas, including nuclear policies, security initiatives, power dynamics, liberal economic principles, and support for free trade where India's National Interest aligned. As a result, there was a need for India to reframe its terminology to maintain the structure of the movement while adapting to new challenges and emerging agendas such as environmental security, food security, healthcare access, counter-terrorism, energy security, poverty alleviation, technology sharing, free trade, and the strategic autonomy of developing nations, among others.

Until recently, the term "Global South"was primarily confined to academic discussions and was not commonly used by diplomats. However, during the G20 meeting held in New Delhi in 2023, there was a notable surge in the discourse around the Global South, particularly as India positioned itself as a potential leader of this group. Nevertheless, this concept of the Global South appeared unfamiliar to former NAM members due to a lack of updates in agendas and coordination among countries from Asia, Africa, and Latin America. The challenge of redefining the concept of the Global South, integrating it within the framework of NAM, and aligning its objectives with India's foreign policy presents a significant challenge for India's political leadership. Nevertheless, India's substantial influence in Global South countries through its humanitarian efforts, multilateral and bilateral engagements, and consistent support for human development has proven to be an effective tool for advancing its geopolitical goals.

India's Engagement with Global South: A Geopolitical Consideration

South–South cooperation (SSC) represents a crucial tenet of the Global South, and it has once again gained prominence in global politics and economics. SSC has served as a central guiding principle and a series of actions aimed at realizing these historical transformations by envisioning mutual advantages and solidarity among the marginalized states within the global system. This concept is closely linked to the idea of the New International Economic Order (Chacko, 2022). This revival has generated fresh enthusiasm for its inherent capacity to transform the global order (Gray & Gills, 2016). The Global South, in its role as a global coalition, is characterized by two major dimensions: the political dimension, encompassing SSC and the G77, complemented by the economic dimension, which includes the United Nations Conference on Trade and Development (UNCTAD). UNCTAD aimed to establish the New International Economic Order (NIEO), which sought to address structural imbalances in trade by promoting "a fair and equitable relationship"between the exports of developing nations and their imports.[2] This approach highlighted the importance of control over natural resources and the authority to nationalize major industries (Biel, 2000).

India has actively engaged in the multilateral arena by establishing several institutions based on the principle of South-South cooperation which includes BIMSTEC, Asian Development Bank, BRICS, IORA (Indian Ocean Rim Association), Mekong-Ganga Cooperation, IBSA, Asian-African Legal Consultative Organization, and many more. India's proactive role in building infrastructure for cooperation among the Global South has produced impressive outcomes as reliance on the North has reduced and the share of the Global South in the flow of capital and finance has increased substantially.

After the end of the Cold War, India's approach to foreign policy was slowly evolving to seek an independent identity by applying the above principles to advocate for a new global order, departing from a collaborative identity like NAM. However, India's nuclear vision played a role in shaping its foreign policy in a way that appeared to be more integrational and cooperative while preserving its geopolitical considerations. Right from the outset, India's foreign policy was rooted in its aspiration to establish an alternative global power equilibrium, with a central emphasis on universal, non-discriminatory disarmament. India's geopolitical ambitions came to the forefront following its nuclear tests in 1998 (Jung, 2017). Nevertheless, India did not actively advocate for its geopolitical goals in global politics. The possession of nuclear weapons significantly shifted India's normative stance on non-alignment. The advent of nuclear technology has significantly altered global security dynamics. Theorists contended that nuclear weapons functioned not as tools of direct warfare, but rather as deterrents and instruments for potential diplomatic leverage. Hence, the fundamental tenet of India's nuclear policy is the conviction that a world free of nuclear arms would enhance the safety of India as well as all countries. (Singh, 1998).

The biggest development in the twenty-first century was India's closer ties with the West which appeared to be a defeat of non-alignment but in India's foreign policy outlook, it marked the beginning of a new era of North–South Cooperation. Thus, Global South rhetoric not only marks South–South Cooperation but marks North-South cooperation without damaging the interests of the North and South (Qasmiyeh & Daley, 2018). Hence Global South is not just an addition to NAM but it is a more constructive development, and seeks to minimize the equity gap in political and economic resources. North–South cooperation refers to a

[2] See UNGA report titled, "Declaration on the Establishment of a New International Economic Order."

comprehensive framework for development collaboration that encompasses political, economic, social, cultural, environmental, and technical aspects between developed countries (the North) and developing countries (the South). This form of cooperation involves two or more developed and developing nations and can occur on bilateral, regional, sub-regional, or interregional levels. Developing countries engage in knowledge sharing, skill transfer, expertise exchange, and resource pooling to collectively work toward achieving their development objectives (UNCTAD report 2014).[3] OECD identifies common regional markets, institutional and regulatory frameworks, custom unions, and inter-state transport and communications networks as common ground of cooperation between North and South.[4]

In current international trends, cooperation has become more evident than conflict as an instrument of exerting greater geopolitical influence. Historically, conflict was a major source of creating alliances that served few powerful nations. But, India through its diplomatic maneuverings has successfully influenced a sizable amount of influence in global institutions and individual countries. India resorted to peaceful means to exert its influence through soft power and meaningful diplomacy (Chacko, 2022).

Undoubtedly, the configuration of the global system has undergone a profound transformation compared to the immediate post-World War II period. As a result, India's capacity for maneuvering has experienced a significant shift. Indian policymakers have shown considerable adaptability in responding to these shifts in the international order. Their responses, spanning economic, political, and strategic realms, have established India as a prominent major power, despite the considerable developmental challenges that it continues to face (Ganguli & Pardesai, 2007).

Conclusion

The idea of the collective Global South might have been inspired by the ideals of the NAM, but it has evolved with some unique characteristics, marked by cooperation rather than the isolationism of the NAM. Ongoing opposition to Northern dominance has fostered shared experiences, common interests, and solidarity among Global South nations. The end of the Cold War's superpower rivalry, the emergence of neoliberal globalization, and the ascent of economically powerful Southern countries have, to a certain extent, weakened Northern control over multilateral institutions which reshaped cooperation among Global South nations.

In the twenty-first century, global interactions have become remarkably intricate, with the movement of people, capital, and knowledge following intricate and multifaceted patterns. The South is no longer a passive entity waiting to be influenced

[3] See UNCTAD report titled "Transfer of Technology and Knowledge Sharing for development: Science, technology and innovation issues for developing countries" published in 2014.

[4] See "Common Ground Between South-South and North-South Co-operation Principles" Issue brief of October 2011.

by modernization from the North, even though Northern institutions continue to make efforts to engage and influence these evolving dynamics. Innovative expressions of transnational unity and collaboration within the South, both among governments and non-government entities, introduce a concept of "Southism" that is not limited by geographical boundaries. This idea more comprehensively captures the complex engagement of both international and local humanitarian personnel in crises, as well as the intermittent importance of national identity in South–South collaboration.

Nations such as India and other former colonies share a common history, which leads to cooperation based on specific issues within a world marked by an anarchic system of world politics. The significance of geopolitics within the Global South makes India's association with this group closely connected to its geopolitical aspirations. India's engagement with the non-aligned movement originally stemmed from its commitment to preserving economic and political independence. Consequently, India adapts its national interests as it progresses and develops in economic, geopolitical, and diplomatic capacities over time.

Through its endorsement of multilateralism, India aims to become a tangible bridge between the Northern and Southern regions by fostering economic, diplomatic, and technological collaboration that preserves the autonomy of all parties involved. On one front, India seeks to overcome historical reluctance that hindered cooperation with Western nations, while on another front, it endeavors to address the challenges posed by its assertive neighbor, China. In pursuit of its national interests, India has transformed the NAM element into the Global South and is continually shifting the global development agenda toward a more "Southernized" approach, reducing the reliance on Northern aid and replacing it with investments in terms of finance, capital, and human resources.

Through active repositioning within regional and global cooperative organizations, India has bolstered its geopolitical ambitions and enhanced South–South Cooperation. India has demonstrated its reliability as a partner in various domains, including trade, finance, capital, human resources, technological support, information technology, informatics, and the sharing of satellite-derived data (applied in agriculture, disaster management, and defense), among numerous other areas of collaboration.

References

Biel, R. (2000). *The new imperialism: Crisis and contradictions in north/south relations*. Zed Books.

Chacko S. (2022) India's trysts with South-South Cooperation. Retrieved Sept 27, 2023, from https://in.boell.org/en/2022/11/30/indias-trysts-south-south-cooperation

Chand, A. (1983). *Nonaligned solidarity and National Security*. Udh Publishers.

Drapac, Vesna. (2010), *Constructing Yugoslavia: A Transnational History, Bloomsbury Publishing*: London

Durkheim, E. (1964 [1893]). *The division of labour in society*. Collier Macmillan.

Featherstone, D. (2012). *Solidarity: Hidden histories and geographies of internationalism*. Zeb Books Ltd.

Frangonikolopoulos, C. A. (1995). The policy and evolution of non-alignment: Past and future. *Paradigms, Taylar and Francis, 9*(1), 62–85.

Ganguli, S., & Pardesai, M. (2007). India rising: What is New Delhi to do? *World Policy Journal, 24*(1), 9–18.

Ganguly, S. (2003). India's foreign policy grows up. *World Policy Journal, Duke University Press, 20*(4), 41–47.

Ghose, A. (1997). Negotiating the CTBT: India's security concerns and nuclear disarmament. *Journal of International Affairs, 51*(1), 239–261.

Gray, K., & Gills, B. K. (2016). South-south cooperation and the rise of the global south. *Third World Quarterly, 37*(4), 557–574.

Halas, E. (2010). Symbolic construction of "solidarity": the conflict of interpretations and the politics of memory. *Polish Sociological Association, 2*(170), 219–231.

Hedrick, B. (2009). *India's strategic defense transformation: Expanding global relationships. Letort Paper* (pp. 11–12). Strategic Studies Institute, 42.

Heyward, J. (1963). Solidarity' and the reformist sociology of Alfred Fouillée, I. *The American Journal of Economics and Sociology, 22*(1), 205–222.

Ikenberry, G. J., et al. (2011). Introduction: Unipolarity, state behavior, and systemic consequences. In G. J. Ikenberry, M. Mastanduno, & W. C. Wohlforth (Eds.), *International relations theory and the consequences of Unipolarity* (pp. 1–32). Cambridge University Press.

International Council for Human Rights Policy report (2010), "*Integrating Human Rights in the Anti-Corruption Agenda: Challenges, Possibilities and Opportunities*", ICHRP, 17 rue Ferdinand-Hodler, CH-1207 Geneva.

Ivana, A. (2017). Belgrade, The 1961 Non-Aligned Conference. Retrieved Sept 5, 2023, from Belgrade, The 1961 Non-Aligned Conference I Global South Studies, U.Va. (virginia.edu).

John Baker, Kathleen Lynch, Sara Cantillon and Judy Walsh (2006), "*Equality: Putting the Theory into Action*", UCD Equality Studics Centre, University College Dublin.

Jung, J. Y. (2017). A path to NSG: India's rise in the global nuclear order. In: Observers Research Foundation. Available via DIALOG. Retrieved from Sept 28, 2023, from https://www.orfonline.org/research/a-path-to-nsg-indias-rise-in-the-global-nuclear-order/

Keethaponcalan, S. I. (2016). Reshaping the non-aligned movement: Challenges and vision. *Bandung: Journal of Global South, 3*, 4.

Kurczewski, J. (2006). Social functions of 'solidarity'. *Polish Sociological Review, 1*(153), 111–127.

Latoszek, M. (2006). Solidarity—A contribution to social movement theory. *Polish Sociological Review, 153*(1), 39–53.

Lüthi, L. (2016). The non-aligned movement and the cold war, 1961–1973. In: Neutrality and nonalignment in world politics during the cold war. *Journal of Cold War Studies, The MIT Press, 18*(4), 98–147.

Mankekar, D. (1980). The role of Media in Nonaligned Movement. In M. K. Saini (Ed.), *Aspect of nonalignment* (p. 76). Kalamkar.

Ministry of External Affairs. (2017). Brief on India-U.S. Relations. Retrieved Sept 25, 2023, from https://www.mea.gov.in/Portal/ForeignRelation/India_US_brief.pdf

Modéer, U. & Eziakonw, A. (2019). Why multilateralism matters. United Nations Development Programme Blog, Issue: September 2019.

Morphet, S. (2004). Multilateralism and the non-aligned movement: What is the global south doing and where is it going? *Global Governance, 10*, 517–537.

Pant, H. (2015). Indian foreign policy responds to the US pivot. *Asia Policy, 19*, 89–114.

Pant, H. (2021). Strengthening global rule-making: India's inclusion in the UN security council. Observer Research Foundation, Issue brief and special report, October 12, 2021.

Qasmiyeh, E. G., & Daley, P. (2018). Introduction: Conceptualising the global south and south-south encounters. In *Handbook of south-south relations*. Routledge.

Raghavan, P. S. (2017). The making of India's foreign policy: From non-alignment to multi-alignment. *Indian Foreign Affairs Journal, 12*(4), 326–341.

Rauch, C. (2008). *Farewell non-alignment? Constancy and change of foreign policy in post-colonial India*. Peace Research Institute Frankfurt.

Rawls, J. (1989). *A theory of justice* (11th ed.). Oxford University Press.

Sladojevic, Ana. (2015), *Slike o Africi/Images of Africa, Belgrade*: Museum of Contemporary Art.

Spaskovska, Ljubica. (2017), *The last Yugoslav generation: The rethinking of youth politics and cultures in late socialism*, Manchester University Press: Manchester.

Singh, J. (1998). Against nuclear apartheid. *Foreign Affairs, Council on Foreign Relations, 77*(5), 41–52.

Tadic, b. (1976), "*the major characteristics and dilemmas of non alignment in contemporary international relations*", India International Centre Quarterly, 3(3), Role of Non alignment in a Changing World (July 1976), pp. 47–55.

The South Commission. (1990). *The challenge to the South: The report of the South Commission*. Oxford University Press.

Yadav, R. S. (1993). NAM in the new world order. *India Quarterly, 49*(3), 47–68.

Strategic Location of Siliguri Corridor and Its Issues

Fidel Rahmati

Introduction

The Siliguri Corridor is a small area in West Bengal, India, surrounded by Nepal, Bhutan, Bangladesh, India's Seven Sister States, and the Chumbi Hills. It's important for politics and diplomacy, but it hasn't seen much practical development yet (Malik, 2015).

The Siliguri Corridor is very important for India because it links the northeastern states to the rest of the country. It's narrow, about 20.8–40 km wide, and sandwiched between Bangladesh and Nepal. Around 100,000 people live there, mostly Hindus (88.16%), with Muslims (9.64%), Christians (0.97%), Buddhists (0.54%), and others (1.20%) making up the rest (Fig. 1).

According to the 2011 records, Siliguri city had a population of 705,579 individuals, with 51 percent being male. In Siliguri, most people are Bengali, but there are also smaller groups such as Gorkhas, Biharis, and Mawaris. In Siliguri, people speak Bengali, Hindi, English, and Nepali, with some also speaking Tibetan languages. The city enthusiastically celebrates various religious festivals like Durga Puja, Chhath Puja, Laxmi Puja, Saraswati Puja, and Diwali.

Siliguri has five main railway stations:

1. Jalpaiguri Town: This is the oldest station in the area.
2. Jalpaiguri Road: Opened in 1944, it serves as the main railway station of the city.
3. Mohitnagar: Located on the Haldibari-New Jalpaiguri line.
4. Raninagar Jalpaiguri Junction: This station serves Jalpaiguri, Raninagar, and nearby areas in the western part of the city.
5. New Jalpaiguri: It is the largest railway junction nearby.

F. Rahmati (✉)
Department of Geography, Osmania University, Hyderabad, Telangana, India

K. Arora, R. R. Anand (eds.), *The Solidarity Approach in Geography*,
https://doi.org/10.1007/978-3-031-86753-8_14

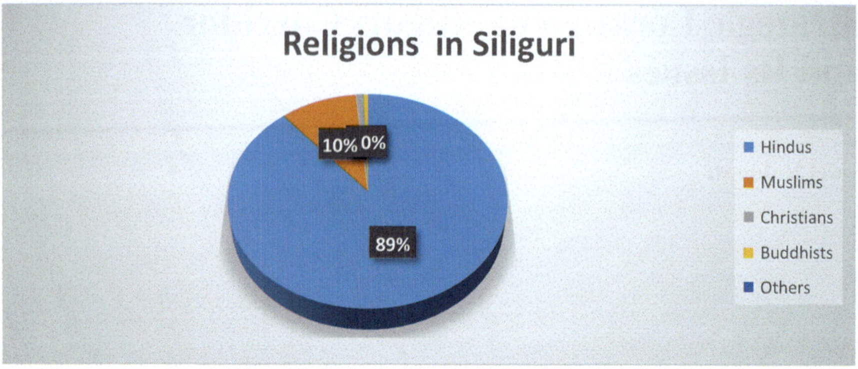

Fig. 1 Religion in Siliguri. (Source: Census of India 2011)

The Siliguri Corridor faces significant security challenges from neighboring countries, including Bangladesh, China, Nepal, and Bhutan. These security concerns emerged following the partition of Bengal in 1947 when India and Pakistan were created. Additionally, Beijing did not officially recognize Sikkim's accession to India until 2003, further complicating the security dynamics in the region.

The Indian central government has proposed a four-lane road project connecting Silchar in Southern Assam to Siliguri as part of its National Highway Development initiatives, highlighting the strategic importance of the Siliguri Corridor as a crucial backbone for India's connectivity and regional development.

Study Area

Jalpaiguri, situated within the Siliguri Corridor, is positioned at coordinates 26.52°N and 88.73°E, with an elevation of 75 m (246 ft.). The Jalpaiguri located at the heart of region within the Siligure Corridor having a strategic location (Fig. 2). The Corridor provides a hup linking northeastern states with rest of the country. Despite its relatively low elevation, Jalpaiguri serves as a vital transportation hub, with its railway stations facilitating the movement of people and goods across the corridor and beyond. Its geographical coordinates underscore its significance in bridging the gap between the northeastern states and the mainland.

Jalpaiguri is situated in the fertile plains at the foot of the Himalayas, enjoying the advantage of many rivers and streams that start in the mountains. Besides the Teesta River, the Karala River, often called the "Thames of Jalpaiguri," runs through the city, splitting it into two parts.

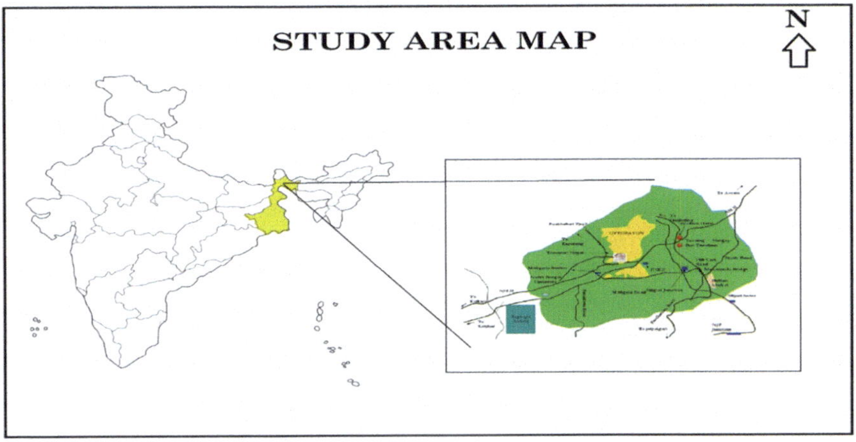

Fig. 2 Study area. (Source: Created by Researcher)

Research Methodology

This study is based on secondary data obtained from a wide variety of sources. The data collection process involved gathering information from reports, books, academic journals, government publications, and statistics from different agencies, online sources, and gray literature. These sources were systematically accessed and reviewed to ensure a comprehensive and robust foundation for the research.

The approach provides a well-rounded perspective and enhances the credibility and depth of the research, ensuring that the conclusions and insights derived from this data are both reliable and informative.

Brief Literature Review

Analyzing existing literature is a fundamental step in research, offering insights, theories, and methodologies while guiding the study's direction. It acts as a compass, ensuring the research contributes to the field and addresses knowledge gaps.

Sun, Y. (1999) conducted a study on multinational corporations (MNCs) choosing entry modes when entering China, focusing on socioeconomic factors. The study found that MNCs with cultural backgrounds more aligned with China tended to have higher foreign equity shares in Foreign Invested Enterprises (FIEs). Conversely, greater socio-cultural differences deterred MNCs from opting for wholly owned subsidiaries. Additionally, a positive relationship was observed between the level of technology in investment projects and foreign equity shares in FIEs. Lastly, the study emphasized that economic conditions and government policies played a pivotal role as determinants in MNCs' choice of entry modes and FIE ownership in China (Sun, Y. 1999).

Davis et al. (2000) conducted a study in the pulp and paper industry, using institutional theory to analyze how U.S. firms chose their entry modes into foreign markets. They looked at factors both within the parent organization and in the host country that influenced these choices, such as wholly owned subsidiaries, exporting, or licensing agreements. Their findings showed that entry-mode decisions were mainly shaped by internal and external institutional pressures. They highlighted the importance of both parent company norms and host country factors in determining international entry strategies. They also noted that wholly owned subsidiaries tended to align closely with internal norms, while other entry modes like exporting or joint ventures were more influenced by external pressures. Mixed entry modes indicated lower levels of pressure to conform (Davis et al., 2000).

In 2015, Hasan Yaser M. conducted a study on the Siliguri Corridor in the Eastern Himalayan Region, shedding light on its pivotal geographic location. Situated in Indian West Bengal, this corridor acts as a vital link connecting Nepal, Bhutan, Bangladesh, India's Seven Sister States, and the Chumbi Hills. Its strategic position renders it economically and politically significant for neighboring nations. However, its development is constrained, and its proximity to China and India's Seven Sister States poses challenges such as smuggling and terrorism. Despite efforts in 2002 to establish a free trade corridor involving Nepal, Bhutan, and Bangladesh, these issues persist, necessitating economic and diplomatic resolutions (Malik, 2015).

Analysis

The Siliguri Corridor occupies a pivotal position at the juncture of Nepal, Bhutan, India's Seven Sister States, and mainland India, establishing it as a central nexus for economic interactions within the Eastern Himalayan Region. Despite its strategic significance, the corridor grapples with persistent challenges, including political instability in the Seven Sister States and limitations on socioeconomic development due to external interests from neighboring countries such as Nepal and Bhutan. These factors impede the region's progress and hinder the realization of its full potential.

One of the primary reasons for the importance of the Siliguri Corridor is its function as a crucial transportation artery within the framework of the Bangladesh–Bhutan–India–Nepal (BBIN) motor vehicles agreement (MVA). This agreement, signed on June 15, 2015, in Thimphu by the four South Asian nations, aims to bolster connectivity for transportation, trade, and tourism. By facilitating smoother movement of goods and people across borders, the corridor underpins regional integration efforts and fosters closer economic ties among the participating nations, thereby unlocking new avenues for growth and development.

Moreover, the Siliguri Corridor's strategic location and its role in the BBIN MVA highlight its potential to serve as a linchpin for broader regional cooperation and connectivity initiatives. By leveraging its geographical advantages and strengthening infrastructural linkages, the corridor can emerge as a catalyst for enhanced regional trade, economic integration, people-to-people exchanges and politico socioeconomic solidarity. However, to fully realize this potential, concerted efforts are required to address existing challenges and foster a conducive environment for sustainable development and cooperation across borders.

Bangladesh

India and Bangladesh boast a significant physiographic boundary stretching approximately 4176 kilometers, making it the world's fifth-largest boundary. Since 1971, these two nations have maintained a strong trade partnership. However, starting in 2002, Bangladesh's trade with China has seen a remarkable surge, surpassing its trade volume with India by a considerable margin (Sahoo, 2011).

India is worried about its deteriorating trade relations and political tensions with China. In 1991, India's exports to Bangladesh were 150% higher than China's exports to Bangladesh. However, by 2010, this gap had shrunk significantly to just 30% (Sahoo, 2011).

Meanwhile, India and Bangladesh face several significant issues and conflicts. A major concern is the illegal immigration to India's Border States, which began after the independence of West Pakistan, now Bangladesh. Another significant challenge is the high levels of crime and human trafficking, particularly in the northern border area of West Bengal. This region is notorious for drug trafficking and other illegal activities. India's Narcotics Control Bureau reports daily seizures of 100–250 km of marijuana and other drugs near Siliguri, West Bengal, indicating the severity of the problem.

Moreover, the area serves as a refuge for Maoist rebels from neighboring countries like Bangladesh and Nepal. Siliguri also acts as a transit point between India and Bangladesh for stolen goods, famously known as "Chor Baazar," where stolen vehicles from all over India are smuggled and sold in Bangladesh. Without strict enforcement measures and regulations, this situation could pose a significant security risk to India.

Nepal

India and Nepal share a deep historical and cultural connection. Nepal's longest boundary, spanning 1751 km, is with India, and historically, Nepal has depended on India for trade and commerce. However, there are geopolitical tensions between the two nations, particularly regarding the Kalapani region in Uttarakhand, which Nepal claims as its own. This disagreement intensified on May 20, 2020, when Nepal released a political map asserting Kalapani as its territory. The root cause lies in

China's increasing influence on Nepal's foreign policy, which India must consider in its long-term national interests. To maintain positive relations with Nepal and counter China's influence, India should contemplate reevaluating its foreign policy approach.

Bhutan

India shares a 699-km international boundary with Bhutan, and historically, Bhutan has been a loyal and friendly neighbor. Until 2020, Indian citizens did not require a visa to visit Bhutan. However, in February 2020, Bhutan imposed a visa fee of INR 1200 due to issues related to unethical behavior and discipline by some Indian citizens, which were causing damage to Bhutan's heritage. Under a 2007 treaty, India has ensured Bhutan's sovereignty and protected it from Chinese military actions, notably at Doklam. Given Bhutan's consistent political, cultural, and economic support, India should maintain its policies and laws to sustain this enduring and positive relationship.

China

Physically, while Siliguri, as part of India, does not share a direct border with China, the Indian Territory extends about 4056 km along the China–India boundary. The Siliguri Corridor holds paramount importance for India, outweighing the interests of neighboring countries such as Nepal, Bhutan, and Bangladesh. The Indian government is actively engaged in advancing the social and economic development of the corridor. However, it proceeds with caution to mitigate any potential disruptions that could affect the stability of its northeastern states, namely Arunachal Pradesh, Assam, Meghalaya, Tripura, Nagaland, Manipur, and Mizoram.

The Siliguri Corridor, or "Chicken's Neck," is strategically vital as it's the sole land link between mainland India and the northeastern states ("Seven Sisters"). Any instability here could harm India's security and economic connectivity in the northeast. The Indian government is cautious in its development efforts to ensure security and uninterrupted connectivity.

Meanwhile, Siliguri serves as a significant gateway for Chinese products entering the Indian market, making it a focal point for trade. In recent years, China has made substantial investments in the neighboring countries of the region. This strategic move is part of China's broader diplomatic efforts in South Asia, where it seeks to play a prominent role and expand its influence on the global stage.

China and India frequently contend over territorial boundaries, engaging in a continuous tug-of-war. Historically, the two nations didn't share frontier regions for thousands of years (Rahmati et al., 2020).

Since its transformation into the People's Republic of China in 1950, China has consistently aimed to exert control over South Asia for economic and strategic advantages, although its efforts have not fully succeeded, largely due to India's progressive stance.

Given that both India and China are significant and growing global powers, they need to prioritize mutual understanding and cooperation rather than engaging in unnecessary conflicts. This approach has the potential to shift global influence back toward Asia from Western countries.

Conclusion

The Siliguri Corridor stands as a critical Special Economic Zone in India, serving as the primary trade route for commerce with Southeast Asian nations. Its strategic significance is bolstered by its unique geographical advantages, yet it remains vulnerable due to its proximity to neighboring countries like Bangladesh, Bhutan, and Nepal. By maintaining positive and constructive relationships with these neighboring nations, India can ensure the security and integrity of the South Asian region against external influences.

China's persistent efforts to expand its influence in neighboring countries have been driven by its pursuit of regional dominance and territorial security in South Asia. In response, India should remain focused on its development and prioritize fostering good relations with its neighbors. By doing so, India can not only safeguard the stability of the Siliguri Corridor but also contribute to regional harmony and progress, ultimately reinforcing the South Asian region against external pressures.

Addressing these challenges requires a comprehensive approach that balances economic development with security concerns and diplomatic engagement, ensuring the long-term stability and prosperity of the region.

Recommendations

Based on a comprehensive analysis of the research topic, the study offers the following recommendations:

1. Develop a Comprehensive Strategy: India should formulate a comprehensive, long-term strategy for the Siliguri Corridor that encompasses both its geoeconomics opportunities and geopolitical challenges. This strategy should prioritize the corridor's strategic interests and proactively manage its potential for enhancing regional stability.
2. Strengthen Regional Relations: Maintaining positive and constructive relationships with neighboring countries such as Bangladesh, Bhutan, and Nepal is essential for safeguarding the Siliguri Corridor's security and South Asian

regional integrity. India should prioritize diplomacy and cooperation with these nations.

3. Focus on Domestic Development: Amidst China's efforts to expand its influence in neighboring countries, India should concentrate on its development and growth. By fostering good relations with its neighbors and emphasizing domestic progress, India can contribute to regional harmony and resilience against external pressures, bolstering the stability of the Siliguri Corridor.
4. Strengthen Border Security: Enhance border surveillance and security measures to curb illegal activities such as smuggling, human trafficking, and infiltration. This could involve deploying additional personnel, utilizing advanced technology for monitoring, and fostering cooperation with neighboring countries to combat cross-border crimes.
5. Promote Socioeconomic Development: Implement targeted initiatives to promote socioeconomic development within the corridor and its surrounding regions. This could include investment in infrastructure, education, healthcare, and employment opportunities to uplift local communities and reduce the incentives for engaging in illicit activities.
6. Foster Diplomatic Engagement: Engage in diplomatic dialogue and negotiations with neighboring countries, particularly Nepal, Bhutan, and Bangladesh, to address territorial disputes and promote cooperation. This could involve establishing joint committees or forums to discuss issues of mutual interest and work toward peaceful resolutions.
7. Enhance Regional Connectivity: Improve transportation and communication infrastructure within the corridor to facilitate smoother movement of goods and people. This could involve upgrading roads, railways, and air links to enhance connectivity between the northeastern states and the rest of India, as well as neighboring countries.

Acknowledgments I would like to extend my sincere gratitude to Dr. Mohd. Akhter Ali, Assistant Professor in the Geography Department at Osmania University, for his continuous encouragement and generous support during the process of writing this paper. His guidance has been invaluable and greatly appreciated.

References

Davis, J., Smith, M., & Lee, R. (2000). Impacts of climate change on agricultural production. Journal of Climate Change Studies, 34(4), 210–220.

Malik, H. Y. (2015). Siliguri: A geopolitical manoeuvre corridor in the Eastern Himalayan region for China and India. *Contemporary Chinese Political Economy and Strategic Relations, 1*(3), 699. https://doi.org/10.1007/s41111-015-0003-4

Rahmati, F., Ali, M. A., & Kamraju, M. (2020). A study on India-China current geopolitical issues and implications. *International Journal of Scientific Research in Engineering and Management (IJSREM), 4*(6), 1.

Sahoo, P. (2011). India losing ground to China on trade with Bangladesh. *East Asia Forum, 20*.

Sun, Y. (1999). A study on environmental changes and their impact. Environmental Science Journal, 45(2), 123–130.

Enhancing Regional Solidarity through HADR: A Case Study of Indo-Nepal Disaster Response

Smrity Upadhyay, Kavita Arora, and Rashmi Rani Anand

Introduction

India and Nepal share a profound and enduring bond, characterized by strong friendship and cooperation, an open border, and widespread people-to-people interactions grounded in shared kinship and cultural heritage (MoEA, India). The bilateral relationship between India and Nepal was established on June 17, 1947, marking the beginning of a strong and enduring partnership. In 2022, the two nations celebrated the 75th anniversary of their official diplomatic ties, commemorating decades of cooperation, shared cultural heritage, and a deep-rooted friendship that has grown stronger over the years (Poudel, 2022). This relationship is demonstrated by the longstanding tradition of unrestricted movement of people across the 1850-km border, connecting Nepal with the northern Indian states of Uttarakhand, Uttar Pradesh, Bihar, West Bengal, and Sikkim (Glaser, 1993). At the heart of this special bond lies the Treaty of Peace and Friendship between India and Nepal, established in 1950, which stands as the bedrock of their diplomatic relations. Under this treaty, Nepalese citizens enjoy privileges and opportunities akin to their Indian counterparts, with nearly eight million Nepalese residing and contributing to India's socioeconomic fabric.

India–Nepal alliance is deeply entrenched in shared customs and traditions and valued as a relationship of "Roti-Beti" (food and marriage) signifying the

S. Upadhyay (✉)
Department of Geography, Delhi School of Economics, University of Delhi, New Delhi, India

K. Arora
Department of Geography, Shaheed Bhagat Singh College, University of Delhi, New Delhi, India

R. R. Anand
Centre for African Studies, School of International Studies, JNU, New Delhi, India

K. Arora, R. R. Anand (eds.), *The Solidarity Approach in Geography*,
https://doi.org/10.1007/978-3-031-86753-8_15

intertwining of familial and cultural connections (Poudel, 2022). Nepal and India are often depicted as the cradles of ancient oriental philosophies, with Hinduism and Buddhism tracing their origins back to these lands. Additionally, both nations share a linguistic connection, with Nepali and Hindi emerging from the common roots of Sanskrit (Karki & Hari, 2020; Poudel, 2022). This deep cultural and linguistic intertwining is exemplified by iconic landmarks such as the birthplace of Lord Buddha in Lumbini, the revered sites of Bodhgaya, Pashupatinath, and the spiritual epicenter of Kashi Vishwanath (Karki & Hari, 2020). These sacred locales serve as enduring symbols of the shared heritage and spiritual continuity between Nepal and India, underscoring the profound cultural bonds that unite the two nations. Furthermore, both nations are united by a common landscape and cultural heritage, reinforced by intermarriages and mythological narratives.

Despite the rich cultural heritage, both India and Nepal grapple with the challenges posed by various natural and anthropogenic disasters. India's diverse geographical and climatic conditions render it vulnerable to an array of calamities, including droughts, floods, earthquakes, landslides, and wildfires, along with pollution incidents (NDMA Report, 2023). Similarly, Nepal, situated within a geophysically dynamic region characterized by youthful mountain ranges and diverse climates, confronts recurring hazards such as floods, earthquakes, landslides, and glacier lake outburst floods (GLOFs) (NDRRMA, 2022). The compounding effect of these multi-hazard exposures often escalates into medium to large-scale disasters.

Methodology

The field of Humanitarian Assistance and Disaster Relief (HADR) is often regarded as a form of soft power disaster diplomacy; however, it is marked by a scarcity of comprehensive literature, which highlights significant gaps in scholarly discourse. This is particularly evident in the lack of long-term case studies within the realm of HADR, presenting a critical limitation in understanding the nuances and complexities of disaster response efforts. Existing literature tends to focus on case studies with narrow temporal constraints, overlooking the broader historical context that shapes observations and outcomes.

This chapter endeavors to address this gap by embarking on a geographically delineated case study spanning the extensive history of HADR between India and Nepal. By delving into the intricacies of this enduring relationship, the study aims to elucidate the potential for comprehending and enhancing HADR practices over the long term, shedding light on evolving dynamics and strategic considerations in disaster response.

To achieve this, the study employs both descriptive and analytical approaches to investigate disaster diplomacy between India and Nepal. Data has been sourced from various government websites and reports. From India, information was collected from the Ministry of External Affairs, the Ministry of Home Affairs, and the annual reports of the National Disaster Management Authority. Similarly, from

Nepal, data was gathered from the Ministry of Home Affairs, the Ministry of Foreign Affairs, and the annual reports of the National Disaster Risk Reduction and Management Authority. Additionally, reports such as the Global Assessment Report (GAR) and the Global Disaster Assessment Report (GDAR), along with content analyses of newspaper articles, were utilized to understand the bilateral mechanisms of disaster response in other countries. Secondary sources, including published research papers, journals, and book references, were also incorporated.

This chapter aims to provide a comprehensive investigation of disaster diplomacy between India and Nepal, highlighting the dynamics of their cooperation and the impact of regional solidarity in disaster management. By exploring the extensive history and strategic considerations in their HADR practices, this study seeks to contribute to the broader understanding and enhancement of disaster response efforts in the region.

Regional Solidarity and Its Importance

Today, India and Nepal are active members of various trans-regional, regional, and sub-regional groupings, reflecting their evolving engagement on multiple fronts. However, delving into history, during the British Raj era, India maintained an "exclusive relationship" with the Kingdom of Nepal. The subsequent partition of the Indian Subcontinent and India's intermittent disputes with neighboring nations has negatively influenced the dynamics of regionalism in the region.

The concept of "region" is interpreted differently across various disciplines, yet scholars generally concur that regions are socially constructed entities (Hettne, 2005). In the specific context of India and Nepal, geographical proximity offers a more focused but significant definition of "region" (Behr & Jokeia, 2011). This is especially relevant when considering the recurrent natural disasters that impact both countries. Studies indicate that disasters do not conform to administrative boundaries (Dutta, 2021; Upadhyay & Arora, 2023); instead, they affect populations living across borders in a state of harmonious coexistence. Therefore, regional mechanisms become crucial in addressing cross-border disaster situations, as they facilitate coordinated responses and cooperation between neighboring nations. Solidarity encompasses various dimensions and is defined by scholars in nuanced ways. Banting and Kymlicka (2017) provided a comprehensive characterization of solidarity, defining it as more than just a concept. It embodies shared acceptance, cooperation, and reciprocal support, particularly in times of adversity or need. They emphasize that solidarity goes beyond mere empathy or sympathy; it entails active engagement and collective action to address challenges and support individuals or groups facing difficulties. Stjernø (2011) further enriches the concept of solidarity by emphasizing its facets of interdependence, inclusion, and safeguarding against social risks. According to Stjerno, solidarity involves recognizing and embracing the interconnectedness among individuals and communities, understanding that their well-being is intertwined. This interconnectedness underscores the importance of inclusivity, ensuring that all members of society, particularly vulnerable or marginalized groups, are actively supported and integrated into social structures. This perspective underlines the interconnected nature of communities and the need to safeguard individuals from various social vulnerabilities which is crucial in terms of the India–Nepal relationship. Ross (2010) adds another layer by describing mutuality as a fundamental component of solidarity. Where mutuality refers to the shared responsibilities and benefits that arise from cooperative actions, reinforcing the notion that solidarity is built on a foundation of joint trust mechanism and support. Solidarity can be also defined as an action or institution through which a country provides aid or relief assistance to another country (Gerhard et al., 2019). Whereas Sangiovanni (2015) provides a slightly different angle by defining a nuanced perspective on solidarity by framing it as a dedication to both sharing and actively pursuing a collective objective aimed at overcoming challenges. Drawing from historical precedent, solidarity has been put into practice to tackle and reduce a broad spectrum of societal risks and challenges, including issues such as marginalization, unemployment, illness, and natural disasters (Ciornei and Ross, 2021). So, the historical perspective illustrates how solidarity has evolved into formal systems and structures designed to protect individuals and communities from a wide range of socioeconomic and environmental threats. Thus, fostering regional solidarity suggests that countries can improve their relations by assisting during crises or natural disasters. This cooperation not only enhances immediate response efforts but also strengthens long-term regional solidarity and collaboration. By supporting each other in times of need, nations can build stronger, more resilient bonds, promoting a sense of mutual trust and interdependence.

India-Nepal relations are very deeply rooted and benefit from its foundation on the principle of "roti-beti" (bread and kinship), fostering a familial bond between the two nations. Additionally, India's role as "big brother" toward Nepal, further facilitates the enhancement of regional solidarity. The Indian approach to humanitarian assistance is deeply rooted in the nation's cultural and spiritual ethos, exemplified by the belief in "Vasudhaiv Kutumbakam" (the world is one family) (Chauhan, 2023), a principle that has been ingrained in Indian society for generations. This ethos, coupled with the principle of non-alignment advocated by Jawaharlal Nehru, has shaped India's humanitarian outlook and commitment to alleviating human suffering. It is observed that the Indian government employs the terms "humanitarian assistance" or "disaster relief" to describe endeavors aimed at mitigating the impact of natural disasters such as cyclones, droughts, earthquakes, floods, or even civil strife (Meier & Murthy, 2011).

Over the past two decades, India's credibility as a provider of humanitarian aid has witnessed a notable enhancement. Guided by its "Neighborhood First" policy, India has demonstrated a steadfast commitment to extending assistance to its South Asian neighbors in times of need. Through concerted efforts, India has emerged as a reliable first responder, playing a pivotal role in addressing humanitarian crises within the region (Canyon, 2021; Chauhan, 2023). This evolution highlights India's growing stature as a significant actor in regional humanitarian endeavors because India's responsibility to support its neighbors, especially during challenging times such as natural disasters, and reflecting its commitment and cooperation among neighboring nations can foster regional solidarity.

Regional Solidarity through Humanitarian Assistance and Disaster Relief (HADR) is an essential framework for enhancing cooperation, resilience, and mutual support among nations within a specific region. The core principle of HADR is that no country should face the aftermath of a disaster alone; instead, neighboring nations should come together to provide aid, share resources, and offer expertise to manage crises effectively. This approach not only improves immediate disaster response but also fosters long-term trust and interdependence, which are crucial for regional stability and development.

The significance of humanitarian assistance particularly in South Asia, gained widespread recognition following the devastating Indian Ocean tsunami in 2004, which prompted a remarkable response from the global community. A staggering 92 countries promptly pledged their support, underscoring the growing awareness and solidarity in addressing humanitarian emergencies (Harmer & Cotterrell, 2005; Parmar, 2012). Before this, at the time of the humanitarian crisis in Bosnia, only 16 countries have come with humanitarian support (Parmar, 2012). This pivotal shift in international dynamics offered the importance of collective action and cooperation in times of crisis among the nations.

Likewise, it is imperative to acknowledge India's evolving role in the realm of aid and development. Historically, India has been a recipient of foreign aid since its independence, relying on external assistance to address socioeconomic challenges (Parmar, 2012). However, in recent years, India's economic ascent and growing global influence have facilitated its transition from an aid recipient to a donor nation. This transformation has not only accentuated India's rising stature on the world stage but has also positioned it as a significant player in international development efforts. Notably, India's status as a non-member of the Development Assistance Committee (DAC) reflects its expanding role and contributions to global humanitarian initiatives (Harmer & Cotterrell, 2005).

HADR to Disaster Response

Understanding what constitutes a disaster and how the academic community defines such events is crucial. Defining a disaster has long been discussed (Quarantelli, 1998; Kelman, 2018). Disaster studies offer various perspectives on the nature of disasters. Some researchers characterize disasters as inherently social phenomena, While others define disasters as events that disrupt society due to unforeseen and often catastrophic occurrences, which can profoundly impact communities, infrastructure, and social systems. These events typically exceed the capacity of affected regions or organizations to cope (Gayan, 2023), necessitating external assistance and coordinated efforts to recover and rebuild (Reinhardt, 2015; Reinhardt & Lutmar, 2022). For instance, an earthquake in a deserted region like Antarctica is not deemed a disaster because it does not disrupt human society. A natural extreme event only becomes a disaster when it leads to significant loss of life and economic assets.

Several organizations argue that there is nothing inherently natural about disasters (UNISDR, 2011). According to Kevin Blanchard, “Disasters are not natural” (UNDRR, 2021) emphasizing that human decisions and actions transform hazards into disasters. Disasters result from sudden, calamitous events that overwhelm communities and exceed their ability to cope with their resources. Raymundo Padilla, an expert on disaster media coverage at Mexico’s Colima University, asserts that “earthquakes do not kill; it is the buildings and bad construction” (UNDRR, 2021). This perspective highlights the critical role of human factors in disaster outcomes. While natural hazards such as earthquakes and tornadoes are unavoidable, the loss of life and economic damage they cause, are not inevitable. Effective disaster preparedness can significantly reduce risks. So basically, disasters are the result of dynamic interactions among hazards, pre-existing local vulnerabilities, and exposure (GAR, 2022). They are influenced by human choices and shaped by the socio-economic, technological, and demographic characteristics of a society (IPCC, 2018; UNDRR, 2019; Gousse-Lessard et al., 2022; GAR, 2022). According to the United Nations, the incidence of extreme weather conditions has surged significantly over the past two decades. This trend was particularly evident in India in 2021, as reported by Business Insider (Sethi, 2021). When compared to the average frequency of major natural disasters over the previous three decades (1991–2020), there was a 13% increase in 2021 alone (GDAR, 2021). Asia, and more specifically South Asia, experiences the highest number of natural disasters globally (GDAR, 2021).

India and Nepal face heightened susceptibility to natural disasters because of their distinctive socioeconomic backgrounds and geo-climatic settings, resulting in multiple recurring disasters each year. The Global Natural Disaster Assessment Report (GDAR, 2021) places both India and Nepal among the top ten countries with the highest death tolls from natural disasters. Furthermore, India is ranked within the top five countries globally in terms of disaster frequency (GDAR, 2021).

When examining the relationship between India and Nepal through the lens of Humanitarian Assistance and Disaster Relief (HADR), it becomes crucial to understand the dynamics of how disasters operate. Disasters evoke empathy and can demonstrate the humanity of rivals, which is essential for creating opportunities for cooperation (Akcinaroglu et al., 2010). Disasters have the potential to facilitate cooperation between nations, but they can also exacerbate rivalries. Some studies propose that despite the human tragedies caused by earthquakes, floods, windstorms, and tsunamis, these catastrophic events can paradoxically provide opportunities for international cooperation (Akcinaroglu et al., 2010). This cooperation may transcend political differences and historical rivalries, as nations come together to provide aid, share resources, and coordinate disaster response efforts. On the contrary, other studies argue that natural disasters have the potential to exacerbate existing tensions and lead to conflict rather than fostering cooperation (Ker-Lindsay, 2000; Bhavnani, 2006; Brancati, 2007; Miguel et al., 2004). These researchers suggest that when resources become scarce, competition for aid and relief efforts may intensify, particularly in regions or between nations with pre-existing political or territorial disputes.

Empirical evidence from various case studies indicates that disaster-related activities typically do not initiate new peace-building initiatives or reduce conflict on their own. Instead, they can catalyze or support pre-existing diplomatic processes, though these benefits are often short-term (Kelman, 2012, 2016; Kelman et al., 2018). Disasters play a significant role in accelerating ongoing interstate diplomacy by intensifying official communications, encouraging solidarity through collaborative recovery initiatives, and promoting humanitarian efforts that defy stereotypes and cultivate environments conducive to cooperation (Quarantelli & Dynes, 1976; Ker-Lindsay, 2000, 2007; Kelman, 2003, 2006; Akcinaroglu et al., 2010). Moreover, humanitarian responses during disasters can bridge divides, fostering empathy and understanding that pave the way for enhanced diplomatic relations and cooperative ventures beyond immediate crisis management.

After a disaster strikes, the process of fostering attitudinal change hinges significantly on the principles of humanitarianism and compassion. Thus, while the relationship between India and Nepal in the context of HADR can be complex, it emphasizes the potential of disaster response efforts to both challenge and enhance diplomatic relations.

HADR at the Time of the Earthquake and Flood

Disaster relief and post-disaster reconstruction in South Asian countries with critical geo-economic and geopolitical interests have increasingly become regular and institutionalized components of India's foreign policy. This strategic approach aims to maintain India's regional dominance and counter China's growing influence, particularly in Nepal. Geopolitically and strategically, Nepal holds immense importance for India. The country's response to natural disasters and calamities in South Asian states has been commendable, as noted by Nandy and Naha (2022). As the core state in the region, India cannot shirk its responsibility to mitigate disaster-induced loss and damage.

India's aspiration to be a global power is heavily dependent on its capacity and acceptance within South Asia. The region frequently experiences disasters such as floods, earthquakes, famines, and industrial accidents. To bolster its standing and counter China's influence, India has effectively employed "relief diplomacy" toward its South Asian neighbors. This approach not only addresses immediate humanitarian needs but also strengthens bilateral relations and regional solidarity.

The importance of Nepal to India's foreign policy is evident from the substantial increase in aid relief following the 2015 earthquake (Table 1). Over the past decade, data shows a significant hike of over 10% in aid provided to Nepal compared to previous years, reflecting Nepal's strategic significance. India's efforts include providing prompt humanitarian aid, engaging in post-disaster reconstruction, and supporting long-term recovery initiatives. These actions underline India's commitment

Table 1 Aid relief fund given by India to Nepal

S. No	Year	Aid in Indian Rupee
1.	2015–2016	3.10 billion
2.	2016–2017	3.20 billion
3.	2017–2018	3.75 billion
4.	2018–2019	
5.	2019–2020	11.98 billion
6.	2020–2021	8.80 billion
7.	2021–2022	7.73 billion
8.	2022–2023	6.8 billion

Source; Indian Budget, 2015, 2016, 2017, 2019, 2020, 2021, 2022, 2023

to regional stability and its role as a responsible global actor in disaster response and humanitarian assistance.

On April 25, 2015, Nepal was hit by a devastating 7.8 magnitude earthquake, which was followed by a powerful 7.4 magnitude aftershock on May 12, 2015, the Government of India demonstrated an exemplary Humanitarian Assistance and Disaster Relief (HADR) response (India-Nepal bilateral relations, 2020). India rapidly mobilized National Disaster Response Force (NDRF) units and specialized aircraft transporting crucial rescue and relief supplies to Nepal within 6 hours of the earthquake's onset. This swift and comprehensive response involved dispatching 16 NDRF teams and conducting 39 sorties using Indian Air Force (IAF) aircraft, delivering a total of 571 tons of essential relief materials. These supplics included food, water essential rescue tools, and medical necessities. India also provided tarpaulins, tents, and blankets by effectively meeting the immediate requirements of those impacted (Embassy of India, Kathmandu, Nepal, 2022).

Additionally, medical teams from India were strategically placed in different regions of Nepal, providing much-needed medical assistance and care to the injured and displaced persons. In addition to the on-ground efforts, India played a crucial role in restoring infrastructure, specifically aiding in the restoration of three power substations in the Kathmandu Valley, which was vital for resuming essential services and stabilizing the affected region.

On June 25, 2015, at the International Conference on Nepal's Reconstruction in Kathmandu, the Indian Government announced a significant relief supply of US$1 billion for the post-earthquake reconstruction. This package included a US$250 million grant and a US$750 million concessional Line of Credit. In February 2016, an agreement was reached on the utilization of the US$250 million grant, allocating US$100 million for the reconstruction of 50,000 private houses and US$50 million each for sectors like the well-being of health, education, and cultural heritage. The collaboration reached new heights when the two governments signed sector-specific agreements on August 24, 2017 (Embassy of India, Kathmandu, Nepal, 2022).

In September 2016, an agreement for a Line of Credit (LoC) totaling US$750 million for post-earthquake reconstruction projects was signed, and it became operational in February 2017. By March 2018, the Government of India had entered into two partnership agreements with the United Nations Development Program (UNDP) and the United Nations Office for Project Services (UNOPS). These agreements aimed to provide socio-technical facilitation for housing beneficiaries in Nepal's Gorkha and Nuwakot districts. Out of the 50,000 earthquake-affected beneficiaries assigned to the Government of India in these districts, more than 45,000 houses supported by India (equating to 90% of the total) have been completed as of now.

The Prime Ministers of India and Nepal remotely observed the progress in the housing reconstruction project, emphasizing India's substantial contributions to Nepal's recovery and rebuilding efforts on January 21, 2020. This collaborative and inclusive approach underlines India's pivotal role as a key regional partner in disaster response and rehabilitation.

India's HADR operations during the 2015 Nepal earthquake serve as a significant reference point for disaster response strategies. It illustrates the importance of rapid mobilization, coordinated efforts, and the provision of multifaceted assistance in the wake of natural calamities. This case exemplifies how timely and well-organized HADR initiatives can mitigate the impact of disasters, support recovery, and strengthen international relations through humanitarian efforts.

Flood

Nepal and India are especially prone to recurrent floods because of their geographical locations. As major riparian nations, more than 6000 rivers originate in Nepal and flow into India, collectively spanning over 45,000 km of length (Khanal et al., 2007; Shrestha & Aryal, 2011). Several rivers originating from the high- or mid-Himalayan regions, primarily feed into the tributaries of the Ganges River system in India these rivers are Mahakali, Karnali, Narayani (Gandak in India), and Kosi (Saptkoshi in Nepal). Flooding is a recurrent natural disaster in both countries, significantly affecting the southern Terai region of Nepal and the northern Indian belts, particularly Uttar Pradesh and Bihar. These floods also lead to a loss in the gross domestic product (GDP) of both India and Nepal (Alfthan et al., 2018). This economic vulnerability is a key reason why the populations in these flood-prone areas remain poorer despite the region's rich natural resources and abundant water (Mishra, 2008).

Floods are defined by various indicators, depending on factors such as severity, timing, and whether they occur in urban or rural settings. According to The Oxford English Reference Dictionary, a flood is described as the overflow or influx of water that exceeds its typical boundaries, especially affecting land areas.

Recent data indicates a heightened incidence and severity of floods in the region, often precipitated by extreme rainfall events (Mishra, 2008; Shrestha et al., 2008; Alfthan et al., 2018). This escalation is linked to intensified seasonal precipitation patterns and glacier retreat caused by global warming, which exacerbates flood risks

(Alfthan et al., 2018). Locally, elevated water levels are denoted by various names such as majarana, baarh, boah, humma, saah, and pralay (Mishra, 2008). Two major rivers, the Gandaki and the Kosi, significantly impact the region, causing flooding almost every year. The extreme monsoon rains are the primary cause of these floods, although the geological setting also plays a crucial role in enhancing the flood conditions. The persistent flooding not only disrupts the lives of the people in these regions but also poses significant challenges to both countries in terms of disaster management and economic stability.

India and Nepal have agreements concerning water resources at both interpersonal and official levels (Bagale, 2020; Ahmed & Gaur, 2020). Due to the recurrent floods during the monsoon season, the construction of barrages and the establishment of hydro-diplomacy between India and Nepal became necessary (Fig. 1), a practice initiated during the British colonial period. Consequently, three major water treaties have been signed between the two nations, focusing on the Kosi, Gandak, and Mahakali rivers.

The Kosi River is infamous in Bihar as the "sorrow of Bihar" due to its frequent changes in course within the region. In 1954, the governments of India and Nepal signed the Kosi Agreement, to jointly manage the different aspects of the river. This agreement led to the creation of the largest irrigation project in Nepal, signifying an early example of bilateral cooperation in water resource management. In December 1959, the governments of both countries signed the Gandak Irrigation and Power Project Agreement, which took effect in 1964 after undergoing amendment. This agreement facilitated the development of irrigation and hydropower projects, benefiting both nations by improving agricultural productivity and energy supply.

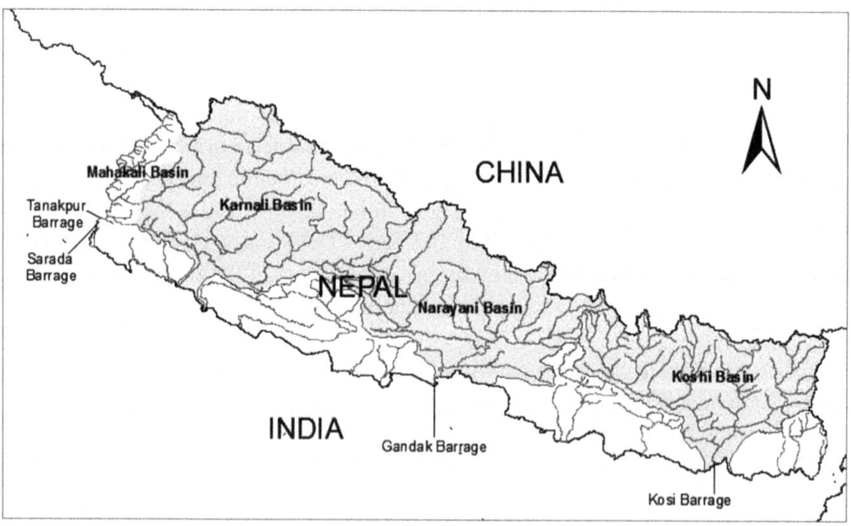

Fig. 1 Major River Basins and Barrages Built on the Rivers. (Source: GoN/Department of Survey)

The Mahakali Treaty was signed in 1996 and came into force in June 1997. This agreement further strengthened the cooperation between India and Nepal. This treaty incorporated an integrated mechanism involving three major power projects: the Sarada Barrage, Tanakpur Barrage, and Pancheshwar Project on the Mahakali River. The treaty also aimed to ensure mutual benefits in irrigation, hydropower, and flood management.

Despite these major agreements and numerous others aimed at the optimal utilization of river water for irrigation and transportation, several issues persist, particularly regarding flood management and cooperation (Siddiqui, 2017). The recurrent flooding during the monsoon season poses significant challenges for both countries. These natural disasters necessitate robust Humanitarian Assistance and Disaster Relief (HADR) efforts to mitigate their impact. Historical instances, like the Kosi floods of August 2008, highlight the ongoing challenges. These floods affected 45,000 people in Nepal and 3.3 million in India, underscoring the need for effective and collaborative flood management strategies. In summary, while the Kosi, Gandak, and Mahakali agreements have laid the groundwork for cooperation in water resource management, the persistent challenges of flood management underscore the need for continued and enhanced HADR efforts.

Effective flood management necessitates a multifaceted approach that includes both structural and non-structural measures. Structural solutions like barrages and embankments play a crucial role in controlling water flow and minimizing flood damage. However, alongside these physical measures, non-structural strategies are equally essential. These include the establishment of early warning systems, that can alert communities well in advance of potential flood events, enabling timely evacuation and preparation. Community preparedness programs educate and empower residents to respond effectively during floods, enhancing resilience and reducing vulnerability. Additionally, effective emergency response coordination ensures that resources and assistance are mobilized swiftly to support affected areas. These measures are essential components of HADR, aimed at reducing the vulnerability of affected populations and enhancing the resilience of both countries to future disasters. By addressing these challenges collaboratively, India and Nepal can strengthen their regional solidarity and improve their collective response to natural disasters.

The Kosi flood of 2008 caused unprecedented damage, marking the highest level of devastation in Bihar in the last five decades of flood history (Kale, 2008). The 2008 Kosi flood had a devastating impact on Nepal. It was described as the most severe flood in the nation's history by the Ministry of Home Affairs and members of the Central Disaster Relief Committee (Shrestha, 2010). During this catastrophic event, transboundary communication between Nepal and India was notably problematic, suffering from significant discontinuities. The maintenance of embankments in Nepal falls primarily under the responsibility of the Government of India. However, local Nepalese officials lacked the authority to directly communicate with the Government of Bihar (GOB) or the Government of Nepal (GON), creating a critical gap in coordination and response efforts. Similarly, the Government of Bihar did not have direct communication mechanisms with the Government of Nepal, further complicating the situation. At the time of the Kosi flood, India was also grappling with its flood crises, which hindered its ability to provide prompt humanitarian assistance to Nepal. This lack of effective transboundary communication and coordination severely hampered the disaster response efforts. Despite the lessons from the 2008 disaster, the 2017 floods showed no significant improvement in Humanitarian Assistance and Disaster Relief (HADR) mechanisms between India and Nepal.

Conclusion

In conclusion, the recurring natural disasters in both India and Nepal stress a critical need for improved communication and collaborative response strategies. The frequent occurrence of such disasters necessitates the establishment of robust and institutionalized Humanitarian Assistance and Disaster Relief (HADR) frameworks. Strengthening these frameworks is essential to mitigate the impact of future disasters and enhance the resilience of affected regions in both countries.

Bilateral treaties and disaster diplomacy between India and Nepal have become increasingly significant, particularly given the limited success of the SAARC Agreement on Rapid Response to Natural Disasters. Political challenges among SAARC nations have impeded the effectiveness of this regional agreement (Canyon, 2021). Consequently, the urgency for direct, bilateral cooperation between India and Nepal in disaster management and response has intensified. Scholars who have examined the bilateral relations between India and Nepal have observed that water treaties and agreements related to the Koshi and Gandak rivers tend to favor India over Nepal. These agreements have contributed to negative perceptions that India seeks to exert significant influence over Nepal's natural resources, particularly water. Critics argue that India has not succeeded in establishing exemplary, mutually beneficial water resource projects between the two nations (Karki & Hari, 2020). The core issue, according to some scholars, lies in the perceived fundamental inequality in the principles underlying these treaties. While India views these treaties as mutually beneficial, critics assert that many agreements, including the Peace and Friendship Treaty of 1950, disproportionately serve India's interests, raising concerns about the equitable distribution of benefits and obligations between the two countries (Karki & Hari, 2020).

Nepal blames India for inundating the Nepalese territory, constructing barrages, roads, and embankments near the international boundary that adversely affects the natural flow of rivers (Dhungel, 2009; Shrestha, 2009), against international principles. Similarly, Indian states, occasionally, blame Nepal for floods in their territory (Mishra, 2008; Dhungel et al., 2009). However, there is huge potential for cooperation between India and Nepal while employing water resources. Adoption of customary international principles of transboundary water sharing can be useful for achieving mutual benefits through water resources cooperation.

India's HADR operations during the 2015 Nepal earthquake serve as a significant reference point for disaster response strategies which can be followed during future Hydro metrological disasters too. India is already extending developmental aid to Nepal, with a focus on building grassroots infrastructure and Nepal stands as the largest recipient of India's humanitarian aid among the SAARC nations.

The relationship between Nepal and India has always been complex and paradoxical (Saran, 2017). To enhance its soft power disaster diplomacy with Nepal and other neighboring countries, India must develop robust, ground-level mechanisms. According to the Global Soft Power Index survey by Brand Finance, India was ranked 29th in 2024, slightly lower than its 28th position in the previous year. India's ranking in soft power indexes indicates that while it enjoys relatively high familiarity and influence among neighboring countries, it holds a lower position in terms of reputation. To improve its soft power, India needs to focus on creating more equitable and mutually beneficial agreements, particularly in areas such as disaster diplomacy and resource sharing. Building trust and demonstrating a commitment to fair and balanced cooperation can help India improve its reputation and effectiveness as a regional leader. Establishing and maintaining strong, equitable partnerships with Nepal will not only bolster regional stability but also enhance India's standing as a responsible and influential global actor.

In summary, addressing the gaps in HADR frameworks and fostering equitable treaties are pivotal for India and Nepal to effectively manage natural disasters and improve their bilateral relations. Enhanced cooperation and trust-building measures will strengthen both nations' resilience to disasters and elevate India's soft power, solidifying its role as a regional leader and responsible global actor.

References

Ahmed, S. I., & Gaur, N. (2020). *India-Nepal Water Cooperation: Issues and Concerns*. China (Tibet), 33(3), 0–3.

Akcinaroglu, S., DiCicco, J. M., & Radziszewski, E. (2010). Avalanches and olive branches: A multimethod analysis of disasters and peacemaking in interstate rivalries. *Political Research Quarterly, 63*(2), 276–287. https://doi.org/10.1177/1065912909358581

Alfthan, B., Gupta, N., Gjerdi, H. L., Schoolmeester, T., Andresen, M., Jurek, M., & Agrawal, N. K. (2018). *Outlook on climate change adaptation in the Hindu Kush Himalaya*. Mountain Adaptation Outlook Series. UNEP/GRID-Arendal/ICIMOD. Retrieved from www.icimod.org.

Bagale, D. R. (2020). *Nepal–India water cooperation: consequences of mutuality or hegemony?*. Water Policy, 22(6), 1098–1108.

Banting, K., & Kymlicka, W. (Eds.). (2017). *The strains of commitment: The political sources of solidarity in diverse societies*. Oxford University Press

Behr, T., & Jokeia, J. (2011). *Regionalism & global governance: The emerging agenda*. Notre Europe.

Bhavnani, R. (2006). *Natural disaster conflicts*. Unpublished manuscript, Harvard University.

Brancati, D. (2007). Political aftershocks: The impact of earthquakes on intrastate conflict. *Journal of Conflict Resolution, 51*, 715–743.

Canyon, D. V. (2021). Solidarity in Europe: From crisis to policy? *Acta Politica, 56*, 209–219. https://doi.org/10.1057/s41269-021-00195-7

Chauhan, A. (2023). *India has emerged as the first responder in humanitarian assistance and disaster relief (HADR) at a global level: CDS general*. PIB Delhi. Retrieved from https://pib.gov.in/PressReleaseIframePage.aspx?PRID=1906761.

Ciornei, I., & Ross, M. G. (2021). *Solidarity in Europe: from crisis to policy?*. Acta Politica, 56(2), 209

Dhungel, D. N. (2009). Historical eye view. In D. N. Dhungel & S. B. Pun (Eds.), *The Nepal-India water relationship: Challenges* (pp. 11–68). Springer. https://doi.org/10.1007/978-1-4020-8402-7

Dhungel, D. N., Pun, S. B., & Adhikari, B. R. (2009). Inundation at the southern border. In D. N. Dhungel & S. B. Pun (Eds.), *The Nepal-India water relationship: Challenges* (pp. 269–293). Springer. https://doi.org/10.1007/978-1-4020-8402-

Dutta, M. (2021). *Disaster and human trafficking*. Springer Nature.

Embassy of India, Kathmandu, Nepal. (2022) https://www.indembkathmandu.gov.in/development-partnership2

GAR 2022 Contributing Paper. *Geneva: United Nations Office for Disaster Risk Reduction*. Retrieved from www.undrr.org/GAR2022

Gayan, U. (2023). Soft power and disaster diplomacy: A case study of India's operation dost. *International Journal of Science and Research (IJSR), 12*, 612. https://doi.org/10.21275/SR23708141534

Glaser, Bonnie S, (1993). China's Security Perceptions: Interests and Ambitions, *Asian Survey*, March.

Gerhard, J., Lengfeld, H., Ignácz, Z., Kley, F. K., & Priem, M. (2019). *European solidarity in times of crisis: Insights from a thirteen-country survey*. Routledge.

Global Natural Disaster Assessment Report. (2021). Ministry of Emergency Management, Beijing Normal University, National Disaster Reduction Centre of China, International Federation of Red Cross and Red Crescent Societies (IFRC). Retrieved from https://reliefweb.int/report/world/2021-global-natural-disaster-assessment-report

Gousse-Lessard, A.-S., Gachon, P., Lessard, L., Vermeulen, V., Boivin, M., Maltais, D., Landaverde, E., Généreux, M., Motulsky, B., & Le Beller, J. (2022). Intersectoral research and multi-risk approaches in Québec: Systemic risk management and its psychosocial consequences. *GAR 2022 Contributing Paper*. Geneva: United Nations Office for Disaster Risk Reduction. Retrieved from www.undrr.org/GAR2022.

Harmer, A., & Cotterrell, L. (2005). *Diversity in donorship: The changing landscape of official humanitarian aid*. Humanitarian Policy Group Research Report 20. Retrieved from http://www.odi.org.uk/resources/download/234.pdf.

Hettne, B. (2005). Beyond the 'new' regionalism. *New Political Economy, 10*(4), 543–571.

India-Nepal bilateral relations. (2020). *Embassy of India, Kathmandu, Nepal*. Retrieved from https://www.indembkathmandu.gov.in/page/about-india-nepal-relations/.

IPCC. (2018). Summary for policymakers. In V. Masson-Delmotte, P. Zhai, H.-O. Pörtner, D. Roberts, J. Skea, P. R. Shukla, A. Pirani, W. Moufouma-Okia, C. Péan, R. Pidcock, S. Connors, J. B. R. Matthews, Y. Chen, X. Zhou, M. I. Gomis, E. Lonnoy, T. Maycock, M. Tignor, & T. Waterfield (Eds.), *Global warming of 1.5°C. An IPCC special report on the impacts of global warming of 1.5°C above pre-industrial levels and related global greenhouse gas emission pathways, in the context of strengthening the global response to the threat of climate change, sustainable development, and efforts to eradicate poverty*. Retrieved from https://www.ipcc.ch/site/assets/uploads/sites/2/2019/05/SR15_SPM_version_report_LR.pdf.

Kale, V. S. (2008). *Himalayan catastrophe that engulfed North Bihar*. Journal Geological Society of India, 72(6), 713–719.

Karki, K. K., & Hari, K. C. (2020). Nepal-India relations: Beyond realist and liberal theoretical prisms. *Journal of International Affairs, 3*, 84–102. https://doi.org/10.3126/joia.v3i1.29085

Kelman, I. (2003). Beyond disaster, beyond diplomacy. In M. Pelling (Ed.), *Natural disasters and development in a globalizing world* (pp. 110–124). Routledge.

Kelman, I. (2006). Acting on disaster diplomacy. *Journal of International Affairs, 59*, 215–240.

Kelman, I. (2012). *Disaster diplomacy: How disasters affect peace and conflict*. Routledge.
Kelman, I. (2016). Catastrophe and conflict: Disaster diplomacy and its foreign policy implications. *Brill Research Perspectives in Diplomacy and Foreign Policy, 1*(1), 1–76.
Kelman, I., Field, J., Surid, K., & Bhat, M. G. (2018). Disaster diplomacy in Jammu and Kashmir. *International Journal of Disaster Risk Reduction, 31*, 21–29. https://doi.org/10.1016/j.ijdrr.2018.02.007
Kelman, I. (2018). *Connecting theories of cascading disasters and disaster diplomacy*. International Journal of Disaster Risk Reduction, 30, 172–179.
Ker-Lindsay, J. (2000). Greek–Turkish rapprochement: The impact of disaster diplomacy? Cambridge Review of International Affairs, 14, 215–232.
Ker-Lindsay, J. (2007). *Crisis and conciliation: A year of rapprochement between Greece and Turkey*. I. B. Tauris.
Khanal, N. R., Shrestha, M., & Ghimire, M. L. (2007). *Preparing for flood disaster: Mapping and assessing in the Ratu watershed*. ICIMOD.
Meier, C., & Murthy, C. S. R. (2011). *India's growing involvement in humanitarian assistance*. GPPi Research Paper No. 13. Retrieved from http://www.gppi.net/fileadmin/media/pub/2011/meier-murthy_2011_indiagrowing-involvement-humanitarian-assistance_gppi.pdf.
Miguel, E., Satyanath, S., & Sergenti, E. (2004). *Economic shocks and civil conflict: An instrumental variables approach*. Journal of political Economy, 112(4), 725–753.
Mishra, D. K. (2008). The Kosi and the embankment story. *Economic and Political Weekly, 43*, 47–52.
Nandy, D., & Naha, A. (2022). *India's Disaster Diplomacy in South Asian Region*. In Gender and the Politics of Disaster Recovery (pp. 54–75). Routledge.
NDMA (2023). https://ndma.gov.in/Resources/AnnualReports
NDRRMA, (2022). https://ndrrma.gov.np/np
Parmar, S. S. (2012). Humanitarian assistance and disaster relief (HADR) in India's national strategy. *Journal of Defence Studies, 6*(4), 47–60.
Poudel, S. S. (2022). *India-Nepal diplomatic relations at 75: Full of contradictions*. The Diplomat. Retrieved from https://thediplomat.com/2022/06/india-nepal-diplomatic-relations-at-75-full-of-contradictions/.
Quarantelli, E. (Ed.). (1998). *What is a disaster?* Routledge.
Quarantelli, E. L., & Dynes, R. R. (1976). Community conflict: Its absence and presence in natural disasters. *Mass Emergencies, 1*, 139–152.
Reinhardt, G. Y. (2015). *Race, trust, and return migration: The political drivers of post-disaster resettlement*. Political Research Quarterly, 68(2), 350–362.
Reinhardt, G. Y., & Lutmar, C. (2022). *Disaster diplomacy: The intricate links between disaster and conflict*. Journal of Peace Research, 59(1), 3–11.
Ross, M. (2010). Solidarity—A new constitutional paradigm for the EU? In *Promoting solidarity in the European Union* (pp. 24–45). Oxford University Press.
Sangiovanni, A. (2015). Solidarity as joint action. *Journal of Applied Philosophy, 32*(4), 340–359.
Saran, S. (2017). *How India sees the world: Kautilya to the 21st century*. Juggernaut Books.
Sethi, V. (2021). *Looking back at the natural disasters that took place in India in 2021*. Business Insider India. Retrieved from https://www.businessinsider.in/science/environment/news/looking-back-at-the-natural-disasters-that-took-place-in-india-in-2021/slidelist/88204897.cms.
Shrestha, A. B., Shah, S. H., & Karim, R. (2008). *Resource manual on flash flood risk management*. Kathmandu: International. Centre for Integrated Mountain Development, ICIMOD
Shrestha, H. M. (2009). River-linking concept of India is viewed from the Nepalese perspective. In D. N. Dhungel & B. S. Pun (Eds.), *The Nepal-India water relationship: Challenges* (pp. 221–242). Springer. https://doi.org/10.1007/978-1-4020-8402-7
Shrestha, R. K., Ahlers, R., Bakker, M., & Gupta, J. (2010). *Institutional dysfunction and challenges in flood control: a case study of the Kosi flood 2008*. Economic and Political weekly, 45–53.
Shrestha, A. B., & Aryal, R. (2011). Climate change in Nepal and its impact on Himalayan glaciers. *Regional Environmental Change, 11*, 65–77.

Siddiqui, S. (2017, May 2). *India-Nepal border plagued by water troubles*. The Third Pole. Retrieved from https://www.thethirdpole.net/en/2017/05/02/india-nepal-border-plagued-bywater-troubles/.
Stjernø, S. (2011). The idea of solidarity in Europe
UNDRR (United Nations Office for Disaster Risk Reduction). (2019). *Global assessment report on disaster risk reduction 2019*. Geneva. Retrieved from https://www.undrr.org/publication/global-assessment-report-disaster-risk-reduction-2019.
UNDRR (United Nations Office for Disaster Risk Reduction). (2021). *Journalists said is no such thing as a natural disaster*. Retrieved from https://www.undrr.org/news/journalists-told-no-such-thing-natural-disaster.
UNISDR (2011). *Annual Report*. Secretariat Work Program.
Upadhyay, S., & Arora, K. (2023). Disaster-induced vulnerabilities and institutional response in indo-Nepal Tarai region. In A. Singh (Ed.), *International handbook of disaster research*. Springer Nature Singapore Pte Ltd. https://doi.org/10.1007/978-981-16-8800-3_44-11

Sikhism, Ad-Dharm Movement, and Ravidasi Identity in the Doaba Province of Punjab

Rajwant Kaur

Introduction

This chapter focuses on the process of change of the Ravidasi community in the Doaba province of Punjab with the influence of the Ad-Dharm movement. This chapter, will look at the growth of caste distinctions among Sikhs through many different processes. While the Sikh religious scriptures thoroughly reject all types of discrimination in society, the practice of untouchability exists among Sikhs. Among the Sikh community, movements developed to eliminate untouchability. Several factors, including historical entrenchment, socio-economic dependencies, political dynamics, insufficient social changes, resistance to change, and the presence of religious institutions, have contributed to the failure of Sikh caste abolition movements.

The Ad-Dharm movement and the rise of caste consciousness in East Punjab and the Doaba region during the 1920s, a well-known movement called Ad-dharma started by Mangoo Ram was successful to mobilize large majority of Chamars from Indian Punjab districts named Jalandhar, Hoshiarpur, Kapurthala and Shaheed Bhagat Singh Nagar (formerly Nawansashar) Doaba region, and played a vital role in changing their identity and social status.

Sikhism and Caste

Sikhism is the youngest religion in India. Indian Punjab is a Sikh preponderance state. The Sikhs comprise 63% of the population in Punjab. Most Sikhs are descended from the landlord Jat caste. Approximately 72% of Sikhs live in rural

R. Kaur (✉)
Centre for the Study of Social Systems, Jawaharlal Nehru University, New Delhi, India

K. Arora, R. R. Anand (eds.), *The Solidarity Approach in Geography*,
https://doi.org/10.1007/978-3-031-86753-8_16

areas. While Sikh doctrine discards the institution of caste, social practices often do not reflect this belief (Puri, 2003). This movement started in the fifteenth century and was an opponent to the Brahminical belief and untouchability. Sikhism completely disapproved of the Brahminical oppressive principles. The Sikh Gurus, primarily Guru Nanak Dev, preached the idea that human beings are equal under God therefore caste systems are meaningless. The sacred book of the Sikhs, *Sri Guru Granth Sahib*, includes the writings of the saints from the lower castes in the Bhakti movement in India.

Reviving the history of Sikh baptismal rite, Guru Gobind Singh announced it to be the phaul, or gate, through which anyone could join the fraternity. He also offered its members prasad, or communion, as a symbol of unity, where people from all four orders of Hindu society would share food from the same dish (Bingely, 1984). Guru Gobind Singh widely attacked all caste distinctions and insisted that everyone would join him. Recognizing that caste-based disunity led to significant national weakness, he declared social equality for all members of the Khalsa. While the higher castes resisted these reforms and many abandoned him, the lower castes embraced his vision and joined his cause in large numbers (Bingely, 1984).

The world-famous historical place of the Sikhs, the Golden Temple, was constructed by Sikh Gurus with four doors. All castes enter through these doors. The holy tank (sarowars) at the gurdwaras of Sikhism is spiritually believed to have healing powers and the powers of purification and to remove sins. All castes and communities can take bath together here, and this has the sociological and psychological potential of negating caste differences. This is in contrast to Hindu pilgrim places where lower castes are not allowed to take bath in the same tank. Similarly, langar (free kitchen) and pangat (congregation) are also important examples of how caste distinctions have been removed. During initiation, everyone sips the Amrit from the same bowl. Further, the creation of the khalsa rejects the caste system. Guru Gobind Singh created the khalsa panth in 1699. Gurus chose the five beloveds from the five different castes to be sacrificed. At that time the guru gave only one title (Singh) to all five different caste persons. But upper caste people do not consider lower caste Sikhs equals. They are referred to as Mazhbi, Ravidassia, and Ramgarhia. Further, the Gurbani (religious scriptures) strongly denounce the caste system. So the Gurus started the langar and pangat institutions and the distribution of karah prasad in the gurdwara sahib gave strength to the idea of equality (Singh, 1933; Sher, 1982).

More than 25 castes, including Ramgarhias, Jats, Khatris, Aroras, Chamars, and Chuaras, are found within the Sikh community, according to Punjab census statistics from 1881 to 1921 (Verma, 2000). There are 37 different scheduled caste communities; currently they are registered under diverse names like Ad-Dharmis and Ravidassias, and the Chuhras are registered as Mazbhi and Balmikis. Despite these egalitarian principles, caste-based discrimination has persisted in practice among Sikhs, mirroring broader societal patterns in India.

The Rise of Anti-caste Movements in Punjab

Punjab's anti-caste movements have a long history and have been significant in transforming the political and social circumstances of the region. The Singh Sabha movement, initiated in 1873 in Amritsar, aimed at revitalizing Sikhism by purging it of superstitions and practices that were perceived as Hindu influences, including caste discrimination. The movement's Lahore faction, which emerged in 1879, was more radical and focused on strict adherence to Sikh principles. However, it still struggled to effectively address caste-based inequalities. The Shiromani Gurdwara Parbandhak Committee (SGPC), established in 1920 to oversee Sikh gurdwaras and promote Sikhism, passed several resolutions (gurmatas) between 1926 and 1933 to address various community issues. While the issue of untouchability was discussed, the SGPC failed to take significant steps to eliminate caste discrimination within gurdwaras. This inaction highlights the persistence of caste biases even within the Sikh community (Singh, 1933; Padam, 1974).

Social Reform and the Singh Sabha Movement

Leaders of this movement wanted to remove untouchability and social evils within the religion. This movement was a reformist movement that made new ideas of practice according to Sikh religious scriptures. The ideology of this movement is based on the Sikh religious scriptures. All Sikhs should follow the tradition which Gurus made. The leaders of this movement compare earlier high-mark traditions and the present which Sikhs mostly followed. Sikhism arose entirely with a new ideology, totally opposed to the fundamentals of other religions. Guru Nanak stressed the oneness of divinity. Guru Nanak openly condemned all religious practices that rejected God's oneness.

In this context of social reforms, Sikh Sabha (associations) were established in Punjab. Examples of the Sabha set up against untouchability are the Amritsar Sabha, Lahore Sabha, and Siri Singh Sabha. The Singh Sabha movement was reformist (Grewal, 1996). Back to the sacred book was the central theme of the Singh Sabha movement (Dhillon, 1973). The primary goals of the Singh Sabha movement were to uphold the Guru Granth Sahib's teachings and the traditional Sikh code of behavior to restore Sikhism to its original purity. If Sikhs do not adhere to the Sikh code of conduct, their distinctive symbols, such as turbans, wrist bangles, and swords, may eventually only be seen in photographs and museums. (Dhillon, 1973, 1989). Guru Nanak emphasized human unity and vehemently denounced societal inequity. According to Guru Nanak, the illusions created in man are the sense of low and high, color, and caste. All Gurus gave the message of brotherhood equality fraternity. Guru Nanak started langar (free kitchen) and pangat (congregation). Guru Amardass made a boli (well). Guru Ramdass made Darbar Sahib. The sixth Guru wore two swords "piri" and "miri" (Holt, 2023). It is important to acknowledge that the concept exists (temporal spiritual and authority) declared by the Sixth Guru, Hargobind (ibid), is the obvious and natural consequence of Guru Nanak's teachings of ensuring justice, equality, and freedom for all men; rejecting asceticism and monasticism; accepting the householder's life and duty and walking the path of charity and sincere service to man (Dhillon, 1989; Singh, 2020; Singh, 2012). The ninth guru, Tegh Bahadur, was martyred because of the cause of justice (Singh, 1922). He made a great sacrifice for the protection of the Hindu religion. Guru Gobind Singh established the Khalsa Panth. The khalsa believes in the brotherhood. The main reason for establishing the above things was that Gurus wanted to remove untouchability and social discrimination within the society (McLeod, 1989).

There is religious diversity within the tradition. The sabha (associations) of intellectuals formulated the powerful discourse of the tat khalsa (pure khalsa) (Oberoi, 1994). Tat khalsa was opposed to the Sanatan Sikhs, those Sikhs who fail to appropriately abide by the code of conduct. According to Census Commissioner Ibbetson, "appearing to be Sikh," or what he called "true Sikh," meant just abstaining from tobacco use and upholding the five visible symbols of the religion. However, everyone who failed to meet these strict requirements was thought to be a non-Sikh (Oberoi, 1994, p. 211). Further Oberoi argues that "Any person, who did not follow his religious code of conduct seriously, is not Sikh. During the colonial period, many Sikhs converted to Christianity. The first person was Daud Singh. Maharaja Ranjit Singh's son Duleep Singh, and Hernam Singh, the nephew of the Maharaja of Kapurthala converted to Christianity. In response the Sabha members spread the awareness that they follow their religion" (Oberoi, 1994, p. 222–223). The Singh Sabha movement also aims to protect the Sikh religion from the influence of other religions.

According to the rigid classification, Sikhs are divided into such groups as Nankpanthi, Ramdasi, Nirmala, and Udasi. The meaning of Sikhism's Sanatan. The sanatanist Sikhs, therefore, held that these traditions, ceremonies, and rituals originated beyond the boundaries of diachronic time, at the beginning of time, as the universe came into existence. It is not implied that Sikhism was in decline or "irrational" because Sikhs participated in the stories, worship, and cults of miracle saints, goddesses, and village gods. These customs were an essential component of a well-rounded way of life and should not be evaluated according to criteria that were created at the turn of the century (Dhillon, 1989; Pall, 2020). The Sabha's leaders aimed to bring Sikhism back to its original, pristine purity. The Sikhs ought to continue to identify as distinct individuals from the caste system. Gaini Ditt Singh and Gurmukh Singh, the two chief leaders of the Singh Sabha movement (Singh, 1922), preached the Sikh religion differently from Hinduism, Islam, and Christianity. Sikhism is an independent religion. A subcommittee was established by the khalsa diwan to recommend strategies for reorganizing the Gurdwaras, which had been taken over by Brahminical priests and special interests. However, because mahants and the pujaris in charge of the gurdwaras had the government's backing, it was unable to take meaningful action. Following this struggle, the mahants and pujaris turned against the leaders of the Singh Sabha (Singh, 1974, p. 68–69).

As mentioned earlier, Dalit Sikhs have not been treated as equals by upper caste Sikhs, despite many thoughtful revolutions passed by the SGPC. It was not surprising, as Harish K. Puri points out. The elimination of untouchability or caste equality was a response to the Jats' inclination toward social dominance and hierarchy, as they made up 70% of Akalis and other upper castes (Puri, 2003).

As mentioned earlier, Dalit Sikhs have not been treated as equals by upper caste Sikhs, despite many thoughtful revolutions passed by the SGPC. As Puri adds, this nature has forced the Dalit Sikhs to institute separate religious places. It has in turn highlighted the Sikhs' pre-existing caste distinctions even more. Furthermore, Dalit Sikhs are forced to look for alternative cultural events in several deras (religious centers) of various religious dominions due to the observance of caste prejudice against them (Puri, 2003). Two different gurdwaras in the same village increase the local politics in the Punjab.

Social Movements and Mobility

Social movements are often driven by attempts to bring about social mobility. They are crucial in not only initiating change but also in shaping the course of social and political transformation. Anti-caste movements emerged both through the colonial and post-colonial phases. The objective of these movements is to achieve political and economic mobilization of participants by invoking their shared heritage and fostering solidarity. Given the severity of the caste system, these movements primarily focused on challenging and opposing it (Natarajan, 1959; ICIMOD, 2024). All social reformists have sought to dismantle the rigidity of social barriers. The core ideology of anti-caste movements was the emancipation from all social evils and practices.

The British historian Malley (1968) argues that the Dalit movement is often depicted as a movement to strengthen the untouchables, initiated and sustained by external forces. These movements focused on preserving or expanding government jobs, welfare programs, and reservations in political offices. However, there has been limited analysis of these Dalit movements nationwide, and no serious efforts have been made to develop a typology of these movements. Shah classifies Dalit movements into two categories: (1) reformative and (2) alternative. Reformative movements aim to address the issue of untouchability by reforming the caste system itself. In contrast, alternative movements seek to establish a new socio-cultural structure through means such as acquiring education, gaining political power, religious conversion, and enhancing economic status. In terms of caste ideology and identity, Shah has restricted Dalit movements into four categories: (1) counter ideology and Dalit identity; (2) Buddhist Dalits; (3) competing ideologies and non-Hindu identities; and (4) movements within culture consciousness. The first is based on the class while the other three are categorized as religious ideologies (Shah, 1990, 2001, 2006).

The Dharma Shastras, a collection of Hindu sacred books, define untouchability. In his expressed introductory speech at the Jaat-Paat, Ambedkar (1891–1956), a messiah of the oppressed and the creator of the Indian Constitution, underlined that untouchability ends when these boundaries are destroyed. The Lahore Society for the Abolition of Caste, or Todak Mandal, was founded in 1936. However, due to the speech's anti-Dharma Shastric nature, Ambedkar was not permitted to give it. Since then, a lot has changed, especially after Independence, but caste-related issues and the curse of untouchability have not entirely driven from heaven. With the establishment of the Indian Constitution in 1950, there has been a noticeable decrease in the overt use of the purity/pollution principle; yet, caste-based activities in politics have taken on a completely new form. With more vocal lower-caste movements changing the political landscape and causing caste violence in some regions of the nation, caste has never played a more significant role in Indian politics than it does now. Despite its longstanding claims to be a caste-free society, Punjab is not a unique situation. Serious caste conflicts have lately broken out in the state's Doaba region in both rural and urban settings. Local politics are made more complex by the recent politicization of caste, which has not only affected Indian political processes but also given rise to diaspora constituencies (Moon, 1989).

Although untouchability and caste are widespread across the nation, they exhibit distinctive characteristics in each area. It is essential to analyze the state of untouchability in Punjab. Additionally, understanding the significance of the Ad-Dharm movement and the increase of Dalit awareness in Punjab is crucial. Movements for social change often arise in response to the deprivation experienced by socially/politically excluded groups.

Ad-Dharm Movement in Punjab

Ad-Dharm, meaning "original religion," refers to a movement that began to uplift Dalits in Punjab. This movement was a significant and effective anti-caste initiative in the region. Emphasizing a native/original/indigenous status, the various Adi movements resulted in further refining the discourse on Dalit cultural heritage as they delved into pre-Aryan oral/folk traditions, cultural heritage, and recollections of the inhabitants. Phule sought to establish a new theistic religion, Periyar advocated for atheism, and Ambedkar embraced Buddhism (Omvedt, 1994). The Mahar movement has played a critical role in shaping contemporary Dalit identity (Jodhka, 2012). In South India, the process of ethnification was strongly promoted by the Adi Karnataka movement in Karnataka and the Adi Dravida movement in Tamil Nadu. Similarly, the Ad-Dharm movement, spearheaded by Mangoo Ram, aimed to awaken Dalits to their organizational capability (majlis), religion (mazhab), and communal pride (qaumiat), which had long been suppressed by the burden of untouchability (Ram, 2004a, 2004b).

These movements symbolized a distinct Dalit identity and supported efforts to regain their lost traditions, culture, music, rich heritage, and language. The

Ad-Dharm movement significantly influenced this development in the area of Punjab (Ram, 2008). The primary goal was to establish a distinct religious identity and socio-cultural structure by creating a separate and respectable space for the oppressed castes through spiritual renewal, political empowerment, and cultural transformation (Ram, 2004a, 2004b). Mangoo Ram, a pioneering leader of the movement in Punjab, fostered a sense of pride among the region's oppressed castes. This movement is widely known by various names, including Ad-Dharm, Ravidassism, Adi-Dharm, Ravidas Dharm, Sant Dharm, Guru Ravidas Dera movement, or Guru Ravidas (Ram, 2011). Mangoo Ram asserted that the untouchables were a quam, a distinctive religious community equivalent to Muslims, Hindus, and Sikhs (Juergensmeyer, 1988).

In the movement's first yearly conference, a poster was dedicated entirely to highlighting the hardships endured by lower castes at the hands of Hindus. The poster also argued that an enduring result of their suffering could only be achieved through the reinstatement of their lost native religion, heritage, and culture (Ram, 2012). Mangoo Ram made a serious plea to all lower castes to come together for their empowerment and liberation. The Chamars are treated as a sub-caste in Punjab. The movement imparted a sense of dignity and confidence to its members, which also translated into a new energy and aspirations for mobility (Jodhka, 2012, p. 122).

Ad-Dharm Movement and Ravidasi Identity

This Dalit movement is recognized by numerous names, such as Ad-Dharm (original religion), Sant Dharm (religion of saints), Ravidas cult, Ravidas Dharm, the Guru Ravidas Deras movement, and Ravidassism. It is deeply rooted in the indigenous, although formerly fragmented, worldview of Punjab's Dalits, who see themselves as the region's original inhabitants and strive for acknowledgment as such.

In other sections of the nation, Dalits often followed the process of Sanskritization, but Punjab took a different path (Ram, 2012). Social reform movements also developed in various parts of India. In Maharashtra, Jotirao Phule (1826–1890) was a leading figure in advancing non-Aryan racial theories and founded the "Sarvajanic Satya Dharma" in 1889, which he identified as the original religion of the Dalits. Meanwhile, in North India, Swami Achhutanand, who transitioned from an Arya Samaji to a Dalit activist, established the "Adi-Hindu Dharma" in 1922 in Uttar Pradesh (Omvedt, 1994). The Adi-Andhra and Adi-Karnataka Adi-Dravida movements in South India are counterparts to the Ad-Dharm movement in Punjab (Juergensmeyer, 1988).

The Dalits are the original natives of India. In Punjab, they are popularly known as Ravidasi Samaj. They have their own rules. Dalits possess a rich cultural heritage, complete with distinct rules, religion, gurus, heroes, folklore, traditions, and symbols. However, the prowling Aryans caged these native populations and established supremacy over them. The Aryans not only disadvantaged the inhabitants of their past and erased their geographies but also pressed their cultural heroes, religion, and gurus into oblivion. Over time, they condensed the original residents to what became scornfully known as the social category of untouchable.

The description further emphasizes that for Dalits to shed their subordination and reclaim their missing identity, heritage, and religion, they must create a separate Dalit individuality, independent of the region's well-known communities—Muslims, Sikhs, and Hindus. It is argued that, unlike in other parts of the country, where Dalits have often sought liberation through conversion or Sanskritization, Dalits in Punjab need to articulate an option and original religious beliefs. This new approach would assist them in constructing a free civilization, breaking away from the harsh binaries of pure/polluted and low/high (Juergensmeyer, 1988; Ram, 2011).

The Dalit philosophy emerged during the Ad-Dharm movement, which began in Punjab. The movement directly opposed the local dominance of Jat landlords and their control over land resources. It emphasized a separate Dalit independence foundation on the philosophy of the fundamental medieval North Indian Bhakti movement. Dalits, especially those from the Chamar caste, worship Ravidas as their Guru, which has extensively contributed to the movement's broad appeal among them. Guru Ravidas is highly regarded among lower castes, largely because 40 hymns and one couplet of his sacred poetry are featured in the Sri Guru Granth Sahib.

Guru Ravidas's battle beside the structure of untouchability, rooted in an enlightened idea of an unrestricted social order (Begumpura), earned him a particular place in the hearts of the Dalit masses, especially at a time when no one else was willing to support the oppressed untouchables. The Ad-Dharm movement strategically leveraged the widespread recognition of Guru Ravidas by using his image as its insignia, understanding his revolutionary manuscripts, and using fables about him to illustrate subordinate-caste power and pride (Omvedt, 2008; Juergensmeyer, 1988). In essence, Guru Ravidas, who had previously started the substitute Dalit plan, was brought to the vanguard in Punjab by the Ad-Dharm movement (Ahir, 1992).

The Ravidas phenomenon sparked the emergence of the Ravidas Deras movement in Punjab. Both the Ravidas phenomenon and the Ad-Dharm movement laid the base for another set of Dalit beliefs in the province. Unlike conversion and Sanskritization, the Ad-Dharm movement and the Ravidas beliefs were initiated by the Dalits themselves. Rooted in the teachings of Guru Ravidass, who was also from the Dalit community, this movement developed distinct cultural symbols that differ from those of Sikhism and Brahminical. The renowned Ad-Dharm movement in Punjab recognized Guru Ravidass's verse and teachings as powerful intellectual instruments for liberation (Ram, 2011; Judge, 2010). The teachings of Guru Ravidass have evolved into a political approach rooted in his dissenter socio-religious ideology and philosophy. This philosophy and ideology have ultimately become the defining characteristic of the emerging alternative.

The Dalit agenda supported by the tradition of Guru Ravidas disagrees with the traditional Hindu Bhakti figures. Rooted in the ethical philosophy of kindness and the dignity of labor, his Bhakti embodied the egalitarian and democratic values of pre-Aryan cultural inheritance. Guru Ravidas redefined the Bhakti movement, transforming it into a powerful technique of social protest against the system of untouchability and the unequal division of wealth and power. He emphasized egalitarianism and the struggle against untouchability practices imposed by the upper castes. Guru Ravidas's faith was centered on a formless God, and he did not shy away from his caste identity, continuing his inherited livelihood of mending shoes.

Omvedt argues that Guru Ravidas uttered his commitment through bani—spiritual philosophy expressed in poetic form. His bani hold worldwide appeal, characterized by fundamental enthusiasm and massive worship for an formless God. It provided a noteworthy podium for expressing hope for a better world and resistance against oppressors, power holders, and religious mistreatment (Omvedt, 2003). Several intellectuals, yet, believe that while his devotional hymns and songs confined the anguishes of the marginalized, they lacked the reformist enthusiasm and straight condemnation of Brahminism and the caste system, which are more apparent in the poetry of Tukaram and Kabir (Dasgupta, 1976).

There is no doubt that Guru Ravidas's entire poetry was against slavery and "untouchability." There is no doubt that Guru Ravidas's entire body of poetry was a strong stance against slavery and untouchability. He asserted that God formed the entire human beings as equals and resides within each of them. Since it is the same God who spread all of humanity, Ravidas argued that it is unwise to segregate society based on caste. In the Sri Guru Granth Sahib, he expressed this sentiment, saying, "Jo ham shehri so meet hamara" (whoever is my fellow citizen is my friend). In this perspective, the perfect vision of a Dalit raj or the city of Begumpura—founded on reason (dhyan) and devotional love (bhakti/bhav)—emerges as an influential icon of the developing Dalit agenda in Punjab expressed through poetry.

Deras and Ravidasi Identity

The report purports to cover the period from January 1926 to April 30, 1931. Its 80 pages contain diverse materials complied with several purposes in mind. One purpose was financial. The main part of the report, however, occupying almost 30 pages, is a general statement of the movement as a new religion, its ideology, its mythic origins, its beliefs and symbols, its hopes for the future, and the demands it makes on its followers (Juergensmeyer, 1988, p. 290–291).

The Ad-Dharm report emphasizes that Dalits were never born as Hindus, so the conception of relinquishing Hinduism is irrelevant. The report further contends that Hinduism was constructed on the remnants of Ad-Dharm, the original faith of Punjab's original citizens. As a result, the Dalits of Punjab are not obligated to follow the cultural practices of a religion imposed upon them by the foreign Aryans. It firmly argues that former untouchables are not Sikhs, Muslims, Christians, or Hindus.

This movement reports undoubtedly show that every Ad-Dharmi embraced their religion with pride. Babu Mangoo Ram Mugowalia played a significant role in anthologizing a separate holy text for the Ad-Dharm religion. This holy book included the philosophy of Guru Ravidas and the writings of other lower-caste saints. Mangoo Ram sought assistance from two prominent Ravidas religious centers of his time—Dera Sachkhand Ballan near Jalandhar and Dera Chak Hakim near Phagwara. Sant Pipal Dass of Dera Chak Hakim was tasked with creating a scripture for Ad-Dharm titled "Sri Guru Ad Prakash Deep Granth." However, this document remained largely symbolic in the movement's history. Meanwhile, Rae Das Ki Bani, a compilation of Guru Ravidas's verses and poetry by Sant Hiran Das, published by the Allahabad-based Belvedere Press in 1908, came to be regarded as the religious text of the Ad-Dharm (Juergensmeyer, 1988).

However, afterward, Sant Isher Das anthologized the religious book of the Ad-Dharmis, titled "Ad Parkesh" (The Original Light). Ravidas deras represent different Dalit identities in their traditions and rituals based on the guru Ravidas's teachings. The phenomena of Ravidas deras have taken new socio-cultural Dalit consciousness to Punjab. This movement, recognized as the Ravidas Deras Dalit or the Ravidassia movement, includes prominent centers like Dera Sachkhand Ballan in Punjab. It represents the latest socio-cultural movement in the region, boosting the confidence of lower castes and socially excluded individuals in Punjabi society, allowing them to take pride in their caste individuality. These religious places are not only spiritual centers but also symbols of a distinct Dalit identity. Their deepest focus is to assert their separation from Hindus and Sikhs, as they face discrimination akin to lower-caste individuals in various aspects of life. Consequently, lower-caste people are encouraged to embrace and openly express their identity. Their uniqueness too stems from the reality that they neither seek asylum in established theologies nor do they mimic the central socio-cultural philosophy of higher-caste civilization. Instead, they pompously set themselves apart from mainstream religious systems and challenge the long-standing dominance of the Sikh holy book over the Dalits of Punjab. The focus of the remainder of this section is on how they differentiate themselves from others and the extent of their success in this endeavor (Ram, 2011, p. 54).

The leaders of the Ravidas deras set themselves apart from Sikh Gurus and Hindu holy places. In Sikhism, just the ten historical Gurus and the Sri Guru Granth Sahib are acknowledged as legitimate Gurus. Within Sikh Gurdwaras, Ravidas is referred to as a bhakta rather than a Guru. This distinction between bhakta and Guru has not merely led to tensions between the Jat Sikhs and Ravidasia community in Punjab and beyond but has also contributed to growing communal division among Dalit and Sikh diasporas (Takhar, 2014; Ram, 2012).

Ravidas deras are separate from the Sikh deras. There is another factor that separates the titles of the last names of the two. For the Sikh gurus, it is Singh but for the Ravidas gurus, the last title of the name is Das. The dressing style of the Sikh guru is different from the Ravidas deras. In the Ravidas deras, some sants maintain unshorn hair and flowing beards and wear turbans, similar to Sikhs, yet they do not identify as Sikhs. However, some Ravidassia deras chiefs are clean-shaven. The symbol of the Ravidasi deras, identified as "Har" (Supreme Being), is regarded as the national emblem of the Ravidassia Society (Fig. 1). It shows the separate identity. In Sikh gurdwaras, the religious symbol is the "Khanda"—a two-edged sword above a quoit, with two crossed sabers below the quoit. A double-edged sword positioned above a quoit, with two crossed sabers placed below the quoit. Hindu temples

Fig. 1 The religious symbols of Ravidasi. (Source: https://www.guru-ravidas-symbol.com)

Fig. 2 Dera Sach Khand Ballan (Jalandhar) Punjab. (Source: https://www.guru-ravidas-symbol.com)

also have their distinct symbols. The areas (prayer) format in Ravidas deras (Fig. 2) differs from those of other religions. Ravidas deras have embraced the greeting "Victory to the Divine Guru" or "Guru Ravidas," with the customary response being "Blessed is the Divine Guru." The phrase "Victory to the Saints" is also widely popular within the community. At the end of prayers, the phrase "Victory to Shri Guru Ravidas, the fearless one" is recited (Ram, 2011).

Solidarity Among the Ravidassis

However, many Dalits have improved their status by leaving their traditional caste-based occupation. They have reinforced their situation through state positive action, diligent efforts, and travels overseas. Several of them set up small-scale businesses. Some of the political movements also spread awareness among the Dalits. The rises of various Dalit-dominated deras are continuing to process lower castes mobility in the state (Ambedkari, 2005). There is a contradiction between the upper castes and lower castes in the state. This has been a power struggle between landlords and landless in the state of Punjab over the past few years. The Vienna attack is an example. The attack targeted the two leading saints of Dera Sachkhand Ballan, headquartered in the village of Bal near Jalandhar city, which is a key center for the Ravidas sect in Northern India. The incident occurred during a sermon on 24 May 2009 at the Guru Ravidas temple in Vienna. Six individuals reportedly carried out the attack but were subdued by approximately 200 devotees present on schedule. Amid the chaos, approximately 16 people were injured. Sant Ramanand, the deputy leader of Dera Ballan, later passed away from his injuries in the hospital (Ram, 2011, Judge, 2010; Jodhka, 2010). This was an important event in the Ravidassia Dalit identity. There was a curfew imposed in the state of Punjab for several days, damage to public property, and thousands of millions of Indian rupees lost due to rioting. The followers of the Dalit community fight for equality but this is not acceptable to the upper castes. One more version links the Vienna incident to a violation of the Sikh code of conduct during the spiritual preaching by the saints of Dera Ballan at the temple. This departure from the "Prakash of Guru Granth Sahib" at Dera Sachkhand Ballan, a practice followed by many Ravidas deras both in Punjab and internationally, is viewed as a contributing factor (Ram, 2011). The last ritual of the Sant Ramanand was completed in the nonappearance of "Parkesh of Siri Guru Granth Sahib." The Vienna attack also contributed to further dividing the society into two parts. The Vienna attack, however, is not an isolated incident. The same case happened in the past between the leading caste and Dalit in Punjab.

It was clear that some Sikh youth had attacked the site, yet the motive was unknown (Judge, 2010, p. 6). However, after this event, the protestors did not attack any Sikh gurdwara during this period. Their violent activities largely remained confined to government property, such as railways and buses, as well as some private vehicles (Judge, 2010, p. 6–7). The Ravidasi religious places not only represent the declaration of a distinct identity in Punjab but also highlight the ongoing development of a substitute Dalit schedule in the state (Singh, 2012).

After all these occurrences the issue remained unresolved and it was, as a matter of fact, a nonissue. The simplest solution to the problem could have been the separate publication of Guru Ravidas's writings and the distinct construction of tradition of reverence of the holy writings. At the end of January, Sant Niranjan Dass announced the final break with the holy text Guru Granth Sahib, which was interpreted by the media in an altogether different manner (Judge, 2010, p. 15). Their sacred text is called Amrit Bani Satguru Ravidas Maharaji. In regions like Punjab, Haryana, and abroad, many Guru Ravidass gurdwaras and deras exclusively read the Amrit Bani. During the 2011 India Census, many individuals from the Chamar community identified their religion as Ravidassia instead of Sikh.

Symbolic Political Assertion of the Ravidasi Community

The Ad-Dharm movement thrived in revitalizing the Chamars in the Doaba province, instilling in them a renewed wisdom of self-assurance. Even though the community was founded primarily for political and practical purposes, its adherents did begin to develop a unique religious vision that set them apart from Sikhism and Hinduism. Ravidasi gurudwaras, Ravidasi associations, and Ambedkar clubs have played a vital function in asserting Ravidasi's identity in the Doaba region of Punjab. Although they do not align with Sikhism, they have looked to it as another source of spiritual and ceremony practices. Dera Ballan, as a religious center, emphasizes universal spirituality and values while vigorously engaging with local politics and Dalit matters. In the broader nationwide framework, Dalit political leaders have commenced connecting across regions, and Dera Ballan remains a significant hub of Dalit political activity in Punjab. Influential politicians from all political parties frequently pay homage to dera heads. Like Prime Minister Narendra Modi, Home Minister Amit Shah, Uttar Pradesh Chief Minister Yogi Adityanath, and Harsimart Kaur Badal are some of the influential persons who paid visits during Guru Ravidass's birth anniversary at Varanasi. Dera Ballan searched for and built the Varanasi temple dedicated to Guru Ravidas.

In 2008, followers of Guru Ravidass from Europe gathered at his birthplace to install a golden palanquin at the Varanasi Janam Sthan Mandir of Guru Ravidas. On that occasion, Mayawati, the leader of the BSP, along with other Dalit leaders, was also present. Significantly, on 16 July 1998, when the Ravidassia diaspora funded the edifice of a splendid gate at the temple doorway, Dr. K.R. Narayanan, then President of India and a Dalit himself, was invited to inaugurate it. Kanshi Ram, a prominent Punjabi leader, individually synchronized the occasion (Jodhka, 2010).

Conclusion

To restore equality within the Sikh community, reformative movements sought to eliminate caste-based hierarchy. However, these efforts were unsuccessful, as Sikhs from higher castes continued to perpetuate discrimination. The above discussion made it clear that the Ad-Dharma movement has played a significant role in the upliftment of the marginalized section of the Doaba region of Punjab. The discussion clearly highlights the religious symbols of the Ravidasis, along with Ravidasi deras and gurdwaras, have played an active role in the socio-cultural and economic development of the community. Apart from assertion, Ravidasi constructed its own religious institutions and health institutions. Ravidasi Diaspora, urbanization, traditional caste-based occupations, and the green revolution destroyed the dependencies of agriculture. They further reinforced their identity with the support of their newly acquired political power and the emergence of Dalit political and economic elites.

References

Ahir, D. C. (1992). *Buddhism in the south-east Asian countries*. South Asia Books.

Ambedkari, N. (2005). Dera Sachkhand Ballan Dae Manjil Whal Vadde Kadam (The surging steps of Dera Sachkhand Ballan towards its destination). *Begumpura Shaher, 2*, 12.

Bingely, A. H. (1984). *Sikhs: Their origin, history, religious customs, fairs, and festivals*. Sumit Publication.

Dasgupta, S. N. (1976). *Hindu Mysticism*. Motilal Banarasidass.

Dhillon, G. S. (1973). *Character and impact of the Singh Sabha movement on the history of the Punjab*. Unpublished PhD diss., Punjabi University, Patiala.

Dhillon, G. S. (1989). *Singh Sabha Movement*. SikhBookClub. Retrieved 16 August, 2024, from https://sikhbookclub.com/Book/Singh-Sabha-Movement.

Grewal, J. S. (1996). The Singh Sabha movement and the development of modern Sikh identity. *Journal of Punjab Studies, 3*(1), 55–72.

Holt, J. D. (2023). *Understanding Sikhism: A guide for teachers*. Bloomsbury Publishing.

ICIMOD. (n.d.). *A comparative study of dalit movements in Punjab and Maharashtra: The Ad Dharm and other movements*. International Centre for Integrated Mountain Development. Retrieved 16 August, 2024, from https://lib.icimod.org/record/files.

Judge, P. S. (2010). *Changing Dalits: Exploration across time*. Rawat Publications.

Juergensmeyer, M. (1988). *Religious rebels in the Punjab: The social vision of untouchables*. Ajanta Press.

McLeod, W. H. (1989). *Who is a Sikh? The problem of Sikh identity*. Clarendon Press.

O'Malley, L. S. S. (1968). The Hindu social systems. In L. S. S. O'Malley (Ed.), *Modern India and the west: A study of the interaction of their civilizations* (pp. 354–388). Oxford University Press.

Oberoi, H. (1994). *The construction of religious boundaries: Culture, identity and diversity in the Sikh tradition*. Oxford University Press.

Omvedt, G. (1994). *Dalit and the democratic revolution: Dr. Ambedkar and Dalit movement in colonial India*. Sage Publications.

Omvedt, G. (2003). *Buddhism in India: Challenging Brahminism and caste*. Sage.

Omvedt, G. (2008). *Seeking Begampura: The social vision of anti-caste intellectuals*. Navayana.

Padam, P. S. (Ed.). (1974). *Rahtnama* (pp. 68–69).

Pall, S. (2020). *Issues of Sikh identity: Sanatanist-Sikh debate*. UC Santa Barbara. Retrieved 16 August, 2024, from https://punjab.global.ucsb.edu/9-Sheena-Pall.

Puri, K. H. (2003). Scheduled castes in the Sikh community: A historical perspective. *Economic and Political Weekly, 38*(26), 2693–2701.

Ram, R. (2004a). Untouchability in India with difference: Ad Dharm, Dalit assertion and caste conflicts in Punjab. *Journal of Asian Survey, 44*(6), 895–912.

Ram, R. (2004b). Untouchability, Dalit consciousness, and the ad Dharm movement in Punjab. *Contributions to Indian Sociology, 38*(3), 323–349.

Ram, R. (2008). Ravidass Deras and social protest: Making sense of Dalit consciousness in Punjab (India). *Journal of Asian Studies, 67*(4), 1341–1364.

Ram, R. (2011). Guru Ravidas, bhakti and his vision of an egalitarian state. *Identity, 2*(6), 18–20.

Ram, R. (2012). Beyond conversion and Sanskritisation: Articulating an alternative Dalit agenda in East Punjab. *Modern Asian Studies, 46*(3), 639–702.

Shah, G. (1990). *Social movements in India: A review of the literature*. Sage Publications.

Shah, G. (Ed.). (2001). *Dalit identity and politics: Cultural subordination and the Dalit challenge* (Vol. 2). Sage Publications.

Shah, G. (Ed.). (2006). *Untouchability in rural India*. Sage.

Sher, S. S. (1982). *Glimpses of Sikhism and Sikhs*. Metropolitan Publication.

Singh, (G.) P. (1933). *Jaat Paat ate Chhoot Chhaat Sabandhi Gurmat Sudhar [Gurmat reform of Casteism and untouchability]*. SGPC.

Singh, G. (2012). Religious transnationalism, Ravidasi identity and the Dera Sant Sarwan Dass, Ballan. *Sikh Formations: Religion, Culture, Theory, 8*(1), 21–36.

Takhar, O. K. (2014). The place of scripture in the trajectories of a distinct religious identity among Ravidassias in Britain: Guru Granth sahib or Amritbani guru Ravidass. *Journal of Contemporary Religion, 29*(1), 105–120.

Verma, A. B. (2000). *The making of little Punjab in Canada: Patterns of immigration*. Sage.

Part V
Climate Change, Sustainable Development, and International Environmental Solidarity

International Solidarity to Address Climate Change through Conservation of Natural Resources: Problems, Prospects, and Indian Perspective

Moushila De and Meenakshi Dhote

Introduction

Climate change is categorically one of the most challenging aspects of these days that is demonstrating its ultimate power across the planet (De & Dhote, 2021). The swift and unprecedented phenomenon of global warming results in a host of effects that touch on environmental, social, and economic aspects. Manifestations of climate change cover increasing temperatures, shrinking polar ice caps, exacerbation of extreme weather events, and noticeable changes in ecosystems (De & Dhote, 2021).

This urgent matter beckons for a consolidated, global reaction. The ramifications of climate change are pervasive, touching all geographical extremities of the Earth. As the window for effective action narrows, there is an imperative to act swiftly to stave off the most crippling impacts of global warming (Chow, 2022; IPCC, 2023). The call for immediate mitigation measures cannot be emphasized enough, given the impending threats of rising sea levels to coastal habitats and the potential disruption to agricultural and water resources, with subsequent implications for food security. The Intergovernmental Panel on Climate Change (IPCC) underscores that without a substantial reduction in the consumption of fossil fuels in the imminent decade, the relentless ascent of global temperatures is poised to inflict catastrophic and irreversible damage to the environment (Mckoy, 2023). To avert this, there is a global mandate to confine global warming to 1.5 °C above the baseline pre-industrial levels by the year 2030, necessitating a nearly halving of current emission levels

M. De (✉)
Department of Environmental Planning, School of Planning and Architecture, New Delhi, India

M. Dhote
Dean Academic, School of Planning and Architecture, New Delhi, India
e-mail: m.dhote@spa.ac.in

K. Arora, R. R. Anand (eds.), *The Solidarity Approach in Geography*,
https://doi.org/10.1007/978-3-031-86753-8_17

(United Nations, 2022). The empirical evidence is irrefutable, necessitating urgent and transformative initiatives (IPCC, 2023).

In addressing climate change, it is recognized as a global conundrum, not confined to specific geographical locales. Acknowledging the shared responsibility, nations have rallied to forge international agreements such as the UNFCCC and the Paris Agreement (UNFCCC, 2015; UNFCCC, 2022). These landmark accords underscore the imperative of collective endeavors in mitigating global warming. Nevertheless, the solution path goes far beyond emission decrease and shift toward renewable power sources. Stewardship of the natural assets of the Earth is fundamental to the solution framework. It is imperative to note the inherent value of forests, oceans, wetlands, and the biodiversity that is found in them considering the key role that they play in the cycling of earth's climate (United Nations, 2021). For instance, forests act as huge carbon sinks and can suck billion loads of carbon dioxide. In addition to protecting in the short term from the effects of climate change, oceans, and wetlands are also carbon-contaminated bodies of water, which provide global temperature stability (Dhote & De, 2021).

Objective of the Study

The main objective of this chapter is to outline in detail the vital role that international collaboration plays in combating climate change through the conservation of natural resources. It aims to underline the universal nature of climate change, representing the global character of the concerted efforts of all countries to handle the negative consequences of global warming. The need for solidarity is therefore at the forefront of international conscience. The chapter calls for the joint actions of nations, utilizing the pooled resources and expertise for sustainable climate adaptation and reduction through nature-based strategies.

Research Methods

This chapter employs a multimethod approach that combines both qualitative and quantitative techniques to assess the effects of climate change. It highlights examples and cases from different geographic locations to illustrate the universal nature of climate change problems and the different ways they are solved. The chapter analyses the results from a variety of international research and includes experienced opinions and findings from environmental institutions to refer to its conclusions. Moreover, it mentions new approaches to the preservation of natural resources which are highly important to expand the adaptation and mitigation tactics in the context of climate change.

Impact of Climate Change on Natural Resources

No doubt, all human-induced activities push the climate to change. These types of changes, which are called climate change, are modifying traditional weather patterns that result in widespread impacts. It is one of the greatest issues connected with these developments: they destroy unique natural resources that cannot be replaced (The Royal Society, 2013).

Climate Change and Its Key Drivers: An Overview

Human society has been overwhelmingly responsible for producing GHGs which in turn have caused the greenhouse effect and the rapid heating of our planet. The main perpetrators are CO_2, CH_4, and N_2O, which are collectively known as the greenhouse gases. The main necessary factors to this phenomenon consist of the burning of fossil fuels, deforestation, and some commercial agricultural practices (NASA, 2023). Two types of factors—natural, such as solar radiation fluctuations or volcanic activity; and artificial, such as higher levels of greenhouse gases—are also influential, but they cannot relate to the fast increase in air temperature observed in the last few decades. The main reason for the increase in the planet's temperature is again the release of CO_2, primarily on account of humans (European Commission, 2020). Hence, the growth in atmospheric emissions interferes with the Earth's natural processes by leading to weather patterns that are out of the ordinary, such as heat waves, droughts, uncommon floods, and powerful storms, all of which could also be said to include a slow rise in sea levels (US EPA, 2023).

Greenhouse gases are very essential in terms of keeping the Earth warm and habitable. However, the over-production of these gases upon human intervention disrupted the thermal balance of our planet (The Royal Society, 2013). Visible impacts of this disruption involve retreating glaciers, diminishing sea ice, increasing sea levels, and altered rainfall patterns. These changes threaten crucial resources such as ecosystems, freshwater reservoirs, and agriculture (NASA, 2023). To address these growing challenges, it is important to reduce greenhouse gas emissions and promote sustainable practices, which preserve the natural environment and support development (European Commission, 2020).

Adaptations in Ecosystems and Biodiversity due to Climate Change

Climate variations are influencing both ecosystems and biodiversity with diverse implications. In response, ecosystems and biodiversity utilize specific strategies to navigate these climatic shifts.

- **Relocation of Species:** Climatic variations prompt species to seek regions with more favorable conditions. Such migrations can redefine ecosystems, altering the dynamics between various species (Malhi et al., 2020).
- **Adaptations to New Climate Patterns:** With shifts in temperature and rainfall, ecosystems are evolving. Observations show plants blossoming sooner during spring, while certain animal migratory patterns are adjusting (US EPA, 2022).
- **Strengthening Ecological Infrastructure:** Implementing ecosystem-centric strategies can safeguard economic interests from the detrimental impacts of climate change. Proactive management of habitats and species, along with hydrological interventions and fortifying ecological frameworks, can bolster the resilience of local species and their habitats (Climate—Adapt, 2023).
- **Preservation and Rehabilitation of Natural Environment:** Prioritizing the safeguarding and rejuvenation of natural territories, and the biodiversity they house, is paramount. The rich tapestry of life they support serves as a formidable natural barrier against the challenges of climate change (United Nations—Climate Action, 2022).
- **Mitigating the fallout of climate change on economic domains** is achievable through ecosystem-centric strategies. The primary focus of these strategies lies in bolstering the resilience of biodiversity and ecosystems (Climate—Adapt, 2023).

Such adaptive measures shed light on the environment's dynamic character and life's incredible capacity to adjust and flourish amidst evolving circumstances. However, the rapidity and scale of climate change currently surpass the adaptability of numerous species and ecosystems, leading to biodiversity reduction and ecosystem degradation (Aurelle et al., 2022; Weiskopf, 2020). To navigate these hurdles effectively, it is essential to curtail greenhouse gas emissions and champion sustainable methodologies that both safeguard natural assets and stimulate economic progression (Climate—Adapt, 2023).

Climate-Induced Stress on Natural Resources

The effects of climate change are directly affecting our most vital natural resources. Water sources are increasingly stressed due to altered hydrological cycles, and agriculture is facing increasing challenges because of temperature extremes, erratic rainfall, and evolving pest dynamics disrupting the delicate balance of food production (National Oceanic and Atmospheric Administration, 2021). Different segments of society are experiencing the detrimental effects of climate change, as its influence reverberates through ecosystems and communities globally (Lindwall, 2022).

Climate Change's Impacts require International Cooperation

Recognizing the universality of climate change is paramount, as its effects are not bound by national or geographical borders—they reverberate globally. Thus, a collaborative international approach is pivotal in navigating its challenges and adapting accordingly. The Paris Agreement stands testament to global efforts, targeting a reduction in greenhouse gas emissions and a cap on global warming.

Yet, many experts are raising alarms, suggesting that nations aren't moving swiftly enough to stave off a potential 1.5 °C rise in global temperatures (Maizland, 2022; Department of Foreign Affairs and Trade, 2018). It underscores the urgency to persistently advocate for collective action and mutual responsibility to safeguard the environment for the coming generations. Addressing the multifaceted threats of climate change demands an unwavering commitment to international cooperation.

One of the Paris Agreement's primary ambitions is to fortify the global stance against climate change, striving to keep the temperature surge well under 2 °C from pre-industrial times and, if possible, to a mere 1.5 °C rise. Beyond just adaptation strategies, the agreement emphasizes fostering climate resilience and transitioning toward a low-emission paradigm.

With the looming threat of a 1.5 °C global temperature hike, nations must expedite their efforts in mitigating climate adversities. In this battle against climate change, unity is strength. While the Paris Agreement lays a foundational framework, there remains much ground to cover. Without swift and concerted action, the global thermostat might rise by 1.5 °C come 2100. The onus is on everyone to collaborate and steward the planet toward a sustainable future.

A Solidarity Approach to Mitigation and Adaptation to Climate Change

Solidarity is not a term but understanding that every country, big or small, rich or developing, has its role in the fight against climate change. This perception highlights the need for countries to join forces, help out where necessary, and work as one. In this consortium, every country plays its part in reducing emissions and adapting to climate change (United Nations, 2023). Major global treaties such as the Paris Agreement, Kyoto Protocol, and UNFCCC provide the foundation. They specify emission targets, and monitoring approaches, and assist in the form of finance to help the developing nations improve (UNESCO, 2019; United Nations, 2018). Institutions such as the United Nations Environment Programme (UNEP) and the Intergovernmental Panel on Climate Change (IPCC) deliver science-based information and policy recommendations that guide the worldwide response. (UNESCO, 2019). However, it is not entirely clear. This spirit of oneness hits impediments not only from economic disparities among nations but also from political debates and conflicts that can hinder progress (United Nations, 2023).

However, there are some bright examples of collective efforts among the world community and common responsibilities in climate actions, joint investments in renewable energy projects, technology transfer, and support for climate resilience programs in some regions that are regarded as vulnerable (Bazzani, 2023; United Nations Climate Change, 2022).

Climate Change Mitigation and Natural Resource Conservation

Natural Resource Preservation and Greenhouse Gas Reduction

The conservation of natural resources and the reduction of greenhouse gas emissions are closely related. Conserving practices are crucial considering that most natural resources are essential for both carbon sequestration and emission reduction (USDA, 2023; EPA, 2016). Deforestation practices result in the release of carbon that was previously stored, while the contribution of greenhouse gases is high in terms of energy production originating from fossil fuels. In addition, the use of natural resources in many situations naturally leads to an increase in emissions (USDA, 2023).

Forests, oceans, wetlands, and biodiversity—collectively representing the tapestry of natural ecosystems—are instrumental in capturing carbon and alleviating the impacts of climate change (NREL, 2012). Implementing sustainable land-use practices, including eco-conscious agriculture and urban planning, aids in the conservation of these invaluable natural resources while fortifying climate resilience (EPA, 2016). In developing conservation strategies, it is essential to factor in the potential alteration climate change may impose on habitats and migratory patterns, as these changes could recalibrate conservation priorities (EPA, 2012). Therefore, conservation approaches that are adaptive to climate shifts should be endorsed, and supported by appropriate policy frameworks and financial instruments (Climatelinks, 2022).

Carbon Sequestration by Forests, Oceans, Wetlands, and Biodiversity

Tropical rainforests are among Earth's most effective carbon sponges. Through the process of photosynthesis, the vast array of trees within these verdant regions actively extracts carbon dioxide from the air. In doing so, they convert carbon dioxide and water into glucose and oxygen. This glucose then transitions into biomass, acting as a solid carbon reservoir (NOAA, 2023).

The significance of oceans and wetlands in carbon capture should be highlighted as well. Coastal wetlands, which comprise ecosystems such as salt marshes, mangroves, and seagrass meadows, have an exceptional ability to remove carbon dioxide from the atmosphere. They store carbon not only in their visible parts such as branches, leaves, and roots but also in their soils (Pew, 2021). Surprisingly, these wetlands have the potential to be ten times more productive in terms of carbon sequestration than mature tropical forests (NOAA, 2023).

In addition, phytoplankton, which is an ocean inhabitant, is a crucial factor in the absorption of atmospheric carbon dioxide. The biodiversity of the ecosystem itself is a buffer to climate change. Highly varied ecosystems have more resilience and adaptability to climatic changes, thus, augmenting their capacity to capture carbon (United Nations—Climate Action, 2022).

Climate Resilience and Preservation of Resources via Sustainable Land Use Approaches

The sustainability of land practices is part of the preservation of our natural assets, as well as the resistance to climatic changes. In light of the dynamic challenges posed by climate change, land stewardship becomes crucial. This section explores the importance of sustainable land practices from a climate-based perspective. It will highlight the essential ecosystem services in climate resilience, the protective role of eco-friendly agriculture in preserving both soil and water resources and in reducing emissions, and the advantages of green urban planning in lowering greenhouse gas outputs and energy consumption while preserving green areas and biodiversity.

- **Ecosystem Services:** The ecosystem provides essential services such as water purification and flood control that are crucial to climate resilience. The provision of these services can be compromised when ecosystems are mismanaged (IISD's SDG Knowledge Hub, 2019).
- **Sustainable Agriculture:** As a result of sustainable agriculture practices, soil health and water resources are protected, and emissions are reduced. There are many examples of sustainable agriculture practices, such as agroforestry and no-till farming (IISD's SDG Knowledge Hub, 2019).
- **Urban Planning:** In addition to reducing emissions from transportation and energy use, sustainable urban planning can preserve green spaces and biodiversity (IISD's SDG Knowledge Hub, 2019). One significant factor is that economically disadvantaged communities, especially those residing in the rural sectors of developing nations, possess limited resources to deal with escalating natural disasters, placing their means of livelihood in jeopardy (United Nations, 2023).

Utilizing renewable land resources for production is imperative, all while safeguarding their enduring productivity and environmental functions. Engaging in sustainable agricultural practices like agroforestry and no-till farming is pivotal, as these methods not only preserve soil vitality and water reserves but also minimize greenhouse gas emissions (FAO, 2023). In addition, services delivered by ecosystems: water cleansing, and flood moderation, among others, are crucial for climate change adaptation. Sustainable urban planning plays a significant part in that regard too, not only in reducing emissions of transport and energy but also in saving green areas and biodiversity.

Incorporation of Climate Knowledge into Conservation Approaches

Climate change is redrawing what conservation efforts seek to protect in terms of the environmental landscapes, affecting habitats, migration patterns, and where species are found. These changes require a review of conservation priorities. A wise approach to conservation is to expect the species to either move or change due to the changing tape of the climate. This section focuses on the necessity of making conservation techniques more sophisticated considering the extensive effects of climate change.

- **Re-evaluating Habitats and Migration:** Climate change does more than just change habitats—it initiates a sequence of changes that forces some species to move to keep thriving. Rising temperatures cause habitats to change and force species to move to higher elevations or toward the poles (Groves et al., 2012). Conservationists should be futuristic in this regard, determining likely movement routes of species and areas that will remain habitable shortly for conservation attention (We Conserve, 2019).
- **Strategizing for Resilience:** Resilient conservation plans are essential. Such planning involves pinpointing crucial habitats and devising tactics that fortify both ecosystems and species against climate change's onslaught. The goal here is to spotlight areas likely to stay viable for diverse species and to channel conservation endeavors accordingly. It is also about recognizing and preparing for regions that climate change will likely affect and formulating strategies to aid species in their adaptation journey (U.S Climate Resilience Toolkit, 2021).
- **Fostering Climate-Adaptive Conservation:** Successful climate-triggered conservation demands suitable policies and proper funding. The authorities and organizations should therefore invest their resources in proceeding with conservation programs that involve climate change considerations. The blueprints and materials include a Climate-Smart Conservation guide as well as procedures developed by the U.S. Fish and Wildlife Service (USFWS). They effectively demonstrate that climate concerns should be integrated gradually into common conservation activity (NWF, 2023; Wilkening et al., 2022).

Innovative Practices in Natural Resource Conservation

Ecosystem-Based Approaches to Climate Change Adaptation

The EbA (Ecosystem-based adaptation) is the most suitable option for the communities as they adapt to the impacts of climate change as the approach involves ecosystem conservation and restoration (UNEP, 2020). EbA is both a financially feasible and effective method of protecting natural habitats and curbing floods, while additionally decreasing vulnerabilities of natural systems and human settlements (U.N. Environment, 2021). The embodiment of the concept of EbA can be seen in programs such as the restoration of coastal mangrove forests to protect against the stormy sea or the preservation of the upland forests to maintain the flow of water to downstream populations (IUCN, 2017).

UNEP claims that one of its programs is to coordinate global efforts and that it is involved in over 45 EbA projects, through the restoration of the estimated 113,000 ha of habitat across the globe, which helps 2.5 million individuals (U.N. Environment, 2021). On the other hand, they are usually more efficient and show some advantages that differ from traditional environmental approaches (Climate Adapt, 2017).

Addressing Climate Change: Embracing Nature-Driven Solutions for Mitigation and Adaptation

Nature-based solutions (NbS) include approaches that take advantage of nature's capabilities to address societal problems which are exemplified by the difficult problem of climate change. Such solutions include actions such as reforestation and the development of green urban areas (World Bank, 2022). By sequestering and storing carbon in addition to minimizing emissions, NbS provides a possible way for climate change mitigation. Secondly, they strengthen communities by providing better flood protection, increased water security, and reinforced resilience (Seddon et al., 2020; Sowińska-Świerkosz & Garcia, 2022). International treaties such as the Paris Agreement and the Convention on Biological Diversity have acknowledged and embedded NbS as a critical element of climate policy. Climate change response requires both mitigation and adaptation, which constitutes an integrated global response (IUCN, 2023).

By safeguarding, responsibly overseeing, and rejuvenating natural ecosystems, nature-based solutions present a dual advantage: they overcome problems such as climate change and food and water shortages, while simultaneously improving human health and biodiversity. An example of this is that coastal trees like mangroves are a barrier to storm surges, which in turn, save lives and assets and, they are home to a wide array of aquatic and avian biodiversity, thus promoting biodiversity (World Bank, 2022). NbS's main objective is to address social needs as well as improve the biodiversity of the planet and the prosperity of its dwellers (IUCN, 2023).

The Paris Climate Agreement sets an ambitious target: limiting the increase in global temperature to 1.5 °C and keeping it at 2 °C or below. The collective actions taken by the global community in the upcoming decade will be pivotal in determining our capacity to mitigate the most severe repercussions of climate change. Among the top strategies identified to curb carbon emissions by 2030, nature-based solutions, especially those focused on forest conservation, restoration, and improved land management practices, hold a prominent place (IUCN, 2022).

Natural Resource Management Using Green Technologies and Eco-Friendly Practices

An effective management of natural resources in a changing climate requires the use of green technologies and eco-friendly practices. In terms of their production process or supply chain, green technologies refer to environmentally friendly technologies (Kenton, 2022). There are a variety of technologies that can reduce environmental impacts while conserving resources, including renewable energy systems, sustainable agriculture practices, and efficient water management systems (Guelfo, 2023; Swallow, 2022). As a result of eco-friendly practices in industries such as forestry, fishing, and mining, ecosystems are minimized, and sustainable resource use is promoted (US Bureau of Labour Statistics, 2012). To promote eco-friendly practices and sustainable resource management, certifications like FSC (Forest Stewardship Council) for timber and organic certification for agriculture are useful tools (Kenton, 2022). Developing energy-efficient technologies, promoting sustainable practices, and minimizing negative environmental impacts are ways to reduce our environmental footprint through the use of green technologies and eco-friendly practices (Seventh & Chomsky, 2023).

Case Studies Highlighting Successful Conservation Initiatives

Case studies provide valuable examples of successful conservation initiatives utilizing innovative approaches to natural resource management in a changing climate. Case studies provide measurable impacts and outcomes of initiatives, such as increased biodiversity, reduced greenhouse gas emissions, and improved community resilience. Wetland restoration initiatives, urban green infrastructure initiatives, or community-based conservation initiatives are examples of case studies from different geographic regions and ecosystems.

The Tongass National Forest, nestled in Southeast Alaska, stands as a testament to effective conservation. Recognized as the planet's most expansive untouched coastal temperate rainforest, it's often dubbed the "lungs of North America." The primary objective of the project centered on the forest is its rejuvenation, with a particular emphasis on addressing the needs of its Indigenous inhabitants. The success of this endeavor leans heavily on community engagement, forming alliances, securing funds, and introducing innovative approaches. Insights from this initiative, including best practices and challenges faced, offer valuable takeaways. These insights, like the crucial role of community participation, forging partnerships, and ensuring adequate funding, can guide and inform upcoming conservation initiatives (Bloom, 2022).

The innovative methods and practical examples discussed offer valuable insights into the role of imaginative and flexible strategies in both conserving natural resources and tackling climate change challenges. Consequently, these approaches highlight the potential for achieving sustainability and resilience in a dynamically changing environment.

The Indian Perspective: Challenges and Opportunities

Climate Change Impact on India

Due to climate change, India is facing multiple challenges due to its vast geographic expanse and diverse geographical characteristics. The following are some of the significant impacts:

- In recent decades, India has experienced a significant increase in average temperatures. It has implications for agriculture since certain crops become less viable in warm weather (The World Bank, 2013).
- Agricultural productivity in India is severely affected by the unpredictable monsoon season. Climate change has resulted in unpredictable monsoon patterns, resulting in either droughts or floods in different parts of the country (The Climate Reality Project, 2022).
- India's coastal areas, especially along the eastern shoreline, face the threat of rising sea levels. Such a rise could potentially displace countless individuals and jeopardize their means of living, especially for those dependent on coastal fishing activities (The World Bank, 2013).
- The Himalayan glaciers are melting at an alarming rate. Water scarcity in the northern plains may result from this impact on rivers originating from the Himalayas (NIC, 2009).
- Changes in weather patterns and habitat loss have resulted in the loss of a great deal of biodiversity (The World Bank, 2013).

In addition, these impacts have a significant impact on India's economy, food security, and well-being. While the nation has already initiated several measures to counteract and adjust to climate change impacts, addressing this universal challenge effectively calls for collective global action.

Conservation Policies and Initiatives at the National Level

India has taken significant steps at the national level to address climate change and safeguard its natural assets. Through its National Action Plan on Climate Change (NAPCC), the country has laid out its blueprint for climate adaptation and fostering a more sustainable developmental trajectory (MOEFCC, 2021). A central aim of the Green India Mission revolves around bolstering forest cover while ensuring the protection, rejuvenation, and enhancement of its dwindling forests (MOEFCC, 2022). The Paramparagat Krishi Vikas Yojana stands as a testament to India's commitment to organic farming and sustainable agriculture (Vikaspedia, 2023). The Jal Shakti Abhiyan initiative underscores the importance of water conservation and efficient water use (Ministry of Jal Shakti, 2021). Complementing these efforts, India is actively working to minimize its reliance on nonrenewable energy sources, setting commendable goals, especially in the domain of solar energy (MOEFCC, 2021).

The implementation of these policies and initiatives is crucial to the sustainable development of India and its resilience to climate change. In addition, there are concerns about the potential negative impacts of certain policies, such as the recent changes made to forest conservation laws, which have been criticized by environmentalists (Bhattacharya, 2023).

The Importance of India's Actions in Addressing Global Climate Change

India plays a pivotal role in global climate change efforts for several compelling reasons. Firstly, being the globe's second most populous nation, any steps India takes reverberate significantly on an international scale (UN, 2023). Secondly, amidst its swift economic expansion, India's approach to aligning growth with sustainability holds valuable lessons for emerging economies (World Bank, 2016). Moreover, with India hosting multiple biodiversity hotspots, its conservation efforts are crucial for global biodiversity preservation (UN, 2023). Finally, as one of the predominant carbon dioxide contributors, India's initiatives in curbing emissions are instrumental for global climate change mitigation (World Bank, 2016). Given that its over 1.3 billion inhabitants constitute approximately 18% of the world's total population, India's status as the third-largest national carbon dioxide emitter underscores the importance of its climate actions (UN, 2023). In India, climate change is already affecting growing seasons, and almost half of Indians work in agriculture and other climate-sensitive sectors, which can lead to significant losses in productivity and health (Leiserowitz et al., 2022). The efforts being made by India to reduce carbon emissions and promote sustainable development are crucial to mitigating the effects of climate change on a global scale.

International Cooperation and Solidarity Opportunities

In promoting clean energy and green initiatives, India has numerous opportunities for international cooperation and solidarity. The following are some of these opportunities: India actively participates in both bilateral and multilateral accords and has also forged partnerships like the International Solar Alliance (ISA) to champion the cause of clean energy, as documented by ISA in 2023. The country's Prime Minister has committed to India to reach net-zero emissions by 2070. Notably, during the G20 summit in 2023, the Director-General of ISA held discussions with Saudi Arabia's Energy Minister, Prince Abdulaziz bin Salman. This meeting led to the inclusion of ISA's Green Hydrogen Innovation Centre in the G20 India summit's Delhi Declaration. Following the UNFCCC, India actively participates in the UNFCCC and commits to the Paris Agreement (Government of India, 2022).

Collaboration with other nations to develop clean technologies can assist India in its transition to a green economy (Phillips et al., 2013). The United States and India maintain a collaborative relationship in the realms of climate change and energy (Mehta & Gerard, 2022). To assist India with its green initiatives, international organizations can provide financial assistance. Green bonds are incorporated into India's climate finance strategy (Hussain & Dill Helena, 2023). It is important to share knowledge with other nations to foster international cooperation. The Climate and Clean Energy Agenda 2030 Partnership between India and the United States aims to enhance climate action plans by mobilizing finance and accelerating clean energy deployment (Mehta & Gerard, 2022). Given India's distinct set of challenges and opportunities, the country holds a pivotal position in both mitigating global climate change and preserving natural resources. Its strategies and actions can offer valuable insights and benefits to other nations, especially those situated in the global south.

Case Study of Conservation in Rajasthan: Efforts in a Challenging Landscape

An Overview of Rajasthan's Climate Change

As a result of its geographical location and topography, the Indian state of Rajasthan is highly susceptible to climate change. Rajasthan is in a region with the highest climate sensitivity, the highest vulnerability, and the lowest ability to adapt to changes in climate (Goswami, 2019, Disaster Management, Relief & Civil Defence Department Government of Rajasthan (2022)). (Figure 1 Map 1 (a–h)) represents the average maximum temperature by district in Rajasthan from 1951 to 2022. From 1951 to 2022, Rajasthan experienced an approximate increase of 0.74 °C in its average maximum temperature. However, this uptick in temperature hasn't been uniform across the entire state. In some districts, the temperature has increased more rapidly than in others. The most affected districts are Bikaner (1.70 °C), Sri Ganganagar (1.58 °C), Nagaur (1.48 °C), Ajmer (1.48 °C), and Jodhpur (1.42 °C).

Figure 2 Map 2 (a–h) represents the minimum average temperature of Rajasthan from 1951 to 2022. Between 1951 and 2022, the average minimum temperature in

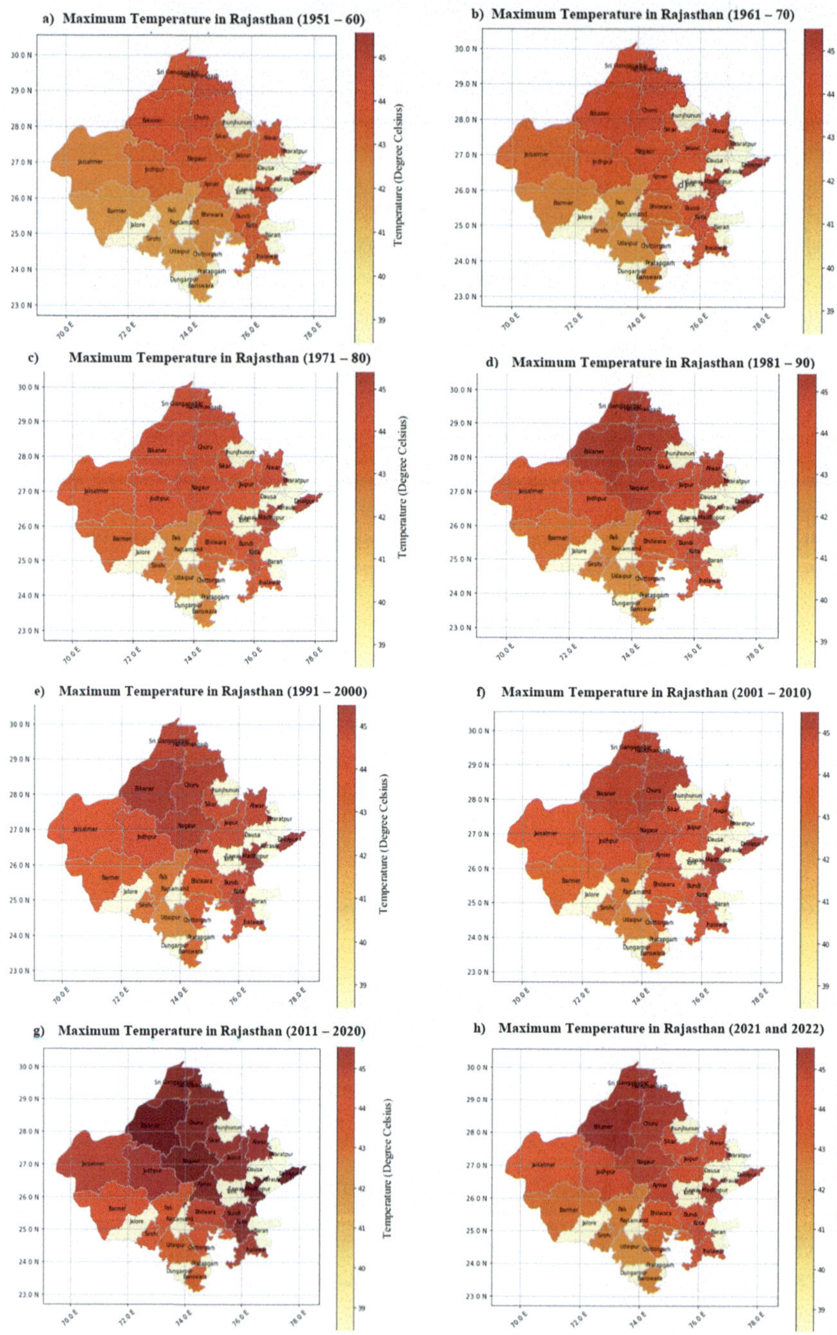

Fig. 1 Map 1 (**a–h**) District-wise maximum temperature of Rajasthan (1951–2022). (Source: Data extracted from Yearly Gridded Maximum Temperature (1.0 × 1.0) data Binary File (IMD))

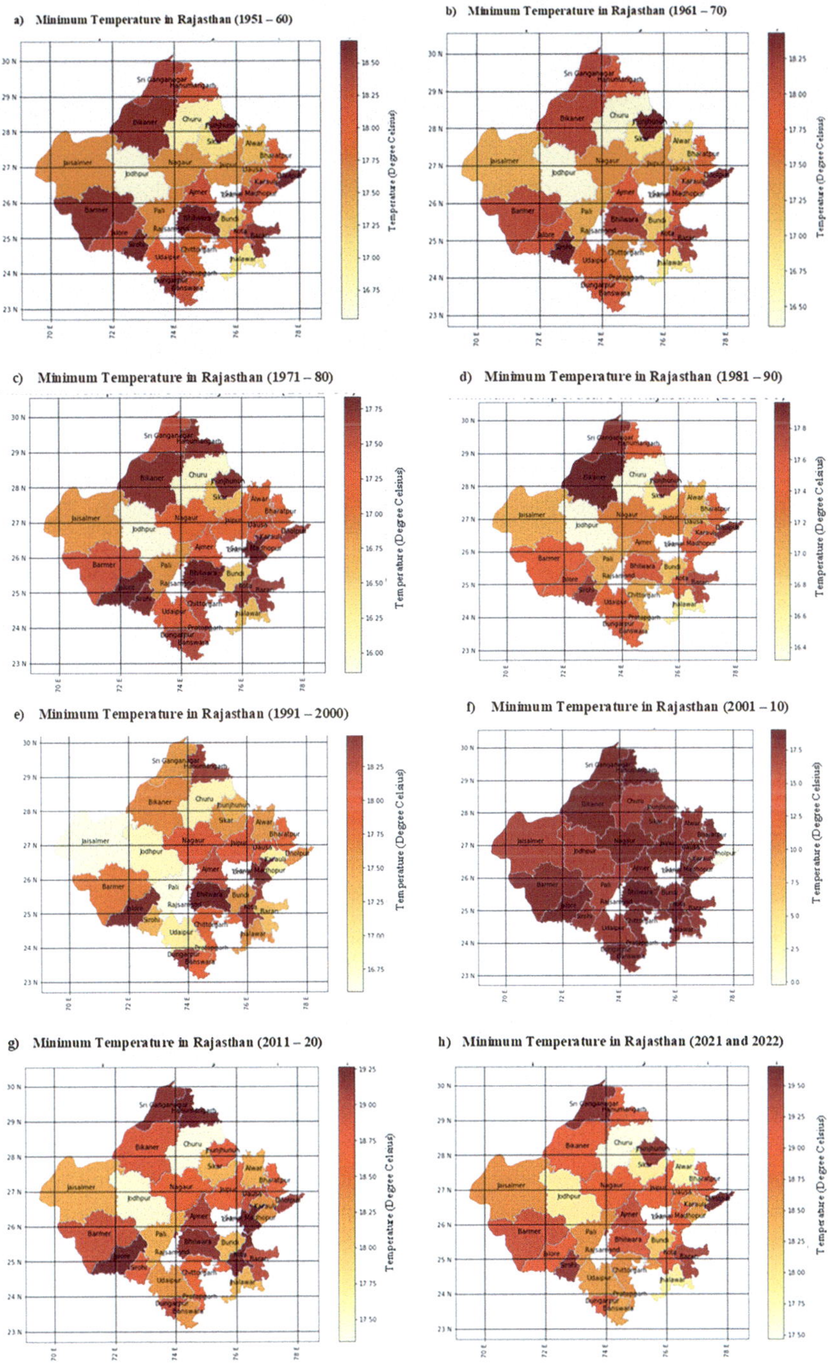

Fig. 2 Map 2 (**a–h**) District-wise minimum temperature of Rajasthan (1951–2022). (Source: Data extracted from Yearly Gridded Minimum Temperature (1.0 × 1.0) data Binary File (IMD))

Rajasthan increased by approximately 0.85 °C. Temperatures are not rising uniformly throughout the state. In some districts, temperatures have risen more rapidly than in others. The most affected districts include Jaipur (1.36 °C), Nagaur (1.36 °C), Jodhpur (1.32 °C), Dausa (1.23 °C), and Bharatpur (1.22 °C), with Jaipur and Nagaur experiencing the highest rises of temperatures

Figure 3 Map 3 (a–e) showcases the simulated surface air temperature readings from 1951 to 2005, as derived from the CCCR-IITM data. Over the last five decades, Rajasthan has seen a steady rise in its average temperature, with an increase of about 1.13 °C. This consistent warming suggests broader changes in the climate. Between the decades 1951–1960 and 1961–1970, there was a temperature rise of around 0.20 °C. The subsequent decade, from 1961–1970 to 1971–1980, saw an increase of about 0.32 °C. The temperature then climbed by around 0.18 °C from 1971–1980 to 1981–1990 and experienced a significant rise of approximately 0.43 °C from 1981–1990 to 1991–2005 (CCCR, 2014).

This study delved into the Representative Concentration Pathway (RCP) 4.5 scenario data, sourced from CCCR IITM, to forecast potential temperature shifts in Rajasthan, India, spanning from 2006 to 2095. The choice to employ the RCP 4.5 data for Rajasthan was influenced by a couple of factors: The CCCR IITM platform exclusively offers the RCP 4.5 dataset for temperature and precipitation projections from 2006 to 2096. India's national commitments align with this, aiming for a net zero target by 2070. Under the RCP 4.5 trajectory, it is anticipated that global yearly greenhouse gas emissions will reach their zenith around 2040, before commencing a decline.

Figure 4 Map 4 (a–i) represents the projected average temperature of Rajasthan from 2006 to 2095. Based on the analysis of temperature data from 2006 to 2095, there is a noticeable change in the average temperature in Rajasthan. The average temperature increased gradually from 27.27 °C in 2006–2015 to 27.48 °C by 2086–2095. It is estimated that the global temperature will rise by approximately 0.21 °C over the next 90 years (Fig. 5). There were some decades during this period that experienced sharper increases than others. As an example, the decade from 2056 to 2065 saw the largest increase in temperature of 0.11 °C over the previous decade. Several factors may contribute to such variations, including natural climate variability and anthropogenic emissions. The whole of Rajasthan is to experience temperature changes; however, particular districts like Sri Ganganagar will suffer from increased average temperatures, making them especially susceptible to the problems of increasing temperature. On the other hand, districts like Udaipur are expected to have more moderate average temperatures

Temperature increase is a factor of influence for many sectors including agriculture, water resources, and human health. Warmer temperatures could lead to higher evaporation rates, which would change soil moisture and possibly influence crop yields. Higher temperatures could worsen the draughts by accelerating evaporation. In addition, it can increase the impact of the El Nio phenomenon leading to the unpredictable and variable nature of the weather patterns.

Although greenhouse gas emissions and climatic fluctuations influence, these temperature changes are in line with the global climate change movement. The RCP

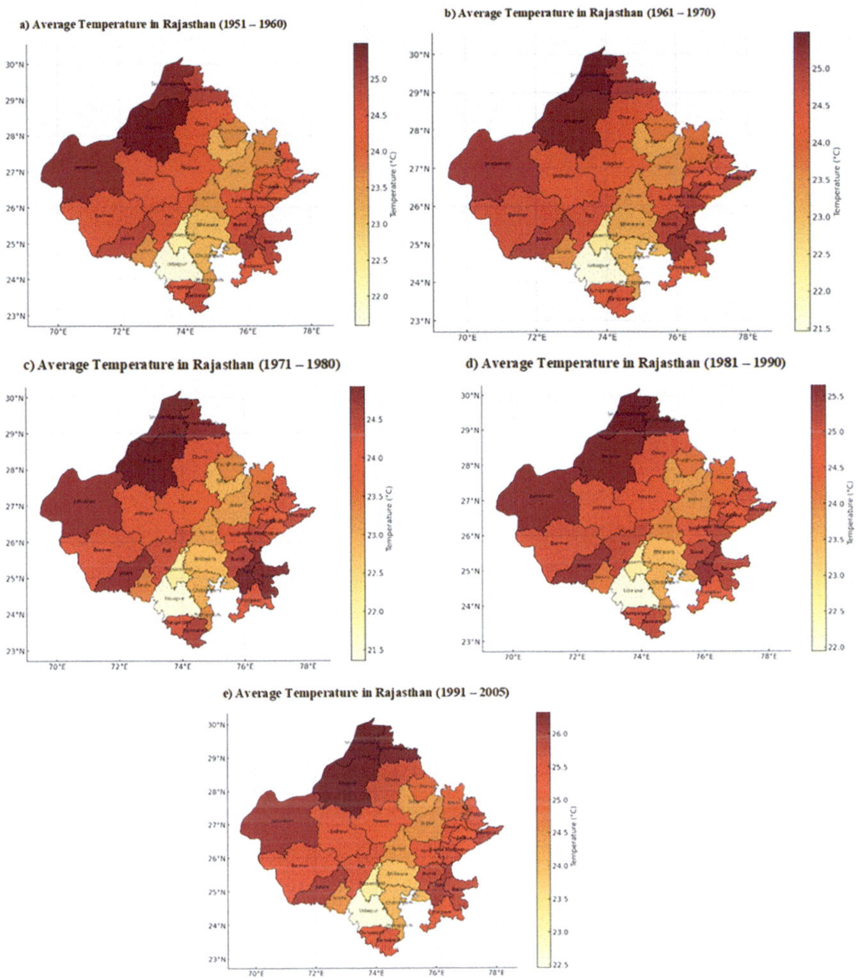

Fig. 3 Map 3 (**a**–**e**) Average temperature of Rajasthan (1951–2005). (Source: Data extracted from Centre for Climate Change Research, IITM historical data (1951–2005))

4.5 sees a route to mitigation though even this is still a significant change in that scenario. The RCP 4.5 scenario considers that the temperature of Rajasthan will probably change considerably in the next decades. Although the temperature rise is small, it can change the climate dynamics of the area. Adaptive strategies, strategic planning, and ongoing monitoring should be employed to successfully address these evolving challenges.

Figure 6 Map 5 (a–e) presents an analysis of historical precipitation patterns in Rajasthan from 1951 to 2005, shedding light on the state's climatic fluctuations. The researcher has examined the historical rainfall data in Rajasthan for five distinct periods: 1951–1960, 1961–1970, 1971–1980, 1981–1990, and 1991–2005, aiming

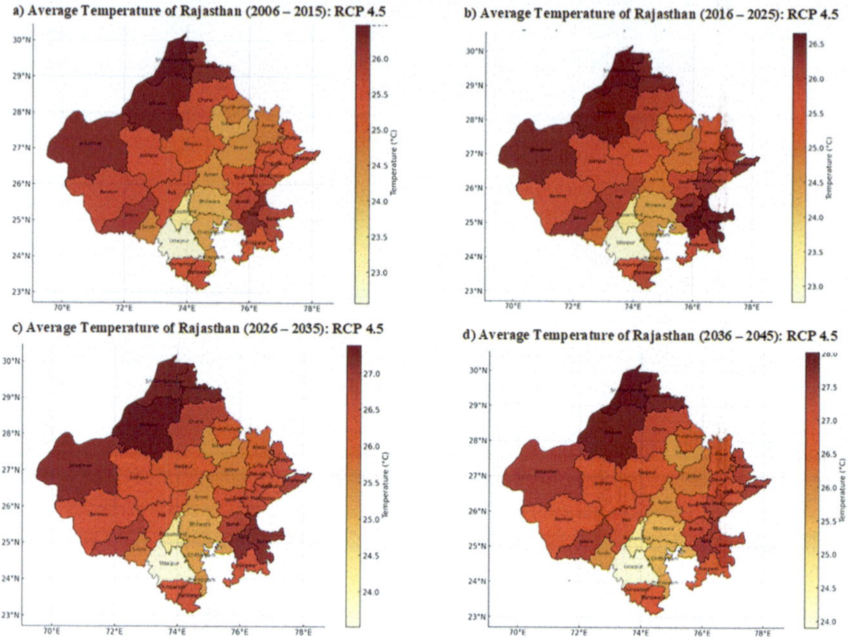

Fig. 4 Map 4 (**a**–**i**) Average temperature of Rajasthan (2006–2095). (Source: Data extracted from Centre for Climate Change Research, IITM: RCP 4.5 (2006–2095))

to discern trends and variations in rainfall. The quantity of rainfall in a region significantly impacts agriculture, water resources, and the overall climate. Notably, during the period from 1991 to 2005, Jhalawar experienced the highest average precipitation, whereas Jaisalmer received the least

Figures 7 Map 6 (a–h) and 8 illustrate the anticipated average precipitation levels in Rajasthan from 2006 to 2095. The average annual precipitation shows an increasing trend from 2006 to 2095 (0.13 mm/day). The decade 2006–2015 recorded the lowest average precipitation at 2.886 mm/day, whereas the decade 2066–2075 is expected to record the highest precipitation at 3.031 mm/day. Over the span from 2006–2015 to 2066–2075, precipitation is projected to rise by an average of 0.145 mm/day, signifying a consistent increase in rainfall over the previous 60 years. After 2075, there will be a slight decline, but the values will remain higher than during the initial decades (Fig. 8) (Srivastava et al., 2009).

It is evident from the trajectory of the RCP 4.5 scenario that mitigation of greenhouse gas emissions is crucial to stabilizing the climate and minimizing adverse impacts. Based on a scenario based on RCP 4.5, the analysis of precipitation in Rajasthan reveals complex patterns over time. The general pattern suggests a subtle rise in rainfall, potentially impacting water resources, agriculture, and the occurrence of extreme weather events. To effectively manage these shifting climate dynamics, it is imperative to maintain ongoing monitoring and implement adaptive strategies.

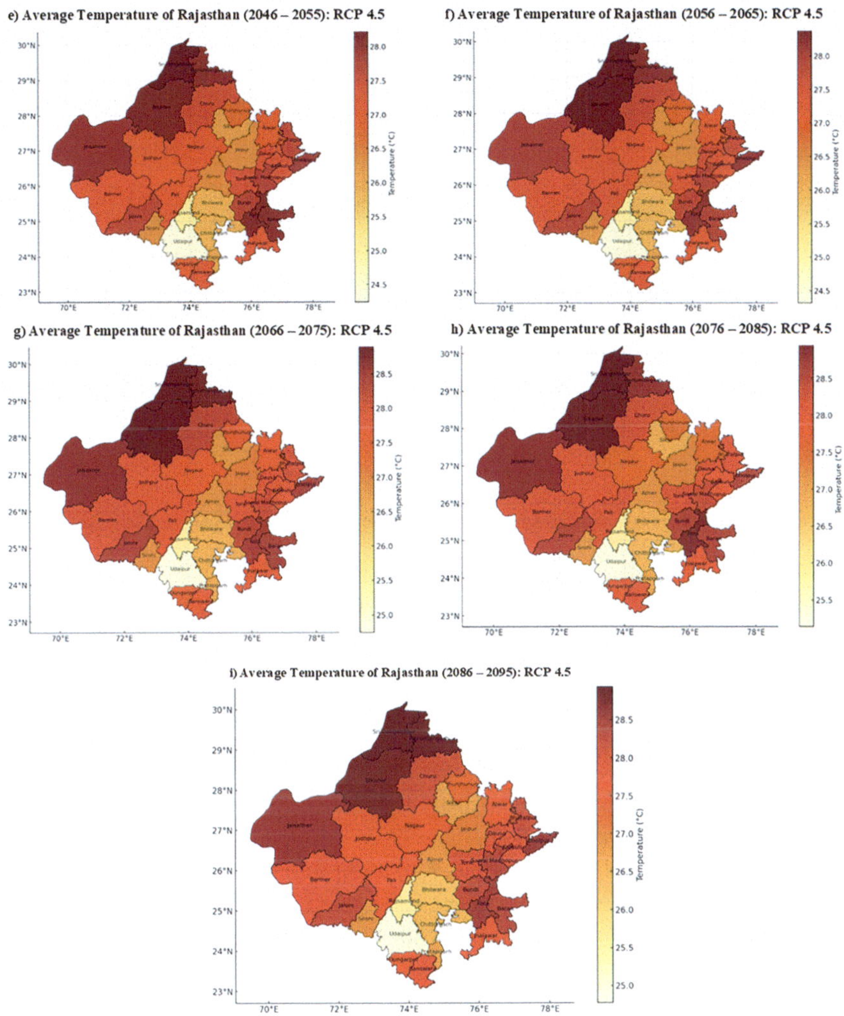

Fig. 17.4 (continued)

Climatic Effects on Natural Resources in Rajasthan: A Focus on Drought

In the state of Rajasthan, India, recurrent droughts pose significant challenges to the well-being of its population. The failure of the southwest monsoon stands out as a primary catalyst for drought conditions across the Indian subcontinent (Khera, 2004). Rajasthan's arid and semi-arid climate, coupled with its geographical features, renders it particularly susceptible to drought occurrences.

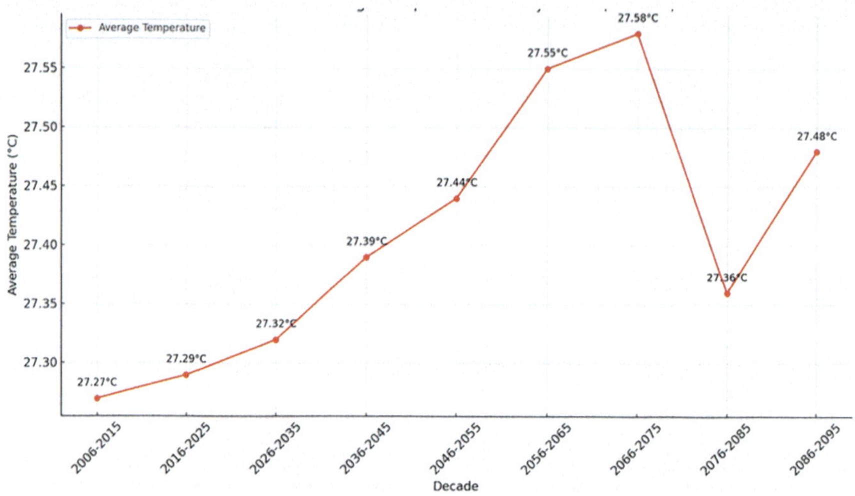

Fig. 5 Decadal average temperatures in Rajasthan: RCP 4.5. (Source: Data extracted from Centre for Climate Change Research, IITM: RCP 4.5 (2006–2095))

It is documented that these droughts have led to extensive crop losses, resulting in scarcities of food, fodder, and water. Rajasthan has witnessed a total of 48 drought years, each varying in terms of severity (Rathore, 2004).

Figure 9 Map 7 (a) illustrates the geographical distribution of drought occurrence across Rajasthan's districts during the years marked by drought between 1901 and 2002. Rajasthan has experienced a significant number of drought years over the past 102 years, with more than half of the districts experiencing drought conditions. The following districts have been particularly affected by drought, with a high frequency of occurrence: With drought conditions prevailing in more than 53% of the years studied, Jodhpur stands out as one of the most affected districts. Among the most affected districts is Sirohi, which has a frequency of over 53%. The arid conditions of Pali make it highly susceptible to drought, which occurs in 52% of the years. In Jalore too, droughts occur approximately 48% of the time. Rajasthan's districts, from north to south, have experienced frequent droughts, highlighting the extent of this climatic challenge

Figure 9 Map 7 (b) represents the analysis of the last decade (2011–2021), which reveals some concerning patterns associated with drought in the region. The map illustrates the number of drought years in each district in Rajasthan for the years 2011–2021. Droughts were more frequent in districts with deeper shades. As can be seen, the drought in Rajasthan is not confined to one region but affects districts throughout the state

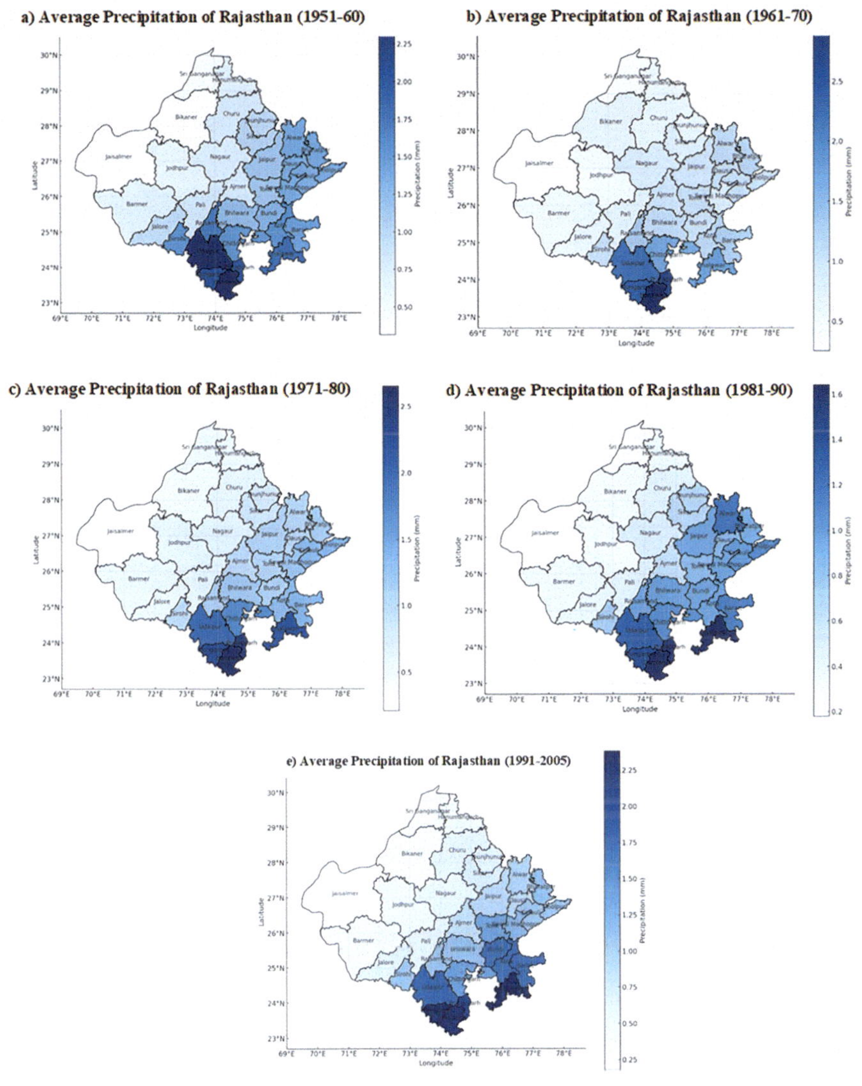

Fig. 6 Map 5 (**a**–**e**) Average precipitation of Rajasthan from 1951 to 2005. (Source: Data extracted from Centre for Climate Change Research, IITM historical data (1951–2005))

Initiatives and Innovations in Conservation in the State

The state of Rajasthan has been a pioneer in adopting innovative conservation approaches. The following are some of the key initiatives and innovations in the state (Fig. 10).

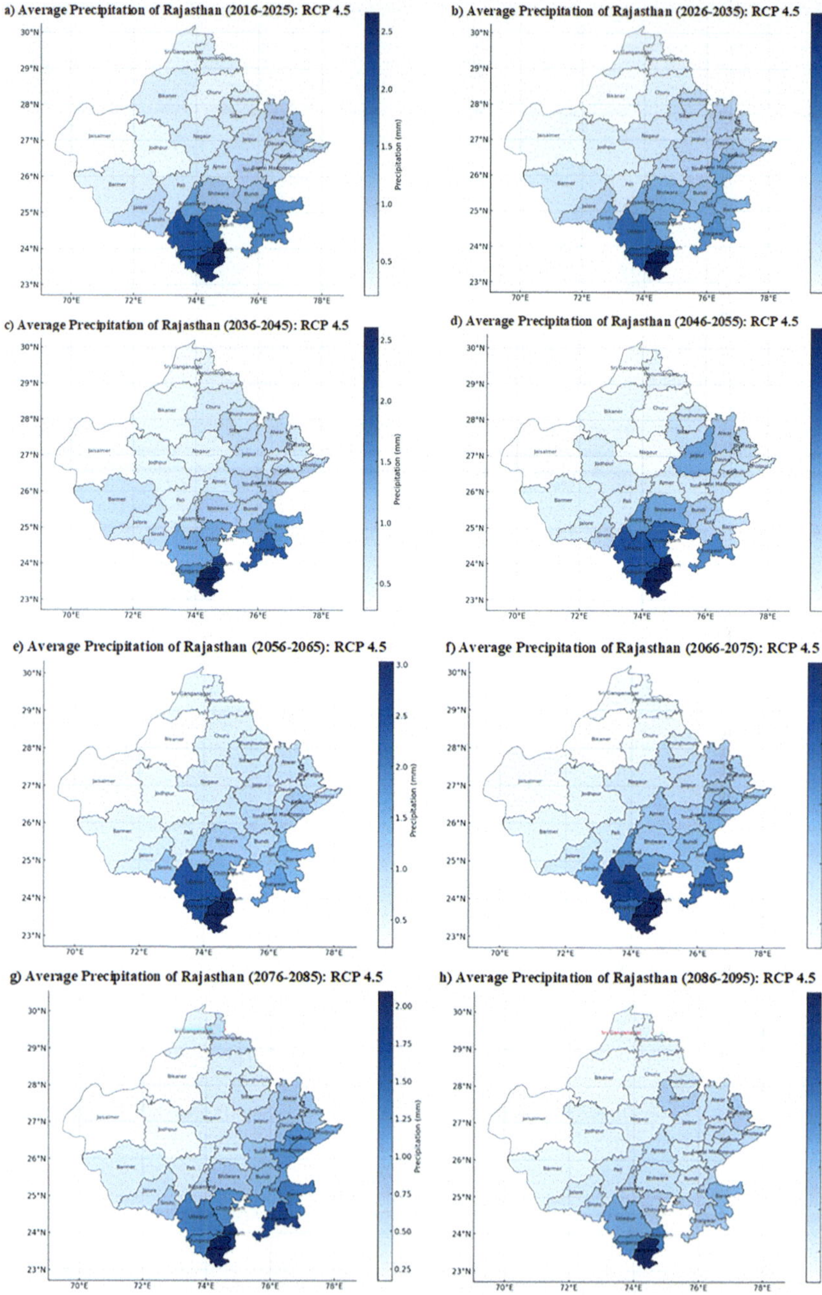

Fig. 7 Map 6 (**a–h**) Average precipitation of Rajasthan from 2006 to 2095. (Source: Data extracted from Centre for Climate Change Research, IITM: RCP 4.5 (2006–2095))

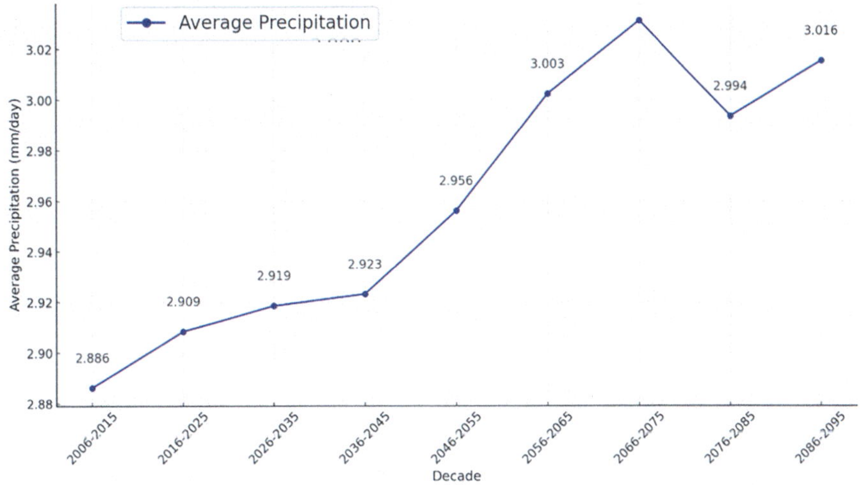

Fig. 8 Decadal average precipitation in Rajasthan: RCP 4.5. (Source: Data extracted from Centre for Climate Change Research, IITM: RCP 4.5 (2006–2095))

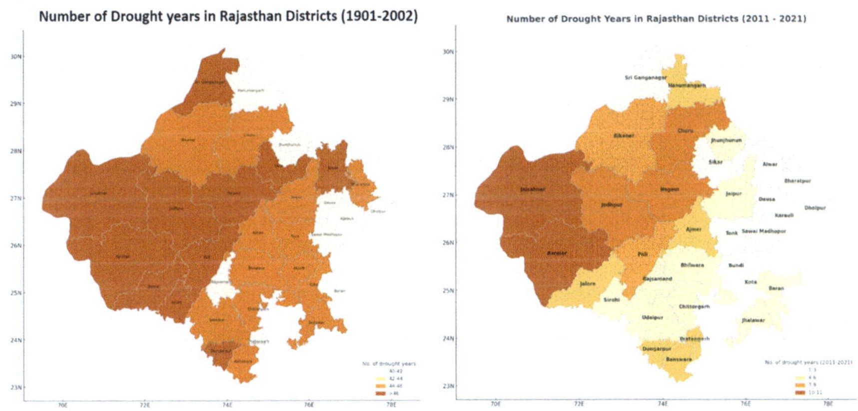

Fig. 9 Map 7 (**a**) Number of drought years in Rajasthan districts (1901–2002); (**b**) Number of drought years in Rajasthan districts (2011–2021). (Data Source: (**a**) State Level Analysis of Drought Policies and Impacts in Rajasthan, India, 2004; (**b**) RTI data from Disaster Management, Relief and Civil Defence Department, Government of Rajasthan, April 2022)

Resilience to Climate Change: Lessons Learned and Implications

Rajasthan's efforts in building climate resilience and adaptation provide valuable lessons for other regions that face similar challenges (Fig. 11). The key lessons learned and their implications are shown in Fig. 11.

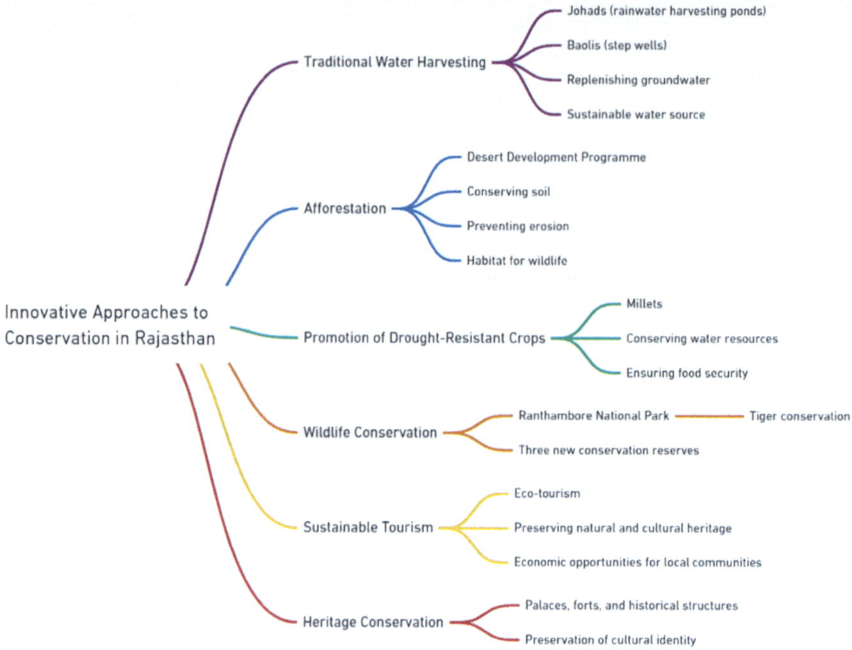

Fig. 10 Innovative conservation approaches taken by government of Rajasthan in the state. (Source: Figure 3 prepared by author from multiple sources)

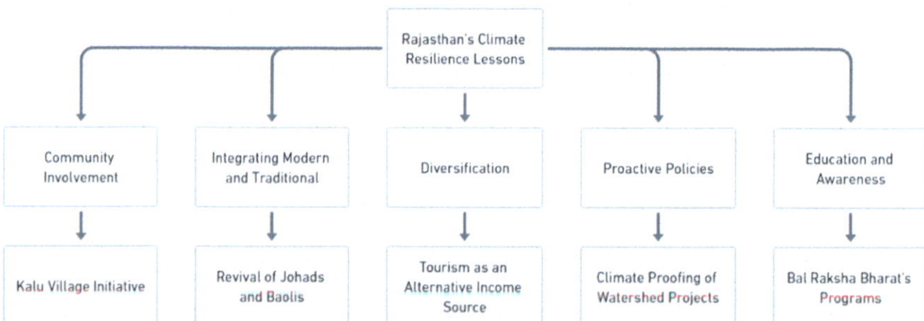

Fig. 11 Rajasthan's Climate Resilience lessons. (Source: Figure 4 prepared by author from multiple sources)

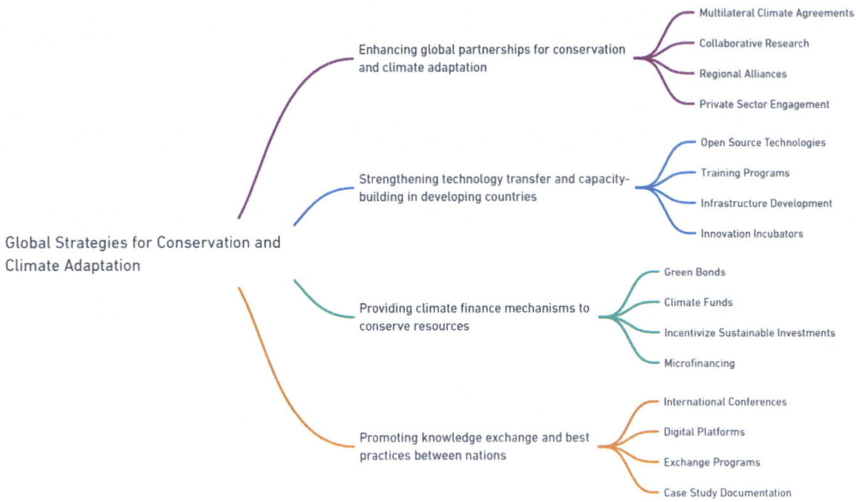

Fig. 12 Recommendations. (Source: Prepared by author)

Promoting International Solidarity for Climate Action: Recommendations

Climate change cannot be addressed by one country alone. For meaningful action to be taken, global efforts must be unified (ISA, 2023).

To conclude, international solidarity for climate action requires a blend of collaboration, technology transfer, financial support, and knowledge exchange (Fig. 12). The global community can effectively address the multifaceted challenges posed by climate change by working together and sharing commitments.

Conclusion

This chapter discusses the multifaceted nature of climate change and its importance from the point of view of India with an additional specific focus on Rajasthan. The chapter further looks into the major impact of climate change across different regions by highlighting the vulnerability of some places that demand specific solutions based on the unique features of each region. Climate change cannot be considered any longer on a national level and is—without a doubt—a global problem affecting each country without border considerations. With its worldwide reach, it is about the recognition that it is a challenge for the entire world which requires everyone's collaboration. Every nation has its part and role, but the primary and above all the necessary element is the international action of the global community.

United and determined efforts are the only way we will manage the severity and complexity of the task ahead.

All stakeholders whether policymakers, environmentalists, business leaders, or individuals must work together. The answer to climate change is our combined effort. Knowledge sharing, resource pooling, and leveraging our strengths can facilitate our quest for sustainable solutions. There is no more time for isolated action—we are moving into an age of cooperation. With the progress in the nation's integration, they will add to the global picture of sustainability and resilience. This vision is attainable; however, it needs strong international cooperation. This chapter should serve not only as evidence of climate change challenges but as a lighthouse of possibilities that can be accomplished together.

References

Aurelle, D., Thomas, S., Albert, C., Bally, M., Bondeau, A., Boudouresque, C., Cahill, A. E., Carlotti, F., Chenuil, A., Cramer, W., Davi, H., De Jode, A., Ereskovsky, A., Farnet, A., Fernandez, C., Gauquelin, T., Mirleau, P., Monnet, A., Prévosto, B., & Rossi, V. (2022). Biodiversity, climate change, and adaptation in the Mediterranean. *Ecosphere, 13*(4). https://doi.org/10.1002/ecs2.3915

Bazzani, G. (2023). Climate solidarity: A framework and research agenda for low-carbon behavior. *Sociological Forum, 38*(2), 352. https://doi.org/10.1111/socf.12885

Bhattacharya, S. (2023). *Why Indian conservationists are wary of proposed changes to the forest conservation act*. Eco-Business. Retrieved 29 September, 2023, from https://www.eco-business.com/news/why-indian-conservationists-are-wary-of-proposed-changes-to-the-forest-conservation-act/.

Bloom, A. (2022) *5 conservation projects focused on long-term success*. The Nature Conservancy. Retrieved from https://www.nature.org/en-us/what-we-do/our-priorities/protect-water-and-land/land-and-water-stories/conservation-projects-long-term-success/.

CCCR. (2014). *Centre for Climate Change Research*. cccr.tropmet.res.in. Retrieved 19 August, 2023, from https://cccr.tropmet.res.in/home/workshop/sept2014/index.html.

Chow, D. (2022). *Immediate climate action is necessary for a "livable future," U.N. report warns*. NBC News. Retrieved from https://www.nbcnews.com/science/environment/urgent-action-climate-change-needed-secure-liveable-future-un-report-w-rcna17764.

Climate Adapt. (2017). *Ecosystem-based approaches*. climate-adapt.eea.europa.eu. Retrieved from https://climate-adapt.eea.europa.eu/en/eu-adaptation-policy/sector-policies/ecosystem.

Climate—Adapt. (2023). *Biodiversity*. climate-adapt.eea.europa.eu. Retrieved from https://climate-adapt.eea.europa.eu/en/eu-adaptation-policy/sector-policies/biodiversity/index_html.

Climatelinks. (2022). *Natural resource management*. www.climatelinks.org. Retrieved from https://www.climatelinks.org/sector/natural-resource-management.

De, M., & Dhote, M. (2021). Impact of Climate Change on Arid Region: Case Study of Rajasthan. International Journal of Environment and Climate Change, 11(12), 474-486. https://doi.org/10.9734/ijecc/2021/v11i1230599

Department of Foreign Affairs and Trade. (2018). *International cooperation on climate change*. Australian Government Department of Foreign Affairs and Trade. Retrieved from https://www.dfat.gov.au/international-relations/themes/climate-change/international-cooperation-on-climate-change.

Disaster Management, Relief & Civil Defence Department Government of Rajasthan. (2022). *RTI data on flood and drought*.

EPA. (2012). *Resource conservation and recovery—A guide to developing and implementing greenhouse gas reduction programs.*

EPA. (2016). *Soils policy: Resource conservation, land, and greenhouse gas reductions.* Office of Solid Waste and Emergency Response (OSWER) | US EPA. archive.epa.gov. Retrieved 28 September, 2023, from https://archive.epa.gov/oswer/international/web/html/201002_resource_conservation_land_and_ghg.html.

European Commission. (2020). *Causes of climate change.* European Commission. Retrieved from https://climate.ec.europa.eu/climate-change/causes-climate-change_en.

FAO. (2023). *Sustainable land management.* | Land & Water | Food and Agriculture Organization of the United Nations | Land & Water | Food and Agriculture Organization of the United Nations. www.fao.org. Retrieved from https://www.fao.org/land-water/land/sustainable-land-management/en/.

Goswami, R. (2019). *Rajasthan to focus on climate change with new action plan.* Hindustan Times. Retrieved 15 August, 2023, from https://www.hindustantimes.com/jaipur/rajasthan-to-focus-on-climate-change-with-new-action-plan/story-X8owvGpMm9a7SmrxhbAv2N.html.

Government of India. (2022). *India's updated first nationally determined contribution under Paris agreement (2021–2030).* UNFCCC. Retrieved from https://unfccc.int/sites/default/files/NDC/2022-08/India%20Updated%20First%20Nationally%20Determined%20Contrib.pdf.

Groves, C. R., Game, E. T., Anderson, M. G., Cross, M., Enquist, C., Ferdaña, Z., Girvetz, E., Gondor, A., Hall, K. R., Higgins, J., Marshall, R., Popper, K., Schill, S., & Shafer, S. L. (2012). Incorporating climate change into systematic conservation planning. *Biodiversity and Conservation, 21*(7), 1651–1671. https://doi.org/10.1007/s10531-012-0269-3

Guelfo, D. (2023). *What is green tech? A comprehensive guide to green technology in 2023.* www.netguru.com. Retrieved 29 September, 2023, from https://www.netguru.com/blog/what-is-greentech.

Hussain, F. I., & Dill Helena, D. (2023). *India incorporates green bonds into its climate finance strategy.* blogs.worldbank.org. Retrieved from https://blogs.worldbank.org/climatechange/india-incorporates-green-bonds-its-climate-finance-strategy.

IISD's SDG Knowledge Hub. (2019). *Sustainable land management critical to combating climate change: IPCC special report* | News | SDG Knowledge Hub | IISD. IISD. Retrieved from https://sdg.iisd.org/news/sustainable-land-management-critical-to-combating-climate-change-ipcc-special-report/.

IPCC. (2023). *Urgent climate action can secure a Liveable future for all—IPCC.* IPCC. Retrieved from https://www.ipcc.ch/2023/03/20/press-release-ar6-synthesis-report/.

ISA. (2023). *International solar alliance.* isolaralliance.org. Retrieved from https://isolaralliance.org/.

IUCN. (2017). *Ecosystem-based adaptation.* IUCN. Retrieved from https://www.iucn.org/resources/issues-brief/ecosystem-based-adaptation.

IUCN. (2022). *Nature-based solutions for climate.* IUCN. Retrieved from https://www.iucn.org/our-work/topic/nature-based-solutions-climate.

IUCN. (2023). *Nature-based solutions.* IUCN. Retrieved from https://www.iucn.org/our-work/nature-based-solutions.

Kenton, W. (2022). *What is green tech? How it works, types, adoption, and examples.* Investopedia. Retrieved from https://www.investopedia.com/terms/g/green_tech.asp.

Khera, R. (2004). *Discussion paper Series-5 drought proofing in Rajasthan: Imperatives, experience and prospects ACKNOWLEDGEMENTS.* ACKNOWLEDGEMENTS.

Leiserowitz, A., Thaker, J., Carman, J., Neyens, L., Rosenthal, S., Deshmukh, Y., Shukla, G., Marlon, J., Sircar, A., & Sekoff, S. (2022). *Climate change in the Indian Mind, 2022.* Yale Program on Climate Change Communication. Retrieved from https://climatecommunication.yale.edu/publications/climate-change-in-the-indian-mind-2022/toc/3/.

Lindwall, C. (2022). *What are the effects of climate change?* NRDC. Retrieved from https://www.nrdc.org/stories/what-are-effects-climate-change.

Maizland, L. (2022). *Global climate agreements: Successes and failures*. Council on Foreign Relations. Retrieved from https://www.cfr.org/backgrounder/paris-global-climate-change-agreements.
Malhi, Y., Franklin, J., Seddon, N., Solan, M., Turner, M. G., Field, C. B., & Knowlton, N. (2020). Climate change and ecosystems: Threats, opportunities and solutions. *Philosophical Transactions of the Royal Society B: Biological Sciences, 375*(1794), 20190104. https://doi.org/10.1098/rstb.2019.0104
Mckoy, J. (2023). *On Climate Change, "the Urgency Is Great"*. www.bu.edu. Retrieved from https://www.bu.edu/sph/news/articles/2023/on-climate-change-the-urgency-is-great/.
Mehta, P. S., & Gerard, J. S. (2022). *Climate technology: US will seek revamp of India's IP regime | policy circle*. Policy Circle. Retrieved 29 September, 2023, from https://www.policycircle.org/environment/india-needs-us-climate-technology/.
Ministry of Jal Shakti. (2021). *Jal Shakti Abhiyan*. jsactr.mowr.gov.in. Retrieved from https://jsactr.mowr.gov.in/.
MOEFCC. (2021). *National Action Plan on climate change (NAPCC)*.
MOEFCC. (2022). *Targets sets under Green India Mission*. pib.gov.in. Retrieved from https://pib.gov.in/PressReleaseIframePage.aspx?PRID=1813175
NASA. (2023). The causes of climate change. In: *Climate change: Vital signs of the planet*. Retrieved from https://climate.nasa.gov/causes/.
National Oceanic and Atmospheric Administration. (2021). *Climate change impacts*. National Oceanic and Atmospheric Administration. Retrieved from https://www.noaa.gov/education/resource-collections/climate/climate-change-impacts.
NIC. (2009). *India: The impact of climate change to 2030: A commissioned research report*. Retrieved from https://www.dni.gov/files/documents/climate2030_india.pdf.
NOAA. (2023). *Coastal blue carbon*. oceanservice.noaa.gov. Retrieved from https://oceanservice.noaa.gov/ecosystems/coastal-blue-carbon/.
NREL. (2012). *NREL-greenhouse gas mitigation through natural resource Management program*. www2.nrel.colostate.edu. Retrieved from https://www2.nrel.colostate.edu/greenhouse-gas-program.html.
NWF. (2023). *Climate-smart conservation*. National Wildlife Federation. Retrieved 29 September, 2023, from https://www.nwf.org/climate-smart.
Pew. (2021). *Coastal "Blue Carbon": An important tool for combating climate change*. In: pew.org. Retrieved from https://www.pewtrusts.org/en/research-and-analysis/issue-briefs/2021/09/coastal-blue-carbon-an-important-tool-for-combating-climate-change.
Phillips, J., Das, K., & Newell, P. (2013). Governance and technology transfer in the clean development mechanism in India. *Global Environmental Change, 23*(6), 1594–1604. https://doi.org/10.1016/j.gloenvcha.2013.09.012
Rathore, M. S. (2004). *State level analysis of drought policies and impacts in Rajasthan, India*. Retrieved from https://www.iwmi.cgiar.org/Publications/Working_Papers/working/WOR93.pdf.
Seddon, N., Chausson, A., Berry, P., Girardin, C. A. J., Smith, A., & Turner, B. (2020). Understanding the value and limits of nature-based solutions to climate change and other global challenges. *Philosophical Transactions of the Royal Society B: Biological Sciences, 375*(1794), 20190120. https://doi.org/10.1098/rstb.2019.0120
Seventh, Q., & Chomsky, R. (2023). *Top 10 green technology innovations*. Sustainable Review. Retrieved from https://sustainablereview.com/top-10-green-technology-innovations-2/.
Sowińska-Świerkosz, B., & Garcia, J. (2022). What are nature-based solutions (NBS)? Setting core ideas for concept clarification. *Nature-Based Solutions, 2*, 100009. https://doi.org/10.1016/j.nbsj.2022.100009
Srivastava, A. K., Rajeevan, M., & Kshirsagar, S. R. (2009). Development of a high resolution daily gridded temperature data set (1969-2005) for the Indian region. *Atmospheric Science Letters, 10*, 249. https://doi.org/10.1002/asl.232
Swallow, T. (2022). *Top 10 green technology innovations*. sustainabilitymag.com. Retrieved from https://sustainabilitymag.com/top10/top-10-green-technology-innovations.

The Climate Reality Project. (2022). *How the climate crisis is impacting India*. Climate Reality Project. Retrieved from https://www.climaterealityproject.org/blog/how-climate-crisis-impacting-india.

The Royal Society. (2013). *Climate change: Evidence and causes | Royal Society*. The Royal Society. Retrieved from https://royalsociety.org/topics-policy/projects/climate-change-evidence-causes/basics-of-climate-change/.

The World Bank. (2013). *India: Climate change impacts*. World Bank. Retrieved from https://www.worldbank.org/en/news/feature/2013/06/19/india-climate-change-impacts.

U.N. Environment. (2021). *Ecosystem-based adaptation*. UNEP—UN Environment Programme. Retrieved from https://www.unep.org/explore-topics/climate-action/what-we-do/climate-adaptation/ecosystem-based-adaptation.

U.S. Climate Resilience Toolkit. (2021). *Climate-smart conservation: Putting adaptation principles into practice*. U.S. Climate Resilience Toolkit. toolkit.climate.gov. Retrieved from https://toolkit.climate.gov/tool/climate-smart-conservation-putting-adaptation-principles-practice.

UN. (2023). *UN DESA Policy Brief No. 153: India overtakes China as the world's most populous country*. Department of Economic and Social Affairs. www.un.org. Retrieved from https://www.un.org/development/desa/dpad/publication/un-desa-policy-brief-no-153-india-overtakes-china-as-the-worlds-most-populous-country/.

UNEP. (2020). *What is ecosystem-based adaptation?* UNEP. Retrieved 29 September, 2023, from https://www.unep.org/news-and-stories/video/what-ecosystem-based-adaptation.

UNESCO. (2019). *A question of international solidarity*. UNESCO. Retrieved from https://en.unesco.org/courier/2019-3/question-international-solidarity.

UNFCCC. (2015). *The Paris agreement*. United Nations Climate Change. Retrieved from https://unfccc.int/process-and-meetings/the-paris-agreement.

UNFCCC. (2022). *Key aspects of the Paris agreement*. United Nations Climate Change. Retrieved from https://unfccc.int/most-requested/key-aspects-of-the-paris-agreement.

United Nations. (2018). *Climate action—United Nations sustainable development*. United Nations Sustainable Development. Retrieved from https://www.un.org/sustainabledevelopment/climate-action/.

United Nations. (2021). *Climate change—United Nations sustainable development*. Sustainable Development Goals. Retrieved from https://www.un.org/sustainabledevelopment/climate-change/.

United Nations. (2022). *About international solidarity and human rights*. OHCHR. Retrieved from https://www.ohchr.org/en/special-procedures/ie-international-solidarity/about-international-solidarity-and-human-rights.

United Nations. (2023). *Disasters, resilience and land management*. United Nations. Retrieved from https://www.un.org/en/climatechange/climate-solutions/disasters-resilience-land-management.

United Nations Climate Change. (2022). *Climate action demands intergenerational solidarity*. Unfccc.int. Retrieved from https://unfccc.int/news/climate action demands intergenerational-solidarity.

United Nations—Climate Action. (2022). *Biodiversity—Our strongest natural defense against climate change*. United Nations. Retrieved from https://www.un.org/en/climatechange/science/climate-issues/biodiversity.

US Bureau of Labour Statistics. (2012). *Green technologies and practices frequently asked questions: U.S. Bureau of Labor Statistics*. www.bls.gov. Retrieved 29 September, 2023, from https://www.bls.gov/gtp/faq.htm.

US EPA. (2022). *Climate change impacts on ecosystems*. www.epa.gov. Retrieved from https://www.epa.gov/climateimpacts/climate-change-impacts-ecosystems.

US EPA. (2023). *Overview of greenhouse gases*. US EPA. Retrieved from https://www.epa.gov/ghgemissions/overview-greenhouse-gases.

USDA. (2023). *Climate*. Natural Resources Conservation Service. Retrieved from https://www.nrcs.usda.gov/conservation-basics/natural-resource-concerns/climate.

Vikaspedia. (2023). *Paramparagat Krishi Vikas Yojana | district south west, government of Delhi | India*. Vikaspedia. Retrieved from https://dmsouthwest.delhi.gov.in/scheme/paramparagat-krishi-vikas-yojana/.

We Conserve, P. A. (2019). *Climate change and conservation: ConservationTools*. conservationtools.org. Retrieved 29 September, 2023, from https://conservationtools.org/guides/173-climate-change-and-conservation.

Weiskopf, S. R. (2020). Climate change effects on biodiversity, ecosystems, ecosystem services, and natural resource management in the United States. *Science of the Total Environment, 733*(733), 137782. https://doi.org/10.1016/j.scitotenv.2020.137782

Wilkening, J. L., Magness, D. R., Harrington, A., Johnson, K., Covington, S., & Hoffman, J. R. (2022). Incorporating climate uncertainty into conservation planning for wildlife managers. *Earth, 3*(1), 93–114. https://doi.org/10.3390/earth3010007

World Bank. (2016). *India*. World Bank. Retrieved from https://www.worldbank.org/en/country/india.

World Bank. (2022). *Climate explainer: Nature-based solutions*. World Bank. Retrieved from https://www.worldbank.org/en/news/feature/2022/05/19/what-you-need-to-know-about-nature-based-solutions-to-climate-change.

Climate Change Impacts Mitigation, and Adaptation through Solidarity Strategies in the Indian Subcontinent

Arabinda Sharma and Chadetrik Rout

Introduction

Climate Change and Its Impacts

Climate change refers to long-term variations in global or regional weather patterns. The prime factor behind it is human activities which increase the greenhouse gases (GHGs) concentration in the troposphere (Wuebbles & Jain, 2001). The geological evidence indicates that climate change is a continuous phenomenon on our planet (Crowley, 1983). Since the Industrial Revolution, human activity has increased the amount of GHGs in the atmosphere, resulting in greater heat retention and surface temperatures (as heat radiation is absorbed by GHGs and re-radiated back to the Earth called the greenhouse effect, Fig. 1). Atmospheric aerosols also affect climate by scattering and absorbing insolation outgoing radiation and by modifying the clouds' microphysical and chemical characteristics (Andreae, 1995).

The Indian subcontinent is more vulnerable to climate change because of its geographical and socioeconomic features. The region has been facing a range of problems including natural resource management, economic losses, frequent and intense disasters, health and productivity, population migration, and environmental problems. The region comprises only 3% of the world's land area, and 21% of the world's population making it a climate risk region (Billah, 2019).

Climate change has made this region vulnerable to numerous climate-related hazards, such as cyclones, heat waves, floods, and droughts. Its coastal areas are

A. Sharma
School of Geography, Gangadhar Meher University, Amruta Vihar, Sambalpur, Odisha, India

C. Rout (✉)
Department of Civil Engineering, Maharishi Markandeshwar Engineering College, Maharishi Markandeshwar (Deemed to be University), Ambala, Haryana, India

K. Arora, R. R. Anand (eds.), *The Solidarity Approach in Geography*,
https://doi.org/10.1007/978-3-031-86753-8_18

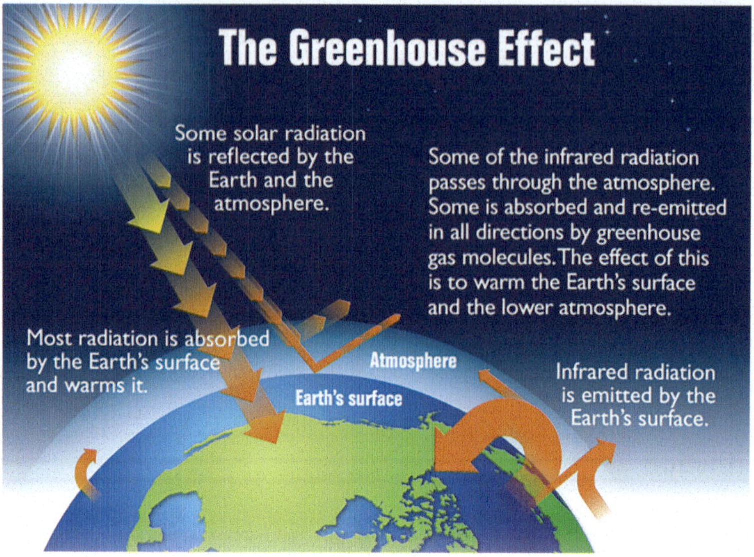

Fig. 1 Earth's greenhouse effect. (US EPA, 2012)

threatened by rising sea levels and the possibility of more frequent and intense cyclones (Mishra, 2017). The glacier melting in the Himalayan region results in water scarcity downstream, affecting a large population relying on Himalayan rivers for domestic, agricultural, and energy generation purposes (Nie et al., 2021). Climate change induced shifting of growing seasons, altered rainfall patterns, and increased pest and disease incidence have threatened crop yields and food security (Muluneh, 2021). The potential disruption of ecosystems due to changing climate has resulted in biodiversity loss, changes in wildlife behavior, and habitat degradation. Coastal coral reefs and mangrove forests are more vulnerable to the changing climate. It is not only the physical environment but also human health affected by it in the form of heat stress, heat waves, disease spread, waterborne diseases, etc. (Ashbolt, 2004). Climate-related impacts can cause community migration and displacement, especially in low-lying coastal areas and regions prone to drought and flooding. This can lead to social and economic problems. Besides these, climate change has affected policy adaptation in the region (Krishnan et al., 2020). It includes initiatives to cut GHG emissions, promote renewable energy, and build climate-resilient infrastructure. Global initiatives, such as the Paris Agreement, focused on keeping global warming below 2 °C over pre-industrial levels (More & Singh, 2016). However, due to the inequitable distribution of limited resources and access to adaptive measures, marginalized communities are disproportionately affected (Nath & Behera, 2011). Thus, it is imperative to recognize climate change and associated issues in terms of the existing capacity to mitigate and adapt to climate change across the Indian subcontinent. The book chapter aims at understanding climate change impacts on the Indian subcontinent and developing solidarity strategies to mitigate and adapt to the changing climate.

To achieve the said objective, the article employed a systematic approach to scholarly literature on climate change impacts, mitigation, and adaptation strategies focusing on solidarity in the Indian subcontinent. The relevant academic databases were searched for keywords like "climate change," "solidarity strategies," "mitigation," "adaptation," "India," and "subcontinent." The selected articles were subsequently subjected to thematic analysis to identify and categorize key solidarity approaches for mitigating climate change impacts and fostering adaptation emphasizing the role of different stakeholders. Finally, the review outcomes were synthesized to explore the effect of the solidarity approach on sustainability and the prospect and challenges of the solidarity approach to address climate challenges in the subcontinent using SWOT analysis.

Need to Address Climate Change Impacts

The need and scope of addressing climate change in the Indian subcontinent are of paramount importance due to the region's vulnerability to the changing climate. This diverse climatic set-up of the region faces several climate-related challenges that require urgent attention and concerted efforts. The significance of tackling climate change on the Indian subcontinent may be attributed to several key reasons such as the large population, high levels of poverty, reliance on agriculture, etc.

The vulnerability of the Indian subcontinent to climate change is threatening agriculture, water resources, coastal communities, and ecosystems of the region. It is home to a significant portion of the global population, and climate change could have profound socioeconomic implications on large populations suffering from poverty and disparities. The region hosts several biodiversity hotspots and unique ecosystems. Loss of biodiversity could have cascading effects on ecosystem services, functions the livelihoods of the local populace, etc. The region is also having several challenges related to food and water security potentially leading to social unrest, migration, loss of livelihood, and conflicts especially due to cross-border sharing of water. Changing climate patterns can further worsen these situations. Besides these, coastal vulnerabilities and an increase in the prevalence of air pollution, heat stress, and vector-borne diseases can affect public health systems and quality of life. Finally, climate action is integral to poverty eradication, gender equality, clean energy access, and resilient infrastructure which are integral parts of Sustainable Development Goals (SDGs).

The scope for addressing climate change in the Indian subcontinent is extensive and covers multiple dimensions including (a) mitigation and adaptation, (b) research and innovation, (c) capacity building, (d) international cooperation, and (e) policy and governance. Thus, addressing climate change in the Indian subcontinent is not only a necessity but also an opportunity for creating a more sustainable and resilient future for the region's diverse populations. To minimize the potential risks and establish a climate-resilient society, governments, stakeholders, civil society, and international partners must all work together in solidarity.

Impacts of Climate Change on Indian Subcontinent

Climate change, population growth, and urbanization have all had significant impacts on agriculture (Nath & Behera, 2011; Nath et al., 2017; Sundar, 2018; Joshi et al., 2021), water resources (Molden et al., 2016; Pani et al., 2021), health (Singh & Purohit, 2014; Jaiswal et al., 2017), biodiversity (Ramakrishnan, 2001; Kumar & Chopra, 2009; Shrestha & Shrestha, 2019), infrastructure (Roy et al., 2023; Mukherjee et al., 2023), monsoon changes (Vinnarasi & Dhanya, 2016), etc. on the Indian subcontinent in the last couple of decades. The impact of climate change is discussed in subsequent sub-headings.

Impact on Agriculture

- Climate change has resulted in altered precipitation patterns, heat waves, and erratic monsoon seasons, all of which harm agricultural productivity.
- Rainfall patterns can shift suitable cultivation areas for various crops, affecting traditional farming practices.
- Increased pest and disease pressures caused by changing climate conditions can reduce crop yields even further.

Impact on Water Resources

- Melting Himalayan glaciers as a result of rising temperatures can result in increased water flow at first, followed by decreased flow as glaciers recede, affecting water availability in downstream regions.
- Changes in precipitation patterns can cause floods and droughts and reduce water availability for various uses.
- Groundwater extraction for irrigation has resulted in declining water tables in some areas, resulting in long-term water scarcity.

Impact on Health

- Because of changes in water availability and quality, changing climate patterns can influence the surge of vector-borne health problems, e.g., dengue, malaria, cholera, etc.
- Heat waves and other extreme weather events can be hazardous to one's health, especially in vulnerable populations.
- Air pollution can cause respiratory and cardiovascular problems due to a variety of sources, including agricultural practices and urbanization.

Impact on Biodiversity

- Natural habitat loss as a result of urbanization, deforestation, and agricultural expansion has resulted in habitat fragmentation and biodiversity loss.
- It can affect the distribution and survival of various plant and animal species, forcing some to migrate or adapt to new conditions.
- Pollution and habitat degradation have the potential to endanger aquatic ecosystems and biodiversity.

However, because climate change challenges are complex and interconnected, they necessitate on-going research, collaboration, and proactive planning to minimize and adapt to their consequences.

Mitigation Strategies

Mitigation refers to the deliberate actions and strategies meant to minimize or thwart the long-term consequences of global warming and its associated effects. These efforts encompass various sectors, including energy, transportation, agriculture, and industry, with the primary goal of decreasing GHG emissions.

Strategies for Reducing GHG Emissions

Mitigation strategies to control GHG release include increased use of green energy sources for electricity generation, improved energy efficiency, afforestation, and reforestation, promotion of electric vehicles (EVs) and efficient public transport, proper waste management practices, cleaner production techniques, carbon pricing mechanisms, investment in research and development (R&D) sector, promotion of sustainable agriculture, international collaboration, adoption of clean cooking technologies, building retrofitting etc. Strategies for successful implementation in the

Indian subcontinent require a comprehensive approach involving government, industry, civil society, and academia.

To be effective, Indian subcontinent climate change mitigation strategies must take into account policy initiatives, technological advancements, socioeconomic conditions, and environmental outcomes. An overview of some important strategies and their potential effectiveness is discussed here.

- **Renewable energy adoption:** The Indian government promotes energy sources like solar, wind, and hydropower, with the National Solar Mission aimed for 450 GW by 2030 (Shah et al., 2022). As of November 30, 2022, it has a cumulative installed capacity of ~61.97 GW of solar power. However, challenges like intermittency and grid infrastructure upgrades persist (Wang et al., 2022).
- **Afforestation and reforestation:** The Green India Mission aimed to increase forest cover by 5 million ha, promoting carbon sequestration and biodiversity but faces challenges like land availability, monitoring, and long-term management (Dogra, 2011). The total forest cover has increased by 154,000 ha, tree cover has increased by 72,100 ha, and total forest and tree cover have increased by 226,100 ha since the previous assessment.
- **Climate policy and international commitments:** India has signed the Paris Agreement and has committed to lower emissions and increase the stake of renewable fuels in its energy portfolio (Sawhney, 2021). Sector-specific strategies are outlined in policies such as the National Action Plan on Climate Change (NAPCC-2008). The effectiveness of these policies is dependent on their implementation and monitoring.
- **Urban planning and sustainable transport:** Rapid urbanization increases pollution and emissions and promotes public transportation and non-motorized modes; sustainable planning reduces carbon footprints (Patil, 2021).
- **Challenges and considerations:** The effectiveness of mitigation strategies can be influenced by socioeconomic factors, energy demand, financial constraints, and the need for technological advancements. Balancing economic growth with emission reductions is a difficult task.

It is noteworthy that the effectiveness of these strategies will differ on how they are implemented, collaborated between governments, industry, civil society, and the changing climatic landscape.

Role of International Cooperation

International cooperation can help to mitigate climate change in the region in many ways such as through shared responsibility, emission reduction targets, technology transfer and capacity building, financial support, etc., and are discussed below.

- **Shared responsibility:** Climate change disregards national boundaries. The actions of one country can cause severe consequences for neighboring countries.

As a result, international cooperation emphasizes countries' shared responsibility to reduce GHG emissions, implement sustainable practices, and mitigate the negative effects of climate change (Khan et al., 2020).

- **Emission reduction targets:** The contributions of Indian subcontinent countries to global greenhouse gas emissions vary. International treaties, such as the Paris Agreement, provide a framework for countries to set emission reduction targets based on their respective capabilities and responsibilities.
- **Technology transfer and capacity building:** Advanced technologies and expertise in renewable energy, sustainable agriculture, and adaptation strategies are frequently found in developed countries. International cooperation can help countries in the Indian subcontinent that transfer these technologies, assisting them in shifting to low-carbon economies.
- **Financial support:** The subcontinent frequently lacks the financial resources needed to carry out climate mitigation and adaptation projects. International cooperation can include providing financial assistance, grants, and concessional loans to assist these countries in investing in clean energy, reforestation, and carbon-reducing infrastructure.
- **Regional collaboration:** Many of the effects of climate change extend beyond national borders, affecting entire regions. Therefore, regional collaboration can aid in addressing common challenges such as managing shared water resources, developing early warning systems for natural disasters, and promoting cross-border adaptation measures (Zuccaro et al., 2020).

Adaptation Strategies

Adaptation strategies refer to a course of action taken to address and mitigate the consequences of climate change. The adaptation strategies for the Indian subcontinent have been a topic of interest due to its vulnerability to climate change impacts.

Strategies to Adapt Climate Change

Several strategies have been proposed and implemented to deal with climate change challenges and are discussed below.

Water Management and Agriculture

- **Rainwater harvesting:** It is capturing and storing rainwater for agricultural and domestic use during dry periods. Some examples of rainwater harvesting initiatives in India are Jal Shakti Abhiyan (launched in 2019), rooftop rainwater harvesting, community rainwater harvesting, traditional water structures, farm pond construction, etc.

- **Drought-resistant crops:** Crop varieties that require less water and are more resilient to changing climatic conditions, such as millets, sorghum, pigeon pea, chickpea, groundnut, and so on are being developed and promoted (Jat et al., 2012).
- **Efficient irrigation:** Drip irrigation, sprinkler irrigation, and micro-irrigation systems are being implemented to optimize water use in agriculture (Daryanto et al., 2017).

Coastal Zone Management

- **Coastal protection:** The Indian government implements coastal protection measures through agencies like MoEFCC, MoES, and state-level departments, including seawalls, breakwaters, and beach nourishment to combat sea-level rise and erosion.
- **Mangrove restoration:** Mangrove forest rehabilitation projects in Bangladesh and India help buffer against storm surges and coastal flooding. The Sundarbans Mangrove Restoration Project in Bangladesh (2021), the Vembanad-Kol wetland restoration project in India (2022), and the Pichavaram Mangrove Restoration Project in Tamil Nadu, India (2001) are some mangrove restoration projects in the Indian subcontinent.

Ecosystem Conservation

- **Protected areas:** Creating and maintaining protected areas to protect biodiversity and natural ecosystems help in flood control and regulate biogeochemical cycles, water purification, and many more ecological services. India currently has 106 National Parks (44,402.95 km^2), 567 Wildlife Sanctuaries (122,564.86 km^2), 105 Conservation Reserves (5206.55 km^2), and 220 Community Reserves (1455.16 km^2) (Bhadury & Ram, 2023).
- **Reforestation:** Planting trees improve carbon sequestration, restore degraded ecosystems, and reduce soil erosion. According to a report, the forest cover of India increased by 3976 km^2 between 2017 and 2019, a 0.56% increase (Sahana et al., 2022).

Infrastructure Development

- **Climate-resilient infrastructure:** To withstand extreme weather events, climate-resilient design principles are being incorporated into infrastructure projects such as roads, bridges, and buildings. In the Indian subcontinent, some examples of climate-resilient infrastructure projects and initiatives include

Chennai's Resilient Urban Development, the Ganges-Brahmaputra-Meghna Delta Plan, Kolkata's Climate-Resilient Transport System, Bhutan's Hydropower Projects, and the Maldives' Coastal Protection Measures.
- **Flood-resilient urban planning:** Cities should improve drainage systems and green spaces to manage stormwater and prevent flooding. Chennai, India, recognized the need for effective flood management and resilience measures after the 2015 floods. Efforts were made to develop flood-resilient urban planning strategies to prevent similar incidents in the future.

Healthcare and Disaster Preparedness

- **Early warning systems:** India launched the Cyclone Warning System in 1985 to provide timely warnings of disasters like cyclones and heat waves. The India Meteorological Department (IMD) has evolved to offer better early warnings and preparedness measures for natural disasters.
- **Healthcare capacity building:** India launched the Ayushman Bharat scheme in September 2018, aiming to improve healthcare systems for climate-related health risks like heat illness and vector-borne diseases. Pradhan Mantri Jan Arogya Yojana (PMJAY) is a scheme that offers health insurance coverage and cashless hospitalization for various medical treatments.

Policy and Governance

- **National climate action plans:** The NAPCC, established in 2008, comprises eight missions focusing on climate adaptation and mitigation strategies. These missions include the National Solar Mission, Sustainable Agriculture, Enhanced Energy Efficiency, Water Mission, Sustaining the Himalayan Ecosystem, Green India, Sustainable Habitat, and Strategic Knowledge on Climate Change.
- **Legal and regulatory frameworks:** Implementing laws and regulations to promote climate-resilient development and environmental protection, including NAPCC, Paris Agreement, renewable energy, state initiatives, carbon market mechanisms, and adaptation strategies.

Monitoring and Evaluation

To evaluate the effectiveness of adaptation strategies and make necessary adjustments, they must be monitored and evaluated continuously (Park et al., 2012). It is important to learn from both successes and failures to improve future strategies. It is also worth noting that effective adaptation frequently requires a combination of strategies.

The Role of Community-Based Adaptation in Resilience Building

Community-based adaptation (CBA) is critical for building resilience in the Indian subcontinent. CBA includes empowering local people to identify, evolve, and implement proposed actions to address the specific challenges they face as a result of changing weather patterns (Dodman & Mitlin, 2013). How CBA can significantly contribute to resilience building in the Indian subcontinent is discussed here.

- **Local knowledge and context:** CBA recognizes the value of indigenous knowledge and practices developed by communities over generations (Temper & Martinez-Alier, 2013). This method ensures that adaptation strategies are rooted in the local context, which increases their effectiveness and sustainability.
- **Tailored solutions:** The Indian subcontinent has a diverse climate, geography, and culture. CBA allows the local community to development of adaptation strategies to their specific circumstances, such as dealing with extreme weather events, water scarcity, or agricultural challenges (Adhikari & Taylor, 2012).
- **Resilience building:** CBA focuses on increasing community resilience (Ensor et al., 2018). This includes addressing not only climate-related risks, but also social, economic, and environmental aspects. Key components include strengthening livelihoods, diversifying income sources, and improving access to resources.
- **Empowerment and capacity building:** Local communities are equipped with the skills, knowledge, and resources needed to manage climate-related challenges as a result of CBA initiatives (Remling & Veitayaki, 2016). This empowers them to respond more effectively to current and future climate impacts.
- **Reducing vulnerability:** CBA contributes to reducing the vulnerability of marginalized groups that are affected most frequently and severely by climate change (Piggott-McKellar et al., 2019). It ensures that their voices are heard and that their needs are factored into adaptation plans.
- **Sustainable practices:** Many CBA projects in India promote environmentally friendly practices, e.g., soil conservation, rainwater harvesting, and agroforestry. These practices not only serve as community adaptation tool but also help in environmental conservation.
- **Long-term vision:** CBA focuses on long-term planning and preparation, assisting communities in developing a vision for their future that takes changing climate conditions into account. This proactive approach lessens the impact of unexpected climate shocks.

Solidarity Strategies

The solidarity approach refers to a collaborative and collective effort among nations, regions, communities, and stakeholders to address the issues raised by climate change. This approach emphasizes shared responsibility, mutual support, and joint initiatives to reduce climate change effects, adapt to its effects, and promote sustainable development. Solidarity in climate action acknowledges that climate change is a worldwide concern that needs coordinated efforts beyond individual countries' capabilities. It embodies the principles of cooperation, equity, and unity in working for a sustainable and resilient future for all.

The most important reasons why solidarity strategies are important in the battle against climate change are as follows:

- **Global nature of the issue:** Climate change affects all countries across the border, irrespective of their level of development. Solidarity strategies acknowledge countries' interconnectedness and shared responsibility for addressing the root cause and consequences (Watson, 2007).
- **Equity and justice:** Solidarity strategies emphasize the values of equity and justice. Despite contributing less to climate change's causes, many developing countries bear a disproportionate share of its consequences. Ayers and Huq (2009) define solidarity as assistance to vulnerable countries and communities in their adaptation to climate change and transitioning to more sustainable practices.
- **Collective action:** To combat climate change, we must all work together to minimize GHG emissions, shift to renewable energy sources, and implement sustainable practices. Solidarity strategies promote international cooperation and coordination, as well as collaboration among governments, industries, civil society, and individuals (Beutler, 2008).
- **Technology and knowledge sharing:** Solidarity strategies encourage cross-national sharing of technological innovations, knowledge, and best practices (Asheim & Coenen, 2005). This allows for the more rapid and effective deployment of sustainable solutions as well as capacity building in areas like renewable energy, sustainable agriculture, and climate resilience.
- **Financial support:** Solidarity strategies offer developing nations monetary assistance and resources to encourage them to reduce and adjust to climate change effects, as agreed upon in international agreements including the Paris Agreement (Falkner, 2016).
- **Political momentum:** Solidarity strategies aid in building political momentum and consensus for bold climate action. When countries work together to achieve common goals, it can spur more ambitious national and international targets and policies (Falkner, 2016).
- **Strengthening international agreements:** The principles of solidarity and cooperation underpin international tie-ups, e.g., the Paris Agreement. Countries can strengthen the implementation and effectiveness of such agreements by working together to achieve common goals (Voigt & Ferreira, 2016).

- **Public awareness and engagement:** Climate change solidarity strategies can increase public awareness and engagement (Howell, 2013). People may be more motivated to adopt sustainable behaviors and support climate change policies if they see their actions as part of a larger, global effort.
- **Resilience building:** Solidarity strategies assist in building resilient communities and countries affected by climate change (Sayre et al., 2017). Nations can better prepare for and respond to the various climate-related challenges by sharing knowledge and resources.
- **Long-term sustainability:** Climate change is a long-term challenge that will necessitate sustained efforts over several decades. Solidarity strategies ensure that countries continue to support each other's efforts and that the commitment to addressing climate change remains strong over time.

Strategies for Addressing Climate Change in Indian Subcontinent

Some key regional solidarity strategies that can be implemented to address climate change are as follows:

- **Multi-stakeholder partnerships:** Collaboration among governments, civil society organizations, local communities, private sector entities, academic institutions, and international organizations should be encouraged. Multi-stakeholder collaborations can help pool resources, expertise, and knowledge to design and implement effective adaptation projects (Gannon et al., 2021).
- **Information sharing and knowledge transfer:** Create platforms for cross-border sharing of climate-related data, research findings, and best practices. Encourage countries and regions to share knowledge to accelerate the adoption of successful adaptation strategies.
- **Capacity building:** Invest in skill development and training at all levels, from local communities to government agencies. Capacity building improves stakeholders' ability to understand, plan, and implement effective adaptation measures (Salafsky et al., 2002).
- **Ecosystem-based approaches:** Encourage collaboration in the protection and restoration of ecosystems that provide critical services for climate resilience (Scarano, 2017). Reforestation, watershed management, and sustainable land use practices are examples of collaborative efforts.
- **Cross-border infrastructure development:** Collaborate on cross-border infrastructure projects such as water management systems, flood defenses, and early warning systems (Wiering et al., 2010). Collaboration can result in more effective infrastructure planning and resource allocation.
- **Disaster preparedness and response:** Cross-border disaster preparedness and response efforts should be coordinated. Climate-related disasters frequently

affect multiple countries, and collaborative planning can speed up response times and resource allocation.

Civil Society and Individuals' Roles in Promoting Solidarity

The efforts of civil society and individuals help to raise awareness, build resilience, and promote long-term practices (Nath & Behera, 2011) for climate change adaptation throughout the Indian subcontinent. Some important ways in which civil society and individuals contribute to promoting solidarity are as follows:

- **Awareness and advocacy:** Civil society organizations (CSOs) run campaigns, workshops, and seminars to educate communities, policymakers, and the general public about climate issues. Their grassroots approach raises awareness and encourages group action.
- **Collaboration and partnerships:** CSOs encourage collaboration among various stakeholders, such as governments, businesses, universities, and other organizations. They also address climate challenges more effectively by pooling resources and expertise.
- **Innovation and best practices:** CSOs frequently pioneer innovative climate adaptation solutions, demonstrating best practices that can be replicated across different regions. They serve as breeding grounds for new ideas and technologies.

Successful Examples of Solidarity Initiatives

Some specific examples of successful climate change adaptation solidarity initiatives in the Indian subcontinent are as follows:

- **People's Science Movement (PSM) in India (1977):** PSM is a grassroots network of environmental organizations, scientists, and activists working to address issues such as climate change. They share knowledge, build capacity, and conduct community-based research to develop sustainable and locally relevant adaptation strategies. .
- **Kudumbashree Mission, Kerala, India (1998):** Kudumbashree is a Kerala-based women's self-help group that has taken steps toward climate adaptation through projects such as waste management, organic farming, and disaster preparedness. These initiatives enable local communities, particularly women, to effectively respond to climate-related challenges.
- **Odisha Community Tank Management Project, India (2006):** This project aims to revitalize and manage traditional community tanks (ponds) in the state of Odisha. It assists local communities in storing water during the monsoon season, reducing the effects of drought and water scarcity during dry periods.

- **Bhutan's National Adaptation Programme of Action (NAPA, 2006):** Bhutan, while not part of the Indian subcontinent, is close by and faces similar climate challenges. Bhutan's NAPA focuses on strengthening vulnerable communities' resilience by funding projects such as climate-resilient agriculture, community forestry, and disaster risk reduction.
- **Bangladesh's Community Climate Change Project (CCCP, 2009):** This project aims to improve the climate resilience of Bangladesh's vulnerable communities. It promotes flood-resistant housing, sustainable agriculture, and community-based early warning systems, all of which contribute to better adaptation and disaster preparedness.
- **Climate Action Network South Asia (CANSA, 2010):** CANSA is a South Asian non-governmental organization network that includes India. It facilitates knowledge exchange, advocacy, and capacity building among its member organizations to promote regional collaboration on climate change adaptation.
- **Indian Network on Ethics and Climate Change (INECC, 2010):** INECC is a platform that brings activists, scholars, and practitioners together to discuss the ethical aspects of climate change. It promotes dialogue, research, and advocacy to promote climate justice and adaptation at different levels.
- **Himalayan Permaculture Center, Nepal (2011):** While not in India, this initiative is close by and demonstrates the potential for solidarity in climate adaptation. Through sustainable agriculture, soil conservation, and water management, the center teaches permaculture techniques that assist Himalayan communities in adapting to changing climate conditions.
- **Ladakh Renewable Energy Initiative, India (2011):** This initiative focuses on providing sustainable energy solutions through solar power and other renewable sources in the high-altitude region of Ladakh. It improves climate resilience and contributes to local development by reducing reliance on fossil fuels.
- **Clean Energy Access Network (CLEAN, 2014), India:** CLEAN is an organization network that promotes clean energy solutions such as solar power and energy-efficient technologies. Their initiatives promote climate-resilient energy, lower GHG emissions, and improve energy access in remote areas.

These examples demonstrate the importance of local engagement, knowledge sharing, and collaborative action in developing resilience to climate effects by highlighting the diversity of solidarity initiatives and projects for solidarity in the Indian subcontinent.

Climate Change, Solidarity Approach, and SDGs

The solidarity approach to climate change aligns closely with the UN SDGs by emphasizing collective responsibility, collaboration, and shared solutions for a more sustainable and equitable world. Table 1 shows how the solidarity approach intersects with specific SDGs:

Table 1 Linking solidarity approach to sustainable development goals

UN SDG	Solidarity aspect	Link to SDG
SDG 1: No Poverty:	Addressing climate change impacts in vulnerable communities through cooperative strategies, ensuring that climate action doesn't exacerbate poverty and inequalities	The solidarity approach recognizes that climate change disproportionately affects the poor and emphasizes the importance of climate-resilient development that lifts people out of poverty
SDG 2: Zero Hunger:	Promoting sustainable agricultural practices and ensuring food security through joint efforts that consider the agricultural effect of climate change	The solidarity approach aligns by encouraging regional collaboration to build climate-resilient agricultural systems that can withstand changing climate conditions
SDG 6: Clean Water and Sanitation:	Climate change can disrupt water availability and quality, affecting communities' access to clean water and sanitation	The solidarity approach involves sharing best practices and innovations for water management, promoting cross-border initiatives to safeguard water resources, and jointly addressing water-related challenges intensified by climate change
SDG 11: Sustainable Cities and Communities:	Climate change impacts urban areas significantly. The solidarity approach promotes collaboration in sustainable urban planning, infrastructure, and disaster resilience	By working together to build climate-resilient cities, countries can ensure that urban centers are environmentally sustainable, safe, and inclusive
SDG 13: Climate Action:	The solidarity approach recognizes that climate change affects all nations, irrespective of their contributions to greenhouse gas emissions. It promotes shared responsibility in mitigating climate impacts and advancing climate resilience	The solidarity approach complements SDG 13 by fostering international cooperation, knowledge exchange, and joint efforts to limit global temperature rise and enhance adaptive capacity
SDG 14: Life Below Water and SDG 15: Life on Land	The solidarity approach recognizes the interconnectedness of ecosystems and promotes joint conservation efforts to address climate impacts on oceans, forests, and biodiversity	By sharing expertise and implementing transboundary conservation projects, countries can protect marine and terrestrial ecosystems from the adversity of climate change
SDG 17: Partnerships for the Goals:	The solidarity approach inherently supports SDG 17 by emphasizing the importance of partnerships and global cooperation in achieving all the SDGs, including those related to climate change	The solidarity approach encourages countries, stakeholders, and organizations to collaborate in implementing effective climate solutions, sharing resources, and fostering mutual support

Table 1 reflects that the solidarity approach to climate change embodies the spirit of international cooperation and collective action, aligning well with several UN SDGs. By linking the solidarity approach to climate change with the UN SDGs, it becomes evident that achieving climate resilience and sustainability requires international collaboration and shared responsibility across various development sectors. By embracing the principles of solidarity, Indian subcontinent can provide promising solutions to mitigate climate change, adapt to its impacts, and advance sustainable development for both current and future generations.

SWOT Analysis of Solidarity Approach

The discussion from the previous sections substantiates the need for actions against climate change in a solidarity approach in the Indian subcontinent. Although the solidarity approach offers numerous opportunities, its adoption is not straightforward due to several factors. In this section, a SWOT (Strengths, Weaknesses, Opportunities, Threats) analysis of the solidarity approach in the Indian subcontinent is carried out to provide insights into the potential factors that can impact the effectiveness of this approach. The outcome of the SWOT analysis is presented in Fig. 2.

Figure 2 shows that the solidarity approach to climate change in the Indian subcontinent has the potential to bring about significant positive impacts by fostering collaboration, knowledge exchange, and integrated solutions. However, it also faces challenges related to political differences, resource allocation, and coordination complexities. By addressing weaknesses and threats and capitalizing on strengths

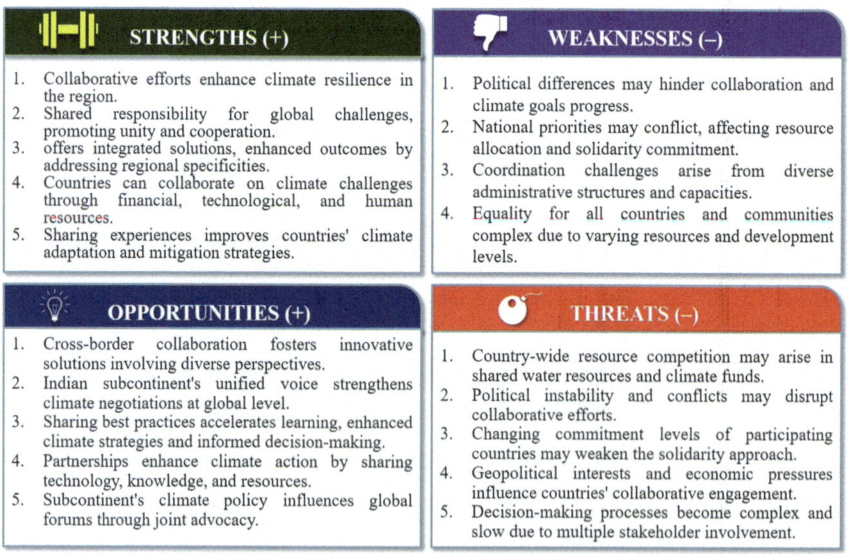

Fig. 2 SWOT analysis of solidarity approach for climate action in the Indian subcontinent

and opportunities, the subcontinent can work toward a sustainable and more resilient future through joint climate action.

Conclusion

The global nature of climate change necessitates coordinated efforts at global and regional levels. In the current scenario, the implementation of solidarity strategies to fight against climate change, and its mitigation and adaptation in the Indian subcontinent is highly imperative. As the region is densely populated, existing social and economic problems climate change can potentially pose serious problems in the region easily. Few countries in the region are acting seriously to meet the SDGs as climate action is inextricably linked to the achievement of SDGs objectives. Therefore, implementing solidarity strategies for mitigation and adaptation of climate change in the Indian subcontinent is vital. At this outset, addressing shared challenges, increasing resilience, promoting sustainable development, and fostering cooperation are key to success. It is also true that there is some regional confrontation among the nations on various issues. To address climate change in the Indian subcontinent requires a multifaceted approach that incorporates mitigation, adaptation, and solidarity strategies. Hence, the future of the region is dependent on its ability to take proactive measures, form partnerships, and implement innovative solutions to protect the environment, societies, and economies for future generations.

References

Adhikari, B., & Taylor, K. (2012). Vulnerability and adaptation to climate change: A review of local actions and national policy response. *Climate and Development, 4*, 54–65. https://doi.org/10.1080/17565529.2012.664958

Andreae, M. O. (1995). Climatic effects of changing atmospheric aerosol levels. *World Survey of Climatology, 16*(06), 347–398.

Ashbolt, N. J. (2004). Microbial contamination of drinking water and disease outcomes in developing regions. *Toxicology, 198*, 229–238. https://doi.org/10.1016/j.tox.2004.01.030

Asheim, B. T., & Coenen, L. (2005). Knowledge bases and regional innovation systems: Comparing Nordic clusters. *Research Policy, 34*, 1173–1190. https://doi.org/10.1016/j.respol.2005.03.013

Ayers, J. M., & Huq, S. (2009). Supporting adaptation to climate change: What role for official development assistance? *Development Policy Review, 27*, 675–692. https://doi.org/10.1111/j.1467-7679.2009.00465.x

Beutler, I. (2008). Sport serving development and peace: Achieving the goals of the United Nations through sport. *Sport in Society, 11*, 359–369. https://doi.org/10.1080/17430430802019227

Bhadury, S., & Ram, L. (2023). An effort to protect the environment and conserve the wild animals in India. *Russian Law Journal, 11*, 612–624.

Billah, M. (2019). South Asian Association for Regional co-operation and its contribution to south Asian politics and economy. *International Journal for Empirical Education and Research, 3*, 21–30.

Crowley, T. J. (1983). The geologic record of climatic change. *Reviews of Geophysics, 21*, 828–877. https://doi.org/10.1029/RG021i004p00828
Daryanto, S., Wang, L., & Jacinthe, P. A. (2017). Global synthesis of drought effects on cereal, legume, tuber, and root crops production: A review. *Agricultural Water Management, 179*, 18–33. https://doi.org/10.1016/j.agwat.2016.04.022
Dodman, D., & Mitlin, D. (2013). Challenges for community-based adaptation: Discovering the potential for transformation. *Journal of International Development, 25*, 640–659. https://doi.org/10.1002/jid.1772
Dogra, A. S. (2011). Contribution of trees outside forests toward wood production and environmental amelioration. *Indian Journal Of Ecology, 38*, 1–5.
Ensor, J. E., Park, S. E., Attwood, S. J., Kaminski, A. M., & Johnson, J. E. (2018). Can community-based adaptation increase resilience? *Climate and Development, 10*, 134–151. https://doi.org/10.1080/17565529.2016.1223595
Falkner, R. (2016). The Paris agreement and the new logic of international climate politics. *International Affairs, 92*, 1107–1125. https://doi.org/10.1111/1468-2346.12708
Gannon, K. E., Crick, F., Atela, J., & Conway, D. (2021). What role for multi-stakeholder partnerships in adaptation to climate change? Experiences from private sector adaptation in Kenya. *Climate Risk Management, 32*, 100319. https://doi.org/10.1016/j.crm.2021.100319
Howell, R. A. (2013). It's not (just)"the environment, stupid!" values, motivations, and routes to the engagement of people adopting lower-carbon lifestyles. *Global Environmental Change, 23*, 281–290. https://doi.org/10.1016/j.gloenvcha.2012.10.015
Jaiswal, A. K., Padmakumari, B., Kumar, N., & Kore, P. A. (2017). An increasing trend in temperature and moisture induced heat index and its effect on human health in climate change scenario over the Indian sub-continent. *Journal of Climate Change, 3*, 11–25. https://doi.org/10.3233/JCC-170002
Jat, R. A., Craufurd, P. Q., Sahrawat, K. L., & Wani, S. P. (2012). Climate change and resilient dryland systems: Experiences of ICRISAT in Asia and Africa. *Current Science, 102*, 1650–1659. http://oar.icrisat.org/id/eprint/6020.
Joshi, D. R., Ghimire, R., Kharel, T., Mishra, U., & Clay, S. A. (2021). Conservation agriculture for food security and climate resilience in Nepal. *Agronomy Journal, 113*, 4484–4493. https://doi.org/10.1002/agj2.20830
Khan, S. A. R., Zhang, Y., Kumar, A., Zavadskas, E., & Streimikiene, D. (2020). Measuring the impact of renewable energy, public health expenditure, logistics, and environmental performance on sustainable economic growth. *Sustainable Development, 28*, 833–843. https://doi.org/10.1002/sd.2034
Krishnan, R., Sanjay, J., Gnanaseelan, C., Mujumdar, M., Kulkarni, A., & Chakraborty, S. (2020). *Assessment of climate change over the Indian region: A report of the Ministry of Earth Sciences (MOES), government of India*. Springer Nature. https://doi.org/10.1007/978-981-15-4327-2
Kumar, V., & Chopra, A. K. (2009). Impact of climate change on biodiversity of India with special reference to Himalayan region-an overview. *Journal of Applied and Natural Science, 1*, 117–122. https://doi.org/10.31018/jans.v1i1.48
Mishra, P. K. (2017). Socio-economic impacts of climate change in Odisha: Issues, challenges and policy options. *Journal of Climate Change, 3*, 93–107. https://doi.org/10.3233/JCC-170009
Molden, D. J., Shrestha, A. B., Nepal, S., & Immerzeel, W. W. (2016). Downstream implications of climate change in the Himalayas. In A. Biswas & C. Tortajada (Eds.), *Water security, climate change and sustainable development. Water resources development and management*. Springer. https://doi.org/10.1007/978-981-287-976-9_5
More, N. S., & Singh, R. P. (2016). COP-21, conference of parties on climate change: Resurrection from Lima to Paris. *Climate Change and Environmental Sustainability, 4*, 231–236. https://doi.org/10.5958/2320-642X.2016.00025.9
Mukherjee, M., Abhinay, K., Rahman, M. M., Yangdhen, S., Sen, S., Adhikari, B. R., Nianthi, R., Sachdev, S., & Shaw, R. (2023). Extent and evaluation of critical infrastructure, the status of

resilience, and its future dimensions in South Asia. *Progress in Disaster Science, 17*, 100275. https://doi.org/10.1016/j.pdisas.2023.100275
Muluneh, M. G. (2021). Impact of climate change on biodiversity and food security: A global perspective-a review article. *Agriculture & Food Security, 10*, 1–25.
Nath, P. K., & Behera, B. (2011). A critical review of the impact of and adaptation to climate change in developed and developing economies. *Environment, Development and Sustainability, 13*, 141–162. https://doi.org/10.1007/s10668-010-9253-9
Nath, R., Nath, D., Li, Q., Chen, W., & Cui, X. (2017). Impact of drought on agriculture in the indo-Gangetic plain, India. *Advances in Atmospheric Sciences, 34*, 335–346. https://doi.org/10.1007/s00376-016-6102-2
Nie, Y., Pritchard, H. D., Liu, Q., Hennig, T., Wang, W., Wang, X., Liu, S., Nepal, S., Samyn, D., Hewitt, K., & Chen, X. (2021). Glacial change and hydrological implications in the Himalayas and Karakoram. *Nature Reviews Earth and Environment, 2*, 91–106. https://doi.org/10.1038/s43017-020-00124-w
Pani, A., Ghatak, I., & Mishra, P. (2021). Understanding the water conservation and management in India: An integrated study. *Sustainable Water Resources Management, 7*, 1–16. https://doi.org/10.1007/s40899-021-00556-2
Park, S. E., Marshall, N. A., Jakku, E., Dowd, A. M., Howden, S. M., Mendham, E., & Fleming, A. (2012). Informing adaptation responses to climate change through theories of transformation. *Global Environmental Change, 22*, 115–126. https://doi.org/10.1016/j.gloenvcha.2011.10.003
Patil, P. (2021). Sustainable transportation planning: Strategies for reducing greenhouse gas emissions in urban areas. *Empirical Quests for Management Essences, 1*, 116–129. https://orcid.org/0000-0001-8747-4679.
Piggott-McKellar, A. E., McNamara, K. E., Nunn, P. D., & Watson, J. E. (2019). What are the barriers to successful community-based climate change adaptation? A review of grey literature. *Local Environment, 24*, 374–390. https://doi.org/10.1080/13549839.2019.1580688
Ramakrishnan, P. S. (2001). Increasing population and declining biological resources in the context of global change and globalization. *Journal of Biosciences, 26*, 465–479. https://doi.org/10.1007/BF02704747
Remling, E., & Veitayaki, J. (2016). Community-based action in Fiji's Gau Island: A model for the Pacific? *International Journal of Climate Change Strategies and Management, 8*, 375–398. https://doi.org/10.1108/IJCCSM-07-2015-0101
Roy, P., Pal, S. C., Chakrabortty, R., Chowdhuri, I., Saha, A., & Shit, M. (2023). Effects of climate change and sea-level rise on coastal habitat: Vulnerability assessment, adaptation strategies and policy recommendations. *Journal of Environmental Management, 330*, 117187. https://doi.org/10.1016/j.jenvman.2022.117187
Sahana, M., Areendran, G., Raj, K., Sivadas, A., Abhijitha, C. S., & Ranjan, K. (2022). Introduction to Forest resources in India: Conservation, management and monitoring perspectives. In *Conservation, management and monitoring of Forest resources in India*. Springer. https://doi.org/10.1007/978-3-030-98233-1_1
Salafsky, N., Margoluis, R., Redford, K. H., & Robinson, J. G. (2002). Improving the practice of conservation: A conceptual framework and research agenda for conservation science. *Conservation Biology, 16*, 1469–1479. https://doi.org/10.1046/j.1523-1739.2002.01232.x
Sawhney, A. (2021). Striving towards a circular economy: Climate policy and renewable energy in India. *Clean Technologies and Environmental Policy, 23*, 491–499. https://doi.org/10.1007/s10098-020-01935-7
Sayre, M., Stenner, T., & Argumedo, A. (2017). You can't grow potatoes in the sky: Building resilience in the face of climate change in the Potato Park of Cuzco, Peru. *Culture, Agriculture, Food and Environment, 39*, 100–108. https://doi.org/10.1111/cuag.12100
Scarano, F. R. (2017). Ecosystem-based adaptation to climate change: Concept, scalability and a role for conservation science. *Perspectives in Ecology and Conservation, 15*, 65–73. https://doi.org/10.1016/j.pecon.2017.05.003

Shah, J. P., Talati, S., & Chetwani, S. (2022). Renewable energy policies and regulation in India. *Water and Energy International, 65*(3), 43–49.

Shrestha, U. B., & Shrestha, B. B. (2019). Climate change amplifies plant invasion hotspots in Nepal. *Diversity and Distributions, 25*, 1599–1612. https://doi.org/10.1111/ddi.12963

Singh, A., & Purohit, B. M. (2014). Public health impacts of global warming and climate change. *Peace Review, 26*, 112–120. https://doi.org/10.1080/10402659.2014.876326

Sundar, I. (2018). Climate change mitigation and adaptation practices in India. *Asian Journal of Multidimensional Research, 7*, 191–200.

Temper, L., & Martinez-Alier, J. (2013). The god of the mountain and Godavarman: Net present value, indigenous territorial rights, and sacredness in a bauxite mining conflict in India. *Ecological Economics, 96*, 79–87. https://doi.org/10.1016/j.ecolecon.2013.09.011

US EPA (2012). Climate change indicators in the United States, 2nd edition, Washington, DC, USA. p.3. EPA 430-R-12-004. https://www.epa.gov/sites/default/files/2016-08/documents/climateindicators-full-2012.pdf (Accessed on 18.03.2025)

Vinnarasi, R., & Dhanya, C. T. (2016). Changing characteristics of extreme wet and dry spells of Indian monsoon rainfall. *Journal of Geophysical Research: Atmospheres, 121*, 2146–2160. https://doi.org/10.1002/2015JD024310

Voigt, C., & Ferreira, F. (2016). 'Dynamic differentiation': The principles of CBDR-RC, progression, and highest possible ambition in the Paris agreement. *Transnational Environmental Law, 5*, 285–303. https://doi.org/10.1017/S2047102516000212

Wang, Y., Xu, C., & Yuan, P. (2022). Is there a grid-connected effect of grid infrastructure on renewable energy generation? Evidence from China's upgrading transmission lines. *Energy and Environment, 33*, 975–995. https://doi.org/10.1177/0958305X211031015

Watson, J. (2007). *Human development report 2007/2008—Fighting climate change: Human solidarity in a divided world, United Nations human development report office*. United States of America. Retrieved 31 August, 2023, from https://policycommons.net/artifacts/2116515/human-development-report-20072008-fighting-climate-change/2871813/.

Wiering, M., Verwijmeren, J., Lulofs, K., & Feld, C. (2010). Experiences in regional cross-border cooperation in river management. Comparing three cases at the Dutch-German Border. *Water Resources Management, 24*, 2647–2672. https://doi.org/10.1007/s11269-009-9572-5

Wuebbles, D. J., & Jain, A. K. (2001). Concerns about climate change and the role of fossil fuel use. *Fuel Processing Technology, 71*, 99–119. https://doi.org/10.1016/S0378-3820(01)00139-4

Zuccaro, G., Leone, M. F., & Martucci, C. (2020). Future research and innovation priorities in the field of natural hazards, disaster risk reduction, disaster risk management, and climate change adaptation: A shared vision from the ESPREssO project. *International Journal of Disaster Risk Reduction, 51*, 101783. https://doi.org/10.1016/j.ijdrr.2020.101783

Climate Change, Urban Challenges, and Collaborative Solutions: Addressing Heat Islands, Urbanization, and Water Crises in South Asian Countries

Anindita S. Chaudhuri

Introduction

South Asia, comprising Afghanistan, Bangladesh, Bhutan, India, the Maldives, Nepal, Pakistan, and Sri Lanka, is a region marked by remarkable diversity in geography, culture, and economic conditions. This dynamic region faces unique challenges and opportunities as it navigates rapid urbanization, significant environmental changes, and socio-economic transformations. With continuous urban growth in the South Asian region, the environmental issues impacting South Asia include urban heat islands. This phenomenon occurs when urban areas experience higher temperatures than their rural surroundings due to human activities and modifications to land surfaces. This has profound implications for public health, energy consumption, and climate resilience. Concurrently, the rapid pace of urbanization in South Asia has led to burgeoning megacities grappling with infrastructural inadequacies, overcrowding, and increased pollution (Urban Resilience Projects—Asian Development Bank, 2020; World Bank, 2020b). Compounding these issues is the region's critical water crisis, characterized by severe shortages, pollution, and the adverse effects of climate change on water availability. Addressing these interconnected challenges requires a concerted and collaborative approach. The countries of South Asia must work together, leveraging their collective knowledge and resources to develop sustainable solutions that can mitigate the impacts of urban heat islands, manage urban growth effectively, and ensure the equitable distribution and sustainable use of water resources (Sustainable Urban Development Projects—Asian Development Bank, 2020). Through regional cooperation and shared strategies, South Asia can

A. S. Chaudhuri (✉)
TERI SAS TERI School of Advanced Studies, Institutional Area, Vasant Kunj, Delhi, India

Department of Geography, Amity Institute of Social Sciences, Amity University, NOIDA, India

K. Arora, R. R. Anand (eds.), *The Solidarity Approach in Geography*, https://doi.org/10.1007/978-3-031-86753-8_19

navigate these complex challenges and foster a more resilient and sustainable future for its diverse populations.

The South Asian region, home to over 1.8 billion people, is one of the most densely populated and culturally rich areas in the world. The diversity in its geography ranges from the snow-capped peaks of the Himalayas to the fertile plains of the Ganges and Indus rivers, from arid deserts to tropical rainforests, and from sprawling urban centers to remote rural villages. This geographical diversity contributes to the complex socio-economic fabric of the region, with each country facing unique challenges and opportunities in its path to development (UNESCAP, 2023).

Figure 1 (Dimitrova et al., 2021; Beck et al., 2018) presents a comprehensive overview of South Asia's climatic zones and related references. It is based on the Köppen-Geiger climate classification, showcasing the diverse climate patterns across India, Pakistan, Nepal, and Bangladesh. The region encompasses various zones, from tropical rainforests to cold deserts. Surrounding the map are boxes for each country, listing key studies on climate-related health effects. These studies are categorized using symbols representing heat effects, cold effects, or both, as well as specific phenomena like heat waves and cold spells. The citations span from 2005 to 2020, indicating on-going research interest in the area. The map not only illustrates the geographical distribution of climate types but also highlights where certain climate effects have been more extensively studied. For instance, heat wave studies are prominent in Pakistan, while Bangladesh shows a focus on both heat

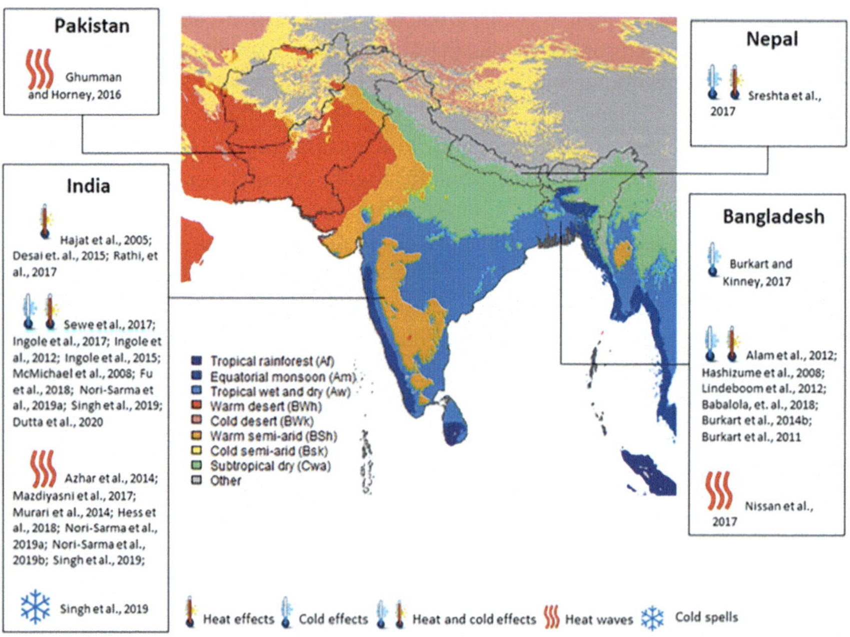

Fig. 1 Map of climatic zones and the reference studies conducted by every country. (Adopted after Beck et al. (2018))

effects and heat waves. This visualization effectively combines climatological data with a summary of regional research efforts, providing insight into both the physical environment and the academic focus on climate-related health issues in South Asia.

The climatic zones and research focus presented in this map have profound implications for urban development in South Asia. Understanding these diverse climate patterns is essential for urban planners to develop climate-adaptive strategies tailored to each region's specific needs. For instance, cities in tropical areas require different approaches compared to those in arid zones, influencing decisions on building materials, urban layout, and infrastructure design. The numerous studies on heat effects and heat waves highlight the urgent need to address urban heat islands through green spaces and innovative building designs. Water management becomes critical in regions with seasonal rainfall variations, necessitating robust drainage systems and water storage facilities. The climate distribution also impacts energy infrastructure planning, with different regions requiring varied heating or cooling solutions (Dimitrova et al., 2021). The focus on extreme weather events in the research suggests a pressing need for resilient urban infrastructure and emergency response systems. Moreover, climate-related health studies indicate that urban health facilities must be equipped to handle a range of climate-specific health issues (UNPD, 2020; Ellis & Roberts, 2016; Climate Change, 2014; Harpham & Tanner, 1995). These climate patterns can also influence migration trends, potentially accelerating urban growth in climatically favorable areas and demanding comprehensive urban planning. The cross-country nature of these climate zones underscores the importance of regional collaboration in urban development strategies. Ultimately, this climatological and research overview provides a crucial foundation for developing climate-smart, sustainable urban policies across South Asia, emphasizing the need for localized, research-informed approaches to city planning and management in the face of climate change challenges (Perera, 2015).

The diverse Köppen climate zones across South Asia, ranging from tropical to polar, create a complex landscape for studying urban heat island (UHI) effects and water crises. This distribution allows for a comprehensive analysis of UHI impacts on human health and environmental pollution across varied climatic conditions (Huang et al., 2022; Oke & Lin, 2023). In tropical and arid zones, intense UHI effects exacerbate heat-related illnesses and air pollution, while in temperate and colder regions, UHI alters local ecosystems and water cycles. The climate diversity also influences water availability and quality differently, from monsoon-dependent areas to glacier-fed regions. This climatic variety makes South Asia an ideal case study for understanding how UHI interacts with different environments, providing crucial insights for developing targeted strategies to mitigate health risks and environmental degradation across various urban settings globally (Santamouris, 2021).

India, the largest humid tropical country in South Asia, is characterized by its vast land area and immense population, making it the second-most populous country globally. India's rapid economic growth has led to significant urbanization, with cities like Mumbai, Delhi, and Bangalore becoming major economic hubs. However, this rapid urbanization has also resulted in severe environmental and infrastructural challenges, including the urban heat island effect (Thapa et al., 2021; Singh &

Ghosh, 2020; Chakraborty & Lee, 2019; Prakash & Venkatesan, 2019). Urban heat islands occur when urban areas experience higher temperatures than their rural counterparts due to the replacement of natural land cover with buildings, roads, and other infrastructure that absorb and retain heat. This phenomenon exacerbates the impacts of heat waves, increases energy demand for cooling, and deteriorates air quality, posing severe health risks to urban populations.

Pakistan, with its diverse topography ranging from mountainous regions in the north to the fertile plains of Punjab and Sindh, also faces significant urban challenges. Cities like Karachi and Lahore are experiencing rapid population growth, leading to increased pressure on infrastructure and resources (Ahmed & Khan, 2023; Hassan & Shahid, 2022). The urban heat island effect is particularly pronounced in these densely populated cities, contributing to higher temperatures and worsening living conditions. Similarly, Bangladesh, predominantly a deltaic country, faces the dual challenges of urbanization and climate change. Dhaka, one of the fastest-growing cities in the world, grapples with severe congestion, inadequate infrastructure, and environmental degradation, including the urban heat island effect (Islam et al., 2022). Sri Lanka, an island nation with a mix of urban and rural areas, is also experiencing the impacts of rapid urbanization. Cities like Colombo are expanding rapidly, leading to environmental challenges such as increased heat and reduced air quality (Jayasinghe et al., 2020; Gunawardena & Weerasinghe, 2023; Ranathunga & Chandana, 2022). Nepal and Bhutan, while primarily mountainous, are seeing urban growth in cities like Kathmandu and Thimphu, respectively. The urban heat island effect is less pronounced in these countries but is becoming a growing concern as urban areas expand. Maldives, a low-lying island nation, faces unique challenges related to urbanization and climate change, particularly the risk of sea-level rise and increased temperatures in densely populated areas like Malé.

Urbanization in South Asia is driven by various factors, including economic opportunities, better living standards, and improved infrastructure and services in urban areas (Seto et al., 2012). However, this rapid urban growth comes with significant challenges. Inadequate infrastructures, such as insufficient housing, poor transportation systems, and lack of basic services like water and sanitation, are common issues in many South Asian cities. Overcrowding, pollution, and environmental degradation further exacerbate these problems, making urban areas increasingly vulnerable to the impacts of climate change (Revi et al., 2014).

The water crisis in South Asia is another critical issue that demands urgent attention. The region's water resources are under severe stress due to over-extraction, pollution, and the impacts of climate change (Revi et al., 2014). Major rivers like the Ganges, Indus, and Brahmaputra are lifelines for millions of people, providing water for drinking, irrigation, and industrial use. However, these rivers are also heavily polluted, with industrial waste, agricultural runoff, and untreated sewage being significant contributors. Additionally, the over-extraction of groundwater for agriculture and urban use is leading to the depletion of aquifers, further exacerbating water scarcity.

The effects of climate change, including altered precipitation patterns, melting glaciers, and increased frequency of extreme weather events, are further straining the region's water resources. For instance, the melting of Himalayan glaciers, which are a crucial source of water for rivers in South Asia, poses a significant threat to water availability in the region (Alam & Rabbani, 2007; Sahu & Prakash, 2023). The changing monsoon patterns are also affecting water availability, with some areas experiencing intense rainfall and flooding, while others face prolonged droughts.

Addressing these interconnected challenges of urban heat islands, urbanization, and water crises requires a collaborative and integrated approach. The countries of South Asia must work together, leveraging their collective knowledge, resources, and experiences to develop sustainable solutions. Regional cooperation is essential to tackle these issues effectively, as the impacts of urbanization, climate change, and water crises are not confined by national borders.

Organizations like the South Asian Association for Regional Cooperation (SAARC) play a crucial role in facilitating regional cooperation and knowledge sharing. SAARC can serve as a platform for countries to share best practices, develop joint strategies, and implement collaborative projects to address these challenges (https://www.saarc-sec.org/). For instance, regional initiatives to promote sustainable urban planning, improve water management, and enhance climate resilience can significantly benefit from shared knowledge and resources. Managing urban growth effectively requires a holistic approach that integrates environmental, social, and economic considerations (IPCC Report on Asia Climate Change, 2022). Countries can collaborate on developing and implementing sustainable urban development strategies that promote compact, inclusive, and resilient cities. Sharing experiences and best practices in urban planning, infrastructure development, and service provision can help countries address the challenges of rapid urbanization more effectively. Regional cooperation can also support capacity building and technical assistance in urban management and governance.

Urbanization Trends and Challenges

Urbanization in South Asia is characterized by rapid growth driven by economic opportunities and rural to urban migration. Each country faces unique challenges and opportunities in managing this growth. Sustainable urban planning, infrastructure development, and regional cooperation are essential to address the environmental and social issues arising from urbanization.

India has experienced a remarkable transition from rural to urban living over recent decades. This shift is characterized by several key aspects. Firstly, rapid urban growth has been notable, with the urban population increasing from around 25.5% in 1993 to approximately 37.0% in 2022. Major cities such as Mumbai, Delhi, Bangalore, and Hyderabad have evolved into megacities, each hosting populations exceeding ten million (Baud & de Wit, 2008) The growth of industries, particularly in the services and IT sectors, has been a significant driver of this urbanization. These cities have become major economic hubs, drawing millions of people seeking better employment opportunities and improved living standards. However, this rapid urbanization has brought about considerable challenges, including inadequate infrastructure, traffic congestion, pollution, and housing shortages. Informal settlements or slums have become common in many cities, often characterized by poor living conditions (Ravindranath & Murthy, 2021). In response to these issues, the Indian government has launched several initiatives aimed at enhancing urban infrastructure and living standards. Programs like the Smart Cities Mission and the Atal Mission for Rejuvenation and Urban Transformation (AMRUT) are designed to address these challenges and promote sustainable urban development.

Pakistan has experienced rapid urban growth accompanied by significant challenges. The urban population has increased from about 29.6% in 1993 to around 40.3% in 2022, with major cities like Karachi and Lahore seeing substantial growth (Qureshi et al., 2010). Economic opportunities in urban areas and the rise of industries and service sectors have driven migration from rural regions, contributing to this urban expansion. However, Pakistan faces considerable challenges, including inadequate infrastructure, traffic congestion, pollution, and informal settlements (Ahmed & Khan, 2023; Hassan & Shahid, 2022). Water scarcity and poor sanitation are also significant issues in urban areas. To address these problems, the government has launched various initiatives focused on urban planning projects, infrastructure development, and housing schemes.

Similarly, Bangladesh has seen rapid urbanization, especially in its capital, Dhaka. The urban population has risen from about 20.1% in 1993 to approximately 39.5% in 2022. Dhaka, one of the fastest-growing cities globally, faces severe congestion and infrastructural pressures. The growth of the garment industry and other manufacturing sectors has driven urbanization, attracting people to cities for better job opportunities and living conditions (Alam & Rabbani, 2007). However, challenges such as overcrowding, inadequate infrastructure, pollution, and informal settlements persist, with Dhaka particularly struggling with severe traffic congestion and pollution (Dhaka City Development Project—Asian Development Bank, 2020; World Bank, 2020a). The government is working on improving urban infrastructure, developing new urban areas, and upgrading informal settlements, with projects like the Dhaka Metro Rail aiming to alleviate traffic congestion to reduce the environmental impact of urbanisation.

In Bhutan, urbanization has been slower compared to its South Asian neighbors but is accelerating. The urban population has grown from about 21.4% in 1993 to around 44.4% in 2022, with Thimphu, the capital, being the main urban center experiencing growth (Dorji & Wangdi, 2021; Tamang & Sharma, 2022). Economic development, particularly in the tourism and hydropower sectors, drives urban growth (Bhutan Green Building Guidelines—National Environment Commission, Bhutan, 2017). Bhutan focuses on sustainable development and maintaining its environmental integrity, managing urbanization to minimize environmental impacts (National Environment Commission Initiatives—National Environment Commission, Bhutan, 2020). The government emphasizes planned urban development, with a focus on sustainability and maintaining Bhutan's cultural and environmental values.

Nepal's urbanization has been influenced by internal migration and economic changes. The urban population has increased from around 8.9% in 1993 to about 22.5% in 2022, with Kathmandu Valley being the primary urban center experiencing significant growth. Migration from rural areas, driven by better employment opportunities and living standards in urban areas, is a key driver of urbanization (Shrestha & Bhatta, 2022; Tamang & Sharma, 2023). However, rapid urbanization has led to challenges such as inadequate infrastructure, traffic congestion, pollution, and informal settlements (Ministry of Forests and Environment, Government of Nepal, 2019). Resilience is also a significant concern. The government is focusing on improving urban infrastructure, developing new urban areas, and enhancing disaster resilience in urban planning.

Maldives presents unique urbanization patterns due to its geography. The urban population has grown from around 25.6% in 1993 to approximately 48.1% in 2022, with Malé, the capital, being heavily populated and facing space constraints. The island's geography limits urban expansion, leading to high population density in urban areas (Hameed, 2023). Climate change and sea-level rise are significant concerns for urban areas. Tourism and fisheries are major economic drivers, contributing to urban growth and offering better services and job opportunities (National Housing Development Program—Ministry of Housing and Urban Development, Maldives, 2020). However, space constraints, environmental sustainability, and climate resilience are major challenges. The government focuses on sustainable urban development and climate adaptation strategies (National Resilience and Adaptation Strategy—Ministry of Environment, Maldives, 2018).

The rise of the urban population in Afghanistan over the past few decades has been significant. Various factors have contributed to this demographic shift, leading to notable changes in the country's urban landscape. Continuous conflict and instability have been primary drivers of urbanization in Afghanistan. Many rural inhabitants have migrated to cities to escape violence, insecurity, and limited economic opportunities in their hometowns (Kamali, 2018; Ehsan, 2017). This has resulted in a substantial increase in the urban population, particularly in major cities like Kabul, Herat, and Mazar-i-Sharif. Urban areas often offer more job opportunities and better access to markets, attracting people from rural regions. Frequent natural disasters, such as droughts and floods, have also played a role in pushing rural populations toward urban areas. Cities are perceived as safer and more stable environments where relief efforts and resources are more readily available. Afghanistan has experienced high population growth rates, which naturally contributes to urbanization. As the overall population increases, the number of people living in urban areas rises correspondingly. According to the World Bank, Afghanistan's urban population has been steadily increasing. In 1980, about 15% of the population lived in urban areas (Mansfield, 2019). By 2020, this figure had risen to approximately 26%. Kabul, the capital, has grown particularly rapidly, with its population ballooning from around 500,000 in the 1980s to over four million in recent years. The rapid rise in the urban population has led to significant challenges (World Bank, 2020c). Urban infrastructure and services are under tremendous strain, with many cities unable to adequately provide housing, water, sanitation, and transportation. Informal settlements and slums have expanded, often lacking basic amenities. Environmental degradation and pollution are also pressing issues.

Figure 2 highlights the substantial population growth in South Asian countries from 1980 to 2020, with the percentage increase over the 40 years providing a clear

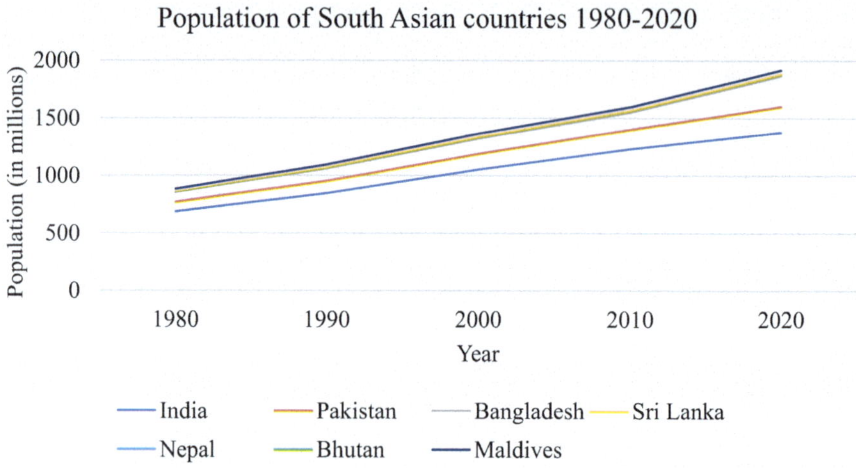

Fig. 2 Population growth pattern in South Asian Countries from 1980 to 2020. (Source: United Nations Population Division, 2020)

comparison. The trends are quite similar for all the countries, the population is continuously increasing, and accommodating the large population is giving rise to other environmental problems. The increasing population is one of the major reasons behind the continuous development of urban areas.

Landuse and Land Cover Distribution in South Asian Countries

Land Use and Land Cover (LULC) in South Asia presents a diverse and dynamic mosaic, reflecting the region's varied geography, climate, and human activities. This vast area has a wide range of LULC types that have significant implications for environmental sustainability, economic development, and human well-being. The map of Fig. 3 has been adopted from the global land use land cover distribution by ESRI (2009–2010) (Bontemps et al., 2009a, 2009b; ESRI) based on a 300 m MERIS sensor on board the ENVISAT satellite mission under the Globe Cover project. It classified the vegetation into 19 categories. A few important LULC categories are discussed in Table 1 which are considered important to fulfill the research objectives. Table 1 highlights the land use and land cover changes over the past four decades in South Asian countries. It reveals a significant increase in urban areas in countries like Pakistan, Nepal, Bhutan, Sri Lanka, and the Maldives, where urban land has doubled. Conversely, there is a reduction in forest cover by approximately 1% in Pakistan, Sri Lanka, and Bhutan, which is unfavorable for mitigating UHI effects. Additionally, Nepal has experienced a decrease in agricultural land, whereas other countries have seen an increase in their agricultural sectors.

The LULC patterns in South Asia are rapidly changing due to factors such as population growth, economic development, and climate change. Deforestation, agricultural intensification, and urban sprawl are leading to significant alterations in

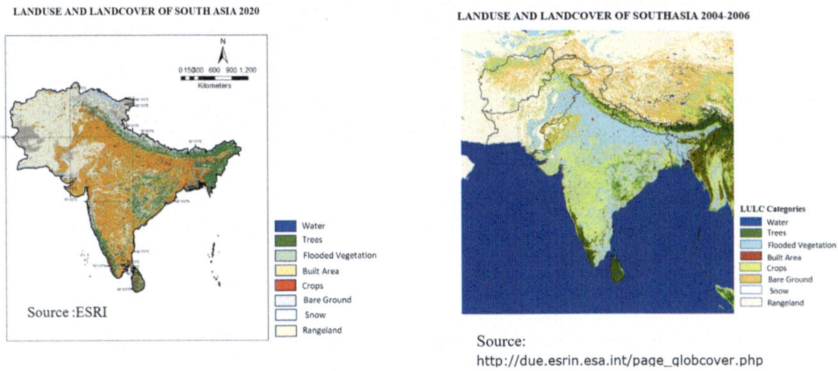

Fig. 3 LULC of South Asia 2006 and 2020. (Source: http://due.esrin.esa.int/page_globcover.php)

Table 1 Distribution LULC in South Asian Countries 1980–2020

Country	Year	Forest (%)	Agricultural land (%)	Urban area (%)	Other land (%)
India	1980	21.0	56.0	2.0	21.0
	2000	22.7	58.6	2.5	16.2
	2020	24.5	60.4	3.2	11.9
Pakistan	1980	2.5	30.0	2.0	65.5
	2000	2.1	32.5	3.0	62.4
	2020	1.9	32.2	4.0	58.9
Bangladesh	1980	14.0	65.2	2.0	19.0
	2000	12.5	68.0	2.5	17.0
	2020	11.1	70.6	3.4	14.9
Sri Lanka	1980	35.0	40.0	0.5	24.5
	2000	31.0	42.0	0.8	26.2
	2020	29.4	43.2	1.2	26.2
Nepal	1980	38.0	30.0	0.5	31.5
	2000	39.0	29.0	0.7	31.3
	2020	40.4	28.8	0.9	29.9
Bhutan	1980	72.0	12.0	0.1	15.9
	2000	71.8	12.5	0.2	15.5
	2020	71.4	13.9	0.3	14.4
Maldives	1980	2.0	8.0	0.5	89.5
	2000	2.5	9.0	0.7	87.8
	2020	3.0	10.0	1.0	86.0
Afghanistan	1980	3.0	30.0	1.0	66.0
	2000	2.0	33.0	2.5	62.5
	2020	1.5	35.0	4.0	59.5

land cover. These changes have profound implications for biodiversity, carbon sequestration, water resources, and local climates, including the exacerbation of urban heat island effects. Understanding and managing these complex LULC dynamics are crucial for sustainable development in South Asia.

It requires integrated approaches that balance economic needs with environmental conservation, urban planning with agricultural preservation, and adaptation to climate change impacts. Monitoring LULC changes through remote sensing and GIS technologies is increasingly important for informed decision-making and policy formulation in the region.

Urban Heat Island Distribution in South Asian Countries

Land Use and Land Cover (LULC) changes in South Asia have a profound impact on Urban Heat Island (UHI) effects, creating a complex interplay between urbanization and local climate patterns (Table 1). As cities expand, natural vegetation and permeable surfaces are replaced with concrete, asphalt, and buildings, significantly altering the urban microclimate. This transformation increases surface albedo, reduces evapotranspiration, and traps heat, intensifying UHI effects (Thapa et al., 2021; Prakash & Venkatesan, 2019; Tshering & Dorji, 2023). The urban heat island (UHI) effect poses a significant challenge across the diverse climatic zones of South Asia depicted in this map (IPCC Report on Asia Climate Change, 2022). India, Pakistan, Bangladesh, and Nepal all grapple with this issue, particularly in their rapidly expanding urban centers. The problem is especially pronounced given the region's predominantly hot, tropical, and subtropical climates. Major cities like Delhi, Mumbai, Karachi, Dhaka, and Kathmandu experience temperatures several degrees higher than surrounding rural areas due to dense construction, lack of green spaces, and heat-absorbing surfaces. The numerous studies cited for these countries, focusing on heat waves and heat-related health impacts, underscore the severity of the UHI effect (Chaturvedi et al., 2020; Wijeratne & Aheeyar, 2019; Hameed, 2023; Tamang & Sharma, 2022; Dorji & Wangchuk, 2021; Wangchuk & Tshering, 2023; Rai & Prasad, 2022; Ravindranath & Murthy, 2021; Sahu & Prakash, 2023; Ghosh & Kumar, 2022).

This phenomenon exacerbates already challenging climatic conditions, leading to increased energy consumption, air pollution, and health risks for urban populations. Addressing the UHI problem requires multifaceted approaches tailored to each city's specific climatic context. These strategies may include increasing urban green spaces, implementing cool roofing technologies, improving urban planning for better air circulation, developing water bodies, encouraging vertical gardens, and enforcing stricter building codes for energy efficiency. The diverse climatic zones highlighted in the map emphasize the need for localized solutions—what works in the tropical rainforest zones of southern India may differ from strategies effective in Pakistan's warm desert regions. The extensive research on heat effects across these countries indicates a growing recognition of the UHI problem and its health impacts, potentially driving policy changes and urban development strategies to combat this pressing issue in South Asian cities.

Table 2 provides an overview of the urban heat island area distribution, and the percentage of urban areas affected by UHIs in South Asian countries over the past

Table 2 Distribution of UHI intensities in various cities of South Asian countries

Country	Major cities	1990		2020		References
		UHI area distribution	Percentage of urban area	UHI area distribution	Percentage of urban area affected	
India	Delhi, Mumbai, Kolkata, Bangalore	Moderate UHI intensity	About 50–60% of major urban areas	Significant UHI effect across urban areas	Up to 70% of the major Urban areas	Thapa et al. (2021), Chakraborty and Lee (2019), Singh and Ghosh (2020), Prakash and Venkatesan (2019)
Pakistan	Karachi, Lahore	Moderate UHI intensity	About 40–50% of urban areas	Significant UHI effect	Up to 60% of the major Urban areas	Thapa et al. (2021)
Bangladesh	Dhaka	Moderate UHI intensity	About 40–50% of urban areas	Notable UHI intensity	About 50–60% of Urban Area	Islam et al. (2022)
Sri Lanka	Colombo, Kandy	Minimal UHI	About 30–40% of urban areas	Moderate UHI effect	Less than 30% of urban Area	Jayasinghe et al. (2020) Wijeratne and Aheeyar (2019)
Nepal	Kathmandu, Pokhara	Minimal UHI	Around 20–30% of urban areas	Moderate UHI effect	Around 30–40% of urban area	Thapa et al. (2021), Bajracharya and Rajbhandari (2019)
Bhutan	Thimphu, Phuentsholing	Limited UHI effect	Less than 20% of urban areas	Limited UHI effect	Less than 30% of urban areas	Dorji and Wangdi (2021)
Maldives	Malé	Minimal UHI	Less than 10% of urban areas	Limited UHI effect	Less than 20% of urban areas	Hameed (2023)
Afghanistan	Afghanistan Kabul, Herat, Mazar-i-Sharif	Limited UHI effect	About 20–30% of urban areas	Moderate UHI effect	About 30–40% of urban area	Chaturvedi et al. (2020), Atasoy (2020)

40 years. The data highlights the growing extent of UHIs in the region due to urbanization and related factors.

Water Crises in South Asian Countries

Groundwater depletion in the South Asian region is a critical and escalating issue, primarily affecting countries like India, Pakistan, Bangladesh, and Nepal. This crisis is characterized by rapidly declining water tables, deteriorating water quality, and increasing water stress across the region. The main driver of this depletion is excessive extraction, primarily for agricultural irrigation, which has been intensified by the Green Revolution's emphasis on water-intensive crops. Rapid urbanization and industrial growth have further increased demand, while climate change impacts are altering recharge patterns (Rodell et al., 2009; MacDonald et al., 2016; Shamsudduha et al., 2012; Wada et al., 2012; Gleeson et al., 2012; Bhanja et al., 2016; Qureshi et al., 2010; Shamsudduha & Panda, 2019; Mukherjee et al., 2015; Shah et al., 2012).

The severity of the Urban Heat Island (UHI) problem varies across South Asia, Fig. 4 illustrates the annual average surface temperature, average soil moisture, groundwater storage, and average annual evaporation rate. These biophysical factors are crucial for understanding groundwater availability in the region. The southeastern part of Afghanistan, along with major portions of Pakistan, Sri Lanka, and India, experiences the highest surface temperatures, ranging from 294 to 305 °F. In contrast, despite high urbanization levels, Bhutan and Nepal do not exhibit intense surface heat development. The maps showing groundwater storage, soil wetness, and evaporation, alongside lower green index values, highlight the impact of rising heat in western South Asia. The greenness index is higher in regions of India, Bhutan, Nepal, and the northern parts of Pakistan and Afghanistan. The average groundwater storage is higher around the Brahmaputra plains (2096 mm), with the Indus-Ganga plain having moderate water storage (337–839 mm). In contrast, regions around Rajasthan and Pakistan have minimal water storage (85 mm). Surface soil wetness, influenced by these parameters, plays a significant role in maintaining surface temperatures. The western part of South Asia is the driest, primarily due to the cold climate and sparse vegetation. For instance, in northwestern India, groundwater levels are declining at rates of up to 1 m/year in some locations. In Pakistan's Punjab province, a similar trend is observed, with water tables dropping by several meters over the past few decades. Bangladesh faces a dual challenge of groundwater depletion in urban areas and widespread arsenic contamination. Nepal's Kathmandu Valley is experiencing rapid groundwater depletion due to urbanization and inadequate water supply infrastructure. The consequences of this depletion are far-reaching, affecting agricultural productivity, food security, rural livelihoods, and urban water supply. It also leads to land subsidence, saltwater intrusion in coastal areas, and the concentration of contaminants like fluoride and arsenic in the remaining groundwater.

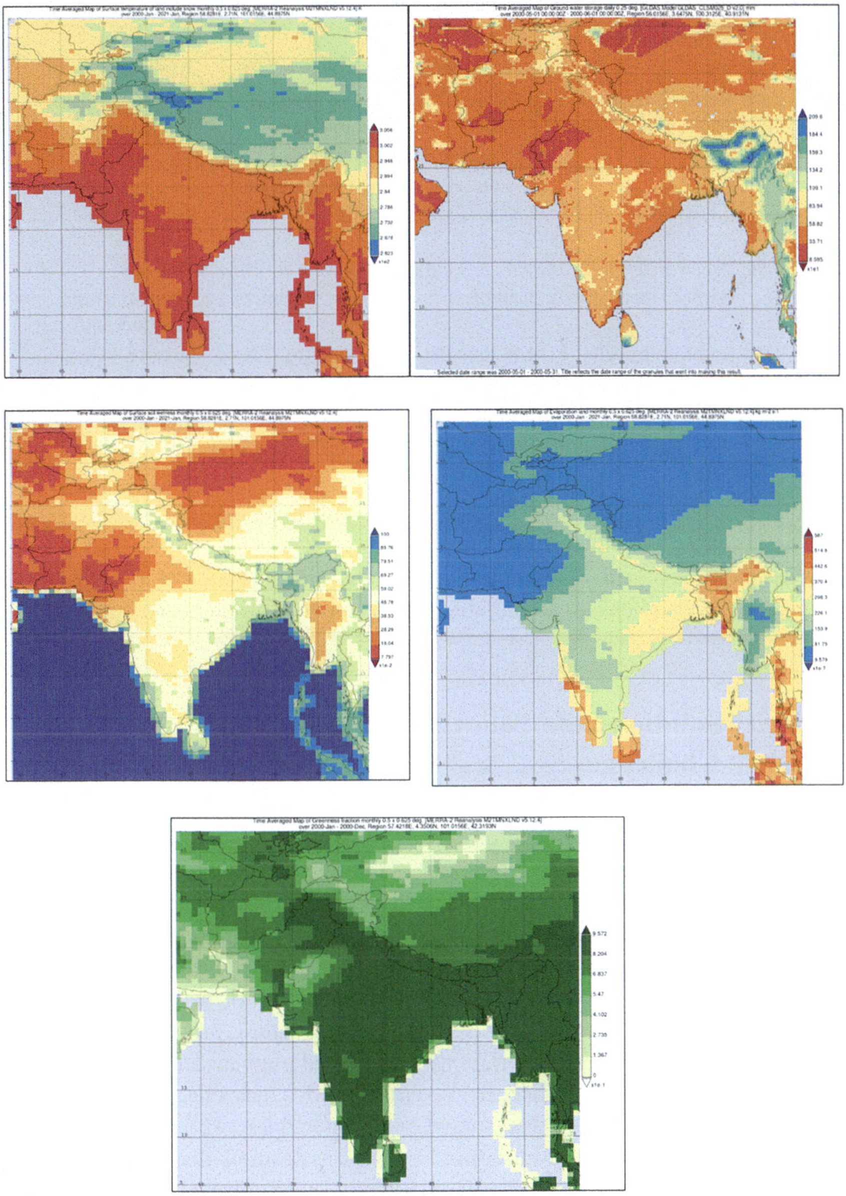

Fig. 4 (**a**) Land surface temperature 2020, (**b**) groundwater storage 2020, (**c**) soil wetness 2020, (**d**) average yearly evaporation rate 2020, (**e**) average green fraction 2020

Table 3 The groundwater depletion situation in major South Asian countries

Country	Major affected areas	Primary causes	Depletion rate	Key challenges
India	Northwestern states, Tamil Nadu	Agricultural irrigation, urbanization	Up to 1 m/year in some areas	Over-extraction, inefficient irrigation
Pakistan	Punjab, Balochistan	Agricultural pumping, drought	About 0.5–1 m/year in critical areas	Energy subsidies, lack of regulation
Bangladesh	Urban areas, Barind Tract	Urban demand, irrigation	About 0.1–0.5 m/year, varying by region	Arsenic contamination, saltwater intrusion
Nepal	Kathmandu Valley	Urban demand, poor infrastructure	About 0.5–2.5 m/year	Rapid urbanization, inadequate recharge
Sri Lanka	Dry zone areas	Agricultural use, droughts	Variable, data limited	Saltwater intrusion, nitrate pollution

Table 3 provides a snapshot of the groundwater depletion issue across South Asia, highlighting the varying rates and causes in different countries. Addressing this crisis requires urgent, comprehensive, and coordinated action at local, national, and regional levels to ensure sustainable management of groundwater resources.

Solidarity Solutions and Policy Recommendations for Urban Challenges

South Asian countries have faced numerous challenges related to rapid urbanization, prompting them to implement various policies and laws in a collective manner to address these issues. Recognizing the need for regional cooperation, many South Asian nations have participated in forums such as the South Asian Association for Regional Cooperation (SAARC) to share best practices and develop joint strategies for sustainable urban development. These collaborative efforts have led to the adoption of policies focusing on improving urban infrastructure, promoting affordable housing, enhancing public transportation systems, and implementing smart city initiatives across the region. Additionally, several countries have enacted regional laws to regulate land use, control urban sprawl, and protect green spaces in and around cities. Some nations have also introduced policies to promote decentralization and strengthen local governance structures to better manage urban growth. While the effectiveness of these measures varies across countries, the collective approach demonstrates a growing recognition of the importance of regional solidarity in addressing common urbanization challenges in South Asia.

SAARC has majorly focused on sustainable urban development through various declarations and action plans. The organization has emphasized the importance of creating liveable cities that are economically productive, socially inclusive, and environmentally sustainable (Sustainable Urban Development Projects—Asian Development Bank, 2020; Bhutan National Urban Development Strategy—Royal

Government of Bhutan, 2010; ADB Reports, 2020). Some key sustainable urban practices, policies, and conferences and meetings promoted by SAARC are as follows:

1. Green building codes: It encouraged member countries to adopt and implement green building codes to reduce energy consumption and promote sustainable construction practices.
2. Urban water management: The organization has supported initiatives for efficient water use, rainwater harvesting, and wastewater treatment in urban areas.
3. Solid waste management: SAARC has promoted integrated solid waste management systems, including recycling and waste-to-energy projects.
4. Sustainable transportation: Efforts have been made to promote public transportation, non-motorized transport options, and electric vehicles in urban areas.
5. Urban agriculture: It has encouraged urban farming initiatives to enhance food security and green cover in cities.
6. Climate resilience: The organization has supported the development of climate-resilient urban infrastructure to help cities adapt to climate change impacts.
7. Knowledge haring: SAARC has facilitated the exchange of best practices and technologies among member countries through workshops, conferences, and training programs.
8. Urban planning guidelines: The organization has worked on developing common guidelines for sustainable urban planning that can be adapted by member countries.

Policies*:*

1. National urban policies: Many SAARC countries have developed national urban policies to guide sustainable urban development. For example, India's National Urban Policy Framework and Nepal's National Urban Development Strategy.
2. Smart cities initiatives: Several countries, including India and Sri Lanka, have launched smart city programs to leverage technology for better urban management.
3. Housing for all: Policies aimed at providing affordable housing, such as India's Pradhan Mantri Awas Yojana and Pakistan's Naya Pakistan Housing Scheme.
4. Urban transport policies: Focused on improving public transportation and reducing congestion, like Bangladesh's Strategic Transport Plan.
5. Urban sanitation policies: Addressing issues of water supply and sanitation, such as India's Swachh Bharat Mission-Urban.

Conferences and Meetings:

1. SAARC summits: Urbanization has been a recurring theme in SAARC summits, with leaders discussing regional cooperation for urban development.
2. South Asian Conference on Sanitation (SACOSAN): A series of conferences focusing on improving sanitation in urban and rural areas.
3. Asian Ministerial Conference on Disaster Risk Reduction: While not exclusively urban-focused, this conference often addresses urban resilience issues.

4. South Asia Urban Knowledge Hub: An initiative supported by the Asian Development Bank to share knowledge on urban issues among South Asian countries.
5. Annual SAARC Planners Meet: A platform for urban planners from SAARC countries to discuss common challenges and solutions.
6. South Asian Cities Summit: Organized by various stakeholders to discuss urban development challenges and opportunities in the region.

These policies and conferences reflect the collective efforts of South Asian countries to address urbanization challenges. However, the effectiveness of these

Table 4 Urban development policies and projects in South Asia: A country-wise overview

India
India has implemented several significant urban policies, with one of the most prominent being the Smart Cities Mission launched in 2015. Here's an overview: **Smart Cities Mission:** – Aim: To promote cities that provide core infrastructure, clean and sustainable environment, and a decent quality of life to citizens through "smart" solutions. – Implementation: 100 cities were selected through a competition to receive funding for smart city projects. – Outcomes: As of 2023, over 7800 projects worth more than $30 billion have been tendered, with about 80% completed or under implementation. Projects include integrated command and control centers, smart water management, solar energy initiatives, and improved urban mobility. **Atal Mission for Rejuvenation and Urban Transformation (AMRUT):** – Aim: To provide basic services like water supply, sewerage, and urban transport to households and build amenities in cities. – Implementation: Covers 500 cities with a population of over 100,000. – Outcomes: Improved water supply to over 2.5 crore people and enhanced sewerage networks benefiting about 1.5 crore people by 2023
Bangladesh
– Urban Sector Development Plan (2015–2035): Aims to manage rapid urbanization and promote sustainable urban development. – Implementation: Focuses on improving urban governance, infrastructure, and services in major cities. – Outcomes: Has led to improvements in urban water supply, sanitation, and solid waste management in cities like Dhaka and Chittagong
Pakistan
– Pakistan Vision 2025: Includes strategies for sustainable urban development. – Implementation: Focuses on developing new urban centers and upgrading existing ones. – Outcomes: Has resulted in projects like the Bus Rapid Transit systems in Lahore and Peshawar, improving urban mobility
Sri Lanka
– Strategic cities development project: Aims to improve urban services and livability in strategic cities. – Implementation: Focuses on cities like Kandy, Galle, and Jaffna. – Outcomes: Has led to improved drainage systems, urban roads, and public spaces in target cities

(continued)

Table 4 (continued)

Nepal
– National Urban Development Strategy (2017): Provides a 20-year vision for urban development. – Implementation: Focuses on developing regional cities and strategic urban clusters. – Outcomes: Has initiated projects for improved urban infrastructure and services, though implementation has been challenged by resource constraints
Bhutan
– Thimphu Structure Plan: Guides the capital city's development with a focus on preserving cultural heritage and environment. – Implementation: Emphasizes green spaces, traditional architecture, and sustainable urban planning. – Outcomes: Has helped maintain Thimphu's unique character while accommodating urban growth
Maldives
– Hulhumalé Development Project: An artificial island project to address overcrowding in Malé. – Implementation: Focuses on creating a sustainable urban environment with modern infrastructure. – Outcomes: Has provided additional housing and reduced pressure on the capital city, though challenges remain in terms of affordability and environmental impact

Source: https://www.saarc-sec.org/

initiatives varies across the region due to differences in implementation capacity, resource availability, and local contexts.
- Outcomes: Improved water supply to over 2.5 crore people and enhanced sewerage networks benefiting about 1.5 crore people by 2023.

These examples show that while each country has its unique approach to urban development, common themes include improving infrastructure, enhancing liveability, and addressing environmental concerns (Table 4). The effectiveness of these policies varies, with factors like financial resources, institutional capacity, and political stability influencing outcomes. It is worth noting that many of these countries face similar challenges, such as rapid urbanization, informal settlements, and inadequate infrastructure. Regional cooperation through SAARC and other platforms has provided opportunities for sharing best practices and lessons learned in urban development.

The solidarity approach is essential for developing comprehensive, effective solutions to urban challenges in South Asia. They allow for more efficient use of resources, foster innovation, and promote the development of regionally appropriate strategies for sustainable urban development. Here are some successful examples of collaborative efforts within South Asia that have addressed urban development challenges:

1. South Asia Urban Knowledge Hub (K-Hub):

- A collaborative platform initiated by the Asian Development Bank (ADB) involving Bangladesh, India, Nepal, and Sri Lanka.
- Focuses on sharing knowledge and best practices in urban planning, water supply, and sanitation.
- Outcome: Has facilitated the exchange of innovative solutions, such as pro-poor water supply strategies and sanitation technologies.

2. South Asian Association for Regional Cooperation (SAARC) Coastal Zone Management Centre:
   - Based in Maldives, it involves all SAARC member countries.
   - Aims to address coastal urban development issues and climate change impacts.
   - Outcome: Has developed regional guidelines for sustainable coastal urban development and facilitated knowledge sharing on coastal protection strategies.
3. Urban Climate Change Resilience Trust Fund (UCCRTF):
   - A multi-donor trust fund supported by the UK, Switzerland, and the Rockefeller Foundation.
   - Focuses on building climate resilience in South Asian cities.
   - Outcome: Has funded projects in Bangladesh, India, and Pakistan, improving urban flood management and climate-resilient infrastructure.
4. South Asia Water Initiative (SAWI):
   - A partnership between the World Bank and the governments of Afghanistan, Bangladesh, Bhutan, India, Nepal, and Pakistan.
   - Addresses water resource management issues in major river basins.
   - Outcome: Has improved transboundary water management, benefiting urban areas dependent on shared water resources.
5. Cities Development Initiative for Asia (CDIA):
   - While not exclusive to South Asia, it has significantly benefited the region.
   - Supports medium-sized cities in preparing sustainable and bankable infrastructure projects.
   - Outcome: Has assisted cities in Bangladesh, India, Nepal, and Pakistan in developing urban infrastructure projects.
6. South Asian Conference on Sanitation (SACOSAN):
   - A government-led biennial convention held on a rotational basis in South Asian countries.
   - Focuses on improving sanitation in the region.
   - Outcome: This has led to increased political commitment and action on urban sanitation issues across South Asia.
7. Greater Mekong Subregion (GMS) Program:

- While primarily focused on Southeast Asia, it includes Bangladesh from South Asia.
- Promotes economic cooperation and addresses transboundary issues including urban development.
- Outcome: Has facilitated knowledge sharing on sustainable urban transport and green city development.

8. International Centre for Integrated Mountain Development (ICIMOD):
   - An intergovernmental organization serving eight countries of the Hindu Kush Himalaya region, including several South Asian nations.
   - Addresses mountain-specific urban issues.
   - Outcome: Has improved understanding and management of water resources in Himalayan urban areas.

Conclusion

The South Asian region has made significant strides in addressing urbanization challenges through collective efforts and regional cooperation. The South Asian Association for Regional Cooperation (SAARC) has played a crucial role in promoting sustainable urban development across member countries. This collaborative approach has led to the implementation of various policies and initiatives focusing on green building, urban water management, sustainable transportation, climate resilience, and knowledge sharing. Individual countries have also developed and implemented their urban development strategies, such as India's Smart Cities Mission, Bangladesh's Urban Sector Development Plan, and Pakistan's Vision 2025. While these country-specific approaches vary, they share common themes of improving infrastructure, enhancing livability, and addressing environmental concerns.

Regional cooperation has proven to be essential in developing comprehensive and effective solutions to urban challenges in South Asia. Successful collaborative efforts like the South Asia Urban Knowledge Hub, SAARC Coastal Zone Management Centre, and the Urban Climate Change Resilience Trust Fund have facilitated knowledge exchange, improved urban infrastructure, and enhanced climate resilience across the region.

However, the effectiveness of these initiatives varies due to differences in implementation capacity, resource availability, and local contexts. Despite these challenges, the collective approach demonstrates a growing recognition of the importance of regional solidarity in addressing common urbanization issues in South Asia.

In conclusion, while progress has been made, continued regional cooperation and knowledge sharing remain crucial for sustainable urban development in South Asia. These collaborative efforts allow for more efficient use of resources, foster innovation, and promote the development of regionally appropriate strategies to address the complex urban challenges faced by South Asian countries.

References

ADB reports ,2020978-92-9262-759-1 (print) 978-92-9262-760-7 (electronic) 978-92-9262-761-4 (ebook).

Ahmed, N., & Khan, A. S. (2023). Urban heat Island effect in major Pakistani cities: Trends and mitigation measures. *Environmental Monitoring and Assessment, 195*(4), 292.

Alam, M., & Rabbani, M. D. G. (2007). Vulnerabilities and responses to climate change for Dhaka. *Environment and Urbanization, 19*(1), 81–97.

Ali, M. M., & Ibrahim, H. (2022). Urban heat Island effect in Malé City: An analysis using remote sensing data. *International Journal of Remote Sensing, 43*(18), 6864–6882. https://doi.org/10.1080/01431161.2022.2082204

Atasoy, M. (2020). Urban heat Island phenomenon in Kabul and its correlation with urbanization. *Urban Climate, 34*, 100685.

Bajracharya, S. R., & Rajbhandari, S. (2019). Urban heat islands in Kathmandu: An analysis of temperature variations using remote sensing data. *Urban Climate, 29*, 100495. This research uses remote sensing data to analyze temperature variations in Kathmandu, emphasizing the impact of urbanization on UHI formation.

Baud, I., & de Wit, J. (Eds.). (2008). *New forms of urban governance in India: Shifts, models, networks, and contestations*. Sage Publications.

Beck, H. E., Zimmermann, N. E., McVicar, T. R., Vergopolan, N., Berg, A., & Wood, E. F. (2018). Present and future Köppen-Geiger climate classification maps at 1-km resolution. *Scientific Data, 5*, 180214.

Bhanja, S. N., Mukherjee, A., Saha, D., Velicogna, I., & Famiglietti, J. S. (2016). Validation of GRACE-based groundwater storage anomaly using in-situ groundwater level measurements in India. *Journal of Hydrology, 543*, 729–738.

Bhutan Green Building Guidelines—National Environment Commission, Bhutan. (2017).

Bhutan National Urban Development Strategy—Royal Government of Bhutan. (2010).

Bontemps, S., Defourny, P., Van Bogaert, E., Weber, J.-L., & Arino, O. (2009a, May). GlobCorine—A joint EEA-ESA Project for Operational Change Detection at Pan-European Scale. In *33rd international symposium on remote sensing of environment (ISRSE), Stresa, Italy*.

Bontemps, S., Defourny, P., Van Bogaert, E., Arino, O., Kalogirou, V., & Perez, J. R. (2009b). GLOBCOVER 2009: Products description and validation report. *ESA Bulletin, 136*, 53.

Chakraborty, S., & Lee, X. (2019). Urban heat islands in Kolkata: A multi-temporal analysis using satellite imagery. *Remote Sensing of Environment, 232*, 111299.

Chaturvedi, R. K., Joshi, J., & Bala, G. (2020). Urban heat Island intensity and its impact on air quality in Kabul. *Atmospheric Environment, 239*, 117747. This research examines the UHI intensity in Kabul and its implications for air quality, emphasizing the role of urbanization.

Defourny, P., Schouten, L., Bartalev, S., Bontemps, S., Caccetta, P., de Witt, A., di Bella, C., Gerard, B., Giri, C., Gond, V., Hazeu, G., Heinimann, A., Herold, M., Jaffrain, G., Latifovic, R., Ling, H., Mayaux, P., Muncher, S., Nonguierma, A., Stibig, H.-J., … Arino, O. (2009). Accuracy assessment of a 300-m global land cover map: The GlobCover experience. In *33rd international symposium on remote sensing of environment (ISRSE), Stresa, Italy*.

Dhaka City Development Project—Asian Development Bank. (2020).

Dimitrova, A., Ingole, V., Basagaña, X., Ranzani, O., Milà, C., Ballester, J., & Tonne, C. (2021). Association between ambient temperature and heat waves with mortality in South Asia: Systematic review and meta-analysis. *Environment International, 146*, 106170. https://doi.org/10.1016/j.envint.2020.106170

Dorji, K., & Wangchuk, D. (2021). Evaluation of urban heat Island intensity and its correlation with land use patterns in bhutanese urban areas. *Environmental Monitoring and Assessment, 193*(10), 643.

Dorji, P., & Wangdi, K. (2021). Impact of urbanization on Thimphu's microclimate and urban heat Island formation. *Bhutan Journal of Natural Resources & Development, 7*(1), 23–35.

Ehsan, E. (2017). Urbanization and its impact on economic development in Afghanistan. *Journal of Central Asian Studies, 24*(2), 123–145.
Ellis, P., & Roberts, M. (2016). *Leveraging Urbanization in South Asia: Managing Spatial Transformation for Prosperity and Livability*. World Bank Publications.
Ghosh, S., & Kumar, K. (2022). Impact of urban heat Island on public health and energy demand in Indian cities. *Journal of Environmental Management, 320*, 116651.
Gleeson, T., Wada, Y., Bierkens, M. F., & van Beek, L. P. (2012). Water balance of global aquifers revealed by groundwater footprint. *Nature, 488*(7410), 197–200.
Gunawardena, K. K., & Weerasinghe, P. A. (2023). Urban heat Island effects in Colombo, Sri Lanka: Evidence from satellite data and ground measurements. *Journal of Urban Planning and Development, 149*(1), 04022048.
Hameed, A. A. (2023). Urban heat Island effects in Male' and Hulhumale': A case study. *Maldives Environmental Research, 5*, 45–58.
Harpham, T., & Tanner, M. (Eds.). (1995). *Urban health in developing countries: Progress and prospects*. Routledge.
Hassan, A., & Shahid, A. R. (2022). Remote sensing and GIS-based analysis of urban heat Islands in Lahore, Pakistan. *Sustainable Cities and Society, 78*, 103680.
Huang, J., Li, C. C. T., & Zhang, J. J. (2022). Urban heat Island effects on local climate and human health: A review of recent advances. *Environmental Research Letters, 17*(4), 043002.
International Labour Office, International Labour Organization (ILO). (2023).
IPCC Report on Asia Climate Change. (2022). Retrieved from https://www.saarc-sec.org/.
IPCC, 2014: Climate Change 2014: Synthesis Report. Contribution of Working Groups I, II and III to the Fifth Assessment Report of the Intergovernmental Panel on Climate Change [Core Writing Team, R.K. Pachauri and L.A. Meyer (eds.)]. IPCC, Geneva, Switzerland, 151 pp.
Islam, A. K. M. S., Choudhury, M. A. I., & Rafiuddin, M. (2022). Impact of urbanization on urban heat Island intensity in major districts of Bangladesh. *Urban Climate, 41*, 101070.
Jayasinghe, P., Jayasooriya, S., & Perera, K. (2020). Impact of urbanization on urban heat Island effects in Colombo, Sri Lanka. *Sustainable Cities and Society, 52*, 101834. This study explores UHI effects in Colombo, examining the impact of urbanization and reduced green spaces on local temperature variations.
Kamali, M. H. (2018). *Urbanization in Afghanistan: Population shift and challenges*. World bank Report.
Kathmandu Valley Development Authority. (2013).
Kathmandu Valley Flood Risk Management Project—World Bank. (2020).
Li, X., & Wang, S. (2023). The role of urban form and land use in urban heat Island intensity: A comparative study of major cities. *Urban Climate, 48*, 101472.
MacDonald, A. M., Bonsor, H. C., Ahmed, K. M., Burgess, W. G., Basharat, M., Calow, R. C., et al. (2016). Groundwater quality and depletion in the Indo-Gangetic Basin mapped from in situ observations. *Nature Geoscience, 9*(10), 762–766.
Malé City Plan—Male City Council. (2018).
Mansfield, D. (2019). *The challenges of urbanization in post-conflict Afghanistan*. World Bank Report.
Ministry of Environment and Forests, Government of Bangladesh. (2009).
Ministry of Forests and Environment, Government of Nepal. (2019).
Ministry of Urban Development, Nepal. (2007).
Mukherjee, A., Saha, D., Harvey, C. F., Taylor, R. G., Ahmed, K. M., & Bhanja, S. N. (2015). Groundwater systems of the Indian Sub-Continent. *Journal of Hydrology: Regional Studies, 4*, 1–14.
National Environment Commission Initiatives—National Environment Commission, Bhutan. (2020).
National Housing Development Program—Ministry of Housing and Urban Development, Maldives. (2020).
National Resilience and Adaptation Strategy—Ministry of Environment, Maldives. (2018).

National Urban Sector Policy—Ministry of Local Government, Rural Development and Cooperatives, Bangladesh. (2006).
Noori, R., & Rahimi, K. A. (2023). Study of urban heat islands in Kabul: Analysis of temporal and spatial variations. *Urban Climate, 49*, 101423.
Oke, T. R., & Lin, T. K. (2023). The urban heat Island effect: Recent insights and future directions. *Journal of Applied Meteorology and Climatology, 62*(1), 67–85.
Perera, N. (2015). *People's spaces: Coping, familiarizing, creating in South Asian Cities*. Routledge.
Population Reference Bureau (PRB). (2023). *2023 World Population Data Sheet*.
Prakash, M. N., & Venkatesan, A. (2019). Study on urban heat islands over Bangalore using remote sensing and GIS. *Sustainable Cities and Society, 47*, 101470.
Qureshi, A. S., McCornick, P. G., Sarwar, A., & Sharma, B. R. (2010). Challenges and prospects of sustainable groundwater management in the Indus Basin, Pakistan. *Water Resources Management, 24*(8), 1551–1569.
Rai, S., & Prasad, K. (2022). Climate adaptation strategies for urban heat Islands in small Himalayan cities. *Urban Climate, 42*, 101089.
Ranathunga, H., & Chandana, W. R. (2022). Evaluation of urban heat Island intensity in Sri Lankan cities using MODIS data. *Environmental Monitoring and Assessment, 194*(3), 160.
Ravindranath, N. H., & Murthy, K. R. (2021). Urban heat Island and its mitigation strategies in Indian cities: A review. *Environmental Science & Policy, 125*, 295–305.
Revi, A., Satterthwaite, D. E., Aragón-Durand, F., Corfee-Morlot, J., Kiunsi, R. B., Pelling, M., Roberts, D. C., & Solecki, W. (2014). Climate Change 2014: Impacts, Adaptation, and Vulnerability. Part A: Global and Sectoral Aspects. In *Contribution of Working Group II to the Fifth Assessment Report of the Intergovernmental Panel on Climate Change* (pp. 535–612). Cambridge University Press.
Rodell, M., Velicogna, I., & Famiglietti, J. S. (2009). Satellite-based estimates of groundwater depletion in India. *Nature, 460*(7258), 999–1002.
Sahu, S. K., & Prakash, M. S. (2023). Assessment of urban heat Island intensity in Indian metropolises: Case Study of Delhi, Mumbai, and Bangalore. *Journal of Urban Climate, 50*, 101500.
Santamouris, M. (2021). Urban heat Island effects and mitigation strategies. *Annual Review of Environment and Resources, 46*, 167–188.
Seto, K. C., Güneralp, B., & Hutyra, L. R. (2012). Global forecasts of urban expansion to 2030 and direct impacts on biodiversity and carbon pools. *Proceedings of the National Academy of Sciences, 109*(40), 16083–16088.
Shah, T., Giordano, M., & Mukherji, A. (2012). The political economy of the energy-groundwater nexus in India: exploring issues and assessing policy options. *Hydrogeology Journal, 20*(5), 995–1006.
Shamsudduha, M., & Panda, D. K. (2019). Spatio-temporal changes in terrestrial water storage in the Himalayan river basins and risks to water security in the region: A review. *International Journal of Disaster Risk Reduction, 35*, 101068.
Shamsudduha, M., Taylor, R. G., & Longuevergne, L. (2012). Monitoring groundwater storage changes in the highly seasonal humid tropics: Validation of GRACE measurements in the Bengal Basin. *Water Resources Research, 48*(2).
Shrestha, N., & Bhatta, K. C. (2022). Urban heat Island effect and its relationship with land use in Kathmandu Valley, Nepal. *Urban Climate, 40*, 100999.
Siddiqui, M., & Ali, M. M. (2022). Impact of urbanization on heat island effect in Afghan cities: A preliminary study. *Sustainable Cities and Society, 77*, 103658.
Singh, R., & Ghosh, M. (2020). Analyzing urban heat Island effect in Mumbai metropolitan region using satellite data. *Environmental Monitoring and Assessment, 192*, 705.
Sustainable Urban Development Projects—Asian Development Bank. (2020).
Tamang, B., & Sharma, R. (2022). Assessing urban heat Island effects in Thimphu City, Bhutan: A remote sensing approach. *Journal of Urban Planning and Development, 148*(3), 04022024.

Tamang, S., & Sharma, B. R. (2023). Mitigation strategies for urban heat Island in Nepalese urban areas: A case study of Kathmandu. *Environmental Science & Policy, 130*, 300–311.
Thapa, R. B., Gupta, M., & Joshi, P. K. (2021). Assessment of urban heat islands in South Asian cities using Landsat imagery. *Journal of Environmental Management, 280*, 111694.
Thimphu Structure Plan—Thimphu City Corporation. (2020).
Tshering, T., & Dorji, P. (2023). Mitigation strategies for urban heat islands in himalayan urban centers: A case study of Thimphu, Bhutan. *Journal of Environmental Management, 323*, 116482.
United Nations Economic and Social Commission for Asia and the Pacific (UNESCAP). (2023).
United Nations Population Division. (2020). *World Urbanization Prospects: The 2018 Revision*. United Nations.
Urban Development and Infrastructure Projects—Asian Development Bank. (2020).
Urban Resilience Projects—Asian Development Bank. (2020).
Wada, Y., van Beek, L. P., & Bierkens, M. F. (2012). Nonsustainable groundwater sustaining irrigation: A global assessment. *Water Resources Research, 48*(6).
Wangchuk, P., & Tshering, J. (2023). Impacts of urban expansion on the urban heat Island effect in Bhutan's cities. *Sustainable Cities and Society, 81*, 103731.
Waste Management Projects—Ministry of Environment, Maldives. (2023).
Wijeratne, S., & Aheeyar, S. (2019). Assessment of urban heat island intensity in Colombo using satellite data. *Journal of Urban Environmental Engineering, 13*(1), 22–32. This research assesses the intensity of UHI in Colombo, using satellite data to analyze the temperature differences between urban and rural areas.
World Bank. (2020a). *Dhaka Urban Development Project*.
World Bank. (2020b). *Urban Resilience Project*.
World Bank. (2020c). World Development Report 2020: Trading for Development in the Age of Global Value Chains. Washington, DC: World Bank.
World Bank. (2024a). *Population Growth for South Asia (SPPOPGROWSAS)*.
World Bank. (2024b). *Population Growth for South Asia (SPPOPGROWSAS)*. Retrieved from FRED, Federal Reserve Bank of St. Louis.

Part VI
Geographies of Care, Responsibility, and Social Mix

Solidarity through Sports: A Geographical Reading

Arnav Anshuman

Introduction

Sport is an expression of human skill, imagination, and performance over space. Sport is like an art form essential to the human experience. The experience of sports on the playing field finds new meaning to it every time a game is in play. Sports are not unique to one group of human beings but a conduit through which people from different identities, classes, races, and social backgrounds express their true beliefs, desires, and emotions. The playing field is where athletes find ways to rid their minds of thoughts of oppression, evil, and animosity. Sports are political entities of their own, and the relationships supersede boundaries, treaties, and alliances and at the same time speak politically and socially. The paper is centered around the political nature of sports as a medium for expanding the dialogue surrounding the solidarity approach in human geography. The paper argues that sports provide a platform for political statements through actions and non-actions alike. Sport has the potential to be a vehicle for political action, where the virtues of sportsmanship and international cooperation reflect the essence of solidarity.

To substantiate the relationships between solidarity, space, and sporting culture, the paper conducts a content analysis of sporting events that are reported in print and social media. The paper looks into sports geographical literature which finds consonance with human and political geography. At the same time, it utilizes literature that spans disciplines to confirm the role of space, place, territory, and scale in the event of sports. The scope of the paper is diverse as the relationship between sports, politics, and solidarity is evident across sports.

The paper has two parts, the first of which examines sports movements worldwide that have shaped political awareness. This section provides a temporal discussion of how sports and politics have come to the fore. In the second section, the

A. Anshuman (✉)
Department of Geography, Delhi School of Economics, University of Delhi, New Delhi, India

K. Arora, R. R. Anand (eds.), *The Solidarity Approach in Geography*,
https://doi.org/10.1007/978-3-031-86753-8_20

focus is on recent developments at the level of nation-states that have enhanced the subtleties of solidarity in sports. This section brings forward the issues of nationalism, race, religion, caste, and class amidst diverse geographical backgrounds to initiate a spatial understanding of the solidarity approach in geography. The paper makes an addition to sports studies and the solidarity approach by extending the conversation of political geography through the lens of sports. The paper brings forward the idea of geographical imageries, space, place, and territory to understand the phenomena of sports and its connection to political movements at the scales of international, national, and regional levels.

Finding Space in Sports

Sport is inherently geographic (Gaffney, 2016). Sports like any human activity happen over space. The nuances of location, landscape, place, human–environment interaction, the city, and global economy become evident when the sport is in play. At the same time, the role of sport in culture is significant and serves as a productive laboratory for furthering geographical understanding (Bale, 1988). The effects of sports activities being carried out in different spaces are vastly different across economic, social, cultural, or political domains, which operate at local, regional, national, or international levels (Ilieş et al., 2014). This space in turn becomes the site of political contestation and acts as a foundation for representation of voices and movements. Issues like nationalism, religion, race, caste, and class find headway into the mainstream through the participation of players and their performances.

The interconnections between sports and geopolitics (Koch, 2016), city (Bale & Moen, 1995), migration (Waite & Smith, 2017), identity (Shobe, 2008), and territory (Bale, 2002) activate the geographical principle of placemaking and the spatial politics associated at varied scales. This paper looks into the geographical literature on sports and brings about the nuances of sports that incite political upheaval movements across the globe. This paper undertakes a content analysis of newspapers and blogs as a methodological tool to understand the varied scales at which the interplay of sports geography and the politics of nationalism, religion, race, caste, and class are manifested. The history of sports events and contemporary issues complement each other to support the claim of sports being a site of political solidarity.

Sports Arena: A Field of Political Consciousness

This section looks at the developments in the history of institutional sports that have effused political consciousness among the players, viewers, and nations. Three significant events in the history of the Olympic games that have set the stage for expansion beyond the stadiums where they occur are the focus of this section. The events include the Olympic Truce movement, the 1936 Berlin Olympics amidst a rising Nazi power, and the "Black Salute" at the 1968 Mexico Olympics. These events showcase that sports have tremendous potential to offer reconciliation, stability, and peace.

Olympic Truce

Before the Ancient Olympic Games in 776 BC, there was a declaration of the Olympic Truce.

The idea was the cessation of hostilities of any form. The aim was to ensure that the host city of the Olympic Games was capable of organizing the event peacefully and providing safe passage for athletes and spectators taking part in the Games. In 1896, the Olympics were revived and the objective was to organize a sports festival that brought together the youth of the world, while also promoting peace, friendship, solidarity, and fair play. Consequently, the United Nations Organization revived the Olympic Truce in 1993 even though the Bosnian War caused the tensest and most volatile circumstances (Loland & Selliaas, 2009).

The Olympic Truce is represented by the dove of peace, which is surrounded by the traditional Olympic flame (Fig. 1). The IOC's ideals and challenges are represented by the peace-dove symbol, which is a symbol of hope and unity in a world that is plagued by conflicts and animosity. The goal is to create a peaceful and better world by utilizing sport and the Olympic concept. The Olympic flame is a symbol of warm friendship for all nations through sharing and togetherness. The flame in the symbol is made up of colorful effervescent elements that symbolize the joy felt

Fig. 1 Symbol for Olympic Truce. (Source: Olympictruce.org)

during the celebration of the human spirit. The Truce is being observed by people of all races, as reflected by the elements (International Olympic Truce Centre—Διεθνές Κέντρο Ολυμπιακής Εκεχειρίας, n.d.).

The Olympic Truce at present is an expression of humanity's desire to construct a world that is guided by fair competition, peace, humanity, and recreation. Moreover, the Olympic Truce acts as an intermediary between the centuries-old and wise tradition and the maintenance of international peace and security. The Olympic Truce offers a reminder that the Olympic principles of respect, peace, and unity are just as relevant today as they were when the original Olympic events were first held in Greece over 3000 years ago.

The Olympic Winter Games PyeongChang 2018, which served as an inspiring example, reinforced the ideals of the Olympic Truce. The event marked the first instance of North and South Korea sharing a platform in the sporting arena. These events demonstrate how sport and the values of Olympism can unite a world that has been divided. The Olympic Truce is a means by which people, athletes, and governments can provide a brief respite from conflict and strife, and sends a powerful message of hope to the international community.

"Unique Friendship" at the Berlin Olympics, 1936

The 1936 Summer Olympics in Berlin sparked worldwide controversy when Jewish athletes were sidetracked from taking part in the events. This was in context to propaganda for National Socialism and the new German state utilized extensively by Hitler and the Nazis (Cody, 2016). The sporting mega-event was entangled with nuances of Aryan's racial superiority, anti-Semitism, and global politics. In this regard, the Berlin Olympics has often been regarded as "Hitler's Olympics" or the "Nazi Olympics" (Glosniak, 2017).

The purpose of the Olympic event was to showcase the superiority of the Aryan race in terms of their genetic makeup rather than athleticism (Hanold, 2012). However, the aspirations of Hitler and the Nazi ideology faced a major setback owing to the performance of Jesse Owens[1] in the competitive events. Owens shattered the myths of Aryan superiority by winning four gold medals in the event (Fig. 2). The tipping point in the ceremony arrived when Hitler refused to "shake hands" with Owens (Bracken, 2019). This showed Hitler's refusal to acknowledge a non-Aryan athlete's ability and reflected his detestation toward Black and Jewish communities.

While on the one hand, the site of Olympic sports was used by Hitler as a medium for the promotion of racial propaganda, a "unique friendship" between Owens and

[1] In 1936, Jesse Owens, also known as James Cleveland "Jesse" Owens, became a Black American track and field athlete who won four gold medals.

Fig. 2 Owens at the Berlin Olympics. (Source: Google Images)

Luzz Long[2] showed that sports can bring individuals together from various backgrounds, including gender, ethnicity, and country, even in the toughest situations (Fig. 3). The latter's technical advice to Owens on the long jump helped the American to secure another Gold Medal for his country. The atmosphere toward Black American athletes reflected in the racial connotation of Nazism, and Luzz's aid to Owens was a major departure in this discourse.

The true act of sportsmanship by Luzz narrates that even the individual response by actors in the sporting field creates a reverberating impact on the sport's political domain. An interesting aspect of the 1936 Olympics was the solidarity shown by India against the colonial and fascist government. The Indian group was among the sole two countries that did not salute as they walked by Hitler. This act was a bold defiance, completely aligned with the core principles of the leading current of Indian nationalism (Majumdar & Mehta, 2020).

Protest on Podium, Mexico Olympics, 1968

Social fields like sports can make it conceivable for racial and ethnic minorities to draw consideration to their cause. Hegemony is reflected in cultural forums and identity politics which eventually become potential sites of struggle. In this vein, the African American Olympic movement used the Olympic podium as a means for

[2] Carl Ludwig "Luz" Long, a distinguished German athlete in the long jump, is remembered for his silver medal victory in the event at the 1936 Summer Olympics in Berlin where he offered expert guidance to his fellow competitor, Jesse Owens, who eventually won the gold medal in the long jump.

Fig. 3 The "Unique Friendship" at the Berlin Olympics 1936. (Source: Google Images)

Fig. 4 Protest at the Mexico Olympics 1968. (Source: Olympictruce.org)

expressing the discrimination against black communities in the US that was a product of the civil rights movement in the 1960s (Hartmann, 1996).

An enduring representation of the silent protest is the image of sprinters Tommie Smith and John Carlos giving the "Black Power Salute" during a medal ceremony at the 1968 Olympics in Mexico City (Fig. 4). The protest, however, extends far beyond the black-gloved fists. The words "Olympic Project for Human Rights" were illustrated on the buttons of all three medal winners, including Australia's silver medal winner Peter Norman.

This demonstration got Smith and Carlos banned from the team but a long-standing tradition of offering political expressions through sport. Norman's demonstration of fortitude despite Australians were confronting racial pressures in their nation was a significant commitment. This act "killed his career" yet forayed a political message across the world.

Olympic events are the highest form of institutional events in sports. Political movements at this level have a large-scale impact on the scales the games are played. These movements are a product of the larger forces that are acting within the society. This section talked about a few important issues in Olympic history that have shaped political consciousness beyond the sporting arenas the game is played. The next section will look into the interplay of these political movements with important geographical concepts like space, place, territory, and scale.

Sports and Solidarity

Insofar we have looked at three key events that have shaped the solidarity movement in Olympic history. However, the world of sports is filled with examples like these. The *Politics of Space and Place* acts as the groundwork at which these events take place at reiterated varied scales. This section will look into the diffused intersections of sports space relationships that reflect the understanding of solidarity in sports. This section will look into the nuances of solidarity through the lens of race, identity, and nationalism that have brought people, players, and governments together in different parts of the world.

Race

As discussed in the previous sections, race has been an intense topic of discussion in sports studies. Racial attributes can never be divested from the arena of sports. The politics of the body is central to the understanding of the way games are played. This has made the sports arena an important site for political contestation through signs, symbols, and performances. Sports have helped shape discourse about race in an unequivocal manner. Drawing inspiration from critical theory and race theory, the movements in sports have transcended regional and national boundaries. Various sporting institutions have taken part in the quest to counter racism. This has been made possible using awareness campaigns through vlogs, punishment of perpetrators of racial abuse for both players and viewers, and usage of signs, symbols, and posters to develop sensitivity(Fig. 5).

The most recent example of sports being a medium to express solidarity with racial movements was seen in the series of events under the tagline *Black Lives Matter*.[3] A movement that gained limelight owing to police brutality in the United

[3] Black Lives Matter is a decentralized political and social movement opposing all racially motivated violence against black people and incidents of police brutality.

Fig. 5 Solidarity against racism at a football game. (Source: FIFA)

States witnessed solidarity from the sporting fraternity across the globe. Players, viewers, and institutions stood united to support the political cause. The movement involved "taking the knee"[4] which became a common sight for demonstration following the events that took place around George Floyd's death (Fig. 6).[5] The movement transcended geographical as well as sporting boundaries with affirmative actions across the globe. In this process, the sports sites offer a medium to represent voices and reinvigorate the political narrative looming over the society at large.

Identity

The degree to which an individual identifies with the role of an athlete and seeks recognition from others is referred to as athletic identity (Good et al., 1993). Participation in itself makes a social statement that is representative of the race, class, caste, community, nation, or religion to which they belong. Sports spaces become sites of political contestations wherein identity is contested and negotiated amidst the power dynamics that surround them. The Māori dance or Haka[6] is a classic example in this case (Fig. 7).

[4] Athletes kneel on one knee to showcase solidarity.

[5] George Perry Floyd Jr. was an African American man who was killed merely based on suspicion of counterfeiting in Minneapolis.

[6] The Haka in Mori culture is a dance or challenge that is performed by a group and involves vigorous movements and rhythmic foot stamps.

Fig. 6 "Take the Knee" protest at a soccer stadium in the United States. (Source: Getty Images)

New Zealand is a country with a disturbing history of Aboriginal' genocide. The colonial rule in the country had subjugated the indigenous Māori[7] people on the island which led to the marginalization of the community. Over time with improved political representation and inclusion, the Māori way of life is finding its roots back in the country. Sports in this case have become yet another site of political representation by showcasing the war dance routine before the beginning of the game. The role of the war dance is to intimidate the opponents before the beginning of the game. Thus, the dance form has found new meanings at the site of sport. The ethnic Māori identity becomes evident through the art form at the venue where the game is played and circulates across the world through the media.

Gender identity has been another area of engagement in sports and cultural studies. The physiological parameters of males and females and the inclusion of the third sex have attracted several debates within academics and the sports field. The case of Serena Williams, a champion tennis player stands as a unique case in this regard. Williams is one of the world's most recognizable athletes and has achieved great success as a female tennis player. Her records in a white-dominant male game transcend boundaries of race, class, and gender. According to (Rios, 2019)

> *...the common mental image of a women's tennis player is white and lithe, in perfect harmony with Western ideals of feminine beauty, while Williams is black and built like a powerlifter. She is an unprecedented affront to our collective notion of the beautiful female athlete.*

However, Williams' femininity is often not just a result of her being a strong female but to some extent, opponents of Williams are defined as weak, fragile, and

---

[7] In New Zealand, Maori are the indigenous Polynesian people who came from eastern Polynesia.

Fig. 7 Haka performance by New Zealand Rugby Players. (Source: telegraph.co.uk)

not-scratchy in physical performances compared to Williams (Dovemark, 2013). At the same time, her dominance on the tennis court, self-confidence, and verbal confrontations categorize Williams as unfeminine and rebellious. Williams' feminine and sexuality are often scrutinized and questioned primarily due to her physique, but also her skin color (Schultz, 2005). Thus, Williams brings the interconnections of race, gender, and sexuality to the center of discussions. African American children, particularly those from low-income areas, have been inspired to participate by her legacy and sister Venus Williams. She is seen as a pioneer who champions the excellence of black women and an activist who persists in her advocacy for women's rights. Her performance on the tennis court challenges the norm of the sport and the normative identity associated with her. Sport settings provide a platform to extend the nuances of identities at the scales where the game is played. Gender, racial, and sexual identities are showcased through performances, chants, signs, and symbols at the site of the sports place.

Beyond Nationalism

The last section looked into the making of community and individual identities within the space of sport. This sub-section extends the debate of identity at the scale of the nation. Sports help in fuelling national identities and vice versa. The interconnections between the two are too hard to ignore. Sports in the globalized world help in bridging the gap between nations. The Olympic Truce, discussed in the previous sections serves as a great example in this regard. Sports aid in nation-building (Koch, 2013) and at the same time bring people from different parts of the world into the same place. The site of sports thus becomes a medium to share bonhomie beyond political, religious, and regional lines.

Both countries divulge in the case of foreign affairs, geopolitical relations, and regional issues. Terrorism and wars have inflicted both nations with vicious scars. The dialogue between the countries is a bare minimum most times of the year. However, cricket emerges as a channel of cooperation between the two nations. The field of play is neutralized when the two teams enter the stadium. With the upsurge in politics, media, and technology, cricket diplomacy between the two countries is a pivot for the diplomatic relationship.

Sports have a connection to political socialization, the creation of political culture, and the advancement of national identity (Houlihan, 1997). Cricket has existed for a considerable time as a means of diplomacy and public relations between India and Pakistan (Fig. 8). Cricket diplomacy, in this regard, has a positive impact on diplomatic talks where people-to-people cricket can significantly reduce hostility between the two sides (Shahid, 2015).

Several instances from the past have validated this issue. Recent examples include the Chennai crowd giving a standing ovation to Wasim Akram's men in the backdrop of the Kargil war. Pervez Musharraf's visit to India in 2005 to watch a cricket match opened the gates for conversation regarding the disputed Kashmir region. The resumption of diplomatic ties, post-2008 terror attacks, after the 2011 World Cup Match in India led to the release of Gopal Das as a goodwill gesture.

Fig. 8 Cricket as a metaphor for nationalism. (Source: BBC)

Fig. 9 Transnational image making through cycling. (Source: cuclingweekly.com)

Matches are played with great sportsman spirit, and the field of play remains neutral during the game. The site of sport extends the dialogue between the two countries beyond the nationalist upsurge evident in both nation-states. The impact of sporting events is seen much beyond the stadiums they are played.

Sports have also been regarded as a "soft-authoritarianism toolkit" that is central to the imagination of a nation (Koch, 2013). The case study of the Asthana Professional Cycling Team illustrates this case of transnational image making of a country through sports. Koch (2013) underscores how in post-Soviet Kazakhstan elite sporting activity enabled ordinary people to support the nation in the sphere of sports (Fig. 9). The performance of the team is seen as something "objective" that fills them with pride and makes a case for their representation on the global stage.

Thus, sports enable the interplay of nationalist identities and politics in multiple ways. The case of India-Pakistan cricket matches talks about the extension of solidarity beyond nationalism lines while Kazakhstan cycling team infuses a sense of solidarity among the people of the country to express their support to the team.

Learning through Solidarity

The solidarity approach is essentially a methodological question in the process of knowledge creation in human geography. This paper tries to understand sports as a medium of political action. Time and again players and sporting institutes have stood for the larger political cause looming over society and public affairs. Through their performances and symbolic iteration, they have transformed the site of sporting activities into a platform for political representation. It is time for geographers to take inspiration from the players and extend the dialogue of solidarity by taking issues of everyday activity like sports. Geographers can substantiate the co-production of sports and space to understand the inequality and marginality in the access to sporting facilities at the scale of everyday life.

In terms of race, identity, and nationalism, these issues come to the forefront through the medium of sports at the scale of the nation-state. Similar nuances of spatial interventions need to be substantiated at the local scale to provide justice for sporting facilities. The importance of sport in the comprehensive development of humans has already been established. However, in a country like India, especially in the urban areas, spaces for recreational activities are minimal. This is an infringement on the basic rights of a large section of society who are deprived of recreation in their daily lives.

Geographers should vouch for these deprivations at the level of everyday life by taking notes from the solidarity approach. There is a dearth of research in this arena. A push for groundwork research in the arena would contribute to sports studies as well as bring new ideas to inculcate solidarity in academics. Sport is an essential commodity of living and should be a basic right of growing children. Most are deprived of sporting facilities, and it should come as a central issue in the study of human geography.

Geographical Understanding of Solidarity in Sports

The paper highlights the instances of solidarity in sports at the scale of the globe. The paper dissects regions and sports and showcases the interplay of sports sites as a standpoint to underline the political representation of individuals, communities, and the nation-state. The Olympic games give insight into how solidarity has been expressed through the historical account of the Olympic Truce, the Berlin Olympics 1936, and the Mexico Olympics 1968. The Olympic Truce narrates the essence of sports that help bring people from different races, religions, backgrounds, and countries together to share a belonging at a common platform. The Berlin Olympics was a standpoint in Olympic history as the bonhomie between the rival athletes stood against the fascist discourse in Germany under Hitler. The Mexico Olympics in 1968 was used as a platform to express solidarity for Americans of African descent who were facing severe repercussions during the violent civil disturbances.

The case of racial struggle at the site of sports is extended through institutional support and the "take the knee" movement in the present context. Sports sites as a medium to express community and individual identity are understood through the Mauri Haka dance routine, and sports and space relations have many wider applications. The site of sport through its geographical nuances of space, place, and territory becomes a medium of political movement in this context we see intersectionality of black, gendered, and sexual identity in the case of Serena Williams. Sport as an important avenue for nationalism is seen through the rivalry that surrounds India–Pakistan cricket matches and cycling as a sport to evoke nationalism in small countries like Kazakhstan. The paper transcends geographical boundaries and presents cases from different parts of the world to narrate the importance of sports sites in inculcating solidarity among people, communities, and nation-states. Toward the end, the paper expresses the lack of access to recreational space in marginalized communities and the need for ground-breaking research to bring out these issues in consonance with the solidarity approach.

Conclusion

The paper gives a spatial and temporal analysis that cuts across different sports to reiterate the role of sports in expressing solidarity. The temporal account presents a timeline of events in terms of the Olympic games to showcase the evolution of political movements in sports. This is understood through the examples of athletics in Berlin and Mexico. The spatial aspects of sport and solidarity bring out instances from countries across the globe to establish sports sites as central to the understanding of the politics of space. The "say-no-to-racism" campaign in football and the "take the knee" movement across sports reflect solidarity against racism. The representation of collective identity is understood through the Haka dance routine in the sport of Rugby in New Zealand. While Serena Williams gives insights into the multi-layered identity that exists in the sports arena, nationalism and sport are understood through the case study of cricket in India and Pakistan and cycling in Kazakhstan. These examples were selected based on an ethnographic approach to bring out instances that highlight the centrality of sports and space in extending the solidarity debate. The paper contributes the sports and solidarity studies by bringing the idea of sport at the helm of affairs to understand the solidarity approach in human and political geography. At the same time, the paper appeals to geographers to analyze the solidarity approach in the context of sporting activities occurring at the local scale to help them gain access to recreation facilities through reforms in public places. It opens up avenues for further research in this regard to bring down the interconnections between solidarity and sports at the level of everyday life.

References

Bale, J. (1988). The place of place in cultural studies of sports. *Progress in Human Geography, 12*(4), 507–524.

Bale, J. (2002). *Sports geography*. Routledge.

Bale, J., & Moen, O. (1995). *The stadium and the city. 347*. Retrieved from https://books.google.com/books/about/The_Stadium_and_the_City.html?id=qUwMAAAAYAAJ.

Bracken, H. (2019). *Was Jesse Owens Snubbed by Adolf Hitler at the Berlin Olympics?* | Britannica. Retrieved from https://www.britannica.com/story/was-jesse-owens-snubbed-by-adolf-hitler-at-the-berlin-olympics.

Cody, N. W. (2016). *The Berlin Olympics: Sports, anti-semitism, and propaganda in Nazi Germany*.

Dovemark, M. (2013). How private 'everyday racism ' and public 'racism denial ' contribute to unequal and discriminatory educational experiences. *Ethnography and Education, 8*(1), 16–30.
Gaffney, C. (2016). Gentrifications in pre-Olympic Rio de Janeiro. *Urban Geography, 37*(8), 1132–1153.
Glosniak, Q. (2017). *The 1936 Nazi Olympic Games; The First Truly Modern Olympiad.*
Good, A. J., Brewer, B. W., Petitpas, A. J., Van Raalte, J. L., & Mahar, M. T. (1993). Identity foreclosure, athletic identity, and college sports participation. *Academic Athletic Journal, 8*(Spring/1).
Hanold, M. (2012). *World sports: A reference handbook*. Bloomsbury Publishing USA.
Hartmann, D. (1996). The politics of race and sport: Resistance and domination in the 1968 African American Olympic protest movement. *Ethnic and Racial Studies, 19*(3), 548–566.
Houlihan, B. (1997). Sport, national identity and public policy. *Nations and Nationalism, 3*(1), 113–137.
Ilieş, A., Dehoorne, O., Wendt, J., & Kozma, G. (2014). For geography and sport, sports geography or geography of sport. *Geosport for Society, 1*(1–2), 7–18.
International Olympic Truce Centre—Διεθνές Κέντρο Ολυμπιακής Εκεχειρίας. (n.d.). Retrieved 13 July, 2024, from https://olympictruce.org/.
Koch, N. (2013). Sport and soft authoritarian nation-building. *Political Geography, 32*, 42–51.
Koch, N. (2016). *Critical geographies of sport: Space, power, and sport in global perspective.* Taylor & Francis.
Loland, S., & Selliaas, A. (2009). *The Olympic truce-the ideal and the reality*. OLYMPIC TRUCE.
Majumdar, B., & Mehta, N. (2020). *Dreams of a Billion: India and the Olympic Games.* HarperCollins.
Rios, T. (2019). *Serena Williams and the Fear of a Dominant Black Woman*. Retrieved from https://www.thedailybeast.com/serena-williams-and-the-fear-of-a-dominant-black-woman.
Schultz, J. (2005). Reading the catsuit: Serena Williams and the production of blackness at the 2002 US Open. *Journal of Sport and Social Issues, 29*(3), 338–357.
Shahid, S. A. (2015). India: Pakistan sports as a tool for peace (cricket diplomacy). *International Journal of Coaching Science, 9*(1), 51–63.
Shobe, H. (2008). Football and the politics of place: Football Club Barcelona and Catalonia, 1975–2005. *Journal of Cultural Geography, 25*(1), 87–105.
Waite, C. A., & Smith, D. P. (2017). Temporary skilled international migration of young professional cricketers: 'Going Down-Under' to move up the career path. *Geoforum, 84*, 70–76.

A Study of Socio-cultural Aspects of Tribal Existence in Association with the Sense of Care and Responsibility

Prajakta Vinod Jadhav

Introduction

Geographers have been interested in the issues related to our moral obligations to care for others for many years. "Care" appears to be nothing new in geography. However, the researcher contends that current societal changes are expanding commercial relations into our relationships and that public support for social welfare is being reduced. These parallel processes have simultaneously marginalized care from public view and made it a more urgent matter in the commercial context. Geographers are in a good position to call attention to these tendencies, and the researcher implored us to consider our obligation to be concerned about these problems and the resulting geographies. In light of (1) market expansions, (2) the current pervasive discourses of personal responsibility (for poverty, inner city decline, unemployment, etc.), and (3) the withdrawal of public support from many crucial arenas emerge an important area to be study and research by geographers. Care ethics draws our attention to the social crises and how it is created via imbalances in power, but it also pushes us beyond criticism and toward the creation of new kinds of institutions, relationships, and actions that promote mutuality and well-being. The researcher thinks about how our academic work, professional training, and professional practices might change in light of care ethics. Incorporating social responsibility into our professional practices, creating spatially extensive connections of interdependence and mutuality, and paying attention to the ways institutional and historical relationships (extension of market relations; famine, unnatural disasters, environmental and cultural destruction) are constructed as all suggested by care ethics.

P. V. Jadhav (✉)
Department of Environmental Studies and Foundation Course, Thakur College of Science and Commerce, Mumbai, Maharashtra, India

K. Arora, R. R. Anand (eds.), *The Solidarity Approach in Geography*,
https://doi.org/10.1007/978-3-031-86753-8_21

The concerns about our moral duties to take care of others have long piqued the interest of geographers. It seems that dealing with the concept of care is not new in geography. The researcher kindly asks that we take into consideration our responsibility to be concerned about these issues and the consequent geographies because geographers are in a strong position to draw attention to these tendencies. The researcher asks all of us to think about our responsibilities as geographers to ask questions in light of (i) market expansions, (ii) the current pervasive discourses of personal responsibility (for poverty, inner city decline, unemployment, etc.), and (iii) the withdrawal of public support from many important arenas.

Care ethics calls our attention to the social crises and how power disparities shape them, but it also encourages us to look past criticism and develop new sorts of institutions, connections, and behaviors that advance welfare and reciprocity. The researcher considered how care ethics might alter our academic endeavors, professional education, and professional practices. Incorporating social responsibility into our professional practices, creating spatially extensive connections of interdependence and mutuality, and paying attention to the ways institutional and historical relationships (extension of market relations; famine, unnatural disasters, environmental and cultural destruction) are developed and constructed as suggested by care ethics. The spread of market relations; hunger, man-made disasters, and environmental issues happen to be the important aspects of this concern.

Objectives of the Study

The objectives of the study are to:

1. Explore the socio-cultural dynamics of tribal communities in India.
2. Investigate the role of the sense of care and responsibility within tribal societies.
3. Examine how cultural practices and beliefs influence caregiving roles and responsibilities among tribal groups.

Research Methods

The study utilized a quantitative research approach, analyzing secondary data obtained from existing ethnographic studies, governmental reports, and academic literature. Secondary data sources provided insights into the socio-cultural aspects of tribal existence and their association with the sense of care and responsibility. Data collection involved a systematic review and synthesis of relevant literature, focusing on caregiving roles, cultural norms, and societal expectations within tribal communities. Statistical analysis was conducted to identify patterns and correlations, shedding light on the nuanced dynamics of care and responsibility in tribal societies.

Hermeneutics of Care

Geographers as a whole have begun to emphasize the dynamic nature of "care" as a concept. "Care" as a practice and the topics it evokes, entangles, and molds are not static or distinct things; rather, it develops over time and space in a variety of moving relationships. In this part, we discuss the implications of geographers grounding their study on care and expanding care beyond just being a subject for study or a yardstick for criticism to inform a disciplinary ethic. We won't say this because we think these are the only circumstances in which geographers should exercise care; rather, we'll say it because we'll argue that how we understand the world and our place in it has a fundamental impact on our ability to practice care. Due to this, our focus will be on methods of research and analysis as well as how geographic data is gathered and "read." To define this fundamental role and to conceive of a potential hermeneutics of care, we turn to work in feminist philosophy and science and technology studies.

Latour (2005) asserts that constructing an account of the world also entails molding it by suggesting a particular arrangement of interactions to which other players would respond. Latour (2004, 2005) asserts that "matters of fact" and "matters of concern" are distinct categories. Matters of fact are things that are generally considered to be common knowledge. Most of us trust traffic signals, we trust that keys unlock doors, and we trust that breast milk is healthy for young children. Conversely, circumstances where the outcome is questionable and less certain are considered topics of concern. On the other hand, context and viewpoint decide if something is true or should be taken seriously. For traffic managers, the frequency of light changes is a constant source of worry. Parents may find that certain facts become concerning when their newborns refuse to feed, have reflux, or are unable to nurse for any reason. Every story has its way of arranging the facts and the problems.

According to Callon and Latour (1981), the establishment of truths that are widely acknowledged and treated as such is frequently the basis for power. When concerns develop, certain performers might give priority to bringing them (back) to reality. On the other hand, different viewpoints on urgent matters are likewise feasible. For example, Puig de la Bellacasa (2017) describes a way of thinking where worries are treated as "matters of care."

Puig de la Bellacasa's demand for seeing factual facts as subjects of concern relies on a longer tradition of thinking about care in feminist philosophy as well as the material-semiotic tradition in which she works. The logic of caring and the logic of control are contrasted by Mol (2008). The logic of control sees the social world as essentially consisting of objective facts, where the same interpretations and actions can produce comparable results in a variety of circumstances.

Insights into the Panchsheel Principles

In contemporary India, tribal communities continue to grapple with numerous challenges, ranging from land dispossession to cultural marginalization. These indigenous groups face persistent struggles for recognition of their land rights, often resulting in displacement due to infrastructure projects and industrialization. Economic exploitation and inadequate access to education, healthcare, and basic amenities further compound their hardships. Despite legal safeguards, tribal populations confront discrimination, and lack of political representation, exacerbating their socio-economic vulnerability. Efforts toward tribal empowerment and inclusive development remain imperative to address these pressing issues and uphold their distinct identities amidst India's modernization drive.

Prime Minister Jawaharlal Nehru established the Panchsheel philosophy for tribal development in India following the country's independence. The five guiding principles of Indian tribal development are known as Panchsheel. Better agriculture, medical facilities, education, and effective communication are all necessary to guarantee total development. However, these developmental paths must be followed within a broad framework that adheres to five essential principles. (1) Nothing should be forced on people; instead, they should cultivate a long line of their talent. Instead, we ought to make every effort to support their customary art and culture. (2) It is important to respect tribal rights over land and forests. (3) We ought to make an effort to build and train a group of our individuals to handle administrative and development tasks. In the beginning, some outside technical staff will be required. However, we shouldn't encroach too many outsiders on indigenous land. (4) We shouldn't overburden these places with several plans or over-administer them. It would be better if we dealt with their social and cultural institutions directly. (5) The quality of the human character that has evolved should be used to evaluate the outcome rather than financial expenditure figures.

As Prime Minister, Jawaharlal Nehru has often noted and elaborated on these concerns, emphasizing the necessity for prudence in developing tribal communities. He draws attention to the catastrophic effects of so-called European civilization on indigenous people around the globe, pointing out how it has destroyed their traditional ways of life and arts and crafts. Nehru says that if the so-called Indian civilization is not properly monitored, it could have a detrimental effect on tribal people's way of existence. The Indian Constitution's Article 46 requires governments to provide special consideration to the educational and economic needs of tribal people, which serves as the foundation for the government's program for tribal development. Special laws should be enacted to protect tribal people from exploitation and to provide social fairness. The protection of tribal interests is a unique duty placed on governors of governments that contain tribal regions. The governor has the authority to amend federal and state laws as they relate to tribal territories and to create rules that safeguard tribal members' land rights and prevent them from being taken advantage of by moneylenders. For this reason, the applications of fundamental rights are modified. Tribal people also have full political rights under the

Constitution. It also makes provisions for Scheduled Tribes, like Scheduled Castes, to have administrative services and seats reserved in the assembly. The Constitution also calls for the creation of Tribal Advisory Councils in each state with a tribal region to offer guidance on issues about the welfare of the tribal people. The President appoints a commissioner for Scheduled Castes and Scheduled Tribes to look into whether the protections afforded to them are being followed. State governments use both executive and legislative measures to stop moneylenders from taking advantage of Native Americans and to stop Native territory from being stolen by non-Natives. The federal and state governments establish unique infrastructure and design unique initiatives for the welfare and economic advancement of tribal communities and their inhabitants, including the development of cottage and village industries and the creation of jobs for them. Huge budgets are created, and enormous amounts are allocated specifically for this reason. After 1971, funding for tribal welfare expanded dramatically. Therefore, the Government of India's tribal policy, which was influenced by Jawaharlal Nehru, is more pertinent to the tribal people of Northeast India. In October 1952, Nehru declared that the Indian people and government should give this entire border region in the northeast particular attention.

The Communal String

The communities in India belong to various cultural attributes. Both the tribe and caste-based communities are a unique part of the Indian sub-continent. Historically, both of them are in interaction with each other's culture, and it is a difficult task to clarify who adopted the culture of whom. It is understood that caste communities were initially tribes and transformation happened through time. The entire journey of transformation is through different stages with modifications, changes, and alterations and finally leads to cultural growth. This also involved constant and continuous interactions between these communities along with the invaders and rulers from other countries. It is also true that whatever changes or modifications happened with the community cultures, these happened with the sustainability of the system that helped the community to continue. The positivity or the suitability of the culture patterns/units/complexes are the most probable factors for adaptation and modifications. The boundaries between the communities are not clear but continuity is featured in most of the spheres of performances and practices.

The continuity of cultural elements in the community while experiencing mobility and changes is the status of the continuum. Usually, the cultures of the communities of tribes are examples of simple societies, and the castes are complex societies in comparison to them. The fundamentals are derived from Robert Redfield's concept of the rural-urban continuum in the "Folk-culture of Yucatan." The socio-cultural elements change but the basic principles of practices remain unchanged. In India, tribal religion has many aspects that are similar to Hindu religion, particularly nature worship and belief in ancestor worship. Annual death rituals are also featured among some tribes like Lanjia Saora of South Odisha (Swain, 2010).

The patriarchal and patrilineal systems are common among the majority of Indian tribes and similar to non-tribal, i.e., Hindu communities. Inheritance of property also follows the descent rules as prevalent in the system. The tribes in India, who are less prone to changes and known as isolated communities, are not at par with non-tribal communities regarding social mobility and changes therein. In some regions, they are considered in lower strata. Thus the members of tribal communities depending upon the social-cultural status also try to move in the social ladder by imitating the culture of caste communities of higher strata. The migrated members of the Santhal community mostly the lifestyles of Hindu culture like worshiping Tulsi herb, observing the ritual calendar of Hindus, and others. The pastoral community of today is integrated into the temple services for their milk products. They observe the sacredness of the buffalo sheds by not allowing females during their menstrual periods and also females are not allowed to prepare milk products supposed to be supplied to the temples.

Numerous times, there was no name given to the collection of native tribe religious customs. The tribal people of northeast India that did not accept Christianity in the second half of the nineteenth century were referred to as "pagans" or simply "non-Christians" because outsiders had given them a label. But the tribes' collective name was of far less importance to them than the beliefs, customs, and rituals that bound them to their sacred beings and guaranteed their long-term well-being. They added more beliefs, rituals, and performances to this storehouse of knowledge by stealing from their neighbors or other people they encountered, believing that doing so would strengthen their culture and increase their control over the environment. Pure classifications did not exist; instead, a type of "mixed-up genre" was in its infancy. Tribal groups have interacted with other religious and sectarian groups as well as Hinduism and later Christianity, but they have managed to preserve some degree of autonomy within their religious systems. The hierarchy of clan and village deities among tribal peoples generally endures despite these continuing connections. The majority of the tribal people's holy specialists are from their communities, with very few coming from other tribes. The schedule of celebrations and festivals is likewise largely unaltered. A second form of interaction is that of purifying them" ("making them Hindu again") or "bringing back people to the Hindu fold," which is significantly different from this kind of interaction and syncretism.

Solidarity and Development

In contemporary India, employing a solidarity approach and integrating care ethics are essential to ensuring the survival and thriving of tribal communities. Solidarity entails fostering a collective commitment to supporting marginalized groups and recognizing their inherent worth and unique cultural contributions. By adopting this approach, governmental and non-governmental entities can collaborate with tribal communities to address their needs holistically, promoting self-determination and preserving their indigenous knowledge systems. Furthermore, care ethics emphasizes empathy, rationality, and responsibility in relationships, advocating for the nurturing of interconnectedness and mutual support. Implementing care ethics within policies and programs involves prioritizing the well-being and dignity of tribal individuals, valuing their perspectives, and safeguarding their rights. This approach entails creating inclusive spaces where tribal communities are actively involved in decision-making processes concerning their land, resources, and socio-economic development. Moreover, it necessitates investing in culturally sensitive healthcare, education, and socio-economic initiatives tailored to their specific needs and aspirations. By embracing both solidarity and care ethics, stakeholders can foster genuine partnerships with tribal communities, promoting resilience, dignity, and sustainable development while respecting their autonomy and cultural integrity in the diverse fabric of contemporary India.

Since the colonial era, the tribal population's political identities have continued to shift. The area the tribes inhabited was designated as a tribal area and given special administrative status by the British overlords in an attempt to isolate the tribes. The tribes were exposed to exploitation by non-tribal moneylenders, zamindars, contractors, and other middlemen because of their unique status and seclusion. Additionally, there was internal exploitation in the form of residents from higher castes taking advantage of others. Throughout the colonial era, there have been numerous tribal uprisings, starting with one in Bihar in 1772 and continuing through numerous others. The British rulers implemented particular measures for the administrative purpose of the tribal lands to quell these uprisings and revolts. Several laws that were created to govern "tribal" areas claimed that "tribals" were a segmented political community that could not be integrated into Indian society at large. For example, the Scheduled District Act of 1874 was founded on the idea that "tribal" territories should not be subjected to general laws. The Government of India Act of 1919 adopted the idea of "Backward Areas," which further muddled the former definition of "non-regulation area."

The Indian constitution was written in the post-independence era, while the country was being reshaped, with the protection of the rights and interests of its tribal population in mind. The Indian constitution included additional specific clauses and granted certain advantages to tribal people. Provisions for "administration and control of Scheduled Area and Schedule Tribe" are found in Article 244 of the Constitution. Several non-tribal populations have been granted a new identity under Article 342 as "Scheduled Tribes," based on standards established by the Indian government.

The Ethics of Care and Responsibilities

Ghurye states that two distinct forces have broken up tribal solidarity: Hinduism and its assimilative process, and British policy. Hindu assimilation leaves tribes feeling less absorbed and more helpless under a sense of brokenness. If a section of a tribe assimilates into the Hindu fold, they are thrust into a very strange social environment. Hindu castes, or at least many of them, have characteristics of tribal society when it comes to the management of internal affairs. Tribal sections that join Hindu society develop an internal organization of caste patterns, and as a result, have been regulating and controlling power within them. Many tribes smugly settle into the Hindu fold. One of the main duties of a welfare state such as India is to provide for every person in the nation. But India's governance practices aren't perfect, which is why not everyone in the nation benefits equally from government assistance initiatives. Right now, the tribal people are the most negatively impacted. The federal and state governments have launched numerous welfare programs, but the residents of tribal territories have not yet benefited from them.

The governments have worked with the India Tribal Care Trust, an NGO that promotes the welfare of the tribal people living in Assam, Jharkhand, Tripura, Maharashtra, Odisha, West Bengal, and Andhra Pradesh, to facilitate the process and provide some measure of respite for the people. It makes sense that every non-governmental group works with the government to secure their backing for the implementation of several social welfare initiatives meant to assist the indigenous people. Both the India Tribal Care Trust and NGOs must comprehend their societal obligations. To be more precise, this organization is well cognizant of its obligations.

Conclusion

When it comes to specific socio-historical characteristics, Indian tribes are distinct. As a result, academics from many fields around the world engage in intense debate and discussion about tribal studies in India. Policymakers, colonial authorities, and Indian sociologists could not agree on how to address the issues and future possibilities of Indian tribes. Three main strategies—the "integration approach," the "assimilation approach," and the "isolation approach"—emerge among Indian groups. As of right now, just three main theories about tribes exist and are widely apparent. Despite efforts being made at many levels to develop theory at every level, their presence or importance is still unknown. This is presumably the reason why the integration, assimilation, and isolation approaches are currently contested and refuted.

The integration approach in Indian groups often fosters a sense of care and solidarity by emphasizing mutual respect and cooperation while preserving cultural diversity. This approach encourages communities to support one another while maintaining their distinct identities and promoting inclusivity and social cohesion.

However, its limitations include the challenge of balancing cultural preservation with integration into the broader society, potentially leading to the dilution of unique cultural practices. The assimilation approach, on the other hand, may prioritize social cohesion at the expense of cultural diversity, undermining the principles of care and solidarity by neglecting marginalized groups' needs. Similarly, the isolation approach, while preserving cultural heritage and autonomy, may limit external solidarity and collaboration, leading to social fragmentation. Understanding the dimensions of these approaches is crucial for fostering inclusive societies that uphold care, solidarity, and cultural diversity.

References

Latour, B. (2004). Politics of nature: How to bring the sciences into democracy (C. Porter, Trans.). Harvard University Press.

Latour, B. (2005). Reassembling the social: An introduction to actor-network-theory. Oxford University Press.

Callon, M., & Latour, B. (1981). Unscrewing the big Leviathan: How actors macro-structure reality and how sociologists help them to do so. In K. Knorr-Cetina & A. V. Cicourel (Eds.), Advances in social theory and methodology: Toward an integration of micro- and macro-sociologies (pp. 277–303). Routledge & Kegan Paul.

Puig de la Bellacasa, M. (2017). Matters of care: Speculative ethics in more than human worlds. University of Minnesota Press.

Mol, A. (2008). The logic of care: Health and the problem of patient choice. Routledge.

Swain, B. (2010). Tribal religion and rituals: A study of Lanjia Saora of South Odisha. New Delhi: Mohit Publications.

Role of Social Networks to Increase Socio-economic Solidarity for Transformative Participatory Development in the Mewat Region, India

Naresh Kumar

Introduction

The use of social networks within the field of social sciences has a robust historical legacy, contributing significantly to the advancement of sociocultural and material conditions of people while informing policy debates. This lineage dates back from the pioneering work of Pierre Bourdieu (1986; Dasgupta & Serageldin, 2000) to the notable contributions of Putnam (1995, 2000, 2007) and Putnam et al. (1993). Social networks have been conceptualized in the context of various forms of capital, such as social capital and cultural capital.

However, in the analysis of social networks, this chapter diverges from the use of the social capital theory due to its inherent limitations. A key limitation of this theory lies in its focus on studying social capital in aggregate terms, neglecting the differentiation between formal and informal social capital, which often existed before any specific project initiation (Krishna, 2002, 2004, 2007). Moreover, it implicitly assumes the absence of social capital before project commencement, a historical oversight that does not align with the dynamic nature of social networks in rural areas of India, where formal organizations are scarce, and local-level informal networks play a crucial role (Chhibber, 2000). Additionally, the chapter seeks to position the "participation" of individuals within development projects within the existing social networks of "community" social groups.

Development practitioners often perpetuate the myth that communities can accomplish anything given sufficient mobilization through local institutions, relying on the presumed hidden potential of these communities for development (Cooke & Kothari, 2001). However, this idealized scenario frequently falls short of expectations. Empirical evidence at the local level (Hickey & Mohan, 2004) often contradicts these claims, revealing severe limitations stemming from inadequate material

N. Kumar (✉)
Post-Doc Fellow, Jawaharlal Nehru University, New Delhi, India

K. Arora, R. R. Anand (eds.), *The Solidarity Approach in Geography*,
https://doi.org/10.1007/978-3-031-86753-8_22

resources and structural constraints that hinder the functioning of "community"-based organizations (CBOs). Furthermore, little is known about how these institutions navigate their interactions with bureaucratic structures, civil society, and political entities, especially in the presence of parallel statutory local institutions at the village panchayat level (Ferguson, 2006; Chhotray, 2007).

The theoretical framework of this study is anchored in the context of ongoing participatory development, a part of a government-sponsored project in collaboration with various stakeholders. According to the Mewat Development Agency (MDA), despite substantial infrastructure development in the region, people's "participation" has been lacking, resulting in prolonged underdevelopment. This perspective emphasizes the choices, aspirations, and roles individuals and communities play in the local development process. This study aims to delve into the participatory development process by critically examining "participation" as a development strategy. It also explores the involvement and influence of state power by scrutinizing the actions of various government departments responsible for implementing the economic components of the project.

The identity of the Mewat region is a complex amalgamation of its historical, cultural, and economic dimensions, clearly defined and further articulated in terms of language and literature. In contrast, its geographical and political-administrative identity is more a product of human invention than a straightforward discovery. This implies that the region's geography shares similarities with its surrounding areas, and there are at least two distinct administrative units, or states, that overlap with portions of the Mewat region. Consequently, the Mewat region in India emerges as a distinctive representation of its own.

This analysis contextualizes socio-economic cohesion within the framework of transformative participatory development by critically examining the notions of "community" and "participation." At the village level, the concept of "community" is far from uniform, encompassing a diverse and often competing set of caste groups and factions. These divisions highlight the challenges of fostering collective action in a socially fragmented environment. Similarly, "participation" in development is a dynamic and contested process shaped by the local power dynamics among various community social groups. It is influenced by the interplay of economic and noneconomic capital(s) possessed by different social groups, which ultimately determines the trajectory and outcomes of local development efforts.

The problematization of "community" and "participation" in this study is closely linked to the conceptual and analytical framework, which emphasizes the complexities of local power dynamics, social networks, and the role of political society in shaping development. By recognizing that "community" is not a cohesive entity but a collection of competing social groups, and that "participation" is a fluid process influenced by various stakeholders, the framework underscores the need for a critical understanding of development in the Mewat region. However, the study's limitations lie in the challenges of fully capturing these intricate dynamics, as local power relations and the impact of informal networks may not be fully observable or measurable within the constraints of this study.

Conceptualizing "Community" and "Participation" in Development

The conceptual framework is underpinned around "participation," "community," and development outcomes. The mainstream development literature offers many insights and highlights the role of local organizations in furthering development agenda. Majid Rahnema (1992) offers a critical overview of the concept of "participation" in development and shows how it is hijacked by various national and international organizations to further their agenda contrary to the promised role that it encapsulates and is expected to deliver to the intended communities. Bastian and Bastian (1996) assert that, in hindsight, nongovernmental organizations (NGOs) often find themselves trapped in a paradox. They aim to please donors, but, inadvertently, they may resort to tokenism and performative actions. This unintended consequence ultimately undermines the original objectives of their projects, leaving the beneficiaries at a disadvantage. Sangeeta Kamat (2002) delves into the complex dynamics of development hegemony in India, with a particular focus on the role of both NGOs and the state. Kamat's work is rooted in ethnographic research conducted among grassroots organizations working with the Bhil "community" in a western Indian state.

Stuart Corbridge et al. (2003) examine the multifaceted role of local governments in rural Bihar, Jharkhand, and West Bengal. They analyze the functions of these governments in development, empowerment, protection, and discipline, providing a comprehensive examination of aspects like "participation," governance, and political society. Their research highlights the intricate and often conflicting relationships among various state agencies, advocating for a closer examination of political society.

Nelson and Wright (1995) delve into the power dynamics of participatory development, seeking to uncover the shifts in power relations required for the successful implementation of participatory development strategies. John Harriss (2001) investigates the depoliticization of the concept of social capital and its detrimental effects on the participatory development paradigm. Anirudh Krishna (2002) delves deeply into the concept of social capital, how it can be applied in the Indian context, and its potential policy implications. Dwaipayan Bhattacharya et al. (2004) interrogate the concept of social capital and its applicability in the Indian context, with their work initially published in a special issue of Economic and Political Weekly in 2001, sparking an ongoing debate.

Craig Jeffrey and Jens Lerche, as featured in Fuller and Benei (2009), explore the state-society interface, shedding light on the often-overlooked public discourse of the state and the reproduction of social groups in rural areas. They examine social relations between wealthy farmers, the police, and access to government jobs. Bill Cook and Uma Kothari (2001) critique the notion of "participation" as a new form of tyranny in development discourse. They not only analyze why "participation" has taken this turn but also propose an agenda that extends beyond "participation" toward deeper empowerment.

Manoj Srivastava et al. (2002) analyze the confluence of rent-seeking behavior, vernacular society, and the Employment Assurance Scheme in Eastern India. Their study uncovers variations in the interpretation of government schemes by English-speaking and Hindi-speaking officials, primarily driven by concerns of potential corruption among junior government officials involved in labor-intensive programs (Corbridge et al., 2003; Gupta, 2012).

An additional significant concern related to "participation" is the lack of clarity surrounding the concept of "community." This concept remains ambiguously defined in academic literature on development, often carrying subjective interpretations (Nelson & Wright, 1995). The paper underscores that ideas about local institutions are frequently based on problematic notions of "community" and "solidarity." The mainstream participatory development theory assumes the existence of a cohesive "community" with clear boundaries (Kothari, 2001). This perception leads to the establishment of local development committees in the name of "community," often replacing traditional institutions. This assumption of a self-evident "community" overlooks the complex, shifting, and subjective nature of real communities as well as the fluidity of social and cultural boundaries.

Moreover, many participatory approaches prioritize village-based solidarity within communities while downplaying processes of conflict, negotiation, inclusion, and exclusion. This model of "solidarity" within "community" tends to ignore social hierarchies, unequal power dynamics, and entrenched religious beliefs, potentially excluding certain "community" members from the benefits of development. In reality, "community" can be both a site of solidarity and conflict, marked by shifting alliances, power struggles, and evolving social structures. Recent research on common property resource management recognizes the role of communities in managing internal conflicts, emphasizing the ever-changing nature of "community" institutions and the power dynamics at play in collective action.

The process flow diagram (Fig. 1) illustrates how participatory development integrates diverse themes for effective outcomes. State power (vernacular and resource market) interacts with existing social networks (civil society and everyday life) to form the foundation for development initiatives. Political networks bridge the gap between bureaucratic and grassroots levels, aligning formal and informal structures. These connections emphasize collaboration among technocrats, political entities, communities, and individuals. The framework facilitates data analysis by integrating quantitative and qualitative methods to evaluate these interactions. The insights gained from this analysis guide policy recommendations, promoting inclusive and sustainable development through coordinated "participation" in the region.

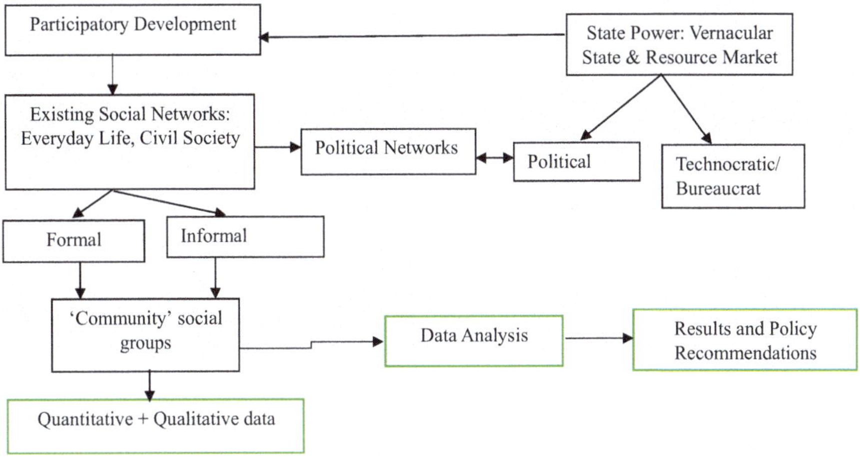

Fig. 1 Conceptual and analytical framework

Materials and Methods

In the initial phase of this research, the crucial information was gathered about the villages through a "community" schedule. This schedule featured a mix of open-ended and closed questions, exploring various aspects of participatory development and the village's social networks. It encompassed inquiries about demographic characteristics, available resources, livelihood opportunities, and involvement in development initiatives.

The key informant interviews, which encompassed both open-ended and closed formats, were conducted with individuals who played pivotal roles in formulating, designing, disseminating, and implementing development schemes at the state, district, block, and village levels.

Next, we carefully selected a subset of villages to undergo detailed surveying, including case studies and focus group discussions in Mewat region (Fig. 2). This selection was carried out through a stratified random sampling method in collaboration with the local "community." The chosen villages were Mohammadpur Ahir in Taoru, Ghasera in Nuh, Mandikhera in Nagina, Indana in Punhana, and Mandkola in Hathin Tehsil.

In each village, we selected "community" social groups across the sample villages. This selection method ensured a balanced representation of both economically disadvantaged and non-disadvantaged sections of society. For the economically disadvantaged group, we applied stratified random sampling to choose households from among the poor Meo, OBC (BCA, BCB), and SC categories. Non-poor households were also selected using stratified random sampling, with input from the local "community." The data collected is both quantitative and qualitative. The quantitative data was analysed with SPSS and qualitative data with MAXQDA. On triangulation, it was found that qualitative data supports the findings of quantitative

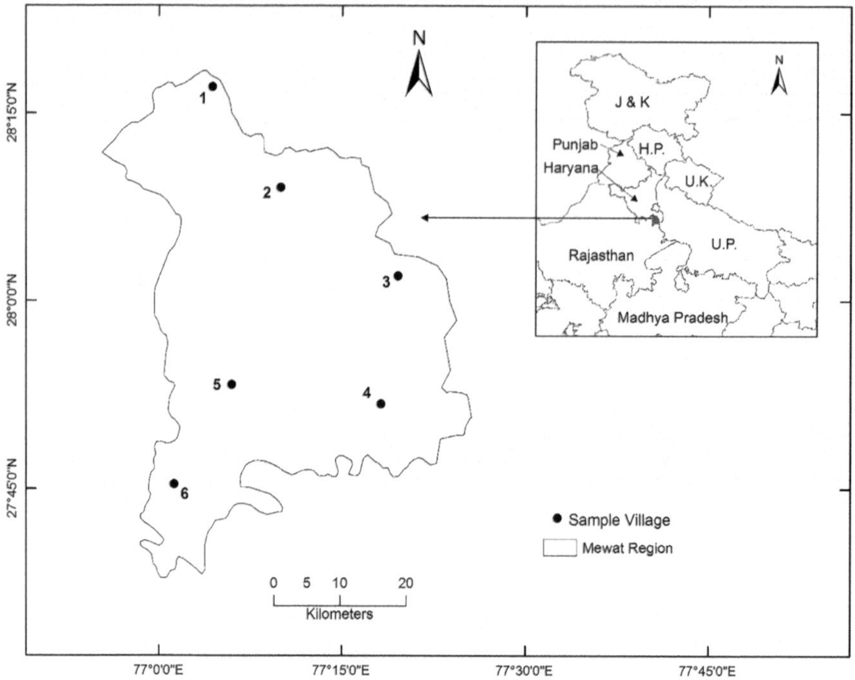

Fig. 2 Location of the study area and sample villages

data. It was found that social and political networks are contextual in nature. One region differs markedly from the other in several aspects. Any empirical measure of these networks has to be properly contextualized in terms of its social and political milieu. There is logic behind using variables in some cases and indicators in other cases meaning thereby that both measures help in understanding variation and magnitude of incremental changes occurring in the society. In analysing economic activities, two sets of methods were considered, (1) division by population (ii) division by households). Division by households was preferred because in rural areas, carrying capacity is a deciding factor on account of variable food supply in relation to population growth. Social segregation is also important as it is not a single unified and homogenous "community" that is the focus of the study but various castes and religion together which are in constant conflict with each other in terms of resource allocation and extraction at the meso and micro level. Economic aggregation is important. If there is no division, then it implicitly means that there is division/multiplication by unit value. In the case of abstract numbers being given as variables, because the population is not considered important here, that is why the number is being given as representative value. Hence in other activities, carrying capacity is not the criteria, availability is the criteria.

Results

Embeddedness of Economic Activities in Social Relations

Economic activities play a pivotal role in shaping social networks and nurturing organic connections within the local "community." These activities encompass a range of endeavors, including small-scale businesses initiated by various village communities. These small-scale businesses in the form of local shops not only facilitate economic transactions but also foster sociocultural interactions with the broader external world. The extent of this exchange is greatly influenced by the existing social and cultural relations among the village residents. Economic activities can be seen as an indirect measure of social cohesion, although their role in strengthening social bonds may be relatively limited. Caste dynamics exert a significant influence on these economic exchanges, with caste often being a determining factor in such interactions (Vijayabaskar & Kalaiyarasan 2014). One of the Sarpanch in the Gram Panchayat told that, "The Chamar community accuses the Jat community of obstructing development work in the village. A close associate of the sarpanch (village head) says that he used to have a tire puncture repair shop on the road. Due to repeated taunts from the Jat community in the village, his shop shut down. His shop was on the road, and now he rarely goes toward the road. The sarpanch's brother, Rakesh, says, "The villagers file an RTI (Right to Information request) or call the vigilance whenever they want." Recently, in an audit, the sarpanch was stripped of his duties due to alleged accusations of fraud in one of the village's development projects" (Field Survey, Mewat, Haryana, 2019).

In Mewat, economic activities predominantly revolve around ancillary functions related to consumption rather than production (Rathee, 1971). Additionally, some small-scale economic initiatives operate within the sample panchayats although their operations extend far beyond the village boundaries. Table 1 provides a breakdown of economic activities in the sample villages, revealing that Mandikhera Panchayat has the highest volume of economic activities, followed by Ghasera and Mohammadpur Ahir. Conversely, Mandkola and Indana report fewer cumulative economic activities compared to the other sample panchayats. These variations in economic activity scores can be attributed to differences in resource endowments among the panchayats. However, they do not fully explain the intricacies of the economic activities taking place.

Diverse economic activities reflect the multifaceted nature of economic life within the "community" and its sociability aspects. Notably, a significant number of sanitation workers come from the Valmiki caste, a subcaste within the SCs. This phenomenon is associated with the higher economic activities observed in the sample panchayats located near bustling economic hubs, industrial model townships, raw material supply centers, and seamless connections to urban centers like Gurugram, Faridabad, and the broader National Capital Region.

In Mewat, a majority of farmers hold marginal or small landholdings. Except for the eastern part near the Yamuna River, the region lacks adequate irrigation

Table 1 Economic activities in the sample village panchayats in Mewat

Type of economic activity	Sample panchayats M. Ahir (BCA)	Ghasera (BCB)	Mandikhera (SC)	Indana (BCB)	Mandkola (Others)
Grocery shop	2.08	0.42	1.25	0.42	0.83
Vegetable shop	0.61	2.44	1.34	0.37	0.24
Meat shop	0.94	0	1.88	0.63	1.56
Milk booth	1.11	0	1.11	1.67	1.11
Cloth shop	1.25	1.56	1.56	0.63	0
Tailoring shop	0.56	1.40	2.53	0.22	0.28
Wash shop	0	5	0	0	0
Cobbler shop	2.08	0	2.08	0.42	0.42
Barbershop	1.23	0.82	0.82	1.64	0.49
Carpenter shop	0.29	1.47	2.65	0.29	0.29
Welding shop	2.00	1.67	0.67	0	0.67
Mason	0.34	0.43	2.80	0.86	0.56
Moneylender	0	0.08	4.17	0.21	0.54
Dhaba	1.79	2.14	0.00	0.36	0.71
Milk seller	2.27	1.16	0.95	0.57	0.05
Agri. input shop	1.88	0.63	0.63	0.63	1.25
Safai workers	1	1.5	1.75	0.25	0.5
Total score	19.43	20.72	26.19	9.17	9.5

Source: Field Survey, Mewat, Haryana (2019)

facilities, limiting the full realization of its agricultural potential. Historical challenges, including recurring famines and sudden droughts, have severely disrupted the traditional agricultural economy. Although the Haryana Government has initiated region-specific schemes aimed at improving the financial condition of small and marginal farmers and alleviating their indebtedness, these efforts have not yielded the desired results due to various policy gaps (Kumar 2016).

The agricultural cycle in the region revolves around two main cropping seasons, Kharif and Rabi, featuring crops such as wheat, mustard, *chana*, *arhar*, onions, *jowar*, and *bajra*. However, due to the absence of proper irrigation facilities, some land becomes infested with termites and is rendered unsuitable for agriculture. This type of land is locally referred to as "*Bhur*" and is typically found in the foothills of the Aravalli Range, where Gujjars and Meos reside. Over time, the economic conditions of Gujjars have improved a lot, while Meos continue to face similar challenges. Although Gram Panchayats possess significant portion of land not suitable for cultivation. One of the Sarpanch told that: "The Gram Panchayat owns around 1014 acres of land out of which 953 acres is in the hills. The remaining 61 acres is illegally occupied by the ITC Company in which they have a golf resort and several farmhouses. Several of these farmhouses are owned by the Non-resident Indian who come only once or twice a year. They have appointed caretakers for upkeep and maintenance. A former Sarpanch of the Gram Panchayat allegedly sold some portion of land to this company. The Gram Panchayat has taken this matter to the court and the case is still going on" (Field Survey, Mewat, Haryana, 2019). Mining is one of the major economic activities and a source of livelihood to people in Mewat. Dumpers can be seen plying on the road near stone-crushing zones. However, very few mines are operating because of the ban imposed by the Supreme Court of India. The Aravalli Mountain Range is facing extinction due to the illegal mining and the uncontrolled use of detonators and explosives used.

Regrettably, the Mewat region still lacks a railway line, despite more than seven decades of independence, with Palwal being the sole exception. There have been sporadic efforts to establish micro and small industrial units by the Haryana State Industrial Development Corporation (HSIDC), but these initiatives have struggled to achieve the necessary economies of scale to stimulate broader industrial activities. The establishment of the *Rojka Meo* industrial model township represents an endeavor to promote industry and employment in the region, addressing some of the economic challenges faced by the local "community." The Golf Course by ITC Company in Taoru also offers employment to the nearby villages especially Mohammapur Ahir.

Civil Society, Social, and Religious Institutions

In a traditional society like Mewat, civil society plays a vital role in raising awareness about the significant challenges concerning education, healthcare, and overall social well-being. These civil society organizations serve as catalysts for empowering the local "community" by providing information and communication strategies, thereby helping residents realize their untapped potential. Moreover, they facilitate the establishment of connections between the "community" and government institutions, which are crucial for holistic development. These organizations also equip the "community" with technical expertise to harness their resources effectively, promoting greater "participation" in "community" development and the overall advancement of the village.

Table 2 provides insights into the number of self-help groups (SHGs) and non-governmental organizations (NGOs) in the sample panchayats and their respective activities. Mohammadpur Ahir, Mandikhera, and Indana each report ten active SHGs in their panchayats, all of which are linked to banks, ensuring their operational effectiveness. Besides offering small loans to their members, SHGs extend financial support for productive group activities such as establishing shops or purchasing sewing machines, fostering economic growth within member households, and inspiring others to follow suit. Many SHGs reported receiving revolving grants, which serve as initial funding for their activities. Organizations like the Institute for Research in Rural Development (IRRD) and S.M. Sehgal Foundation have played pioneering roles in water harvesting and awareness campaigns like Jal Chetna Yatra in Mewat (India Water Portal, 2010).

Table 3 displays the number and locations of religious centers in the sample panchayats, with mosques predominating due to Mewat's Meo population (Fig. 3).

Table 2 SHG and NGOs active in the village panchayats in Mewat

Name of the panchayat	Total SHGs	Bank linked (yes = 1, no = 0)	NGO active	Focus area	A common problem faced by NGOs
M. Ahir (BCA)	10	1	Spandana, Deepalaya	Education, healthcare, social well-being	Lack of government support
Ghasera (BCB)	6	1	Savera, Development Support Team	Health and sanitation	Inadequate funding
Mandikhera (SC)	10	1	Mewat Development Society (MDS), Sehgal Foundation	"Community" development, water management	Presence of "middlemen"
Indana (BCB)	10	1	MSEDS, Sehgal Foundation	Rural development, water management, "community" media	Lack of representation of NGOs in policy and decision-making structures
Mandkola (others)	8	1	MDS	"Community" development	Low "community" outreach

Source: Field Survey, Mewat, Haryana (2019)

Table 3 Number and location of religious centers in the village panchayats in Mewat

Religious center	Location of religious center	Total number of religious places			Social meeting place (*Pathak*)	
		Temple	Mosque	Others	Panchayat Bhawan	Marriage hall
M. Ahir	BCA	5	5	1	1	1
Ghasera	Meo (BCB)	1	21	0	0	0
Mandikhera	SC	5	5	1	0	0
Indana	Meo (BCB)	4	7	1	0	0
Mandkola	Others	0	5	0	0	0

Source: Field Survey, Mewat, Haryana (2019)

Additionally, local deities such as *Kheda Devta*, *Devi Ma*, *Lord Shiva*, *Shani Dev*, *Jaharveer Baba*, *Geetanand* Maharaj, *Chirshu*, *Dadi Chavand*, and *Baba Bhai* are worshipped in the Aravalli hills. Notably, Mohammadpur Ahir is the only village with a Panchayat Bhawan and a marriage hall, also known as a "community" *dharamshala*.

As per the latest report from the Aspirational District Programme of Niti Aayog, Mewat district ranks at the lowest rungs of development. The region grapples with inadequate infrastructure, low female literacy rates, larger household sizes, and a shortage of skilled manpower. The perception among the locals is that government

Fig. 3 Badi Masjid in Indana, Punhana, and Ghasera, Nuh, respectively

teachers and healthcare officials do not readily serve this underdeveloped area, as it holds limited political influence and stands geographically distant from the state capital. Mewat constitutes only 4% of Haryana's population. The local administration, often following political directives, does not inspire much confidence among the local population. Insufficient staffing and inefficiency further compound the challenges faced by the region.

Mewat's underutilized human resource potential, exacerbated by limited literacy levels, has hindered its developmental progress. High dependency ratios contribute to unemployment and underemployment, leading to severe economic hardships. For these reasons, the formation and growth of civil society in the region face impediments. Meos constitute the majority "community" in Mewat, followed by scheduled castes (SCs) and other communities. Over time, Meos have experienced persecution at the hands of various historical regimes, and they are not much liked by the bureaucrats, police, and local authorities (Amir Ali 1971; Maheshwari 2003). Even self-proclaimed "community" and political leaders of Mewat have at times betrayed their unique hybrid identity. This history of betrayal has left the "community" wary of the local administration's actions, often suspecting hidden motives. Memories of the 1947 communal tensions still linger in their minds. Tragically, some members of the media, law enforcement, and civil society harbor suspicions that certain members of the "community" have taken the wrong path to petty crime and dacoity (Kumar, 2016). Consequently, during episodes of communal tension in the region, the "community" is often the first to face scrutiny and targeting by officialdom and law enforcement agencies.

Political and "Community" Institutions

The significance of local "community" institutions in driving development outcomes is a well-established theme in the development literature (Jeffrey Craig & Jens Lerche 2000). In the context of Mewat, one such traditional "community" organization is the *Khap*[1] Panchayat, often referred to as the "*Badi* Panchayat," while the statutory Gram Panchayat is known as the "*Chhoti* Panchayat." The *Badi* Panchayat commands immense respect and reverence within the village. Its decisions are arrived at through consensus and are binding on all "community" members. Led by a Meo *Chaudhary*, the Badi or Pal Panchayat holds sway over decisions concerning various aspects of social life. However, with the advent of democratic elections and the establishment of statutory Gram Panchayats in villages, the influence of traditional institutions like the *Khap* has diminished. Additionally, the local bureaucracy often fails to address the concerns of communities regarding development bottlenecks.

Panchayati Raj Institutions (PRIs) have played a pivotal role in driving the development process in rural India, fostering "participation" in socio-economic activities. A noteworthy aspect is their role in offering opportunities to disadvantaged and marginalized sections to contest elections and become agents of social change. While some commentators argue that this democratic process can lead to factionalism, it also promotes effective opposition, exerting pressure on the Gram Panchayats to fulfil its duties more faithfully and work toward village betterment. However, a key challenge is the lack of sufficient panchayat funds to support development programs. The final authority to grant sanction of funds to Gram Panchayats rests with the Block Development Officer (BDO), and delays in fund disbursement can impede progress. Streamlining this process to expedite fund allocation is imperative.

Many of the sampled PRIs lack the proper infrastructure to conduct development activities and often resort to holding panchayat meetings at the Sarpanch's residence. In cases where the Sarpanch is a Dalit, upper-caste women panchayat members may hesitate to attend such meetings. If a Dalit or upper-caste woman holds the Sarpanch position, her husband frequently assumes her duties. This unequal treatment extends to her exclusion from Gram Sabha and other panchayat meetings with male members, perpetuating traditional social customs that conflict with her official role. This contradiction highlights a significant challenge in women's empowerment, as the government claims to empower them to participate equally in society, which does not always align with reality.

Table 4 pertains to "community" institutions, which play a crucial role in the lives of the local population and contribute to increased "participation" in the local development process. These institutions gain importance because they are formed by the people themselves, enabling them to better understand their needs and aspirations.

This table illustrates the structure and composition of Sarpanch and panchayat members in the sample Gram Panchayats, along with their associated

[1] It is the congregation of all caste panchayats in the region and elsewhere in North India. The members of these panchayats claim that they represent all castes. It is claimed that even members of the SCs are invited at these meetings. But it is doubtful whether they have any voice in decision-making or not. They are also known as Pal Panchayats.

Table 4 Structure and composition of sample Gram Panchayats in Mewat

Gram Panchayat	Presence of Panchayat Bhawan[a]	Panchayat members (*Panch*)									Sarpanch		
		SC		OBC (BCA)		Meo Muslim (BCB)		Others		Total	M	F	Education
		M	F	M	F	M	F	M	F				
M. Ahir	1	3	1	5	2	2	0	0	0	13	OBC		12th
Ghasera	0	2	1	0	1	11	3	0	0	18	Meo		10th
Mandikhera	0	0	3	0	2	10	1	0	0	16	SC		8th
Indana	0	2	3	1	0	3	2	0	0	11	Meo		8th
Mandkola	1	1	3	3	0	0	0	10	3	20		Others	10th

Source: Field Survey, Mewat, Haryana (2019)
[a] Indicates Yes = 1, No = 0

characteristics. Only two out of five panchayats report having a Panchayat *Bhawan*. Adequate representation of scheduled castes (SCs) and other backward classes (OBCs) is observed in Gram Panchayat bodies, and women are also well represented. Despite lower literacy rates, women are on par with men in terms of representation. However, it should be noted that many of the women's responsibilities are assumed by their husbands due to constraints related to meeting times and locations. In one of the Gram Panchayats, the Sarpanch (female) has been suspended pending an enquiry. The brother of her husband told that, "the villagers, especially individuals from the Jat community, are not allowing the *Sarpanch* to carry out development work. The *Chamar* community appears uncomfortable and intimidated in front of the dominant *Jat* community in the village. The gram panchayat has currently completed three and a half years, with one and a half years still remaining. According to some village elders, "The auditor's raid has affected several gram panchayats, and duties have been stripped from many. So, this hasn't happened here only" (Field Survey, Mewat, Haryana, 2019). In another Gram Panchayat, The Sarpanch told that, "under the government's welfare scheme, every Harijan in the village who does not own land has been allotted 100-square-yard plots from the panchayat land. The sarpanch claims that the blank registries of these plots have been made in the names of the Harijans. However, the mutation (*dakhilnama*) has not yet been done. When asked why the mutation hasn't been completed, the sarpanch says that the *Patwari* keeps delaying it by saying, "They will include it in this plan or that plan." According to him, people from the *Jat* community in the village are pressuring the *Patwari* not to carry out the mutation. Therefore, excuses are being made to buy time" (Field Survey, Mewat, Haryana, 2019).

Social Assets

Social assets play a pivotal role as they provide the "community" with the resources necessary to meet their fundamental needs, thus elevating their standard of living and overall social well-being. Table 5 offers insights into the social assets available in the sample Gram Panchayats across various communities in the region. Ghasera, Indana, and Mandkola Gram Panchayats report the presence of cooperative societies that provide credit facilities and facilitate the purchase of seeds and fertilizers. However, M. Ahir and Mandkola note that the cooperative society is located in nearby towns, requiring residents to commute for these facilities.

All the sample panchayats have access to primary, secondary, and senior secondary schools within their boundaries, ensuring educational opportunities for the local population. Ghasera stands out with the highest number of Anganwadi centers,

Table 5 Social assets of the panchayat villages in Mewat

	Availability (yes = 1, no = 0)				
Assets	M. Ahir (BCA)	Ghasera (BCB)	Mandikhera (SC)	Indana (BCB)	Mandkola (Others)
Cooperative society	B	1	A	1	1
Primary	1	1	1	1	1
Middle	1	1	1	1	1
Secondary	1	1	1	1	1
Senior secondary	1	1	0	0	1
Anganwadi center*	1 (5)	1 (11)	1 (5)	1 (3)	1 (4)
Post office	1	1	1	1	1
Fair price shop (depot)	1 (OBC, *Sarpanch* father)	1 (Meo)	1 (Meo)	1 (Meo)	1 (other)
Bank (nearest)	1	b (5–10 km)	a (0–5 km)	a (0–5 km)	c (10–15 km)
"Community" toilet	1	0	0	0	0
Common pastures	1	0	0	0	0
Drinking water	1	2	3	1	1
Primary health center	1	0	1	1	1
Railway station (nearest)	Haileymandi (25 km)	Palwal (25 km)	Gurugram (60 km)	Hodal (10 km)	Palwal (15 km)

Note: Figures inside the parenthesis () indicate the number of Anganwadi centers. Tap water supply = 1, Kunda = 2 (*Kunda* is a type of well constructed to store and preserve water for household use in Mewat.). Tap water supply and Kunda and *Mayur Jug* = 3 (*Mayur Jug* is a commercial water jug manufacturer company based in Delhi and having extensive operations in the Mewat region. They have submersible pump sets in select villages and supply water to the nearby villages.). A = 0–5 km distance. B = 5–10 km distance

Source: Field Survey, Mewat, Haryana (2019)

followed by Mohammadpur Ahir and Mandikhera. In contrast, Mandkola and Indana report relatively fewer centers compared to other sample panchayats.

Anganwadi centers[2] are categorized into two types based on caste composition: general and other (SC, BCA, BCB). These centers are staffed by Anganwadi workers and Anganwadi helpers and are typically housed in various types of buildings, including Panchayat *Bhawan*, one-room sets, and primary schools. Their primary functions include providing supplementary nutrition to adolescent girls, and pregnant and lactating mothers, as well as organizing awareness meetings on topics such as adolescent growth, menstruation, reproductive health, and nutrition. Immunization programs are also periodically conducted for the intended beneficiaries.

All the sample panchayats have access to post office facilities, which are vital for government communication and citizen-centric services, allowing them to send and receive letters and other items over long distances. Fair price shops, also known as ration shops, serve the needs of marginalized and deprived sections of society, particularly those belonging to SCs and OBCs. Notably, none of the individuals from the SC "community" and only one from the OBC category own fair price shops. These shops are typically licensed to be run by influential Meos and individuals from the general category. Senior secondary schools are available in only BCA and BCB, which shows that students especially girls may drop. Tap water supply is available to BCA, BCB, and others "community" group categories. There is no railway facility available in the region. "Community" pastures are declining day by day. Primary health centers are available to all, except BCB group.

Participation of "Community" and Political Society in MADP

The plan component of MADP includes various activities that encompass economic initiatives, components related to social development, and project management (Kumar, 2021). The economic activities within the program encompass soil and water conservation, canal irrigation, arable crops, horticulture development, livestock development, and off-farm enterprises.

For the success of any participatory project, it is crucial to examine both the intentions and goals of the scheme as well as the generation of project ideas. The initial step involves understanding how officials at various levels and in different state departments interpret the goals and intentions of the MADP. These interpretations, as presented by the local and state bureaucracies, often diverge in substance and subsequent actions. This contrast in perceptions between local and state

[2] It was reported by respondents in the field that there is separate Anganwadi for general category children and others like SCs and their location also differs as per caste. Both cannot mix with each other, and they have separate staff also.

bureaucracies has been highlighted in Akhil Gupta's influential study (2012) on Indian bureaucracy and its interactions with the general public.[3]

The second step revolves around the generation of project ideas, determining how the project should be structured in terms of design and execution. To foster the emergence of a project structure, it is imperative to engage in consultations and discussions at the "community" level, involving all segments of society. It is essential to emphasize the concept of "community" as a diverse and heterogeneous entity. Additionally, it is crucial to ensure that meetings take place, addressing whether there was proper attendance and "participation" in these meetings.

Before these meetings, it is necessary to inform the "community" about the meeting time and location to facilitate their attendance. Adequate dissemination of information is pivotal for project design and implementation. Furthermore, it is essential to consider whether these meetings are closed or open. Closed meetings tend to exclude "participation," while open meetings offer the potential for fair and transparent proceedings. It is important to determine whether these meetings resulted in the generation of ideas or merely followed a ritualistic and stage-managed approach.

Another critical aspect of such meetings is evaluating whether the ideas generated during these gatherings contributed to project design and implementation. In other words, did the meetings significantly influence the project? At a subsequent level, the dynamics involving political, administrative, and contractor roles in implementing project activities come into play. This also involves understanding the interactions between these stakeholders at various project levels, both locally and at the state level in the field. During the execution of project activities, contractors often employ various political linkages to expedite project tasks and sometimes exploit the situation for personal gain.

Impact of MGNREGA on Livelihood Security

The Mahatma Gandhi National Rural Employment Guarantee Act (MGNREGA) was introduced in 2008 in India as a rural employment and social security initiative. Its primary objective is to offer livelihood security to rural households by providing 100 days of employment for unskilled manual labor. However, the implementation of MGNREGA work in the study area has encountered numerous challenges, which have been extensively documented through interviews with various respondents. These issues have been incorporated into the field questionnaire, focusing on the availability of employment opportunities over the last 5 years, including during off-seasons.

[3] Even within the bureaucracy, the difference between the English-speaking and the vernacular bureaucracy is so stark and apparent. Moreover, majority of the officers like Deputy Commissioner, Superintendent of Police, and District Magistrate are not from the region, and there are frequent transfers and postings here. Mewat is considered a punishment posting.

Furthermore, the disbursement of wages under the program has faced criticism in recent years due to persistent delays in data entry of muster rolls and the apparent apathy of local administrations. Another crucial aspect is whether payments were made promptly after the completion of the work, considering that the workers often depend on these daily wages for their sustenance and other needs. How payments were disbursed, either separately or at a central location, is another aspect to investigate. Additionally, the conditions under which workers engage in MGNREGA work in mining and quarry areas within the region, particularly about environmental factors, warrant exploration.

The subsequent set of variables related to MGNREGA centers on its impact on beneficiaries (Table 6). These impacts encompass food security, protection against extreme poverty, household functioning, reduced distress migration, alleviated indebtedness, economic empowerment of women, and increased purchasing power capacity for beneficiary households. Assessing the well-being of the "community" in terms of living conditions is another essential facet. The condition of the primary village roads leading to the homes of the "community" holds significance as it reflects socio-economic interactions within the "community" and its external networks.

Various catalysts influence the well-being of the "community," such as increased agricultural production, job opportunities, and entrepreneurial activities. The relative economic status of the "community" can be categorized as wealthy, well-to-do, average, poor, or very poor. Additionally, trust levels among "community" members can be evaluated based on lending and borrowing activities. Informal sources of

Table 6 Impact of MGNREGA on beneficiary households in Mewat

Attributes	Responses of sample Gram Panchayats (yes = 1, no = 0)						
Food security	M. Ahir (BCB)	Ghasera (BCA)	Mandikhera (SC)	Indana (BCA)	Mandkola (Others)	Total	Beneficiaries' response
Protection from extreme poverty	1	0	0	1	0	2	1. Only partly 2. Somehow managing 3. Yes, but must migrate a small distance when work is not available 4. Late payment
Running family	1	0	1	1	1	4	
Reduction in distress migration	1	1	1	1	1	5	
Reduction in indebtedness	0	0	0	0	0	0	
Economic independence for women	0	0	0	1	1	2	
Purchasing power	1	1	1	1	1	5	

Note: Yes[1] and no[2] denote the number of Gram Panchayats saying yes or no to the said variable
Source: Field Survey, Mewat, Haryana (2019)

lending may include borrowing from local moneylenders in the village or *arhtiyas*[4] in the grain market where farmers sell their produce.

This social security and livelihood guarantee scheme plays a vital role in supporting landless households and those living at subsistence levels, especially those without robust social support systems. It serves as a lifeline for landless and marginal farmers, enabling them to earn additional income to sustain their families. In regions like Mewat, proper implementation of this program could significantly improve living standards. However, experiences from sample villages suggest that the program faces many challenges. Local Gram Panchayats unanimously acknowledge a shortage of sufficient funds and cite high labor costs as a significant challenge.

Some Gram Panchayats report providing only 26 days of wage labor in a single financial year, with work coming to a halt due to high demand and labor costs. Consequently, there is a heightened demand for accountability and transparency from "community" social groups, especially those in opposition to the Gram Panchayat regarding work allocation and social audits of the scheme. Confusion prevails regarding the functioning and meeting of Gram Sabha in the village, where they can demand work from the Gram Panchayat.[5] The lack of awareness among villagers about the provisions of the scheme complicates matters further.

Housing and Well-Being

In the past 5 years, determining whether the roads leading to the houses of the communities have improved, deteriorated, or remained unchanged is crucial. Housing availability within the "community" can be a critical variable, with deficiency or adequacy depending on the residents' living standards.[6] Evaluating the quality of housing, including whether it has improved, worsened, or stayed the same, is essential. These parameters collectively contribute to an assessment of whether the overall quality of life within the "community" has improved, worsened, or remained static compared to 5 years ago.

The state of "community" housing and the well-being of residents in sample Gram Panchayats reveal several key insights and challenges. Firstly, the type of roads leading to "community" houses shows that all Gram Panchayats in the sample

[4] *Arhtiyas* are commission agents or middlemen in the grain market or *mandi(s)* who facilitate the buying of produce of farmers to government agencies or wholesale agents under the Agricultural Produce Market Committee Act.

[5] Majority of the community social groups report that Gram Panchayat does not inform about the time and place of meeting of Gram Sabha. It is only in the meeting of Gram Sabha that they can demand work in MGNREGA.

[6] Although the state plays a crucial role in providing housing to the marginalized communities at subsidized rates under various schemes, it still remains a critical challenge with intergenerational change and fluctuating agricultural farm income in the region.

have concrete roads, except for Mandikhera and Indana, which have half-concrete roads. Notably, all respondents have reported road improvements in the last 5 years. In terms of housing, some Gram Panchayats, such as M. Ahir and Ghasera, have adequate housing, while others, including Mandikhera, Indana, and Mandkola, report deficiencies. One person from the Chamar community in Mandikhera, who is an associate of the *Sarpanch*, points toward the Harijan *Chaupal* and says that the work is still incomplete due to lack of funds. It remains unfinished. The main gate of the *chaupal* has not been installed yet, and the floor is still unpaved.

However, most Gram Panchayats have seen improvements in housing quality over the past 5 years, except for Mandikhera, where it has remained the same. The overall quality of life in these communities varies: none are considered wealthy, but Ghasera is regarded as well-to-do, M. Ahir as average, Mandikhera and Mandkola as poor, and Indana as very poor. In summary, these insights indicate that the conditions of housing and roads and the perceived quality of life in the surveyed Gram Panchayats in Mewat are mixed and call for government policy support.

Discussion

This chapter discusses various aspects of economic activities, social institutions, livelihood security through the MGNREGA, and overall well-being in the Mewat region of India. It offers a comprehensive understanding of the economic, social, and political landscapes of the area. In terms of economic activities, the chapter highlights the pivotal role these activities play in shaping social networks and fostering connections within the local "community." It emphasizes that these activities, often small-scale businesses initiated by village communities, not only facilitate economic transactions but also promote sociocultural interactions with the broader external world. The caste and religion dynamics in the region significantly influence these economic exchanges.

The economic activities are primarily centered on consumption rather than production, with limited resources and challenges such as inadequate irrigation facilities affecting agricultural potential. It also affects animal husbandry as it is a critical source of income. The government schemes aimed at improving the financial condition of small and marginal farmers have not yielded the desired results due to policy gaps.

Regarding MGNREGA, the program's impact on livelihood security and its challenges are highlighted. Delays in daily wage payments and issues related to project implementation are rampant. Civil society organizations have a crucial role and influence in raising awareness and empowering the local "community." In terms of social assets, the chapter provides insights into the presence of various institutions such as self-help groups (SHGs) and nongovernmental organizations (NGOs) in the sample panchayats. It discusses their roles in providing financial support and technical expertise for "community" development.

It also delves into the presence of religious centers in the sample panchayats and their significance in the region's social fabric. In the political sphere, the role of institutions, including traditional "community" organizations like the *Khap* and statutory Gram Panchayats as part of PRIs, is examined. The chapter highlights challenges such as inadequate infrastructure as well as support for PRI activities and issues related to gendered "participation."

Finally, the overall quality of life in the surveyed Gram Panchayats is discussed by taking into account factors like housing conditions and social infrastructure. It provides a detailed overview of the socio-economic and political landscapes in Mewat, shedding light on the challenges and opportunities faced by the communities. Overall, this chapter presents an analysis of the economic, social, and political dimensions of the Mewat region, highlighting the complexities and nuances of development and "community" well-being in this context.

Conclusion and Policy Recommendations

It can be concluded from the above discussion that the Mewat region grapples with complex socio-economic challenges, demanding a holistic regional approach that integrates economic development with social cohesion and solidarity. Limited agricultural productivity, historical issues, and resource constraints hamper progress. The MGNREGA scheme requires streamlining to ensure timely wage payments and efficient execution while fostering socio-economic solidarity. Civil society organizations play a pivotal role in raising awareness and connecting communities with government institutions, necessitating support for capacity building. Equitable access to vital social assets is paramount for inclusive development. Empowering women by giving them due recognition in local governance bodies is essential, and infrastructure improvements, especially road connectivity, are critical for socio-economic solidarity. Robust monitoring, evaluation, and "community" engagement are fundamental to drive socio-economic progress.

In light of the intricate socio-economic landscape in Mewat, a comprehensive set of policy support is vital to address the region's multifaceted challenges and promote socio-economic solidarity. First and foremost, agricultural support schemes should be tailored to the specific needs of small and marginal farmers, focusing on irrigation infrastructure and diversification to enhance agricultural productivity.

To strengthen the role of civil society in promoting awareness and connecting communities with government institutions, capacity-building support should be provided to these organizations because CBOs outside the region are not familiar with the complex socio-economic realties of the region. Equitable access to critical social assets should be prioritized, extending their reach to remote areas to promote socio-economic solidarity. Women's "participation" in local governance should not only be actively encouraged but also be facilitated, and fund allocation to Gram Panchayats expedited to bolster socio-economic solidarity efforts.

Investment in road infrastructure is essential, as improved connectivity facilitates economic interactions and promotes socio-economic solidarity across the region. To ensure that development programs effectively contribute to the well-being of the "community," a robust monitoring and evaluation system should be established. Finally, active "community" engagement through open meetings and participatory processes will be instrumental in facilitating inclusive development and enhancing socio-economic solidarity in Mewat. These policy recommendations, when implemented comprehensively, can pave the way for a more equitable and prosperous future for the region.

Acknowledgments This research is the result of my postdoctoral research work funded by the ICSSR, India, at Jawaharlal Nehru University, New Delhi, India, from 2018 to 2020 period. I sincerely thank my supervisor and other faculty members at JNU for providing feedback during the postdoctoral seminar in 2019. All the due research ethics were undertaken in the completion of primary fieldwork, and data was anonymized to protect the identity of the respondents.

References

Ali, Hashim Amir., Mohammad Rafiq Khnan., & Om Prakash Kumar (Eds.) (1971) The Meos of Mewat: Old Neighbours of New Delhi. Oxford university Press.

Bastian, S., & Bastian, N. (Eds.). (1996). *Assessing participation: A debate from South Asia*. Konark Publishers.

Bhattacharya, D. (Eds.). (2004). Interrogating social capital: *The Indian experience*. Sage.

Bourdieu, P. (1986). The forms of capital. In J. G. Richardson (Ed.), *Handbook of theory and research for the sociology of education, translated from the French by Richard Nice* (pp. 241–258). Greenwood.

Chhibber, A. (2000). Social capital, the state, and development outcomes. *Social Capital. A Multifaceted Perspective*, 296–309.

Chhotray, V. (2007). The 'anti-politics machine' in India: Depoliticization through local institution building for participatory watershed development. *The Journal of Development Studies, 43*(6), 1037–1056.

Cooke, B., & Kothari, U. (Eds.). (2001). *Participation: The new tyranny?* Zed books.

Corbridge, S., Williams, G., Srivastava, M., & Veron, R. (2003). Making social science matter-I: How the local state works in rural Bihar, Jharkhand and West Bengal. *Economic and Political Weekly, 38*(24), 2377–2389.

Dasgupta, P., & Serageldin, I. (Eds.). (2000). *Social capital: A multifaceted perspective*. World Bank Publications.

Ferguson, J. (2006). The anti-politics machine. *The Anthropology of the State: A Reader*, 270–286.

Fuller, C. J., & Benei, V. (2009). *The everyday state and society in modern India*. Social Science Press.

Gupta, A. (2012). *Red tape: Bureaucracy, structural violence, and poverty in India*. Duke University Press.

Harriss, J. (2001). *Depoliticizing development: The World Bank and Social Capital*. LeftWord.

Hickey, S., & Mohan, G. (2004). *Participation: From tyranny to transformation*. Zed Books.

India Water Portal (2010). *Jan Chetna Yatra, S M, Sehgal Foundation, Mewat, Haryana*. Retrieved 12 September, 2023, from https://www.indiawaterportal.org/events/jal-chetna-yatra-s-m-sehgal-foundation-mewat-haryana.

Jeffrey Craig and Jens Lerche. (2000). Dimensions of dominance: Class and state in Uttar Pradesh. In C. J. Fuller & V. Bénéï (Eds.), *The everyday state and Society in Modern India* (pp. 91–114). Social Science Press.

Kamat, S. (2002). *Development hegemony: NGOs and the state in India*. Oxford University Press.

Kothari, U. (2001). *Power, knowledge and social control in participatory development. Participation: The new tyranny?* pp. 139–152.

Krishna, A. (2002). *Active social capital: Tracing the roots of democracy and development*. Columbia University Press.

Krishna, A. (2004). Understanding, measuring and utilizing social capital: Clarifying concepts and presenting a field application from India. *Agricultural Systems, 82*, 291–305.

Krishna, A. (2007). How does social capital grow: A seven year study of villages in India? *Journal of Politics, 69*(4), 941–956.

Kumar, N. (2016). *Social identities and cultural capital among Meos and other social groups in Mewat, Haryana*. PhD. Thesis submitted to Jawaharlal Nehru University, New Delhi.

Kumar, N. (2021). *Participatory development and state: A case study of beneficiaries under Mewat area development project, Haryana*. Final post-doctoral report submitted to Indian Council of Social Science Research, New Delhi.

Maheshwari, B. (2003). The Meos of Mewat: *perspectives on ethnicity and nation building*. Social Change, 33(1), 81–91.

Nelson, N., & Wright, S. (1995). *Power and participatory development: Theory and practice*. ITDG Publishing.

Putnam, R. (1995). Bowling alone: America's declining social capital. *Journal of Democracy, 6*(1), 65–78.

Putnam, R. (2000). *Bowling alone: Collapse and revival of the American community*. Simon & Schuster.

Putnam, R. D. (2007). E pluribus Unum: Diversity and community in the 21st century the 2006 Johan Skytte prize lecture. *Scandinavian Political Studies, 30*(2), 137–174.

Putnam, R. D., Leonardi, R., & Nanctti, R. Y. (1993). *Making democracy work: Civic traditions in modern Italy*. Princeton University Press.

Rahnema, M. (1992). Participation. In W. Sachs (Ed.), *The development dictionary: A guide to knowledge as power* (pp. 116–131). Zed Books.

Rathee, C. R. (1971). *Meos of Mewat: An economic analysis*. Mainstream Annual.

Srivastava, M., Corbridge, S., Veron, R., & Williams, G. (2002). Making sense of the local state: rent-seeking, vernacular society and the *Employment Assurance Scheme in eastern India*. Contemporary South Asia, 11(3), 267–289.

Vijayabaskar, M., & Kalaiyarasan, A. (2014). *Caste as social capital*: The Tiruppur story. Economic and Political Weekly, 34–38.

Where and in Which Aspect to Care? Insights from Spatial Variation and Levels of Human Development in Bardhaman District of West Bengal

Tamali Halder

Introduction

Regional disparity is the outcome of unbalanced regional development that varies from region to region depending upon several characteristics—economic, sociocultural, and demographic. Spatial unit of various scales impacts the level of development and its response to different developmental measures, as each unit has its own cultural and socio-economic framework. It is observed that certain regions have an edge on specific parameters over other areas owing to the development of human resources and capital. Still an unfolding concept, development refers to a change where there is an upward movement of an entire social system, which means overcoming persistent poverty and achieving human development that is supposed to improve the lives of people. Coccia (2019) discusses that the pivotal elements of development in a society are the growth rate of wealth, improvement in the health of the society, and encouraging and ensuring the growth of technology and knowledge in society. From the economic framework, the concept focuses on either capital accumulation, diminishing returns, or appropriation of a nation's wealth. From the social point of view, the term "development" is understood in a social framework where the needs of the people from different strata of society are satiated by the proper use of available resources in such a manner that there is equity and equality in the distribution of the facilities like healthcare, housing, education, transportation, and other infrastructure. Nothing can be seen in isolation, and so is the concept of human development, under study.

T. Halder (✉)
Shahid Matangini Hazra Government General Degree College for Women,
Purba Medinipur, West Bengal, India

Centre for Regional Studies, University of Hyderabad, Hyderabad, Telangana, India

K. Arora, R. R. Anand (eds.), *The Solidarity Approach in Geography*,
https://doi.org/10.1007/978-3-031-86753-8_23

The late Pakistani economist Mahbub ul Haq in collaboration with the economist and Nobel laureate Amartya Sen devised human development, which ushered in a new way of capturing development. They aimed at capturing the social and economic facets of development. The purpose was to view it from a different perspective other than compartmentalizing it within the framework of income levels or economic growth rate. As per the UNDP 1990 reports, HDI was anchored upon the premise that human development revolves around three fundamental aspects of human life, primarily health, defined by the longevity of life, knowledge, and decent living standards. The year 1980 witnessed a change of development from being represented by the accumulation of capital (during the 1960s by dependency theory) to capacitation, that is, enlargement of people's choice, as shown by Amartya Sen (1985) in his report on human development. The report focuses on making the index show levels of various facets of human development.

Human Development Index (HDI), as explained from 1990 to 2010, comprised life expectancy at birth, which indicates a long and healthy life; knowledge characterized by two aspects: the adult literacy rate and the combined gross enrolment ratio for primary, secondary, and tertiary schools; and GDP per capita indicating the decent standard of living which in turn measures the purchasing power parity measured in terms of US dollars. From 2010, the parameters for measuring the three indicators changed. Life expectancy at birth was used to measure long and healthy life, which was similar to what was defined in 1994. Mean years of schooling of the population of youth aged 25 and over and expected years of schooling for children were used to measure knowledge. Gross national income (GNI) per capita adjusted by purchasing power parity (PPP) was used to measure the standard of living.

The major criticism of the formulation of HDI is that very little information or knowledge is present about its application at the sub-national level (Mohanty & Dehury, 2012) because the data of the sub-indices used to measure the three primary parameters is unavailable at the micro level or the community development block level. Instead, data on other sub-indices are available, which are equally essential to be considered. In this context, an attempt has been made to demonstrate the spatial pattern of development levels and recognize the inter-block disparity in Bardhaman district using the methodology of HDI.

Objectives

The main objectives of the current work are:

1. To investigate the block-wise disparity in terms of social well-being, relative prosperity, and livelihood pattern that make up the human development
2. To locate the highly, moderately, and lowly developed CD blocks

Study Area

Bardhaman district comprises two main divisions. The western part of the district, or the Paschim Bardhaman, is famous for urbanization and industrialization. In contrast, the eastern part, Purba Bardhaman, is primarily agricultural (District Human Development Report, 2011). The level of analysis of the present research is at the community development (CD) block because it is the primary administrative setting that can effectively implement any plan, and any spatial unit that is smaller in extent like Gram Panchayats or more extensive area like districts as a whole might create trouble in gathering the data and implementing the desired plan. Bardhaman district stands in fifth position regarding the Human Development Index (Government of West Bengal, 2004, p. 13). A detailed analysis has been done at the block level to understand the spatial variation of the Human Development Index, which further gives an insight into the areas lagging in health, education, and standard of living (Fig. 1).

Database and Methodology

The entire work surrounds the secondary source extracted from the District Statistical Handbook, Bureau of Applied Economics and Statistics, Government of West Bengal, of the period 2015–2016. The district of Bardhaman (or the then-called Burdwan) was officially bifurcated into the east or Purba Bardhaman and

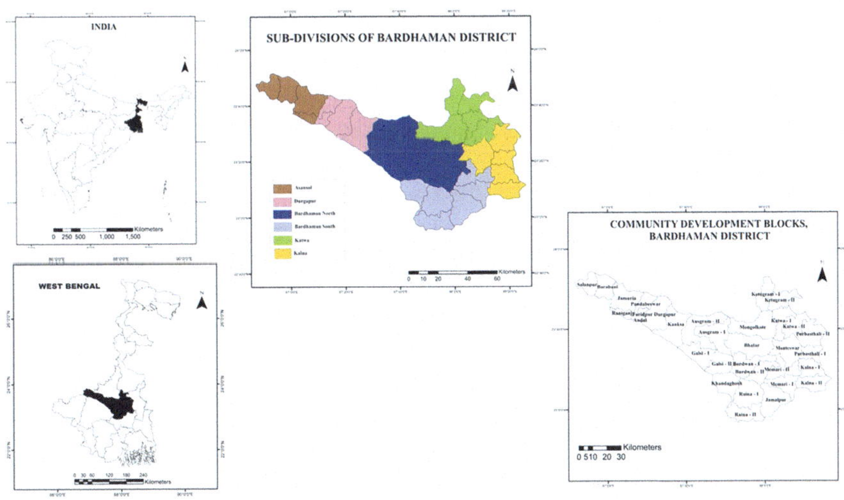

Fig. 1 Location map of the study area

west or Paschim Bardhaman on 7th April 2017. The latest data available is that of 2015–2016 and herein lays the limitation of the work as the present work is done based on the years 2015–2016. The latest data of Paschim Bardhaman district is yet to be published on the grounds of administrative reasons after the district's bifurcation. So, the current work entails working on the whole of the district using the data of 2015–2016. The quantitative method has been used to surmise and interpret the different levels of development in the district. A range method of normalizing values similar to the one employed by the UNDP in capturing HDI has been applied. A brief description of the same is required. A country's or region's average achievement is reflected in the average composite or index values derived from using an established minimum and a maximum value for each selected dimension. These chosen highest and lowest values are termed "goalposts." Such benchmark values help to locate a country or region's position compared to other areas. The final values derived from standardizing each variable representing health, standard of living, and education range from 0 to 1. The current work employs the same technique to see the HDI at the CD block.

Analysis

Health Care Facility Index (HCFI)

Since HDI focuses on three main aspects, health, knowledge, and income, this work also focuses on these issues at the CD block level. However, the parameters used to reflect these three aspects vary from what has been summarized in the UNDP report. For example, life expectancy is used as the parameter to explain the healthy development of any nation, but at the CD block level, such data is unavailable. But other dimensions have been taken up for analysis. Six dimensions, that is, bed population ratio (BPR), health care institution population ratio (HCIPR), doctor population ratio (DPR), doctor health care institution ratio (DHCIR), bed health care institution ratio (BHCIR) (Basak & Siddique, 2020, p. 171, 172), and health centre density (Ashraf et al., 2013, p. 144), have been taken up for analysis. The description of each of these dimensions, the reason behind choosing it, and the formulae used to reach the health care index needed are as follows:

Bed population ratio (BPR): In many developed and developing countries, hospital capacity planning is an important issue that is being dealt with due to the surge in the costs of inpatient care, limited resources, and growing demand of the increasing population for hospital care (Ravaghi et al., 2020). In many countries, indices like bed occupancy and ratio of beds to population are considered essential to determine the extent of inpatient care. The number of beds indicates the capital stock of the hospital (Rechel et al., 2010). In any case, both the excess and shortage of beds in a hospital can lead to a crisis. When there is a shortage of beds, the functioning of the hospital comes to a standstill where the surgeries are postponed or new patients (both critical and non-critical) are not admitted, thereby lowering the productivity of the institute (Green, 2002). In a scenario with excess beds, the resources are underutilized, and additional costs have been borne, which could have been diverted to other urgent procedures. Therefore, population is considered the primary factor in planning the hospital bed capacity. In the current work, all types of hospital beds, whether public or private, have been considered for analysis. Keeping this in mind, the ratio of beds per 10,000 populations in a particular spatial unit (in this case, a community development block) has been extracted. The following formula then represents it:

$$\text{Bed population ratio}\left(\text{BPR}\right) = \frac{\text{Number of beds}}{\text{Total population}} \times 10{,}000$$

Health care institution population ratio (HCIPR): A health care facility is an essential factor in determining an individual's health. Their longevity and quality of life determine a person's health. Hospitals play a vital role in ensuring that the person lives a happy and long life, as observed in the cases of severe or critical patients dealing with life-threatening injuries and diseases. It is here that the hospitals in an area play a crucial role. Under situations of limited hospitals in areas during critical times as evincible from the COVID-19 pandemic, scenarios like overcrowding emerge, and the staff often neglects many patients in critical conditions to tend to overtly critical patients (Yagma et al., 2020). Therefore, the dimension of HCIPR has been taken for analysis because strategic planning of several hospitals at the regional level is essential when accurately predicting long-term regional needs (Ravaghi et al., 2020). In this case, all types of health care institutes, private or public, have been considered or studied, and the formula to represent the same is

$$\text{Health care institution population ratio}\left(\text{HCIPR}\right) = \frac{\text{Number of health care institutes}}{\text{Total population}} \times 100{,}000$$

Doctor population ratio (DPR): For developing nations, the inadequacy of health workers was considered a significant hurdle in "reducing under-5 mortality rate by two-thirds and maternal mortality rate by three-quarters" between 1990 and 2015 as per the Millennium Development Goals (Deo, 2016). The number of doctors per population plays a crucial role in the longevity of the life of a people. In an economy, if the doctors are fewer, then the issues of the patients are not attended to, resulting in the patient's untimely death without proper diagnosis and prognosis. Literature suggests that a low physician population ratio contributes to the short life expectancy, as evincible from the case of Haiti (Fournier & Dodard, 1997). On the other hand, the state health ranking of the USA improved under better conditions of doctor population ratio (Bigbee, 2008). Such a condition is prevalent in Scandinavian countries as well. In Norway, the average life expectancy of people is 81.53 years, which is significantly correlated with the ideal number of physicians. In Europe, the highest densities of physicians are found in Norway (Ringard et al., 2013). Since the availability of doctors bears a significant effect on the correct and timely treatment of patients, the availability of doctors per 100,000 population has been considered by using the following formula:

$$\text{Doctor population ratio}\left(\text{DPR}\right)=\frac{\text{Number of doctors}}{\text{Total population}}\times 100,000$$

Doctor health care institution ratio (DHCIR): The availability of doctors in the institutes (public or private) per a defined population set constitutes DHCIR. It is of great concern when a hospital is understaffed with doctors or other medical practitioners. When there is a shortage of doctors, the functioning of the hospital comes to a standstill where the surgeries are postponed or new patients (both critical and non-critical) are not admitted, thereby lowering clinical efficiency (Green, 2002). In India, the public health care system is dismal due to low funding and constrained manpower (Aggarwal, 2019, p. 221). Also in private hospitals, to curtail the cost, sometimes fewer doctors are hired, and it increases the burden on the limited doctors who work with overflowing cases of patients. The adequate availability of doctors in an institute can go a long way as medical students are tutored under different doctors with different specializations and the probability of a huge influx of patients getting care increases (Dandavino et al., 2007). Thus, the formula used to represent DHCIR is

$$\text{Doctor health care institute ratio}\left(\text{DHCIR}\right)-=\frac{\text{Number of doctors}}{\text{Number of medical institutes}}$$

Bed health care institution ratio (BHCIR): This dimension measures the available bed capacity in hospitals. As discussed in the case of BPR, the number of hospital beds also indicates clinical productivity and efficiency. The availability of beds increases the probability of inpatient care, and there will be fewer instances of patients waiting in a queue outwards (Ravaghi et al., 2020). The formula used to calculate BHCIR is as follows:

$$\text{Bed health care institute ratio}\left(\text{BHCIR}\right)=\frac{\text{Number of beds}}{\text{Number of medical institutes}}$$

Health centre density (HCD): Quality of life depends on health to a considerable extent which is dependent on the location of hospitals, clinics, and health care centres. In critical cases, it is observed that the nearby location of medical institutes plays a key role in saving critical patients as they can get admitted on time followed by proper treatment (Roy, 2009, p. 81). For our purpose, number of health care institutes per 10 km^2 has been taken up for analysis, and the formula is

$$\text{Health centre density}\left(\text{HCD}\right)=\frac{\text{Number of health care institutes}}{\text{Area of the CD block}}\times 10$$

Health care index (HCI): The health care index is a composite of the above-mentioned six dimensions—bed population ratio (BPR), health care institution population ratio (HCIPR), doctor population ratio (DPR), doctor health care institution ratio (DHCIR), bed health care institution ratio (BHCIR), and health centre density (HCD). In each of these dimensions, the range methods proposed by the UNDP have been employed to reach a scale-free number. Upon reaching a scale-free number, the average of all six dimensions has been calculated, and the answer is the health care index in Paschim Bardhaman district. The formula used to calculate HCI is

$$\text{Health care index}\left(\text{HCU}\right)=\frac{\text{BPR}+\text{HCIPR}+\text{DPR}+\text{DHCIR}+\text{BHCIR}+\text{HCD}}{6}$$

The HCI has been taken as an average of six dimensions' normalized or standardized values. HCI calculated for different CD blocks shows an uneven pattern. The values of HCI along with the standardized values of each table have been listed in Table 1.

The overall scenario of health care facilities is mostly lowly developed, and significant regional variation is found within the district. Suppose one delves into the scenario of the HCI at the subdivision level. In that case, Asansol subdivisions occupy the first position with an average (mean of HCI values prevailing at the CD blocks falling under the Asansol subdivision) HCI value of 0.3758. Bardhaman Sadar North subdivision occupies the second position, which means on average, the health care facility development stands at 33.15% as the HCI value is 0.3315. Katwa subdivisions acquire the lowest position indicating that the condition of the bed population ratio, health care institution population ratio, doctor population ratio, doctor health care institution ratio, bed health care institution ratio, and health centre density is dismal in the CD blocks falling under it. The western portion of the district is more developed than the eastern part. At the block level, the highest HCI value is prevalent in the CD block of Burdwan I about 0.7921, located on the eastern side of the district. In contrast, the dismal condition prevails in Pandabeshwar, located on the western side or a part of the district now known as Paschim Bardhaman district (with the official bifurcation of the whole of Burdwan district in the year 2017).

The implication of the result (Table 1) is straightforward: Burdwan I located in the south-eastern corner of the district in the Burdwan North subdivision has achieved approximately 80% of development in the health arena with 20% still left to work on. Such success in the CD block of Burdwan I can be attributed to its first position in dimensions like health care institution population ratio, bed population ratio, and doctor population ratio. Likewise, very highly developed condition prevails in Burdwan II in the same subdivision, Salanpur (occupies the highest rank in doctor health care institution ratio), and Andal is located on the western side (Fig. 2). The lowest value of HCI (0.575) is found in the CD block of Pandabeshwar falling under the Durgapur subdivision as it occupies the last rank owing to poor development in case of bed population ratio, health care institution population ratio, doctor

Table 1 Health care index values at CD block level, Bardhaman district, 2015–2016

Subdivisions	CD blocks	X1_HCIPR	X2_BPR	X3_DPR	X4_DHCIR	X5_BHCIR	X6_HCD	Health care index	Rank
Asansol subdivision	Salanpur	0.354	0.493	0.643	1.000	0.875	0.237	0.600	3
	Barabani	0.473	0.089	0.182	0.209	0.093	0.195	0.207	10
	Raniganj	0.907	0.061	0.082	0.029	0.000	1.000	0.347	6
	Jamuria	0.962	0.108	0.372	0.212	0.031	0.412	0.349	5
Durgapur subdivision	Andal	0.300	0.262	0.358	0.640	0.529	0.419	0.418	4
	Faridpur-Durgapur	0.328	0.024	0.365	0.604	0.023	0.107	0.242	9
	Pandabeswar	0.000	0.000	0.000	0.173	0.172	0.000	0.057	31
	Kanksa	0.578	0.117	0.019	0.000	0.101	0.177	0.165	17
Burdwan North subdivision	Galsi I	0.237	0.103	0.117	0.259	0.248	0.064	0.171	16
	Burdwan I	1.000	1.000	1.000	0.594	0.676	0.483	0.792	1
	Burdwan II	0.534	0.807	0.736	0.793	1.000	0.222	0.682	2
	Ausgram I	0.316	0.052	0.165	0.281	0.083	0.054	0.158	18
	Ausgram II	0.389	0.049	0.225	0.317	0.053	0.045	0.179	15
	Bhatar	0.280	0.072	0.119	0.227	0.142	0.063	0.151	19
	Galsi II	0.243	0.123	0.130	0.281	0.291	0.056	0.187	14
Burdwan South subdivision	Memari I	0.193	0.124	0.200	0.518	0.363	0.115	0.252	8
	Memari II	0.237	0.049	0.077	0.173	0.113	0.078	0.121	24
	Jamalpur	0.233	0.071	0.188	0.419	0.172	0.114	0.199	11
	Raina I	0.376	0.145	0.287	0.419	0.223	0.112	0.260	7
	Raina II	0.463	0.025	0.125	0.142	0.002	0.143	0.150	20
	Khandaghosh	0.234	0.062	0.194	0.432	0.148	0.060	0.188	13

Subdivisions	CD blocks	X1_HCIPR	X2_BPR	X3_DPR	X4_DHCIR	X5_BHCIR	X6_HCD	Health care index	Rank
Katwa subdivision	Mongalkote	0.192	0.039	0.119	0.317	0.113	0.043	0.137	22
	Ketugram I	0.138	0.010	0.045	0.173	0.053	0.036	0.076	30
	Ketugram II	0.416	0.157	0.103	0.129	0.220	0.146	0.195	12
	Katwa I	0.196	0.009	0.042	0.119	0.023	0.093	0.080	29
	Katwa II	0.182	0.017	0.086	0.245	0.053	0.057	0.107	27
Kalna subdivision	Purbasthali I	0.152	0.029	0.049	0.173	0.113	0.116	0.105	28
	Purbasthali II	0.201	0.028	0.065	0.173	0.077	0.109	0.109	26
	Kalna I	0.152	0.029	0.104	0.335	0.113	0.093	0.138	21
	Kalna II	0.274	0.031	0.111	0.216	0.053	0.130	0.136	23
	Monteswar	0.221	0.038	0.072	0.173	0.093	0.065	0.110	25.000

Source: Computed and compiled by the author

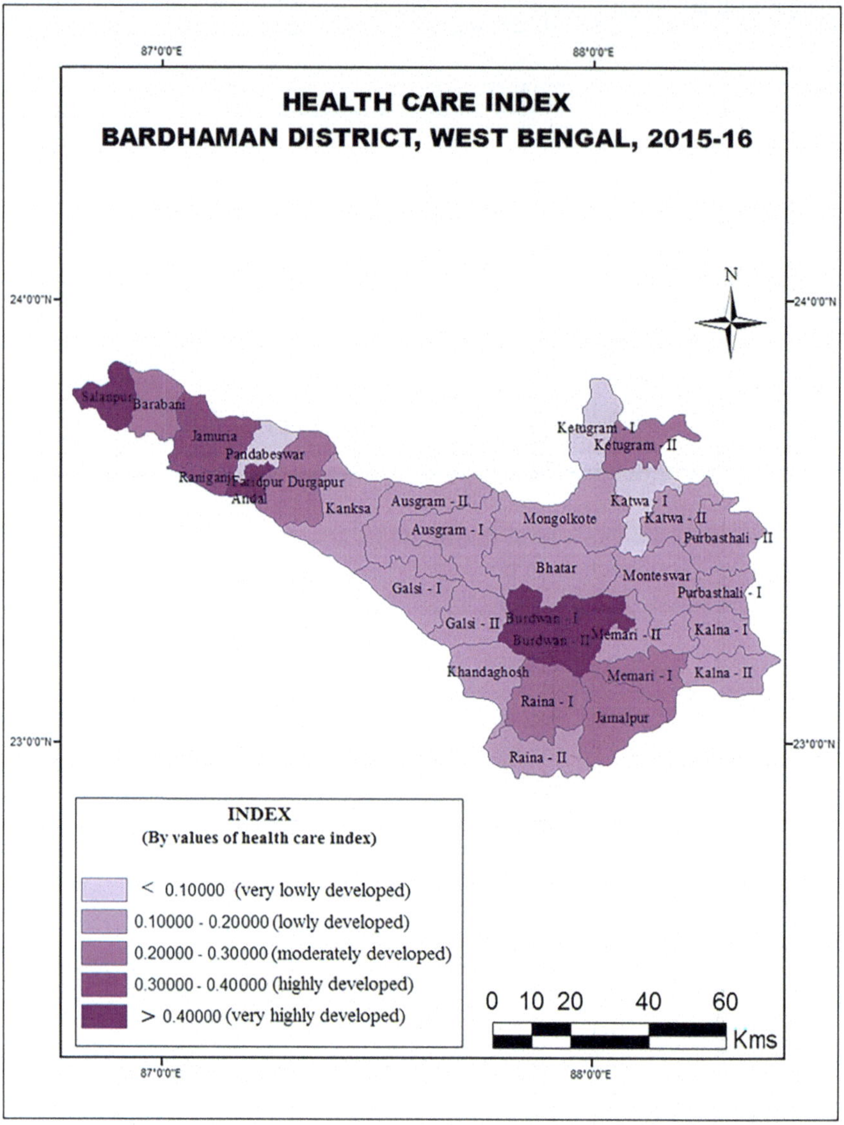

Fig. 2 Health Care Facility Index

population ratio, and health centre density, followed by Ketugram I and Katwa I of Katwa subdivision. Apart from Raniganj, no other CD block has even one health care institute within a vicinity of 10 km^2. The Planning Commission of India has suggested that the DPR's benchmark or ideal situation should be 1:1000 (Zodpey et al., 2018). Still, at the district level, it can be inferred that none of the CD blocks have achieved this critical limit of DPR. Kanksa, located within the Durgapur subdivision on the district's western side, occupies the lowest position in terms of DHCIR. Similarly, Raniganj located within the Asansol subdivision occupies the lowest rank in BHCIR.

The CD blocks have been classified into five groups for the sake of understanding of different levels of development in HCI, which are as follows:

Very lowly developed (below 0.10000): Katwa I, Ketugram I, Pandabeswar
Lowly developed (0.10000–0.200000): Jamalpur, Ketugram II, Khandaghosh, Galsi II, Ausgram II, Galsi I, Kanksa, Ausgram I, Bhatar, Raina II, Kalna I, Mongalkote, Kalna II, Memari II, Monteswar, Purbasthali II, Katwa II, Purbasthali I
Moderately developed (0.200000–0.30000): Raina I, Memari I, Faridpur-Durgapur, Barabani
Highly developed (0.30000–0.40000): Jamuria, Raniganj
Very highly developed (more than 0.40000): Burdwan I, Burdwan II, Salanpur, Andal

Educational Attainment Index (EAI)

Coming to the education aspect, the UNDP employs the dimensions of mean years of schooling of the population of youth aged 25 and over and expected years of schooling for children to measure the levels of development of knowledge. But at the CD block level, these two dimensions' data are unavailable. Therefore, the Educational Attainment Index (EAI) has been calculated from the other available block-level parameters describing the educational scenario of the district. A total of five dimensions are used to reach at Educational Attainment Index (EAI). So an explanation of each dimension followed by the reason for considering it has been detailed in the ensuing paragraphs. UNDP methodology has been used to evaluate adult literacy and enrolment indexes, which in turn have been combined to formulate the education attainment index. For the analysis of data at the CD block, the data of the percentage of students enrolled in primary, middle, secondary, and senior (higher) secondary schools were already documented in DSHB, and the same has been used for analysis.

Percentage of students enrolled in primary schools: Universally, primary education is considered and accepted as the basic or the foundation stage for paving further levels of education worldwide. "It provides the mini-structural framework on which the quality of other levels of education is anchored" (Etor et al., 2013).

Percentage of students enrolled in middle schools: The major purpose of middle school is to meet the needs of the early adolescent and prepare them for the next level of secondary education. Therefore, middle school is considered to be the "most widely accepted configuration between the elementary and secondary levels" (Smith, 1992a, 1992b, p. 2).

Percentage of students enrolled in secondary school: In a globalized country like India, there is a huge demand for highly skilled labour, leading to indirect demand for secondary education. Secondary education will act as a foundation by preparing the students as a labour force for the globalized economy. The importance of secondary education is manifold. It develops the critical thinking power of the student and improves their problem-solving skills in abstract problems and formal reasoning (Lapuz & Fulgencio, 2020). A pool of skilled labour is good for a national economy and a global economy (Lewin & Caillods, 2001). So, for faster economic growth in a nation, primary education is insufficient. As per the World Bank report of 1993, a huge public investment is required in this field that has paid rich dividends in East Asia. The economic and social returns from secondary education are more than primary education. So in this work, the percent of students enrolled in secondary standards has been considered.

Percentage of senior/higher secondary students: Higher secondary schooling not only helps the student to acquire more skills and attitudes but also metamorphoses the child in such a manner that they develop job-oriented skills and start taking control of their lives (Mingat & Tan, 1996; Lewin & Caillods, 2001). With the help of such education, not only are the students skilled enough to earn more income, but poverty is also reduced over time (Mingat & Tan, 1996). As education is a continuing process from childhood to adult life, primary and secondary education should be available to all children (UNESCO, 2000).

After calculating the scale-free values of each of the above-mentioned education dimensions, combined enrolment index has been extracted from it by taking the average of these parameters. So the formula is

$$\text{Combined enrolment index} = \frac{\begin{array}{c}\%\text{of Students enrolled in primary schools} + \%\text{of Students enrolled in middle schools} + \\ \%\text{of Students enrolled in secondary schools} + \%\text{of Students enrolled in higher secondary schools}\end{array}}{4}$$

To calculate EAI, the next step was to consider the two-third weightage of adult literacy rate and one-third weightage of the combined enrolment index (Government of West Bengal, 2004, p. 19), the results of which are documented in Table 2. Attention was paid while combining all these dimensions in the index and attaching proper weights to them. Two-thirds of the weightage has been assigned to total literacy index and one-third of the weightage to the combined enrolment index (Roy, 2009). The formula to calculate EAI can be summarized as

$$\text{Educational attainment index} = 2/3\,\text{Total literacy index} + 1/3\,\text{Combined enrolment index}$$

At the district level in terms of education, the eastern side is relatively well developed than the western counterpart with the highest value of EDI (0.78) located in Raina II of Burdwan South Sadar subdivision and the lowest value found in Ketugram II of Katwa subdivision (Table 2). It is symbolic that blocks like Raina II, Raina I, and Purbasthali which fall into the very highly developed blocks have successfully attained 78%, 77%, and 73% of educational attainment. In contrast, CD blocks like Jamuria, Katwa II, Ketugram I, Ausgram I, Ausgram II, and Ketugram II have been lagging as they fall in the category of the very lowly developed CD

Table 2 Educational development index of Bardhaman, 2015–2016

Subdivisions	CD blocks	Combined enrolment index	1/3rd combined enrolment index	2/3rd total literacy index	Educational attainment index	Rank
Asansol subdivision	Salanpur	0.275	0.092	0.550	0.642	5
	Barabani	0.319	0.106	0.155	0.262	25
	Raniganj	0.183	0.061	0.339	0.400	17
	Jamuria	0.231	0.077	0.149	0.226	26
Durgapur subdivision	Kanksa	0.413	0.138	0.446	0.583	8
	Andal	0.502	0.167	0.485	0.652	4
	Faridpur-Durgapur	0.133	0.044	0.351	0.396	19
	Pandabeswar	0.291	0.097	0.303	0.400	18
Burdwan North Sadar subdivision	Galsi I	0.295	0.098	0.297	0.395	20
	Burdwan I	0.565	0.188	0.434	0.623	7
	Burdwan II	0.458	0.153	0.351	0.503	13
	Ausgram I	0.126	0.042	0.147	0.189	29
	Ausgram II	0.296	0.099	0.088	0.186	30
	Bhatar	0.622	0.207	0.240	0.447	16
	Galsi II	0.291	0.097	0.176	0.273	23
Burdwan South Sadar subdivision	Memari I	0.429	0.143	0.350	0.493	15
	Memari II	0.444	0.148	0.371	0.519	12
	Jamalpur	0.683	0.228	0.349	0.576	10
	Raina I	0.464	0.155	0.612	0.766	2
	Raina II	0.344	0.115	0.667	0.781	1
	Khandaghosh	0.419	0.140	0.486	0.626	6
Katwa subdivision	Mongalkote	0.566	0.189	0.086	0.275	22
	Ketugram I	0.320	0.107	0.088	0.194	28
	Ketugram II	0.106	0.035	0.000	0.035	31
	Katwa I	0.250	0.083	0.189	0.272	24
	Katwa II	0.234	0.078	0.137	0.216	27
Kalna subdivision	Purbasthali I	0.699	0.233	0.500	0.733	3
	Purbasthali II	0.548	0.183	0.189	0.371	21
	Kalna I	0.469	0.156	0.423	0.579	9
	Kalna II	0.361	0.120	0.442	0.562	11
	Monteswar	0.566	0.189	0.306	0.495	14

Source: Computed and compiled by the author

blocks. Most blocks are moderately developed, except Jamuria, Ausgram II, Ausgram I, Katwa II, Ketugram I, and Ketugram II.

The calculation of the educational attainment index has been done based on five dimensions, the data of which was available from the Census and DSHB. These indicators or dimensions have been used to evaluate the enrolment and literacy (of the adult) indices, which in turn have been combined to formulate the education attainment index using the methodology of the UNDP. Percentage of students enrolled in primary, middle, secondary, and senior secondary classes have been combined to reach at enrolment index. The percentage of literate people in a particular CD block has been used as the sole indicator to reflect the adult literacy index. Enrolment index was found to be the lowest in Ketugram II and highest in Purbasthali I. Adult literacy index is found to be the highest in Raina II and lowest in Ketugram II. On average, in terms of EAI, the Katwa subdivision occupies the lowest position, while the Burdwan South Sadar subdivision occupies the highest position (Fig. 3).

For the sake of analysis, five classes have been taken to interpret the values of EAI, which are as follows:

Very lowly developed (less than 0.250000): Jamuria, Katwa II, Ketugram I, Ausgram I, Ausgram II, and Ketugram II
Lowly developed (0.25000–0.40000): Raniganj, Pandabeswar, Faridpur-Durgapur, Galsi I, Purbasthali II, Mongalkote, Galsi II, Katwa I, and Barabani
Moderately developed: Memari II, Burdwan II, Monteswar, Memari I, and Bhatar
Highly developed: Andal, Salanpur, Khandaghosh, Burdwan I, Kanksa, Kalna I, Jamalpur, and Kalna II
Very highly developed: Raina II, Raina I, and Purbasthali I

Standard of Living Index (SOLI)

According to the UNDP, SOLI captures gross national income (GNI) per capita adjusted by purchasing power parity (PPP). As a proxy for the standard of living of an average person in a region, usage of the average per capita income of the region has been proposed by the UNDP. However, the problem arises while looking at the meso or micro level. "No official block or municipality level statistic for a variable that may be taken to be a sufficiently close approximation to the per capita income is available" (Government of West Bengal, 2009, p. 21). Since no such secondary data is available at the block level, different dimensions were used to capture the scenario of SLI—percentage of mouzas with drinking and electricity facilities, food grain productivity index, percentage of total working population, and main workers. The data from DSHB was available in absolute form, so the first percentage of selected variables has been extracted. In congruence with the previous method, to ensure that the values lie between zero and unity, the raw data of all the variables have been converted into values free from the units and then standardized regarding maximum and minimum values of the particular dimension. The formulas and the justification behind taking up selected variables for the study are as follows:

Percentage of mouzas that have been electrified: According to Masuduzzaman (2012) and Nhalur, Josey, & Mandal (2018), household residential electricity consumption refers to the volume of electricity consumed by the people in the households in their mundane life upon which the standard of living is anchored. From

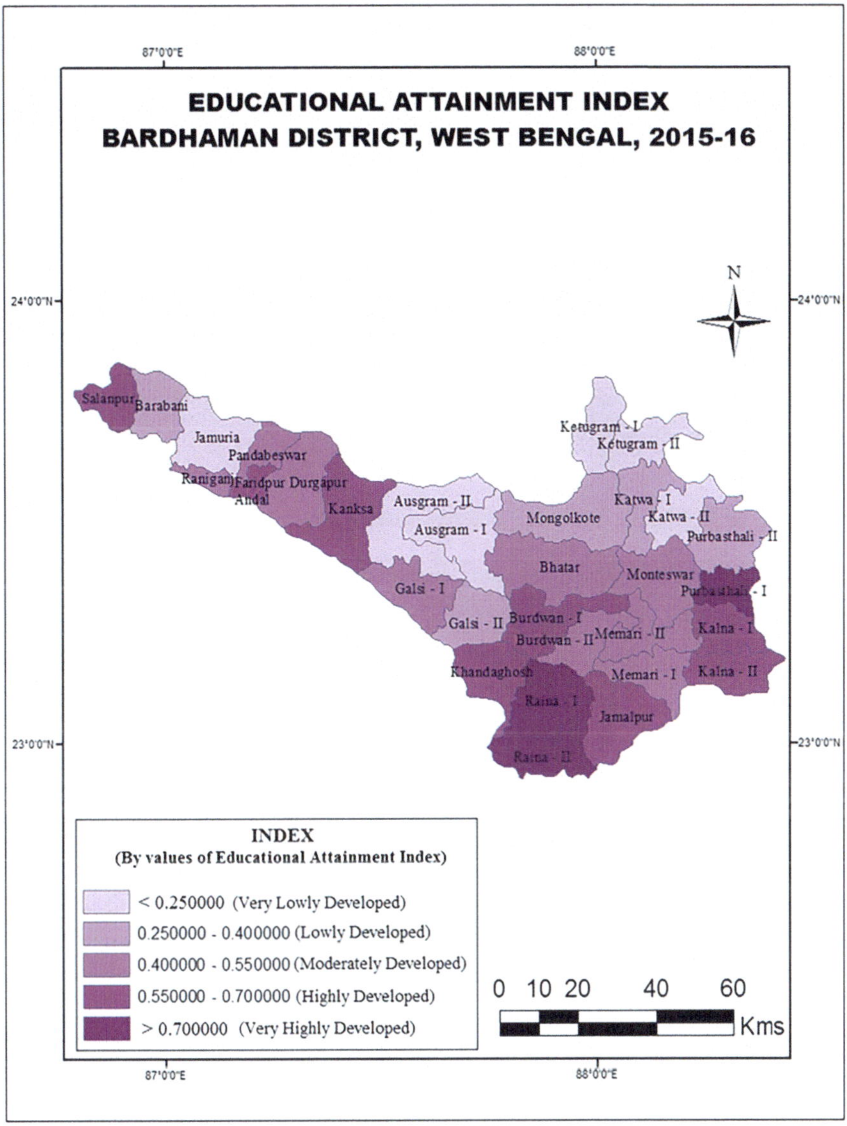

Fig. 3 Educational attainment index

DSHB, the original value is represented in absolute number so it has been converted into percentage. In our analysis for employing the range method of the UNDP to reach scale-free numbers, in case of the denominator, the highest and lowest values have been expressed as 100 and 10, respectively. A value of 100 has been considered the maximum and 10 as the lowest value for the electrified mouzas (Ashraf et al., 2013). So the formula used to calculate this index is

$$\%\text{of mouzas with electricity index} = \frac{\text{Observed value} - 10}{100 - 10}$$

Percentage of mouzas with drinking water facilities: Drinking water is a basic requirement for life and an important determinant of the standard of living. "A certain quantity of water is essential for life and maintenance of personal hygiene, absence of which results in health problems like dehydration, skin diseases, etc." (Puttaswamaiah, 2002, p. 8). From DSHB, the original value of this variable is represented in absolute number, so it has been converted into percentage. In our analysis for employing the range method of the UNDP to reach at scale-free numbers, in case of the denominator, the highest and lowest values have been expressed as 100 and 10, respectively. The value of 100 has been considered maximum and 10 as the lowest value for the mouzas with drinking water facilities (Ashraf et al., 2013). The formula used to compute the index is

$$\%\text{of mouzas with drinking water facilities index} = \frac{\text{Observed value} - 10}{100 - 10}$$

Index of net per capita *value of food crops*: Agriculture is the backbone of India's economy, providing about 50% of the employment to the people (Arjun, 2013). Such a scenario is applicable at the district level of Bardhaman as well where the major source of the income of the people is from the primary sector, although the western portion of the district is well known for its industrial sector (Government of West Bengal, 2011). Due to the absence of data at the CD block level, proxy variables like net per capita value generated from the main crops have been used up for analysis (Ashraf et al., 2013). Crops like rice (aman), wheat, and potato have been considered the basic food crops required for sustenance. Other crops like pulses, grams, mustard, and maize have not been considered as their production is negligible in the study area. The value of production of crops was available in metric tonnes, and the wholesale price of the food grains was available in quintals. It is to be noted that the amount of wholesale food grains is sold at different marks. And the wholesale price varies per market. To maintain consistency, an average of the prices at different markets is taken. To represent it in terms of the same unit, the value of food grain production has been converted from metric tonnes to quintals (1 metric ton = 10,000 quintals). For further calculation, the value generated by the crops has been divided by the population to reach net per capita income (Ashraf et al., 2013). The formula is

$$\text{Step 1: Value of a particular food grain} = (\text{Total production} * \text{Wholesale price})$$

$$\text{Step 2: Per capita value of a particular food grain} = \frac{\text{Value of a particular food grain}}{\text{Total population of a particular CD block}}$$

Step 3 : Net per capita value generated by different foodgrains in a CD block
= Per capita value of Aman rice + Per capita value of wheat + Per capita value of potato

The average district per capita income of Bardhaman is 17,537.98 (Government of West Bengal, 2004, p. 79). Assuming that as the maximum and Rs. 1000 as the minimum, the range method of normalizing the values has been employed.

% of workforce participation: Any country's economy rejuvenates when a huge population works in different sectors. It increases the economic productivity of the nation and raises the disposable income from the job sector. The earnings accrued from working can then be utilized for improvement in standard of living. Keeping that in mind, such a variable has been taken to depict the scenario of economic livelihood, the formula of which is

$$\%\text{of working population} = \frac{\text{Total workers}}{\text{Total population of a particular CD block}}$$

% of main workers: Main workers are those who had worked for the major part of the reference period which is 6 months, unlike the marginal workers. As these people can log in more time for work in a year, it can be assumed that they also get paid more and can invest it behind their lifestyle, which automatically impacts their standard of living (Halder and Patel, 2023). That is why the percentage of main workers has been chosen as the variable to capture the concept of economic livelihood (Khan et al., 2015). The variable will give an insight into the SOLI, and the formula for calculating it is

$$\%\text{of the main working population} = \frac{\text{Number of main workers}}{\text{Total workers of a particular CD block}}$$

The variables like the percentage of households with drinking and electrical facilities and the percentage of working and main workers have been standardized based on normative goalposts using the range method. Upon calculating the index value of each of the variables after standardizing it, the average of the same has been taken to finally arrive at SOLI, the formula of which is

$$\text{Standard of living index}\left(\text{SOLI}\right) = \frac{\begin{array}{c}\text{Index value of mouzas with electric facilities} + \\ \text{index value of mouzas with drinking water facilities} + \\ \text{net per capita value generated by food grains index} + \\ \text{working population index} + \text{main workers index}\end{array}}{5}$$

Although there is a moderate level of development in the district, there is a wide regional variation within the district, when one delves deep into the aspects of standard of living. A clear distinction between the district's eastern and western sections of the district is visible. The eastern half of the district is prominently well developed in terms of SOLI than the western counterpart (Fig. 4). The highest standard of living is found in Jamalpur of Burdwan South Sadar subdivision and the lowest in Pandabeshwar with a value of 0.518 (Table 3) indicating that about 50% of the development is yet to happen in Pandabeshwar. Five variables have been combined to compute the SOLI. If one looks into subdivisions, the Burdwan South Sadar subdivision, Kalna, and Burdwan North Sadar subdivision have performed quite impressively in this aspect. In contrast, the Asansol subdivision occupies the lowest rank. For the sake of interpretation, the values of derived SOLI have been divided into five classes which are as follows:

Very lowly developed (less than 0.550000): Jamuria, Pandabeswar and Raniganj
Lowly developed (0.550000–0.65000): Andal, Barabani, Salanpur, Faridpur-Durgapur, and Ketugram I
Moderately developed (0.65000–0.75000): Kanksa, Galsi I, Monteswar, Purbasthali II, Katwa I, Ketugram II, Burdwan I, Burdwan II, Ausgram I, Purbasthali I, Katwa II, and Mongalkote
Highly developed (0.75000–0.85000): Raina II, Ausgram II, Raina I, Kalna I, Khandaghosh, Memari II, Galsi II, Memari I, and Bhatar
Very highly developed (more than 0.85000): Kalna II and Jamalpur

Human Development Index

The Human Development Index gives a clear picture of the basic sustenance of humans in terms of health, educational attainment, and standard of living. Table 4 provides a display of the values of HDI and depicts how it is closely related to the dimensions. The conditions of all three dimensions are comparatively well off in only two CD blocks—Burdwan I and Burdwan II. In most of the CD blocks like Memari I, Memari II, Jamalpur, Raina I, Raina II, Khandaghosh, Purbasthali I, Purbasthali II, Kalna I, Kalna II, and Monteswar, conditions of two dimensions are good like in case of educational attainment and standard index, whereas health care index is relatively low. This points to the fact that to ensure improved HDI, the respective block development officers must ensure developments in the aspect of health facilities. In the rest of the CD blocks, although the index values of standard of living are appreciable, these areas have failed to make significant progress as evincible from the dismal situation of educational attainment and health care index, a situation that is evident in most of the CD blocks like Barabani, Raniganj, Jamuria, Galsi I, Faridpur-Durgapur, Pandabeswar, Kanksa, Ausgram I, Ausgram II, Bhatar, Galsi II, Mongalkote, Ketugram I, Ketugram II, Katwa I, Katwa II, and Purbasthali II. Most of such CD blocks falling under either Katwa subdivision or Durgapur subdivision are in dire need of policies to be aimed for development in the sectors lagging in education and health facilities.

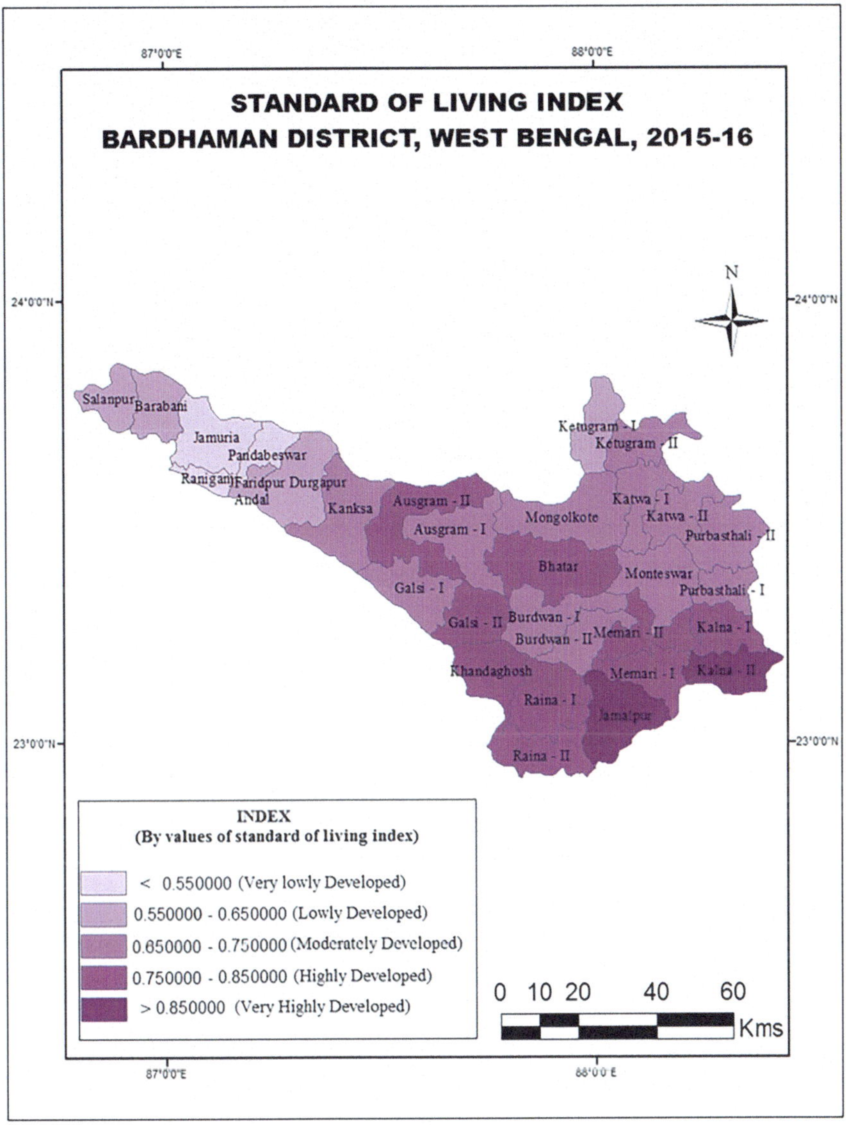

Fig. 4 Standard of living index

To interpret the data of the Human Development Index and the map, the index scores have been divided into five classes to analyse the levels of human development in different corners of the district (Fig. 5):

Very lowly developed: The CD blocks scoring less than 0.400000 fall in this category. A total of ten blocks represent very low development in the aspects of HDI altogether, like Ketugram I, Ketugram II, Ausgram I, Pandabeswar, Barabani, Katwa

Table 3 Standard of living index, Bardhaman 2015–2016

Subdivisions	CD blocks	X1_Mouzas with drinking water facilities	X2_Mouzas electrified	X3_Workforce participation	X4_Main workers	X5_Agr-cultural index	Standard of living index	Rank
Asansol subdivision	Salanpur	1.000	1.000	0.034	0.735	0.021	0.558	26
	Barabani	1.000	1.000	0.178	0.432	0.179	0.558	27
	Raniganj	1.000	1.000	0.036	0.681	0.007	0.545	29
	Jamuria	0.928	1.000	0.169	0.462	0.145	0.541	30
Durgapur subdivision	Kanksa	1.000	1.000	0.755	0.323	0.335	0.683	23
	Andal	1.000	1.000	0.055	0.648	0.023	0.545	28
	Faridpur-Durgapur	1.000	1.000	0.249	0.386	0.337	0.594	25
	Pandabeswar	1.000	1.000	0.000	0.586	0.005	0.518	31
Burdwan North subdivision	Galsi I	1.000	1.000	0.812	0.213	0.427	0.690	22
	Burdwan I	1.000	1.000	0.618	0.616	0.356	0.718	17
	Burdwan II	1.000	1.000	0.735	0.546	0.313	0.719	16
	Ausgram I	0.945	1.000	1.000	0.000	0.706	0.730	15
	Ausgram II	1.000	1.000	0.972	0.170	0.718	0.772	10
	Bhatar	1.000	1.000	0.773	0.697	0.670	0.828	3
	Galsi II	1.000	1.000	0.893	0.298	0.882	0.815	5
Burdwan South subdivision	Memari I	1.000	1.000	0.970	0.676	0.460	0.821	4
	Memari II	1.000	1.000	0.790	0.582	0.671	0.809	6
	Jamalpur	1.000	1.000	0.883	0.675	0.792	0.870	1
	Raina I	1.000	1.000	0.572	0.545	0.775	0.779	9
	Raina II	1.000	1.000	0.572	0.643	0.576	0.758	11
	Khandaghosh	1.000	1.000	0.630	0.611	0.704	0.789	7

Subdivisions	CD blocks	X1_Mouzas with drinking water facilities	X2_Mouzas electrified	X3_Workforce participation	X4_Main workers	X5_Agr-cultural index	Standard of living index	Rank
Katwa subdivision	Mongalkote	1.000	1.000	0.439	0.653	0.656	0.750	12
	Ketugram I	1.000	1.000	0.216	0.623	0.392	0.646	24
	Ketugram II	1.000	1.000	0.365	0.813	0.411	0.718	18
	Katwa I	1.000	1.000	0.308	0.921	0.339	0.714	19
	Katwa II	1.000	1.000	0.349	1.000	0.382	0.746	13
Kalna subdivision	Purbasthali I	1.000	1.000	0.642	0.913	0.129	0.737	14
	Purbasthali II	1.000	1.000	0.664	0.764	0.125	0.710	20
	Kalna I	1.000	1.000	0.691	0.660	0.577	0.786	8
	Kalna II	1.000	1.000	0.803	0.750	0.721	0.855	2
	Monteswar	0.938	1.000	0.357	0.787	0.372	0.691	21

Source: Computed and compiled by the author

Table 4 Human Development Index, Bardhaman, 2015–2016

Subdivisions	CD blocks	Educational attainment index	Health care index	Standard of living index	Human Development Index	Rank
Asansol subdivision	Salanpur	0.642	0.600	0.558	0.600	4
	Barabani	0.262	0.207	0.558	0.342	28
	Raniganj	0.400	0.347	0.545	0.431	17
	Jamuria	0.226	0.349	0.541	0.372	24
	Kanksa	0.583	0.165	0.683	0.477	14
Durgapur subdivision	Andal	0.652	0.418	0.545	0.539	7
	Faridpur-Durgapur	0.396	0.242	0.594	0.411	20
	Pandabeswar	0.400	0.057	0.518	0.325	29
	Galsi I	0.395	0.171	0.690	0.419	19
Burdwan North subdivision	Burdwan I	0.623	0.792	0.718	0.711	1
	Burdwan II	0.503	0.682	0.719	0.635	2
	Ausgram I	0.189	0.158	0.730	0.359	25
	Ausgram II	0.186	0.179	0.772	0.379	23
	Bhatar	0.447	0.151	0.828	0.475	15
	Galsi II	0.273	0.187	0.815	0.425	18
	Memari I	0.493	0.252	0.821	0.522	10
Burdwan South subdivision	Memari II	0.519	0.121	0.809	0.483	13
	Jamalpur	0.576	0.199	0.870	0.549	6
	Raina I	0.766	0.260	0.779	0.602	3
	Raina II	0.781	0.150	0.758	0.563	5
	Khandaghosh	0.626	0.188	0.789	0.534	8
	Mongalkote	0.275	0.137	0.750	0.387	22
Katwa subdivision	Ketugram I	0.194	0.076	0.646	0.305	31
	Ketugram II	0.035	0.195	0.718	0.316	30
	Katwa I	0.272	0.080	0.714	0.355	27
	Katwa II	0.216	0.107	0.746	0.356	26
	Purbasthali I	0.733	0.105	0.737	0.525	9
Kalna subdivision	Purbasthali II	0.371	0.109	0.710	0.397	21
	Kalna I	0.579	0.138	0.786	0.501	12
	Kalna II	0.562	0.136	0.855	0.518	11
	Monteswar	0.495	0.110	0.691	0.432	16

Source: Computed and compiled by the author

II, Ausgram II, Katwa I, Mongalkote, and Jamuria, with most of it located on the north-eastern side under Katwa subdivision followed by Burdwan North and Asansol subdivisions.

Lowly developed: The CD blocks with HDI values of 0.40000–0.45000 have been classified as administrative units which are lowly developed but are relatively better off than the ones falling in the very lowly developed category. Such units are Faridpur-Durgapur, Purbasthali II, Galsi I, Galsi II, Monteswar, and Raniganj. This

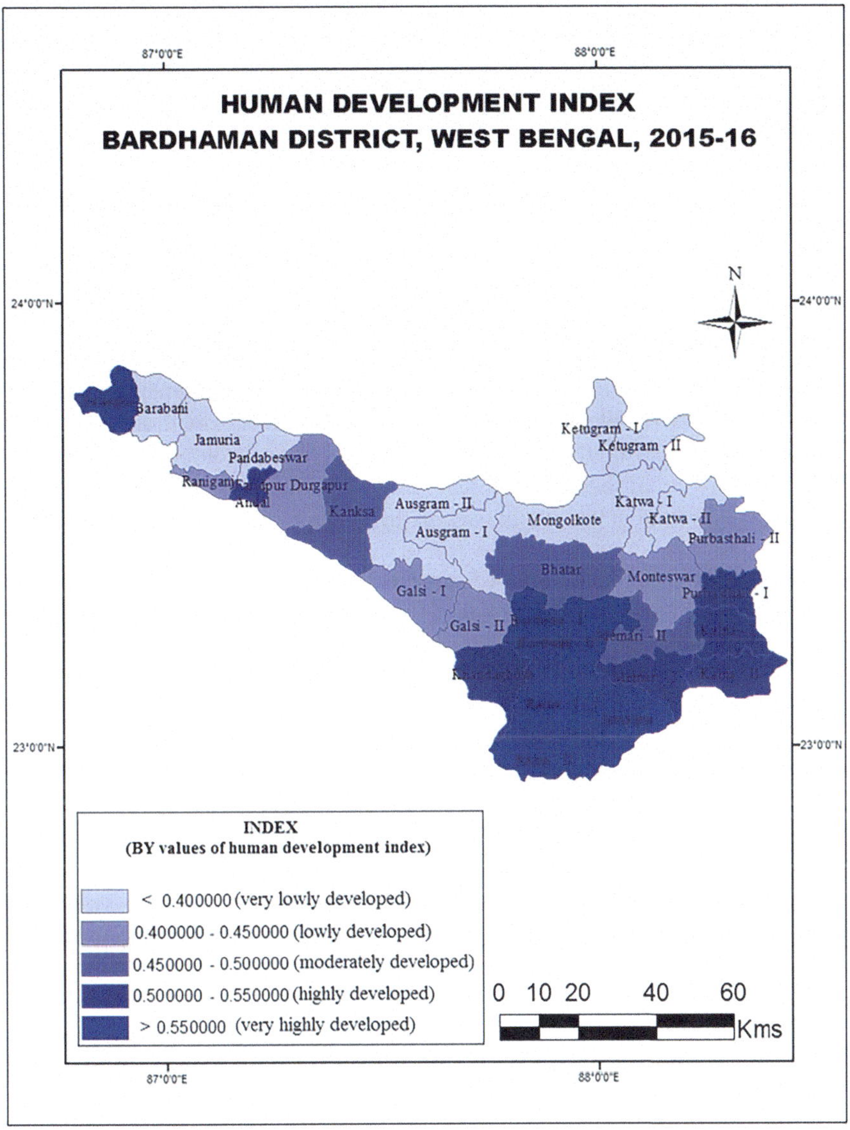

Fig. 5 Human Development Index

class does not form any regional pattern, that is, it is scattered across the districts, with each such unit falling under three different subdivisions. In this category, Monteswar with an HDI value of 0.431 scores the highest, owing to its better performance in educational attainment and standard of living index. However, its progress in the health care index stands very low. It is probably the reason behind its poor overall status in HDI.

Moderately developed: The CD blocks with an HDI value ranging from 0.45000 to 0.50000 have been classified as moderately developed. The administrative units falling in this category are caught up in the transition between lowly and highly developed CD blocks, owing to their moderate progress in educational attainment, health care, and standard of living. Three CD blocks, namely Bhatar, Kanksa, and Memari II, fall in this category.

Highly developed: The CD blocks with a value of HDI ranging from 0.50000 to 0.55000 fall in this category. Six CD blocks fall into this category: Khandaghosh, Memari I, Purbasthali I, Kalna I, Kalna II, and Andal. In this category, Andal occupies the first position with an HDI value of 0.54.

Very highly developed: In this category, the CD blocks have achieved all-round development in all three aspects forming the HDI, with a value ranging more than 0.60000. Such values are found in only six such units, namely Jamalpur, Raina II, Salanpur, Raina I, Burdwan II, and Burdwan I, most of which are located in Burdwan's south and north subdivisions. Such a better condition of HDI in this subdivision can be attributed to the fact that the district headquarters of Paschim Bardhaman is located in this subdivision in the location called Burdwan, which has all the amenities in comparison to the rest of the CD blocks that have positioned in this area under very highly developed category.

Geographies of Care and Responsibility

Space, scale, and territory play a crucial role in solidarity. A region or place determines the location where different socio-economic dynamics play out. In contrast, scale and territory determine how we perceive solidarity by bounding it within a particular geographical location, at the nation, state, district, CD block, or village level. The place is inherently a social construct (Pierce et al., 2011). According to Pierce et al. (2011, p. 54), place-making is "the set of social, political, and material processes by which people iteratively create and recreate the experienced geographies in which they live." It can be surmised that people not only identify themselves with the place but also engage in sociopolitical struggles that emerge out of inequalities in basic amenities of life provided to them, like education, health, and a dignified standard of living. Scale and territory are important concepts while studying it since they are "connected with how we operationalize and bound solidarity" (Schuermans et al., 2013, p. 10). It becomes essential to bind the above discussion with territoriality as it helps one to perceive a bounded space and study in detail the phenomena taking place within the bounded space and helps to differentiate it from the surrounding regions (Creswell, 1997). Along these lines, this work has been carried out to explore the regional variation in the basic amenities provided to the citizens of the CD block of Burdwan district of West Bengal by the Government of India. Mapping out the contours of development not only gives an idea about the presence of inequalities but also chalks out the geographies of care (Lawson, 2007) and geographies of responsibility (Massey, 2004) that should be shouldered by the government to reduce the disparity in the provision of the basic amenities of life. Geographical delineation and spatial arrangements of the CD blocks allow the researcher to perceive deeper into the aspect of HDI (Khan et al., 2015). Figure 5 depicts the composite scenario of health, education, and standard of living index. Three indices have been combined by providing equal weightage to health, education, and standard of living to depict the scenario of human development in the study area. Burdwan I occupies the highest position in terms of HDI with a value of 0.711, followed by Burdwan II with a value of 0.635 and Raina I (0.601). On the other hand, Ketugram I occupies the lowest position with a value of 0.305, followed by Ketugram II (0.316) and Ausgram I (0.359). From the analysis, it has been identified that the CD blocks falling under the Katwa subdivision need urgent attention in all three dimensions, whereas the CD blocks falling under the Asansol and Durgapur subdivision of Paschim Bardhaman district need specific attention in the arenas of educational attainment and standard of living of people so that it can be enhanced to improve human development in these areas.

Conclusion

From the methodology section, one might criticize the variables taken up for study as the standard of living could not be properly addressed owing to a lack of data from DSHB at CD block level. But exercising utmost attention, some variables have been taken up as a proxy of SOL (to the best of our knowledge) with justification, which has given a glimpse of the same. This work confirms that a wide regional variation exists in health care facilities, educational attainment, standard of living, and therefore Human Development Index at the district level of Bardhaman. On average, there is an inclination of development towards the south-eastern side or the central part of the district, where the community development blocks are located under either Bardhaman south or north subdivision. If one looks at the disaggregated level, then the scenario is slightly different. In terms of both educational attainment and standard of living, Purba Bardhaman has a better position than Paschim Bardhaman.

In contrast, in terms of health care facilities, along with the well-developed facilities in the Purba Bardhaman, with it being concentrated more in Burdwan I and Burdwan II of Bardhaman north subdivision, it is also very well developed in most of the CD blocks of Paschim Bardhaman side. It can also be concluded that the CD blocks located in the central portion of the undivided Burdwan district or close to the headquarters of Burdwan have witnessed a better scenario in terms of HDI. This study aimed to identify the regional disparity in HDI to ensure that proper policies can be framed at the block level to fill up the cavity in development. Within the Purba Bardhaman district, there is a disparity. From the analysis, it has been identified that the CD blocks falling under the Katwa subdivision need urgent attention in all three dimensions, whereas the CD blocks falling under the Asansol and Durgapur subdivisions of Paschim Bardhaman district need specific attention in the arenas of educational attainment and standard of living of people so that it can be enhanced to improve human development in these areas.

Acknowledgments I am thankful to Dr. Surendra Kaur Rawal (ICSSR Postdoctoral Fellow) for giving valuable insights while framing certain indexes.

References

Aggarwal, S. K. (2019). Healthcare in India and violence against doctors: The missing links. *Indian Journal of Nephrology, 29*(4), 221–231.

Arjun, K. M. (2013). Indian agriculture—Status, importance, and role in Indian economy. *International Journal of Agriculture and Food Science Technology, 4*(4), 343–346.

Ashraf, S. W., Ahmed, S., & Rawal, S. K. (2013). Status of human development in Murshidabad District of West Bengal. *Indian Journal of Human Development, 7*(1), 141–160.

Basak, A., & Siddique, G. (2020). An analysis of health Care Facility Systems in Purba and Paschim Bardhaman Districts, West Bengal. *India. Journal of Global Resources, 6*(1), 170–178.

Bengal, G. O. (2004). *West Bengal human development report*. Development and Planning Department.

Bengal, G. O. (2009). *District human development report: South 24 Parganas*. HDRCC, Development & Planning Department, Government of West Bengal.

Bengal, G. O. (2011). *District human development report: Bardhaman*. Kolkata: Government of West Bengal. Retrieved April 14, 2023, from https://www.indiastatpublications.com/District_Factbook/West_Bengal/Barddhaman#:~:text=Main%20source%20of%20income%20in,150.14%20for%20the%20year%202021.

Bigbee, L. J. (2008). Relationships between nurse- and physician-to-population ratios and state health rankings. *Public Health Nursing, 25*(3), 244–252.

Coccia, M. (2019). Theories of development. *Global Encyclopedia of Public Administration, Public Policy, and Governance, 6*(2), 1–7.

Creswell, T. (1997). Weeds, plagues, and bodily secretions: A geographical interpretation of metaphors of displacement. *Annals of the Association of American Geographers, 87*, 330–345.

Dandavino, M., Snell, L., & Wiseman, J. (2007). Why medical students should learn how to teach. *PubMed, 29*(6), 558–565.

Deo, G. M. (2016). Time to revisit recommendations on doctor to population ratio in India. *Journal of the Association of Physicians of India, 64*, 72–73.

Etor, C. R., Mbon, U. F., & Ekanem, E. E. (2013). Primary education as a Foundation for Qualitative Higher Education in Nigeria. *Journal of Education and Learning, 2*(2), 155–164.

Fournier, A., & Dodard, M. (1997). The health care delivery crisis in Haiti. *Family Medicine, 29*(9), 666–669.

Green, L. V. (2002). How many hospital beds? *INQUIRY A Journal of Medical Care Organization, Provision and Financing, 39*(4), 400–412.

Halder , T., & Patel, P. P. (2023). An appraisal of housing quality of living in Durgapur: A spatial approach. *Transactions, 45*(2), 207–222.

Khan, A., Chatterjee, S., Mistri, K., & Das, A. (2015). Level and pattern of human development in Paschim Medinipur District, West Bengal, India. *Journal of Human Ecology, 49*(3), 219–229.

Lapuz, A. M., & Fulgencio, M. N. (2020). Improving the critical thinking skills of secondary school students using problem-based learning. *International Journal of Academic Multidisciplinary Research, 4*(1), 1–7.

Lawson, V. (2007). Geographies of care and responsibility. *Annals of the Association of American Geographers, 97*(1), 1–11.

Lewin, K., & Caillods, F. (2001). Financing the development of secondary education in developing countries. *Prospects, 31*, 61–72.
Massey, D. (2004). Geographies of responsibility. *Geografiska Annaler Series B, 86*(1), 5–18.
Masuduzzaman, M. (2012). Electricity consumption and economic growth in Bangladesh: Co-integration and causality analysis. *Global Journal of Management and Business Research, 12*(11), 46–56.
Mingat, A., & Tan, J.-P. (1996). The full social returns to education: Estimates based on countries' economic growth performance. *Economics Education*, 1–17.
Mohanty, S. K., & Dehury, B. (2012). Human development indices: Old and nee. *Artha Vijnana, LIV*(1), 19–38.
Nhalur, S., Josey, A., & Mandal, M. (2018). Rural electrification in India from 'connections for all' to 'power for all'. *Perspectives, 53*(45), 31–37.
Pierce, J., Martin, D. G., & Murphy, J. T. (2011). Relational place-making: The networked politics of place. *Transactions of the Institute of British Geographers, 36*(1), 54–70.
Puttaswamaiah, S. (2002). *Drinking water supply: Environmental problems, causes, impacts, and remedies—Experiences from Karnataka*. Gujarat Institute of Development Research.
Ravaghi, H., Alidoost, S., Mannion, R., & Bélorgeot, V. D. (2020). Models and methods for determining the optimal number of beds in hospitals and regions: A systematic scoping review. *BMC Health Services Research, 20*(186), 1–13.
Rechel, B., Wright, S., Barlow, J., & Mckee, M. (2010). Hospital capacity planning: From measuring stocks to modeling flows. *Bulletin of the World Health Organization, 88*(8), 632–636.
Ringard, A., Sagan, A., Saunes, I. S., & Lindahl, A. K. (2013). Norway: Health system review. *Health Systems in Transition, 15*(8), 1–162.
Roy, A. (2009). Status of human development in Puruliya District. *Geographical Review of India, 70*(1), 81–86.
Schuermans, N., Spijkers, F., & Loopmans, M. (2013). Solidarity in human geography: Responsibility, care, place and encounter. *KU Leuven*, 1–42.
Sen, Amartya (1985). *Commodities and Capabilities*. Oxford University Press India.
Smith, J. A. (1992a). *Dissertation. Effective middle school teaching: Factors that promote and maintain it*. Portland State University.
Smith, J. A. (1992b). *Effective middle school teaching: Factors that promote and maintain it*. Portland State University.
UNESCO. (2000). *World education report: The right to education; towards education for all throughout life*. UNESCO.
Yagma, N. M., Donkor, A. G., & Gokler, M. E. (2020). Management of Hospitals during the COVID-19 pandemic. *Medical Research Reports, 3*(1), 155–161.
Zodpey, S., Sharma, A., Zahiruddin, Q. S., Gaidhane, A., & Shrikhande, S. (2018). Allopathic doctors in India: Estimates, norms and projections. *Journal of Health Management, 20*(2), 151–163.

"Conflict in Solidarity of Society": An Assessment through SDG 5.4

Robin Raj and Anjali Srivastava

Introduction

Solidarity is necessary for building cohesive and harmonious societies and is closely linked to values like empathy, social justice, and inclusivity. It embodies people's demands for both personal development and a feeling of belonging, and it links individual freedom with social justice, economic efficiency, equitable resource distribution, pluralism, and universal principles for solving all issues. It is the outcome of a democratic endeavor to forge a national identity, social peace, and economic vigor with the goals of constructing a just system, sustaining the momentum of unbridled economic growth, and averting social rifts.

The social construction of gender roles often impacts the successful functioning of solidarity in society. These gender roles compel people to "see what is expected, act and react in expected manner, and learn what is expected," which creates conflict in solidarity as one has to compromise with their independence which in turn results in gender disparity in society (Pillay-Naidoo & Vermeulen, 2023). The United Nations Conference on Sustainable Development introduced sustainable development goals which objective was to develop a set of global goals that address the pressing political, economic, and environmental issues the world is currently experiencing. Target 5.4 of these goals, which focuses on gender equality, asks for the acknowledgment, reduction, and redistribution of unpaid family care work—that is, the everyday care and cleaning done in households, largely by female family members.

R. Raj (✉)
S. J. Mehta School of Management, Indian Institute of Technology Bombay, Mumbai, Maharashtra, India

A. Srivastava
India Child Protection, Okhla Phase-II, New Delhi, India

K. Arora, R. R. Anand (eds.), *The Solidarity Approach in Geography*,
https://doi.org/10.1007/978-3-031-86753-8_24

UN Women's Policy Brief No. 5 on maintaining quality care services and UN Women's Progress of the World's Women 2015–2016 both define "unpaid care and domestic work" as "the work that is done in the home by family members for maintenance and well-being of family members without pay." It also entails doing volunteer community service. Since it is theoretically possible to pay a third party to execute these tasks, they are classified as work. Time-use surveys are a tool that measures the unseen role of women's domestic and unpaid labor. They highlight variations between the sexes and provide a vital example of how gender roles influence the allocation of tasks within a home (Medina-Hernández et al., 2021). The International Labour Organization (ILO) reports that, in 2018, women completed 76.20% of all unpaid domestic and care labor hours, which is representative of a threefold increase over men.

In the context of Indian society, gender inequality is a prominent problem. Despite several decades' worth of laws and a constitution guaranteeing equal rights for men and women, some deeply ingrained gender inequality has a terrible toll on women's lives. The International Labour Organization (2018) reports that whereas men only spend approximately 31 min providing unpaid care, Indian women perform this labor for about 297 min. Sepúlveda (2013) argues that when women undertake unpaid caregiving in proportionately higher amounts than men, it violates their rights. Additionally, this discrepancy prevents women from being economically empowered. Global Gender Gap Report (WEF, 2023) places India at 127th place out of 146 nations in terms of gender disparity, attributing this unequal distribution of caring tasks to discriminatory societal structures and gender role norms.

In this context, this study investigates the gender distribution in the time use of unpaid domestic and caregiving activities in India, using time-use data. Additionally, this study intends to identify the factors influencing the time allocated to UDCW.

With the evolution of society and challenging traditional gender norms, there needs to be a shift in the relationship between the sex dimension and the solidarity approach. A woman's time could be better spent engaging in activities connected to the market or developing her educational and professional abilities for every minute she spends on unpaid caregiving. Thus, a holistic public policy regarding the improvements in the quality of care without decreasing women's well-being and autonomy is the need of the hour.

Review of Literature

The conventional breadwinner model, which classified the public sphere as primarily composed of paid labor, civil society, politics, and family as the "men's sphere" and the domestic sphere as comprised of the home, family, and unpaid labor, as the domain of women, has evolved into a more gender-neutral form in recent decades (Gornick & Meyers, 2005).

Stiglitz et al. (2007) characterize unpaid care labor as a vital component of the economy and an indispensable factor that enhances the well-being of people, their families, and society. According to Sepúlveda (2013), the unequal allocation of unpaid care labor between men and women violates women's rights and hinders their potential to become economically empowered.

A report published by the ILO (2018) states that in Asia and the Pacific, women engage in unpaid caregiving 4.1 times more frequently than men. It is primarily this unpaid care job that keeps women out of the workforce and hinders their ability to advance in it.

One of the studies by Singh and Pattanaik (2020) highlights the fact that, from an Indian perspective, a significant percentage of women perform unpaid domestic work due to three main reasons: career (low opportunity cost of performing unpaid job in the market), choices (failure of market and states to offer needed provisioning), and constraints (social and religious).

The Oxfam India Report on the state of employment in India (2019) revealed that in India, women who provide unpaid care work spend 312 min a day in urban regions and 291 min a day in rural areas. This imbalance is caused by the patriarchal and casteist nature of our society, which while undervaluing care work considers it as an exclusively women's job.

Jayati Ghosh, in the book Never Done and Poorly Paid (Ghosh, 2009), draws attention to the difficult conditions that women in the Indian subcontinent have to endure as a result of the interaction of speculative and structural forces that serve to reinforce patriarchy. Internationally, it has been found that when a couple starts a family, it is usually the woman who is obliged to reduce outside working hours or even give up work altogether (Palomba & Sabbadini, 1995).

A report published by the UN in 1980 states that women make up almost half of the world's population, work about two-thirds of all working hours, earn one-tenth of all revenue, and possess less than one-tenth of all property worldwide (United Nations, 1980).

Consequently, a great deal of research has been conducted highlighting the importance of unpaid caregiving and household tasks. The unpaid care and household labor, as well as how it influences societal solidarity, will be highlighted in this study using existing literature.

Materials and Methods

The data used in our analysis comes from the Time Use Survey of India, a large-scale sample survey organized by the National Statistical Office (NSO), Ministry of Statistics and Programme Implementation (MoSPI), Government of India. The survey's objective is to gather data regarding the activities that people engage in and how long they last. The data was gathered from sample households between January and December 2019 and includes information on time utilization for individuals aged 6 and above.

The survey collected information from 138,799 households (rural: 82,897 and urban: 55,902), which listed 447,250 persons of age 6 years and above (rural:

Table 1 Classification of major activities as per ICATUS 2016

Classification	Activity
1	Employment and related activities
2	Production of goods for own final use
3	Unpaid domestic services for household and family members
4	Unpaid caregiving services for household and family members
5	Unpaid volunteer, trainee, and other unpaid work
6	Learning
7	Socializing and communication, community participation, and religious practice
8	Culture, leisure, mass media, and sports practices
9	Self-care and maintenance

Source: TUS Activity Classification, National Statistical Office (Government of India, Ministry of Statistics and Programme Implementation, 2020)

273,195 and urban: 174,055). The survey included males, females, and transgenders, but due to the smaller sample size of transgender, this analysis has been restricted to males and females only. Thus, our analysis sample is of 445,170 persons, which comprise 226,644 males and 218,526 females.

The Time Use Survey of India (2019) has classified the activities of the population based on the International Classification of Activities for Time Use Statistics (ICATUS) 2016, prepared by the United Nations Statistics Division. The ICATUS 2016 is a three-level hierarchical classification (composed of major divisions, divisions, and groups) of all the activities that an individual may engage in within 24 h (Table 1). It is made up of major divisions, divisions, and groups. Its objective is to provide a uniform framework for time-use statistics based on meaningfully grouped activities. The ICATUS 2016 has classified the activities into nine major divisions which are given in Table 1.

Definition and Measurement of Dependent Variables

Our study emphasizes Sustainable Development Goal (SDG) 5 Target 4, i.e., "Recognize and value unpaid care and domestic work through the provision of public services, infrastructure, and social protection policies and the promotion of shared responsibility within the household and the family as nationally appropriate." According to the UN, the indicator to measure SDG 5.4 is defined as the proportion of time spent in a day on UDCW, by sex, age, and location (United Nations, 2015). The term "UDCW" describes actions about the supply of services for household members' or relatives' personal usage when residing in different homes. These activities are listed in ICATUS 2016 under the major divisions "3. Unpaid domestic services for household and family members" and "4. Unpaid caregiving services for household and family members" (Economic Commission for Latin America and the Caribbean (ECLAC), one of the regional commissions of the United Nations).

Table 2 Unpaid domestic and caregiver services for family members and the household include the following activities

Major activity	Divisions of activity
Unpaid domestic services for household and family members	Food and meals management and preparation Cleaning and maintaining of own dwelling and surroundings Do-it-yourself decoration, maintenance, and repair Care and maintenance of textiles and footwear Household management for own final use Pet care Shopping for own household members Traveling, moving, transporting, or accompanying goods or persons related to unpaid domestic services for household members Other unpaid domestic services for household members
Unpaid caregiving services for household and family members	Childcare and instruction Care for dependent adults Help to nondependent adult household members Traveling and accompanying goods or persons related to unpaid caregiving services for household members Other activities related to unpaid caregiving services for household members

Source: TUS Activity Classification, National Statistical Office (Government of India, Ministry of Statistics and Programme Implementation, 2020)

Accordingly, this study has also merged these two major groups of activities into one and named it "UDCW."

In the study, the dependent variable is the total time spent in a day expressed as min/day in UDCW (Table 2).

Definition and Measurement of Explanatory Variables

- **Location**: Based on demographic and geographic factors, the area is divided into two categories: rural (reference) and urban.
- **Employment Status**: A person's type of labor force participation influences how much time they spend performing unpaid caregiving and household tasks. Based on the participants' participation in gainful economic activity over the 365 days before the survey day, the survey gathers data on the typical major activity (main paid activity) of those who are 6 years and above. Five types of an individual's employment status are taken into consideration in the study: self-employed (own account worker, employer, worked as a helper in household enterprise), salaried or regularly paid (worked as regular salaried/wage employee), contract labor (worked as casual wage labor: in public works and other types of work), not in the labor force (attended domestic duties; engaged in free collection of goods; sewing, tailoring, weaving, etc. for household use; rentiers; pensioners; remittance recipients; not able to work due to disability; others such as begging and prostitution), and unemployed (did not work but was seeking and/or available for work).

- **Household Income**: As a stand-in for household income, the typical monthly per capita consumption expenditure (MPCE) of the household is used. The usual MPCE is divided into five quintiles—poorest (reference), poorer, middle, richer, and richest.
- **Type of Dwelling Unit**: Dwelling unit refers to the kind of dwelling/house individuals are residing in. In the study, the following types of dwelling structures have been considered—kutcha house (reference), semi-pucca house, pucca house, and no dwelling.
- **Source of Cooking Fuel**: The kind of cooking fuel used by the household influences the time spent in unpaid domestic work. In this study, the source of cooking fuel is categorized into three—nonsolid fuels (reference), solid fuels, and others. Nonsolid fuels consist of LPG, other natural gas, kerosene, gobar gas, and biogas, whereas solid fuels comprise traditional cooking fuels like firewood, dung cake, coke/coal, and charcoal. The category others includes households using any other fuel or households with no cooking arrangement.
- **Electricity**: It describes the individuals living in households with electricity (reference) and households without electricity.
- **Level of Education**: The following categories comprise the educational hierarchy: (a) illiterate, (b) primary, (c) secondary, and (d) 12th and above.
- **Marital Status**: It records an individual's current marital status. The study classifies the variable as never married (reference), currently married, widowed, and divorced or separated.
- **Number of Young Children (<6 Years) in the Household**: It includes the total number of children below 6 years of age in a household.
- **Number of Children (6–14 Years) in the Household**: It includes the total number of children aged 6–14 years in a household.
- **Number of Working-Age Persons (15–60 Years) in the Household**: It includes the total number of working-age persons, i.e., of those aged 15–60 years in a household.
- **Number of Senior Citizens (Above 60 Years) in the Household**: It includes the total number of senior citizens of age above 60 years in a household.
- **Age**: Age entails the number of years lived by an individual.
- **Religion**: The study divides religion into three major categories: Hindu (reference), Muslim, and Others (Christian, Buddhist, Jain, and others).
- **Caste**: Caste is the traditional and hereditary social identity of an individual. It has been categorized as scheduled caste or scheduled tribe (reference), other backward class, and others.

Methodology

The Time Use Survey's unit-level data was used to estimate each person's daily total time spent on "UDCW" for the analysis. The entire amount of time participants spend in an activity divided by the total number of participants yields the average duration of time spent by an individual. Similarly, the average amount of time that men and women spend on unpaid care and household labor is determined by dividing the entire amount of time that they spend doing so by the total number of men and women, respectively. The same method has been used to determine the average amount of time that men and women in various age groups, in both rural and urban locations, spend performing unpaid domestic and caregiving tasks.

Furthermore, OLS regression is applied individually for males and females to determine the factors affecting the time spent on unpaid domestic and care duties.

The basic form of OLS regression is

$$Y_i = \beta_0 + \beta_1 X_1 + \beta_2 X_2 + \ldots + \beta_n^* X_n,$$

where Yi is the dependent variable depicting minutes spent on UDCW. X_1, X_2, …, X_n represent socioeconomic and demographic characteristics of the individuals such as location, employment status, household income, dwelling unit, source of cooking fuel, electricity, level of education, marital status, number of young children in the household, number of children in the household, number of working-age persons in the household, number of senior citizens in the household, age, religion, and caste as independent variables.

Results and Discussion

Descriptive Statistics

Figure 1 represents the average time allocated to UDCW by sex, location (segregated by sex), and age groups (segregated by sex). The figure reveals that in India, males spend an average of 47.3 min on UDCW, whereas females spend significantly more time on it with an average of 362 min. The findings suggest that females spend considerably more time on UDCW than males. This highlights the substantial imbalance in the distribution of unpaid domestic and care responsibilities among males and females.

The situation is almost the same in rural and urban areas where females spend significantly more time on UDCW than their male counterparts. In rural areas, males allocate an average of 49.5 min to it, whereas females spend an average of 365.2 min on it. Similarly, in urban areas, females spend an average of 356.2 min, while males spend 44 min on UDCW.

Figure 1 depicts that this gender disparity is persistent across all age groups where females spend significantly more time on UDCW than males. The time allocated to UDCW by females in her life cycle resembles a pyramid-like pattern. The females in the 6–15 age group allocate an average of 42.2 min, which escalates to its zenith in the 31–45 age group with an average time of 487.6 min and then declines to an average of 152.8 min in the age of 60 and above. In the life cycle of males, the young males in the 6–15 age group allocate an average of 13.2 min to

(i) Sex

(ii) Location segregated by Sex

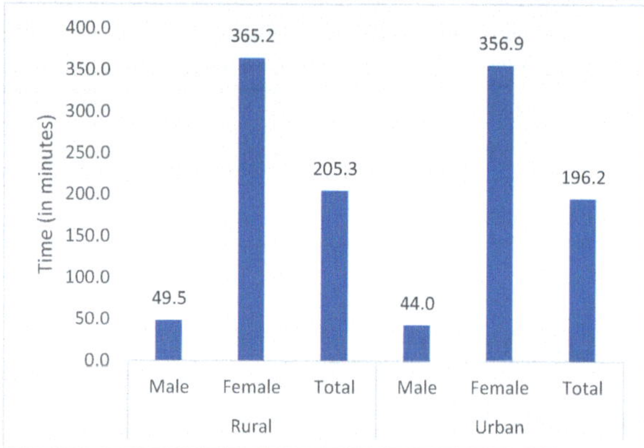

(iii) Age Groups segregated by Sex

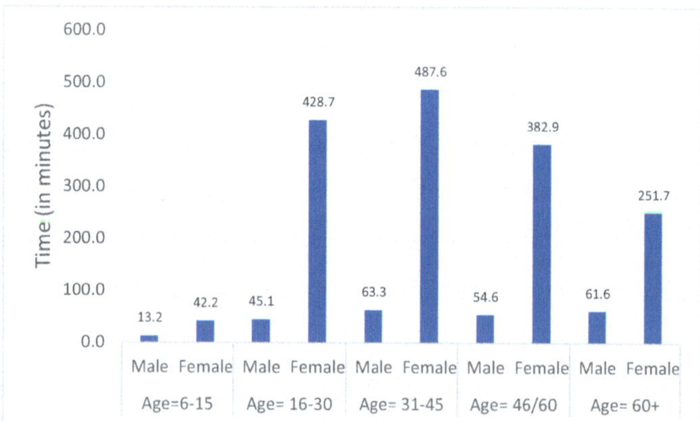

Fig. 1 Average time spent in UDCW by (1) Sex. (2) Location segregated by sex. (3) Age groups segregated by sex

UDCW, and the males above the age of 15 spend approximately 45–65 min on these activities, as shown in the figure. The results indicate that gender disparities originate early in life, often during childhood. Families are often viewed as the initial educational environment for children, and within Indian society, gender stereotypes are deeply ingrained within family dynamics. Typically, household chores are assigned to females, while tasks outside the home are designated for males. This underscores the necessity for a shift in attitudes and beliefs within families to promote greater gender equality.

Regression Analysis

Table 3 displays estimates for time spent on unpaid domestic and caregiving tasks derived from ordinary least square analysis. Individuals' time spent performing unpaid household and caregiving tasks is influenced by their socioeconomic status and demographic background.

Table 3 OLS regression coefficient in the models predicting the impact of socioeconomic and demographic backgrounds of the individual on time spent on unpaid domestic and care activities

Variables	Male	Female
Location		
Rural®		
Urban	−3.91***	−3.71***
Employment status		
Not in the labor force®		
Self-employed	1.63***	−89.85***
Regular wage/salaried	0.04	−144.23***
Casual labor	−0.20	−113.92***
Unemployed	27.89***	61.99***
Household income		
Poorest®		
Poorer	1.08*	5.43***
Middle	0.82	8.83***
Richer	1.51**	3.49**
Richest	5.56***	−14.04***
Type of dwelling unit		
Kutcha house®		
Semi-pucca house	−4.94***	1.89
Pucca house	−13.27***	−14.81***
No dwelling	−10.15	−7.15
Sources of cooking fuel		
Nonsolid fuels®		
Solid fuels	2.73***	16.64***
Others	−23.55***	−101.15***
Electricity		
Household with electricity®		

(continued)

Table 3 (continued)

Variables	Male	Female
Household without electricity	−0.48	6.79***
Level of education		
Illiterate®		
Primary education	3.31***	32.42***
Secondary education	8.62***	76.93***
12th and above	12.65***	82.32***
Marital status		
Never married®		
Currently married	20.88***	480.20***
Widowed	72.78***	340.76***
Divorced/separated	64.13***	340.54***
Number of young children in the household (<6 years)	11.25***	15.94***
Number of children in the household (6–14 years)	−6.80***	−23.64***
Number of working-age persons in the household (15–60 years)	−7.70***	−16.31***
Number of elderly in the household (above 60 years)	−7.03***	−25.48***
Age	0.14***	−3.50***
Religion		
Hindu®		
Muslims	1.32**	30.90***
Others	9.65***	19.01***
Caste		
SC/ST®		
OBC	−3.32***	−15.67
Others	−1.92***	***-1.15

Note: Significance level: ***$p < 1\%$, **$p < 5\%$, *$p < 10\%$
Source: Calculations of data done by the authors based on NSO Time Use Survey, 2019

Taking location into account, it can be said that, on average, urban men and women spend 3 min less each day on unpaid care and household duties than their rural counterparts.

An individual's time availability to indicate unpaid care and household duties depends on their employment position. From the standpoint of time availability, substantially less time is left over for domestic chores because full-time employees spend the majority of their time in the workforce. The table demonstrates that regular salaried women devote the least amount of time to UDCW, i.e., 144 min less than the ones who are not in the labor force. This is followed by women who are in the casual labor force (spending 133 min less) and self-employed (spending 89 min less). The unemployed men and women spend 27 min and 61 min more on UDCW compared to those not in the labor force. Thus, the overall scenario states that women who are employed tend to spend less time on unpaid domestic and care activities and unemployed (did not work but was seeking and/or available for work) men devote more time to unpaid domestic and care activities in comparison to the ones who are not in the labor force (attended domestic duties; engaged in free collection of goods; sewing, tailoring, weaving, etc. for household use; rentiers;

pensioners; remittance recipients; not able to work due to disability; others such as begging and prostitution).

It is believed that households in the higher income quintiles can engage outside domestic helpers based on their household income. However, this does not hold in the case of males in the study as the amount of time spent on UDCW by males belonging to the richest income quintile is 5 min more in comparison to males belonging to the lowest income quintile. Nonetheless, compared to women in the lowest income quintiles, women in the upper income quintiles spend 14 min less on these activities.

Considering the dwelling unit, it can be stated that the individuals residing in the pucca house spend less time in UDCW in comparison to the ones residing in the kutcha house. Women and men tend to spend 14 and 13 min less time in comparison to the one who is residing in a pucca house.

When one considers sources of cooking fuel, one finds that those who use solid fuels typically devote more of their time to unpaid caregiving and household tasks. Compared to females who do not use solid fuels, females who use solid fuels dedicate 16 min more to unpaid care and household labor. The finding suggests that the females in the household that uses solid fuels may spend their time in the collection and gathering of firewood or making cow dung cakes.

Considering the household without electricity, the female individual spends 6 min more on UDCW in comparison to the females who are residing in a household with electricity. Thus, the data suggest that proper households with basic amenities have a significant effect on time allocation for unpaid domestic and care activities.

Education is a significant factor in the amount of time that both males and females spend performing unpaid caregiving and household tasks. Generally, males who have completed 12 years of school or more spend around 12 min more on unpaid household and caregiving tasks than illiterate males do, and females who have completed 12 years or more spend about 82 min more.

In terms of marital status, there is a correlation between being married and spending more time providing unpaid care for females at home as married females spend almost 8 h managing the household chores in comparison to never-married females. Married males spend 20 min more on these activities in comparison to never-married males.

The table also shows that as the number of young children (<6 years) increases, the time devoted by individuals to UDCW also increases. This demonstrates that with the increase in one young child or birth of one child in the family, the time spent on UDCW requires 15 min of women's daily time and 11 min of men's daily time. However, the same does not apply to children (6–14 years), working-age persons (15–59 years), and senior citizens (60 and above). The table shows that apart from young children, with every additional member in the household, i.e., children (6–14 years), working-age person (15–59 years), and senior citizens (60 and above), the amount of time spent by men and women on UDCW decreases. The analysis also shows that with the increase in the age of individuals in the household, women in the family tend to devote 3 min less to unpaid domestic and care activities.

Religious practices and beliefs are frequently linked to gender-based standards. Islamic women tend to spend 30 min more on UDCW in comparison to Hindu women, whereas in the case of other (including Christian and others) religions, women spend 19 min more in comparison to Hindu women. Across all the religions, the contribution of men belonging to other religions is significantly high for UDCW.

In analyzing the social group, i.e., caste, it was found that in contrast to other social groups, scheduled caste and scheduled tribe men and women spend more on unpaid domestic and caregiving duties. Women belonging to other backward classes (OBCs) spend 15 min less in unpaid domestic and care activities in comparison to women of the SC/ST category. Therefore, compared to women in other social groups, it may be said that women from marginalized and vulnerable groups devote a greater amount of their time to unpaid caregiving and household chores.

Limitations of the Study

The study is based on Sustainable Development Goal 5.4 (SDG 5.4), which lacks specific, quantifiable targets or metrics for tracking progress. As a result, this work focuses solely on the analysis of time allocation between males and females for UDCW, making it challenging to gauge the extent of progress achieved in addressing this issue.

The analysis in this study relies on time-use data from India, as surveyed in 2019. Unfortunately, there is no prior national-level time-use data available for reference. The absence of historical data limits the ability to track changes and progress over time, making it challenging to determine the effectiveness of policies and interventions aimed at addressing UDCW.

Certain occupations, such as agriculture or under the Mahatma Gandhi National Rural Employment Guarantee Act (MGNREGA), exhibit cyclical patterns due to seasonal variations in work demand. These variations can significantly influence the time allocation patterns of individuals regarding UDCW. This study does not analyze the impact of seasonal fluctuations, potentially missing important insights into how these patterns change over a year.

The time-use patterns of individuals may vary between weekdays and weekends due to differences in work schedules, school attendance, and other factors. However, this study does not take into account these variations in time allocation for UDCW. This limitation can hinder a comprehensive understanding of how these patterns change depending on the day of the week, potentially overlooking valuable information for policymakers.

In summary, this study presents several limitations that stem from the lack of clear targets in SDG 5.4, the absence of historical data for comparison, the failure to address seasonal variations, and the neglect of differences in time allocation between weekdays and weekends. These limitations emphasize the need for future research that accounts for these factors for a more comprehensive understanding of UDCW in India.

Recommendations

Time-use surveys serve as an essential instrument for collecting data on how individuals allocate their time, shedding light on the disparities in unpaid work between genders and different demographic groups. They help us comprehend the time allocations made to unpaid domestic and care activities, which can further guide us in the formulation of policies. Thus, it is required that time-use surveys are conducted periodically to understand the situation and to track the progress. Based on our study, the following are some of the recommendations that might bring a change in society with more inclusive gender equality.

The reform is needed from the very foundational stone of learning, which is the school curriculum. The school curriculum must incorporate gender studies and raise awareness about unpaid domestic and care activities by teaching gender equality, life skills, and socioeconomic importance of such work. It may foster empathy, challenge traditional gender roles, and encourage students to engage in discussions and community projects that promote a more equitable and responsible approach to these vital responsibilities.

There is also a need to quantify the UDCW. Thus, the government must establish a framework for calculating GDP losses attributed to unpaid domestic and care activities. This initiative will not only promote gender equity by recognizing the undervalued contributions of caregivers but also facilitate data-driven policy decisions that address the economic and societal significance of unpaid work.

There should also be some women-centric policies like providing affordable and accessible childcare services to alleviate the caregiving burden, with a particular focus on women. Such initiatives not only empower women to participate in the workforce but also promote economic and gender equality, facilitating a more balanced and inclusive society.

Conclusion

Time-use statistics are essential for measuring the financial contributions of unpaid domestic work. The purpose of the study is to look into how time is spent on unpaid caregiving and household tasks by gender. It also seeks to comprehend the social, economic, and demographic aspects that account for how much time each gender spends on unpaid caregiving and household duties in India as well as how this impacts societal cohesion. The result of this study signifies that women in India are devoting large amounts of time to unpaid care and domestic work while men are spending less than an hour. Furthermore, the amount of time spent on these activities varies greatly depending on the socioeconomic status and demographics of households and individuals. The findings highlight the fact that urban individuals, both men and women, spend less time in unpaid domestic and care activities in comparison to their rural counterparts. Considering their employment status, working women tend to devote less time to UDCW. Self-employed (own account worker, employer, worked as a helper in household enterprise) and unemployed men (did not work but was seeking and/or available for work) dedicate more time to these pursuits in comparison to men who are not in the labor force (attended domestic duties; engaged in free collection of goods; sewing, tailoring, weaving, etc. for household use; rentiers; pensioners; remittance recipients; not able to work due to disability; others such as begging and prostitution). A higher level of household income is positively related to unpaid domestic and caregiving activities for men but is adversely associated with women belonging to higher-income households. Individuals residing in pucca houses devote less time to unpaid domestic and care activities in comparison to those residing in kutcha houses. Further, those using the solid source of cooking fuel devote more time to caregiving activities in comparison to nonsolid fuels. Women facilitated with electricity in households devote less time to caregiving activities in comparison to women in households without electricity. An increased degree of education is positively related to time spent by men and women in unpaid domestic and care activities. Married women invest more time in these pursuits in comparison to unmarried women, and married men also devote more time to unpaid domestic and care activities in comparison to unmarried men. An increase in the number of young children (below 6 years of age) in a household requires an additional amount of time to be devoted by both men and women towards unpaid domestic and care activities. However, a negative relation has been observed with the increasing number of children (6–14 years), working-age population (15–60 years), or senior citizens (60 above) with the time allocated to unpaid domestic and care activities in the household. Islamic women and women from marginalized communities (SC/ST) usually take more time to perform unpaid caregiving and household duties.

The distribution of time by gender as seen through the intersectionality lens shows the increasing invisibility of women's labor in the public sector and supports Talcott Parsons' theory of functionalist perspective that views society as an intricate system which components interact to foster stability and solidarity through gender inequality with a view where predefined gender norms complement each other like men support the family financially, while women look after the home. Moreover, this view is also supported by classical scholars like Plato and Aristotle who believe that men are made up of gold, silver, and iron and are chosen by nature to take up different occupations and oppose granting even citizenship rights to women.

In a nation like India, where cultural norms have a significant impact on women's activity status and minimize their time spent on work-related activities, it is clear that women perform the majority of unpaid household and caregiving tasks. This unpaid work that women are involved in is usually not quantified and therefore is considered to have little, if any, value resulting in gender imbalance and hence creating conflict in social solidarity for getting equal opportunity to develop. SDG 5.4 seeks to address this gender gap and uphold social cohesion by valuing unpaid care and domestic labor and promoting shared duties within the family and home via the provision of public services, infrastructure, and social protection laws. Thus, to reduce gender inequality and promote equitable involvement of men and women in both the public and commercial sectors of the Indian economy, unpaid domestic and caregiving responsibilities must be redistributed within the household and business.

References

Ghosh, J. (2009). *Never done and poorly paid: Women's work in globalizing India*. Women's Unlimited.

Gornick, J., & Meyers, M. (2005). *Families that work: Policies for reconciling parenthood and employment*. Industrial & Labor Relations Review, 58, Article 84. https://www.researchgate.net/publication/37148578.

Government of India, Ministry of Statistics and Programme Implementation. (2020). Time use survey (January–December 2019). National Statistical Office.

International Labour Organization. (2018). *Care work and care jobs for the future of decent work*.

Medina-Hernández, E. J., Fernández-Gómez, M. J., & Barrera-Mellado, I. (2021). Analysis of time use surveys using co-statis: A multiway data analysis of gender inequalities in time use in Colombia. *Sustainability (Switzerland), 13*(23). https://doi.org/10.3390/su132313073

Oxfam India. (2019). *Mind the gap: The state of employment in India in 2019.* .

Palomba, R., & Sabbadini, L. (1995). *Different times-how men and women use their time in Today's Italy*. Istituto Poligrafico e Zecca dello Stato.

Pillay-Naidoo, D., & Vermeulen, C. (2023). Seeking support through solidarity: Female leader's experiences of workplace solidarity in male-dominated professions. *Frontiers in Psychology, 14*. https://doi.org/10.3389/fpsyg.2023.1119911

Sepúlveda, M. (2013). *Report of the Special Rapporteur on extreme poverty and human rights: Unpaid care work and women's human rights*. SSRN Electronic Journal. https://doi.org/10.2139/ssrn.2437791

Singh, P., & Pattanaik, F. (2020). Unfolding unpaid domestic work in India: Women's constraints, choices, and career. Palgrave. *Communications, 6*(1). https://doi.org/10.1057/s41599-020-0488-2

Stiglitz, J., Sen, A., & Fitoussi, J.-P. (2007). *Report on the commission on the measurement of economic performance and social Progress*. Commission on the Measurement of Economic Performance and Social Progress, Paris, Paris. http://www.stiglitz-sen-fitoussi.fr/.

United Nations. (1980). *World conference of the United Nations decade for women: Equality, development and peace, Copenhagen, 14 to 30 July 1980: Report*. United Nations.

United Nations. (2015). *Transforming our world: The 2030 agenda for sustainable development*. United Nations.

World Economic Forum (WEF). (2023). *Global Gender Gap Report 2023*.

Part VII
Hazard, Disaster, and Integrated Crisis Management

Creative Leadership and Solidarity in the Management of the COVID-Related Lockdown Situation in a Grama Panchayat

Neena Joseph

Introduction

This chapter is an attempt to look at the creativity with which the response to COVID-19 was planned and implemented. During 2021–2022, the entire world was living through the experience of COVID-19. This being a pandemic of unprecedented nature with no models to emulate, tremendous creativity was required to contain and manage the pandemic, to fashion our lives and engagements to confront it, and to live with it. Grossman et al. (1990) believes that there are multiple alternative ways of solving a problem, and history bears witness to this fact; whatever the domain of human endeavor is taken as an example, it would be beneficial to examine the application of creativity to solve problems and face the challenges thrown up by this pandemic in a grama panchayat in Kerala, India. This chapter gives the lived experiences of creative governance in the said turbulent period and contains live lessons of creativity, which can be adapted across not only lockdown-like situations, but also any crises. The success of this crisis management can also be attributed to the solidarity that existed among the citizens.

Creativity has been theorized and defined differently as a product, a human capacity, and a process. "Creativity implies the production of a totally or partially novel identity" (Stagner & Karwoski, 1973). "Creativity is the capacity of a person to produce compositions, products or ideas which are essentially new or novel and previously unknown to the producer" (Drevdahl, 1956). According to Anderson (1992), creativity is "nothing more than going beyond the current boundaries, whether those are boundaries of technology, knowledge, current practices, social norms, or beliefs." Cropley (2011) states that creativity is not confined to arts, but is present in various domains of life including business, technology, medicine, administration, education, and architecture, and goes beyond encompassing ideas,

N. Joseph (✉)
Institute of Management in Government (Retd), Faculty, XLRI/LIPI, Kochi, Kerala, India

K. Arora, R. R. Anand (eds.), *The Solidarity Approach in Geography*,
https://doi.org/10.1007/978-3-031-86753-8_25

processes, and services. Creative people identify potential in any situation, and they are "opportunity thinkers" (Grossman et al., 1990). Creativity has utilitarian value.

The attempt is to examine the responses of local self-government in Kerala to the COVID-19 pandemic. Kerala is internationally acclaimed for the efficient management of local self-governments. Isaac (2020) notes that Kerala excelled in COVID management based on recovery vis-à-vis the national performance: recovery (50 vs. 11%), mortality (0.5 vs. 3.4%), and prevention of primary transmission (0.4 vs. 2.6%). Roy and Dav (2020) observes that Kerala's achievement is the result of its unique history of decentralized planning and participatory governance.

With the commencement of Kerala's Peoples Plan Campaign in 1996, many democratic spaces were thrown open for citizens. The opportunity unleashed the creative potential of the citizens while partnering in local governance. In India's three-tiered decentralized system of governance, gram panchayats are the basic self-contained sub-system of self-governance at the ground level. During a pandemic or any disaster, it is the local community and the local self-governments that are the first respondents. Hence, for this paper, a gram panchayat is chosen for study. Kodakara panchayat in Thrissur district in Kerala is selected since Thrissur is the first place in India where a COVID patient was admitted. In Kerala, the stories of the management of COVID were shared extensively through several ongoing webinars organized by the Kerala Institute of Local Administration (KILA) under the name "Orumayude Pradhirodha Kadha" which translates to "The Heroic Stories of the United Resistance." Kodakara had contributed many unique and creative good practices, many of which were subsequently accepted by the state government to be replicated across the state.

This is an exploratory study. The objectives of this study are to examine the significance of creative leadership in managing crises at the level of grassroots governance and to explore the relevance of the already existing solidarity among the citizens in such situations.

The case study method is adopted. The information for this chapter was collected from the YouTube of KILA and by interviewing the Panchayat President (PP) of Kodakara. Due to the COVID-related restrictions, it was not possible to conduct FGDs and to interview more individuals. The study was done during the third and fourth weeks of October 2020.

Case Study of COVID Response in Kodakara Grama Panchayat: Creativity in Action

With an area of 21.29 sq. km, Kodakara Grama Panchayat (KGP) has a predominantly agrarian economy with a population of 33,245 and a literacy rate of 95.8%. KGP has 19 wards with 10,600 households. The Panchayat President (PP) was a head load worker turned elected representative. Right from 2000 till date with a gap of 5 years in between (when the presidentship was reserved for women), he has been the PP. In 2005, he was elected unopposed. During this interval, he managed to pass the tenth standard equivalent examination. He is a voracious reader and has read the Malayalam translation of many classics across the globe. Going through the KILA presentation, it was learned that he is an excellent public speaker with fabulous content and inspiring delivery skills. The political profile of the panchayat samiti (committee) was Left Democratic Front (LDF) (10), United Democratic Front (UDF) (6), and Bharatiya Janata Party (BJP) (3), during the COVID period.

The first COVID patient in the country was hospitalized in Thrissur district during the last week of January 2020. Right from that time, in KGP, exploration commenced about the practical implications of the response to this virus and the panchayat's role in resisting it. From the outset, KGP had the support of health and police departments and that of the district administration in planning the COVID-19 resistance activities (CRAs). The departments of health and police played key roles. CRA activities at the grama panchayat (GP) level were led by the PP and his team. The month of February was utilized for widespread awareness generation. Innumerable meetings were held at the ward level with the involvement of voluntary organizations and many civil society groups to plan and implement the awareness programs. Twenty thousand notices containing the necessary information and the required precautionary measures against COVID-19 were printed and distributed among the households, shops, and markets and in KGP. This was a well-orchestrated campaign with the partnership of Jana Maitri police, social organizations, NGOs, cooperative societies, and business community during the beginning of March.

By March, the number of COVID-affected persons increased. On March 1, 17, and 19, PP held meetings with KGP members, health department officials, and representatives of private hospitals to plan and implement CRA. From the 19th onwards, daily meetings started to be held to evaluate the day's work and to plan the next day's work.

On March 21, to avoid widespread panic, in KGP, posters were printed and exhibited with the slogan, "No need for panic, vigilance is enough." Within 3 days, in all the wards, meetings were organized with GP members, officials from health and other departments, and representatives of various citizen groups. In each ward, a committee of 20 volunteers was constituted to undertake CRA.

The state government proclaimed a lockdown on March 23, 2020, and the national government declared it on March 24. By the time lockdown was announced, systems and structures were in place in KGP to face this public health crisis. There was no delay in making decisions and to forge ahead. Starting on March 24, KGP launched the program of sanitizing public spaces. A disaster management squad of 20 persons was constituted including the cream of elected representatives and officials from KGP, officials of health, police and revenue departments, and citizens. This squad worked synergistically to sanitize the selected spots such as town areas and other crowded places. Various clubs and libraries coordinated these activities. KGP was the pioneer in the state for this activity.

On March 25, KGP allocated Rs. 20 lakhs in the annual plan for CRA.

On March 25, the Chief Minister instructed all the panchayats to organize community kitchens. But on March 24 itself, the preparations of the community kitchen commenced in KGP. On March 26, the first community kitchen in the district started functioning in KGP and began to feed the destitute and the vagrants. These kitchens were bustling with activities, and this became possible by the liberal contribution of shramadhan (voluntary work) by the citizens. Residential associations donated rice and vegetables. Posters were printed and distributed with the caption "anna dhanam mahadhanam," meaning donation of food is the greatest donation. Member of Legislative Assembly (MLA) visited the community kitchens in Kodakara whenever he could and extended all sorts of help. The education minister visited the community kitchen on April 3. Leading newspapers published the news about the activities. Community kitchens were run on donations in the form of labor (mainly by Kudumbashree members), provisions, and vegetables. The panchayat fund was not utilized for this.

A unique program of KGP was to make free travel arrangements for patients who needed chemotherapy and dialysis. KGP got the services of the ambulance from Perambra church, free of cost, and also one from Ekalavya, a social organization at a reduced rate. During COVID time, KGP was able to provide ambulance services, 20 times during the lockdown period alone.

The citizens owned up to the entire program. PP used to post daily events on Facebook. The citizens responded to the Facebook posts, enthusiastically. The donation of Rs. 1200 by a senior lady who subsisted on a welfare pension was specifically noteworthy.

The volunteers were busily engaged in various CRA engagements. They were also concurrently evaluating the activities and making the needed interventions taking instructions from PP and his team. The contribution of doctors, health inspectors, and Accredited Health Activists (ASHA workers) under the supervision of the health department was invaluable. Irrespective of the diverse political affiliations, all the panchayat members cooperated in all activities. The contribution of Kudumbashree members in terms of the continuous toil in the community kitchens was commendable.

From March 27, under the leadership of PP, daily meetings were held at 4 p.m. The team comprising GP members, and the staff of the departments of police, health, and revenue, evaluated the day's CRA and planned the CRA of the succeeding day.

An innovative program, viz. "vilikkoo, veettilethikkam" ("call, your stuff will be delivered at home"), was launched successfully. This was for the delivery of provisions and medicines. This was a godsend, especially for the senior citizens staying alone. Wide publicity was given through multiple channels including posters. They succeeded in distributing medicines to 1500 persons and vegetables to 750 houses with the help of the ward-level volunteers, Jana Maitri police, and the selected staff of the health department. This started on March 30. This solidarity was laudable.

In Thrissur district, the maximum number of guest workers was in Kodakara—1140. Their full data was collected. Meetings were organized to discuss the matter of safeguarding their rights. It is well known that this was the segment of the population that was devastated by the sudden declaration of lockdown, loss of livelihood, and deprivation of even food and shelter. Many employers refused to take care of them. From among them, only 500 persons came through contractors. The KGP took up the responsibility of protecting the rest. Community kitchens were set up. The MLA visited KGP whenever possible, and community kitchens had the privilege of his frequent visits. The district collector also gave all possible support in this matter. KGP apprised Kerala State Local Self Government Department minister and the district administration on this issue. On March 29, the minister came down to the panchayat office, and a meeting was convened with the Superintendent of Police, Revenue Divisional Officer, and District Labour Officer. Through interventions, the identified employers were directed to provide them with food, shelter, and basic needs. Based on the directions of the High Court, a group comprised of district judges and sub-judges visited the migrant laborers and interacted with them since KGP was the panchayat having the maximum number of guest laborers. The guest laborers conveyed their utmost contentment with being taken care of and protected very well by the KGP. This thoughtful gesture is in glaring contrast with the fate of many guest laborers who, from many other parts of India, walked several hundred kilometers to reach their homeland. Some of them were crushed dead on the railway track where they laid their head to capture a wink of sleep to ease their fatigue. Some died, and some others gave birth on the way. The term guest workers was coined in Kerala to refer to the laborers who migrated to Kerala from northern and eastern states to better their economic prospects through physical labor. Kodakara set the trend for the special care given to the guest laborers.

To retain the citizens within their homes during the lockdown and also to make them stay at home during the quarantine period, arrangements were made to deliver books to their doorsteps from the KGP library. KGP library is Wi-Fi enabled and houses 30,000 books. From March to October, 1000 books were delivered to doorsteps. The slogan was "Home can be made library—stay at home, stay safe." This door delivery service commenced in the first week of April.

Another citizen owned three drones, and he happily let KGP utilize them for surveillance to detect and prevent the breach of COVID protocols. This was used for 3 days during the second week of April. Some groups of men who were playing cards were identified, and such crowding could be prevented.

As the cases were increasing, the GP arranged the Ayurveda hospital as the isolation ward with 30 beds on April 4. During April, 326 persons were quarantined in 7 isolation wards. Health workers in the health department were assigned the duty of visiting quarantined persons daily.

During May, poor families who did not come under any welfare pension schemes of the government were given Rs. 1000 each from KGP.

COVID-19 threw everything in disarray. The Government of Kerala declared the Subhiksha Keralam project to achieve self-sufficiency in vegetables for the households. In 10,000 households, a packet of vegetable seeds was delivered. Classes on vegetable cultivation were already organized in KGP before January. Forty thousand saplings of various vegetables and also those of banana (1000) and rambutan (1000) were distributed. These started on April 3. The slogan was "veettil oru pachakkarithottam," i.e., a vegetable garden in each house. The idea was that each household member develop the habit of vegetable cultivation and would become self-sufficient in this aspect. This started in June after the relaxation of the lockdown. 11.5 tons of ginger, 6 tons of chena, 6 tons of chembu, and 3.5 tons of turmeric seeds were distributed.

The ration was provided by the government regularly, and welfare was taken care of by KGP. The wives at home toiled hard and met major family expenses by borrowing money from Kudumbashree. During the lockdown, many men slipped into the habit of not going out for work, while women folk toiled tirelessly both at home and in community kitchens. A very short film was produced using a mobile phone, depicting the theme of overburdened women and the irresponsibility of men. This was a nil-budget scheme utilizing the artistic talent of a citizen who volunteered to do this. The idea emerged during discussions, while PP was overseeing the functioning of community kitchens, where a few leaders and volunteers used to congregate on a daily basis for evaluation and planning of CRA. on a daily basis.

KGP received generous donations which were inspired by the faith and conviction of the people in the dedicated services rendered through CRA. So much so that, KGP was able to donate Rs. 5 lakhs to Chief Minster's disaster relief fund.

A few days before each marriage in KGP, a team under the leadership of a ward member would visit the house, give a small gift, and present a mangala pathram. This contained tips for a happy family life along with the protocols to be observed while conducting the marriage function. This practice started in the first week of August. This was an ingenuous way to implement a compliance mechanism without coercive control.

One noteworthy practice in KGP was the daily publishing of the COVID statistics in the panchayat with ward-wise, age-wise, and gender-wise details without mentioning the names. The source of infection was also mentioned. This transparency helped in motivating the citizens to take precautionary measures. This started in the first week of September.

In response to the CM's exhortation for volunteerism, in May, 150 persons responded positively. GP started to avail their services as well for their CRA.

Discussion

It can be seen that many meetings were held extensively and there was daily evaluation and planning of activities. Some aspects of brainstorming might have happened, especially for collecting divergent ideas and then convergently evaluating them and choosing appropriate useful and viable alternatives. We can see that the political composition is very precarious—ten LDF and nine UDF members in the KGP samiti. Still, COVID could be managed in team spirit. The continuous political term of PP for 15 years shows his acceptance among the citizens. The success of CRA would not have been realized if the opposition members were not part of the team right from the early planning stages and in all the preliminary discussions.

It can be seen that the success story is the result of inputs from a wide spectrum of persons from all levels including the district collector and ministers. Many cross-learnings might have happened during the KILA webinars. We can also see the convergence of talents and the creative use of the talents for CRA activities, be it the use of drones for surveillance, awareness generation, prevention of COVID-19-spreading activities during weddings, and timely production of a short film to put down unhealthy tendencies. Entities with diverse talents and expertise contributed to the cause, and it is this solidarity which undergirded the success of CRA in the panchayat. Creativity is the ability to take anything and make something from it (Grossman et al., 1990).

According to Khandwalla (1984), "novelty and relevance" are hallmarks of creativity. Creative responses displayed in KGP are indisputably relevant and have been replicated across the state through government orders. A few examples are volunteers, mangalaparam, and "vilikkoo veettilethikkaam." The national daily, Indian Express, had reported these innovative processes. Chief Minister in his daily broadcast press meeting mentioned these creative steps and exhorted other local bodies to follow suit. The figures for the number of beneficiaries for all types of door delivery services indicate that all possible attempts were made to delay the spread of the virus to the community. Door delivery was arranged for provisions, medicines, books, saplings, and what not. This was a very effective feat for preventing the rapid spread of COVID-19 in the initial days and thus delaying the peak of the pandemic.

Khandwalla elaborates on six abilities for creativity—fluency, flexibility, originality, sensitivity, ability to identify the causes, and elaboration. COVID times threw up the challenge of retaining citizens inside their homes. Fluency is exhibited by the multipronged approach to solving the problem through door delivery of many types of services of and by engaging the citizens inside their homes with vegetable cultivation. Originality can be observed in the practice of visiting the houses that are planning a marriage, giving a gift, presenting a manageable pathra, and discussing protocol issues. This ensures compliance with COVID protocol. Sensitivity is evidenced by how the tendency to evade work and shirk the responsibility of men during COVID-19 days was identified. Making a short film, portraying the problem, and suggesting solution are quite original ideas. This issue of shirking responsibility by men came up in the course of casual conversation on the premises of the community kitchen, and one solution which came up was the production of a short film. The presence of all the six qualities mentioned above can be seen in the initiatives. KGP Samiti was sensitive to the needs of the neediest among the poor and could go deep down to the possibility of them getting shattered in the eventuality of a medical emergency in a family during COVID-19 times. The ambulance scheme was born out of this sensitivity and so also the scheme to distribute Rs. 1000 to those who did not have any welfare pensions.

The rapidity with which the response measures were implemented was amazing. According to Khandwalla, the blocks to creativity fall into seven categories. The PP did not have a fear of failure, since there was strong public support. Again, the learnings from the management of the 2018 floods in Kerala came in handy. There was no allergy to ambiguity—it is seen that any good ideas were readily accepted by the team. There was no touchiness. Support was sought through posters. Facebook announcements of schemes attracted donations. Community kitchens were run under the full sponsorship of people at all levels. Even poor people contributed financially. Residential associations also cooperated wholeheartedly. Resource myopia is another block to creativity. But this was absent in the case of KGP—drones, libraries, and talents to produce films all came in handy during the COVID days to resist and contain COVID-19. There was no dearth of material resources that could be garnered—be it vegetables, books, provisions, or medicines. There was no issue of starved sensibilities because of the wide reading habits, political experience, and habit of mingling freely and conversing without reservation—all these habits enabled PP to empathize with the citizens and to develop a superb level of sensitivity to the problems. In KGP, there already was the practice of brainstorming for ideas, and KGP benefited from the wisdom of a wide variety of persons. Cross learnings happened. There was no rigidity. KGP rose to the occasion and geared up to face the COVID. Khandwalla was describing these blocks at the level of an individual. PP, being a voracious reader himself, was eager to encourage the citizens to broaden their horizons through reading. The case study does not give evidence of such blocks which could have raised formidable barriers in the response to the pandemic. The gradual buildup of trust over the past two decades as a people's representative came in handy in the management of COVID and was indeed an enabling factor. The volunteerism offered to PP by the citizens was crucial for its success.

The leadership rendered during COVID was being discussed through the lens of creativity.

While going through the activities in KGP, it at once becomes clear that all this creativity could not have been possible, if there was not an already prevailing firm foundation of solidarity. Functions, funds, and functionaries could not have been coordinated and streamlined with this amazing rapidity in the absence of such a background. During COVID days, the already citizen-centric KGP became a space that could be approached and accessed by the citizens around the clock. The solidarity was visibilized and fortified during the 2018 floods in Kerala. There was splendid and seamless coordination in managing the flood which affected 600 families. The 600 families were given shelter in rapidly put-up 11 camps. Citizens with sufficiently large homes offered space in their homes for their fellow citizens who lost their houses or who could not live any longer in the flood-affected houses. KGP had a glorious history of managing the camps through local philanthropy. KGP was able to incorporate this experience and the learnings in the management of the pandemic.

The dictionary meaning of solidarity is: "Solidarity is an awareness of shared interests, objectives, standards, sympathies creating a psychological sense of unity of groups or classes." Richards (2001) argues that neither egalitarianism nor hierarchy alone can ensure the cohesion of social groups in the long run. It can be seen that the PP adopted a blended approach with strict surveillance with the help of drones during the lockdown period. At the same time, he merged with the crowd and maintained continuous communication with the citizens. For example, the issue of irresponsible males caught his attention amid his direct communication in the community kitchen where women were toiling from dawn to dusk to feed the migrant laborers and the senior citizens living alone. The idea of Managala Pathra was implemented uncompromisingly and meticulously while at the same time making the enforcement palatable and compliance-ready by winning their hearts through gifts and family visits. In all their endeavors, he offered an opportunity for the citizens to flourish—the youth prepared a script and acted it out for the short film to reform the errant husbands.

Conclusion

This case study demonstrates the efficacy of decentralized governance and its paramount significance in responding to grassroots-level needs, especially during a disaster, be it a flood or a pandemic. The first response in a disaster comes from the local population. This response can be made effective and efficient if there is local creative leadership undergirded with a rock-solid foundation of socially, culturally, and historically constructed base of solidarity. Resources could be easily marshaled from multiple sources in an emergency if there is solidarity within the community. Decentralized governance offers plenty of opportunities for the citizens to partner with the local government to solve their problems creatively, nurture social capital, and forge solidarity. This type of local body can set trends and create models that can be replicated and up-scaled.

Recommendation

Decentralized governance has to be strengthened in the country. Solidarity has to be consciously and deliberately created, nurtured, and developed, which is the most powerful base that can be leveraged for synchronized teamwork which is a lifeline to deal with any kind of disaster and unfavorable situation. We need to select leaders with a democratic spirit—who are sensitive to the needs of the citizens, especially those of the vulnerable sections; who mingle with the citizens; who identify their expertise and talents; and who utilize them to fulfill the needs of the citizens. Leaders selected should have the capability to secure the support of the top echelons of governance to realize the needs of the people.

Acknowledgements I acknowledge Mr. Madan Mohan, the CEO of Panchayat Presidents' Association in Kerala, with whom I had made many rounds of discussion before finalizing the selection of this panchayat. I also acknowledge with gratitude the time and energy spent by Mr. PR Prasadan, the then President of Kodakara Panchayat in Kerala. Due to COVID constraints, I could make only telephone conversations with him. Without his wholehearted support and time spared for several interviews, this chapter would not have materialized. I also thank the several citizens and a few members of the panchayat with whom I spoke to triangulate the information.

References

Anderson, J.V. (1992) 'Weirder than fiction: The reality and myths of creativity', The Executive, 6(4). p. 41: http://search.proquest.com.ezproxy.waikato.ac.nz/docview/210534761?pq-origsite=summ on.

Cropley, A. J. (2011). Definitions of creativity. In M. A. Runco & S. R. Pritzker (Eds.), *Encyclopedia of creativity* (pp. 511–524). Academic Press.

Drevdahl, J. E. (1956). Factors of importance for creativity. *Journal of Clinical Psychology, 12*, 22.

Grossman, S. R., Rodgers, B. E., & Moore, B. R. (1990). *Unlocking creativity in the workplace*. Jaico Publishing House.

Isaac, TM Thomas. (2020). *What nation can learn from Kerala: Lockdown is not enough: Preparedness, dencentralisation, are key*. Retrrieved from Apr 17, 2020, from https://indianexpress.com/article/opinion/columns/coronavirus-covid-19-kerala-curve-6365935/

Khandwalla, P. N. (1984). Fourth eye excellence through creativity

Richards. (2001). Social solidarity. *International Encyclopedia of Social and Behavioural Sciences*

Roy, A., & Dav, E. S. K. (2020). Analysing Kerala's response to the COVID-19 pandemic when people and governments come together. Economic and Political Weekly, 55(18).

Stagner, R., & Karwoski, T. F. (1973). *Educational psychology* (p. 314). Eurasia Publishing House.

Use of ICT for Disaster Relief Distribution Across Hostile Terrain

Bhoomi Gupta and Sachin Gupta

Introduction

Disaster risk management (DRM) is at the forefront of humanitarian efforts in the twenty-first century due to the increasing frequency and severity of both natural and man-made disasters. DRM has evolved into a critical field of study and practice, aiming to mitigate disaster impact, save lives, and aid affected communities in their recovery. It encompasses a range of activities from risk assessment and preparedness to response and recovery strategies.

The importance of disaster risk management cannot be emphasized enough, as disasters continue to disrupt communities, strain resources, and challenge our ability to respond effectively. In this context, information and communication technologies (ICT) have become invaluable tools in enhancing the effectiveness of DRM. ICT, which includes mobile applications, data analytics, and communication networks, has revolutionized disaster management by enabling real-time information exchange, coordination, and resource allocation.

An ongoing challenge in DRM is the need for innovative relief distribution solutions. Traditionally, the distribution of essential supplies, especially in challenging terrains and remote disaster-affected areas, has been a complex and resource-intensive task. Conventional means of transportation often become inaccessible or inadequate after a disaster, leaving affected populations vulnerable and isolated. This chapter explores a novel approach that leverages ICT and unmanned aerial vehicles (UAVs) to address this specific challenge.

B. Gupta (✉)
Department of ITE, MAIT, GGSIPU, Delhi, India
e-mail: bhoomigupta@mait.ac.in

S. Gupta
Department of CSE, MAIT, GGSIPU, Delhi, India
e-mail: sachin.gupta@mait.ac.in

K. Arora, R. R. Anand (eds.), *The Solidarity Approach in Geography*,
https://doi.org/10.1007/978-3-031-86753-8_26

Given the escalating disaster risks and the growing impact of climate change, innovative solutions are imperative. The integration of ICT into relief distribution holds promise in improving the speed, precision, and effectiveness of response efforts. This chapter focuses on the development and application of a mobile-based relief distribution application, with a focus on its potential to bridge the gap in disaster-affected areas where conventional means of transportation and communication are limited. We consider a scenario in which rescue and relief teams are equipped with basic smartphones and have access to UAVs capable of delivering essential supplies over short distances.

Disaster Risk Management and Its Challenges

Understanding Disaster Risk Management

Disaster risk management is a comprehensive approach aimed at lessening the adverse effects of disasters on communities, economies, and the environment. It is a dynamic and evolving field that covers preparedness before a disaster, immediate response during a disaster, and recovery efforts post-disaster. DRM is firmly based on the understanding that disasters, whether triggered by natural hazards or human activities, are intricate occurrences influenced by a variety of factors.

DRM involves proactive measures to recognize, evaluate, and alleviate risks. This encompasses analyzing vulnerabilities, evaluating potential hazards, and formulating strategies to bolster resilience (Munawar et al., 2022). The primary objective is to reduce the loss of life, property, and livelihoods while ensuring a prompt and effective recovery of affected areas.

Challenges Encountered in Traditional Relief Distribution

Traditional relief distribution confronts significant challenges, especially in harsh terrains and disaster-affected regions. These challenges can significantly obstruct timely and efficient response efforts, and some of the challenges are listed below:

1. *Inaccessibility*: Disasters frequently render roads impassable, make airports unusable, and disrupt traditional supply chains. Harsh terrains, like mountains or flood-prone areas, exacerbate these challenges.
2. *Limited resources*: In disaster-affected areas, essential resources such as food, water, and medical supplies become scarce, generating a high demand for swift distribution. Conventional methods may struggle to keep up with this demand.
3. *Communication disruptions*: Communication networks may be damaged or overwhelmed during disasters, impeding coordination among relief teams and affected communities.

4. *Logistical complications*: Orchestrating the movement of relief supplies, personnel, and equipment in chaotic post-disaster environments is a logistically intricate and time-consuming task.
5. *Security challenges*: Hostile terrains can introduce security risks for both relief workers and the affected population, further complicating the distribution process.

DRM: ICT Opportunities

Information and communication technologies (ICT) has emerged as a one-stop solution to address the multitude of challenges inherent in conventional relief distribution. ICT presents several key advantages over the traditional approaches:

1. *Real-time data*: ICT enables the swift collection and dissemination of real-time data, allowing relief agencies to promptly assess the situation on the ground, identify critical needs, and allocate resources efficiently.
2. *Remote coordination*: Communication technologies facilitate remote coordination, allowing relief teams to collaborate across geographical distances. This is especially crucial when accessing hostile terrains poses challenges.
3. *Data analysis*: ICT tools such as geographic information system (GIS) assist in analyzing disaster impacts, identifying vulnerable populations, and optimizing resource allocation.
4. *Mobile applications*: Applications designed for mobile devices provide accessible platforms for information sharing, enabling affected individuals to request assistance or report their needs directly to relief agencies.
5. *UAVs for relief delivery*: Unmanned aerial vehicles (UAVs or drones) equipped with ICT capabilities can surmount geographical barriers by delivering essential supplies to remote and hostile terrains.

The subsequent sections of this chapter dig deeper into the multiple dimensions of ICT's role in disaster relief distribution.

Advantages and Constraints of ICT Solutions

The integration of ICT in disaster risk management (DRM) not only magnifies the speed and efficiency of relief distribution but also enhances situational awareness and reduces response times, resulting in the preservation of lives as discussed further.

Advantages

1. *Heightened situational awareness*: ICT furnishes real-time data and situational awareness, facilitating faster decision-making.

2. *Enhanced coordination*: Communication tools enable seamless collaboration among relief agencies and stakeholders.
3. *Accessibility*: Mobile-based applications simplify the process for affected populations to request assistance and access information.

While the advantages of ICT in disaster relief are substantial, it is imperative to recognize the constraints as well:

Limitations

1. *Dependency on infrastructure*: ICT solutions are contingent on functional communication networks and power sources, which may be disrupted during disasters.
2. *Digital disparities*: Not all affected populations have access to ICT devices or possess the necessary digital literacy to use them effectively.
3. *Data privacy and security*: Safeguarding the privacy and security of sensitive data poses a concern in the utilization of ICT.
4. *Maintenance and training*: ICT systems necessitate continual maintenance and training for relief personnel.

UAVs as an Emerging Use Case of ICT in Disaster Relief

Among the multiple solutions being offered by ICT for DRM, an emerging and notably impactful application for hostile terrain lies in the use of unmanned aerial vehicles (UAVs) or drones. These technologies, integrated with ICT capabilities, have brought about a paradigm shift in disaster relief efforts (Messaoudi et al., 2023).

Introduction to UAVs and Their Capabilities

Unmanned aerial vehicles (UAVs), commonly known as drones, have emerged as a transformative force in disaster relief operations. These remotely piloted aircraft, equipped with diverse sensors and technologies (as shown in Fig. 1), have expanded beyond their initial military usage to become indispensable tools in humanitarian endeavors. UAVs come in various sizes, ranging from compact quadcopters to more substantial fixed-wing aircraft, each tailored to specific relief tasks.

The capabilities of UAVs are extensive and encompass the following:

1. *Aerial surveillance*: UAVs excel in capturing high-resolution images and videos of disaster-affected areas, offering real-time situational awareness to relief agencies. This aerial perspective is particularly valuable in challenging terrains where ground access is constrained.

Fig. 1 UAV monitoring in a hostile terrain. (Source: Authors)

2. *Swift assessment*: UAVs facilitate rapid and precise damage assessments by capturing detailed imagery of affected infrastructure, enabling relief teams to effectively prioritize response efforts.
3. *Aid delivery*: Certain UAVs are engineered for cargo transportation, capable of conveying essential supplies, medical equipment, and even lifesaving medications to remote or inaccessible areas. This functionality is especially critical in hostile terrains where conventional transportation faces hindrances (Gupta and Gupta, 2019).

The Transformation of Relief Distribution Through UAVs

UAVs are fundamentally reshaping the landscape of relief distribution, presenting several exclusive advantages, some of which have been presented below:

- **Speed and agility**: UAVs can be swiftly deployed, ensuring a timely response to disaster-affected areas. They demonstrate exceptional agility in navigating challenging terrain, reaching isolated communities more expeditiously than traditional ground-based methods.
- **Risk mitigation**: Particularly in hostile terrains where physical access poses significant risks to relief workers, UAVs operate with minimal human exposure, enhancing safety for both relief personnel and the affected population.
- **Precision in delivery**: UAVs excel in delivering aid to precise locations, guaranteeing that essential supplies reach their intended recipients with accuracy. This precision minimizes waste and optimizes the impact of relief efforts.
- **Data acquisition**: Beyond aid delivery, UAVs gather valuable data on disaster impacts and response needs. This data enriches decision-making processes, enabling relief agencies to allocate resources with precision.

Case Studies Demonstrating UAV Utilization in Disaster Relief

To demonstrate the tangible impact of UAVs in disaster relief distribution, notable case studies underscore their efficacy in navigating hostile terrains:

- **Nepal Earthquake (2015)**: Following the devastating earthquake in Nepal, UAVs were instrumental in assessing damage to heritage sites, road networks, and remote villages. They played a pivotal role in guiding relief efforts and ensuring that aid reached isolated communities amidst the rugged Himalayan terrain.
- **Puerto Rico Hurricane Maria (2017)**: UAVs were deployed to assess damage to critical infrastructure, such as power lines and bridges, in the aftermath of Hurricane Maria in Puerto Rico. They facilitated rapid response and assisted in restoring essential services in challenging and remote locations.
- **COVID-19 Response**: Amid the COVID-19 pandemic, UAVs were utilized to deliver crucial medical supplies, test kits, and personal protective equipment to remote and inaccessible areas. Their deployment mitigated the risk of infection transmission and ensured timely access to healthcare.

These case studies underscore how UAVs have become indispensable tools in disaster relief, especially in scenarios where conventional methods struggle to meet the demands of hostile terrains and remote locations. UAV technology continues to advance, presenting new possibilities for its application in disaster response and recovery efforts.

Augmenting ICT Effectiveness in DRM Using Mobile Apps

As a logical progression, integrating a mobile app into disaster risk management (DRM) strategies can significantly enhance ICT effectiveness, particularly in challenging terrains. The modality involves creating a comprehensive and user-friendly mobile application that facilitates real-time communication, data collection, and coordination among stakeholders involved in disaster relief efforts. This app should encompass the following key features:

1. **Real-time communication**: Enable instant and seamless communication between relief teams, local authorities, and affected populations. This could include text messaging, voice calls, or video conferencing functionalities. The application should also be able to use near-field communication or other peer-to-peer infrastructure-less methods in case the network service is patchy.
2. **Geolocation and mapping**: Integrate GPS capabilities to provide accurate location data, aiding in resource allocation and swift response to critical areas.
3. **Data collection and reporting**: Allow for the collection of essential data directly through the app, enabling immediate reporting of disaster impacts, relief needs, and resource availability.

4. **Resource coordination**: Establish a platform for efficiently coordinating resources, including volunteers, supplies, and equipment, by matching available resources with the areas in urgent need.
5. **Information dissemination**: Disseminate vital information about the disaster, relief efforts, safety protocols, and available assistance through the app to both relief workers and affected individuals.
6. **Multilingual support**: Include multilingual support to ensure accessibility and understanding for a broader range of users, considering linguistic diversity.

Scenario for Augmenting UAV Potential Through the Mobile App

In a hypothetical scenario following a severe earthquake in a remote, mountainous region, a mobile app designed for disaster relief is employed to enhance UAV potential:

1. **App deployment**: Relief teams quickly deploy the mobile app, ensuring that all team members have it installed on their smartphones (Gupta and Gupta, 2017). The app is programmed to interface with UAVs and display relevant UAV-collected data.
2. **UAV coordination**: When UAVs are deployed for aerial assessments and surveillance, the app establishes a live link with the UAVs, enabling real-time streaming of imagery and data directly to the relief teams on the ground.
3. **Data analysis and decision-making**: Relief personnel use the app to analyze the UAV-collected data, assessing damage severity and identifying areas needing immediate assistance (as shown in Fig. 2). This analysis aids in prioritizing relief efforts.
4. **Resource allocation**: The app assists in swiftly allocating resources, including medical supplies and emergency kits, based on the information gleaned from UAV imagery and on-ground reports.
5. **Aid delivery coordination**: UAVs equipped for cargo delivery use the app to receive instructions on where to deliver aid, ensuring precise and timely delivery to the most affected and inaccessible areas.

By combining the capabilities of UAVs and a purpose-built mobile app, disaster relief efforts are significantly optimized, especially in hostile terrains, enabling faster and more effective response and recovery operations.

Fig. 2 Mobile application for transportation modeling. (Source: Authors)

Network Infrastructure Failure and NFC

During a disaster, network infrastructure failure is a critical concern. In such situations, peer-to-peer communication using near-field communication (NFC) or similar technologies becomes crucial:

1. **NFC for communication**: NFC allows devices to establish a direct communication link within a close range, even without an internet connection. Relief workers can use NFC-enabled devices to share critical information and coordinate and exchange data.
2. **Offline data sharing**: NFC facilitates offline data sharing, essential when Internet connectivity is compromised. Relief teams can share vital updates, reports, and assessments directly through their NFC-enabled smartphones (Nefros et al., 2022).
3. **Simplified data exchange**: NFC simplifies the exchange of specific information, such as medical records, emergency contacts, or resource availability, directly between devices, aiding in efficient disaster response.
4. **Redundancy in communication**: Integrating NFC alongside conventional network-dependent communication ensures redundancy in communication channels, enhancing the resilience of the communication system in disaster scenarios (Sharifi and Allam, 2023).

Considerations for Context and Usability

In the development and implementation of the relief distribution mobile app, it is imperative to address contextual and usability factors (Tarhan et al., 2016):

- **Cultural sensitivity**: The app must be culturally sensitive, respecting local customs and practices. Incorporating multiple language options and culturally appropriate content is vital for effective communication and user engagement.
- **Accessibility**: Ensuring the app's accessibility to all is crucial, including individuals with disabilities or limited digital literacy. User interface design should be intuitive and easy to navigate.
- **Data security and privacy**: Prioritizing robust data security measures is essential to safeguard sensitive information. Maintaining user privacy should be of utmost concern throughout the app's design and deployment.
- **Scalability**: Designing the app with scalability in mind is vital, allowing it to handle increased user loads during large-scale disasters without compromising performance. This scalability ensures continued effectiveness during times of heightened demand.
- **Training and support**: Providing comprehensive training and ongoing support for relief personnel and affected individuals is necessary to ensure the effective and efficient utilization of the app. Training should encompass app functionalities, disaster protocols, and troubleshooting guidance.

Implementation Challenges

Implementing the proposed relief distribution app integrating UAV usage poses several challenges, which may be related to the characteristics of the UAV on the one hand and integration with the existing solutions as discussed below.

Challenges in UAV Usage

1. **Integration complexity**: Integrating the app seamlessly with UAV operations, ensuring real-time data exchange, and enabling UAV control through the app require intricate technological integration.
2. **Technical expertise**: Relief personnel need training to effectively operate the UAV functionalities within the app. Acquiring the necessary technical expertise can be time-consuming and resource-intensive.
3. **Cost constraints**: Acquiring and maintaining UAVs and the app infrastructure can be expensive, especially for organizations with limited budgets. Funding constraints may hinder optimal implementation.

4. **Regulatory compliance**: Adhering to aviation and disaster relief regulations, especially regarding UAV operations in disaster-stricken areas, presents legal and regulatory challenges that need to be navigated.
5. **Battery life and range limitations**: UAVs have limited battery life and operating range, potentially restricting their coverage and effectiveness in reaching remote or disaster-affected areas.
6. **Weather conditions**: Adverse weather conditions like strong winds, heavy rain, or storms can impede UAV operations, affecting their ability to assist effectively during disasters.

Challenges in Practical Implementation and Integration

Infrastructure readiness: One of the primary considerations during implementation is the readiness of the infrastructure. Reliable mobile networks and access to power sources are essential prerequisites for the application to function effectively. In areas prone to disasters, ensuring the resilience of these infrastructure elements is a critical task.

Training and capacity building: Relief personnel and affected individuals need training to utilize the application efficiently. Training programs should cover not only the technical aspects of the application but also its broader implications for disaster relief coordination. Capacity-building efforts should extend to local communities, ensuring that users at all levels can actively participate in the relief process.

Integration with existing systems: In many cases, the relief distribution application will need to integrate with existing disaster management systems and databases. This integration ensures seamless data flow and coordination between the application and other relief efforts. Compatibility and data interoperability become critical considerations (Han et al., 2023).

User engagement and feedback: Engaging with users, including both relief workers and disaster-affected populations, is essential. Regular feedback loops should be established to gather insights, identify areas for improvement, and ensure that the application remains responsive to evolving needs.

Future Prospects and Conclusion

The inception of the relief distribution application and its integration with information and communication technologies (ICT) for disaster risk management (DRM) mark the initial stages of an exciting evolution in humanitarian response (Khan et al., 2020). Anticipating future progress, numerous prospects and advancements in this domain come into view:

- **Enhanced predictive analytics:** Accurate predictive modeling of disaster impacts will become pivotal. This will empower relief agencies to proactively position resources, strategize for potential scenarios, and enhance risk mitigation.
- **5G and satellite connectivity:** The ongoing deployment of 5G networks and advancements in satellite connectivity will significantly bolster the reliability and speed of communication in disaster-affected areas, encompassing even the most challenging terrains. This will enable seamless coordination and real-time data exchange.
- **IoT and sensor integration:** The integration of the Internet of Things (IoT) and sensor technologies will assume a more prominent role in disaster monitoring and response. These devices will furnish valuable real-time data concerning environmental conditions, infrastructure integrity, and safety of affected populations.
- **Humanitarian UAV innovations:** The continuous development of specialized UAVs tailored for disaster response, such as cargo-carrying drones and long-endurance surveillance UAVs, will persist. These advancements will further expand the capabilities of relief agencies, facilitating access to remote and hostile terrains (Mohd Daud et al., 2022).

Concluding Remarks

The utilization of ICT in enhancing disaster relief distribution across hostile terrains signifies a remarkable shift in the humanitarian sector, where technology acts as a catalyst, amplifying response efforts. Through the utilization of ICT tools, real-time data, and UAV technologies, relief agencies can surmount geographical barriers, curtail response times, and ensure that vital supplies reach those in need, even in the most arduous environments.

Looking ahead, the integration of ICT in disaster risk management (DRM) will continue its evolution, offering vast opportunities for further innovation, efficiency, and effectiveness. The relief distribution application and the case studies discussed in this chapter illuminate a path of hope, showcasing the potential to save lives and alleviate suffering in the face of disasters. In a world where the frequency and intensity of disasters are escalating, the ongoing advancement of ICT for DRM stands as a testament to our unwavering commitment to fostering a more resilient and responsive global community. It is a beacon guiding us toward a future where technology plays an increasingly vital role in safeguarding and supporting our fellow humans during their most vulnerable moments.

References

Messaoudi, K., Oubbati, O. S., Rachedi, A., Lakas, A., Bendouma, T., & Chaib, N. (2023). A survey of UAV-based data collection: Challenges, solutions, and future perspectives. *Journal of Network and Computer Applications, 216*, 103670. https://doi.org/10.1016/j.jnca.2023.103670

Mohd Daud, S. M. S., et al. (2022). Applications of drone in disaster management: A scoping review. *Science & Justice, 62*(1), 30–42. https://doi.org/10.1016/j.scijus.2021.11.002

Khan, A., Gupta, S., & Gupta, S. K. (2020). Multi-hazard disaster studies: Monitoring, detection, recovery, and management, based on emerging technologies and optimal techniques. *International Journal of Disaster Risk Reduction, 47*, 101642. https://doi.org/10.1016/j.ijdrr.2020.101642

Gupta, S., & Gupta, B. (2019). Performance modeling and evaluation of transportation systems using analytical recursive decomposition algorithm for cyclone mitigation. *Journal of Information and Optimization Sciences, 40*(5), 1131–1141. https://doi.org/10.1080/02522667.2019.1638003

Munawar, H. S., Mojtahedi, M., Hammad, A. W. A., Kouzani, A., & Mahmud, M. A. P. (2022). Disruptive technologies as a solution for disaster risk management: A review. *Science of the Total Environment, 806*, 151351. https://doi.org/10.1016/j.scitotenv.2021.151351

Sharifi, A., & Allam, Z. (2023). Chapter 10—Contributions of smart technologies to disaster resilience. In A. Sharifi & A. R. Khavarian-Garmsir (Eds.), *Urban climate adaptation and mitigation* (pp. 219–240). Elsevier. https://doi.org/10.1016/B978-0-323-85552-5.00010-5

Gupta, B., & Gupta, S. (2017). ICT implementation- GIS and Smartphone application for disaster evacuation. In *Proceedings of International Conference on Information, Communication and Computing Technology* (pp. 5–16). Edupedia Publications Pvt Ltd.

Han, D. C., Takabatake, T., Esteban, M., & Shibayama, T. (2023). Development and implementation of a smartphone application for tsunami evacuation. *International Journal of Disaster Risk Reduction, 96*, 103915. https://doi.org/10.1016/j.ijdrr.2023.103915

Nefros, C., Kitsara, G., & Loupasakis, C. (2022). Geographical information systems and remote sensing techniques to reduce the impact of natural disasters in smart cities. *IFAC-PapersOnLine, 55*(11), 72–77. https://doi.org/10.1016/j.ifacol.2022.08.051

Tarhan, C., Aydin, C., & Tecim, V. (2016). How can be disaster resilience built with using sustainable development? *Procedia-Social and Behavioral Sciences, 216*, 452–459. https://doi.org/10.1016/j.sbspro.2015.12.059

From Adversity to Unity: Women, Pandemic, and Social Solidarity

Maya Kumari, Pushpanjali Kumari, and Shubham Kumar Sanu

Introduction

The gendered impacts of pandemic outbreaks have not been addressed by any policy, public health efforts, or specialists. Recently, the Gender Gap Index Report 2022 was published, in which India's rank is 135 among 146 countries. It is also the worst performer in health and survival in the world (The Hindu, 2022). The repercussions of the pandemic are not uniform, and its impact is intricately woven into the preexisting fabric of inequality, hierarchy, and discrimination. Historical data from past epidemics in Indiac, such as Ebola, indicates that women and children were disproportionately vulnerable, facing heightened risks of violation and discrimination (Anand et al., 2023). Since the COVID-19 pandemic is still unfolding, many challenges are arising, mainly in the socio-economic context, and many of the negative consequences, particularly for women, may remain unknown to the general public. Women have been disproportionately affected by the crisis in many ways (Sharma et al., 2023a, b). During times of crisis, women's lack of control over resources such as education, land, finance, and technology generally translates into lower incomes, fewer options, and increased risk (Krishnakumar, A. and Verma S., 2021).

Emerging data and reports from those on the front lines reveal that all forms of violence against girls and women, including domestic abuse, have increased since

M. Kumari
Delhi School of Social Work, University of Delhi, Delhi, India

P. Kumari
Department of Sociology, Banaras Hindu University, Varanasi, Uttar Pradesh, India
e-mail: pushpanjalipathak@bhu.ac.in

S. K. Sanu (✉)
Department of Geography, Indraprastha College for Women, University of Delhi, Delhi, India

K. Arora, R. R. Anand (eds.), *The Solidarity Approach in Geography*,
https://doi.org/10.1007/978-3-031-86753-8_27

the beginning of COVID-19. Even before the COVID-19 pandemic, the UN women's data declares that 1 in 3 women worldwide had experienced physical or sexual violence, the majority of which was committed by an intimate relationship. During the global crisis outbreak, domestic violence against women has reached epidemic proportions in India, which the UN has termed a "shadow pandemic" (The Hindu, 2022). In developing countries like India, disease outbreaks enhance the responsibilities of girls and young women in caring for old and ill family members as well as out-of-school siblings. In fulfilling familial responsibilities, they usually ignore their education and health (Sharma et al., 2022). However, this pandemic increased their vulnerability more because early and forced child marriage, girls' education dropout, early pregnancy, multiple violence, and sexual exploitation were on the top. That is also visible in the Gender Gap Index Report 2022, which declares that approx. 46 million women were pushed into poverty, and also India is the worst country in the world in terms of health and survival.

This study aims to unravel the intricate web of challenges faced by women during the COVID-19 pandemic. As we explore the multifaceted impacts, it becomes evident that adversity has not only heightened vulnerability but has also given rise to remarkable stories of unity and resilience among women. Through examining social solidarity, this study seeks to shed light on how collective strength and support have emerged as crucial factors in empowering women to navigate the adversities imposed by the pandemic.

COVID-19 and Rural Women

India's healthcare is still underdeveloped, despite the establishment of a national rural health mission and the government's commitment to providing accessible, equitable, affordable, and quality healthcare to all. More than 50% of Indian people residing in rural India are involved in agricultural activity (Anand et al., 2022). During the initial stages of lockdown, everything was closed. In such a situation, those who are already suffering from lots of problems such as unemployment, poverty, illiteracy, malnutrition, and inadequate resources are at the highest risk. Because of illiteracy and unawareness, they are not aware of their health and well-being, and even if they are aware, because of poverty and lack of resources, they cannot do anything. That is why the second wave of the great humanitarian challenge was the most challenging for India, destroying rural India badly.

The situation of women during this time was extremely difficult; there was a significant increase in domestic violence cases, and home deliveries by untrained midwives were again started, resulting in an increased number of maternal deaths and health hazards. Many women are working as maids in rural India, and their entire livelihood and family are dependent on that income. However, due to social distancing, many people avoided having a maid in their homes. In that situation, their lives became difficult for them, and they suffered from starvation and malnutrition for months. During vaccination, women were unable to register themselves for COVID-19 vaccination due to technological backwardness, and even if they were able to do so with the help of family members, the next challenge they faced was that, due to sociocultural aspects, they could not go to the vaccination center alone; they must go in a group. So, women became more vulnerable at the time of the global crisis.

Children, too, faced a crisis as e-learning initiatives were promoted while schools remained closed. Disregarding socio-economic disparities, the assumption that every child had access to mobile phones led to psychological distress for those without such access (Sharma et al., 2022). Parents, in turn, were compelled to sell their properties to meet this unexpected demand. Additionally, essential school programs providing sanitary pads to adolescent girls and midday meals to children were temporarily halted, resulting in several health issues.

In such adverse situations, social solidarity and community-level cooperation play a measurable role in the survival and well-being of women and, to some extent, alleviating the impact of a pandemic on rural women.

Impact on Health

Women's health in Indian society varies from place to place, according to economic condition, family type, and cultural practices. Women's health is not independent; it is dependent on multiple indicators. So, the global Gender Gap Index Report of 2022 declares India's rank as 146 in health and survival among 146 countries. It is the worst performer in health and survival in the world. Therefore, it is very clear from the beginning that the pandemic has not affected every individual, especially women, in the same way. It does depend upon several indicators. A study done on rural women declares that women's mental health became worse during COVID-19. The depression rate increased by 38%, anxiety rate increased by 44%, and exhaustion increased by 74%. The household head's monthly income was also reduced by 8625–3584; on average, 50% of monthly income has been declined (Kannan S., 2020, Times of India, 2022).

The proportion of direct maternal deaths also increased from 45.6% to 52.5%, approximately a 7% increase. Even utilization of maternal healthcare services during the pandemic also reduced by socially vulnerable groups (Table 1) (Thakur et al., 2022a, b).

The utilization of maternal healthcare services was 27.1% in 2019, whereas in 2020, it was 15.5%. Women belonging to the BPL category who utilized maternal healthcare services was 38.3%, and in 2020, it was only 20.8% (ibid, 2020). In the above table, it can be seen that the proportion of utilization of maternal healthcare services significantly increased by the socially advantaged group, whereas a high

Table 1 Comparison of utilization of maternal healthcare services (2019–2020)

		2019 ($n = 214$)	2020 ($n = 211$)	p-value
Educational qualifications	Literate	156 (72.9%)	178 (84.5%)	0.004**
	Illiterate	55 (27.1%)	18 (15.5%)	<0.0001**
Economic status	Not BPL	132 (61.7%)	167 (79.2%)	<0.0001**
	BPL	72 (38.3%)	17 (20.8%)	<0.0001**
Average hospital stay		16.5 ± 2.1 days	126 ± 6.3 days	<0.0001**
Time taken to reach hospital from illness to current hospital		27 ± 3 h	36 ± 4 h	<0.0001**
ICU admission		67 (31.3%)	49 (23.2%)	0.061369
Residual morbidity at discharge		17 (7.9%)	25 (11.85)	0.177313

Source: Thakur et al. (2022a, b)

decrease can be observed in the socially disadvantaged groups. COVID impacted the health of women in several ways, and the less utilization of health services is directly related to the life of the mother and her unborn child.

Impact on Education

The impact of the COVID-19 pandemic on education has been substantial, particularly affecting women's access to learning. According to the UNESCO estimates, approximately 11 million girls may not be able to return to school due to the disruptions created by the pandemic. Despite preexisting gender discrimination and societal norms surrounding female education, the closure of schools and the promotion of online learning during the pandemic exacerbated existing disparities. A study highlighting the situation in Bihar, a state with a high poverty rate, revealed that only 28% of girls had access to mobile phones. Additionally, the study indicated that a mere 23% of households in India had Internet facilities for remote learning (Rana, 2021). These stark realities contributed to a widening educational gender gap, reflected in India's ranking of 107 in educational attainment among 146 countries, as per the Global Gender Gap Index Report of 2022. The pandemic has not only underscored the importance of addressing existing gender-based educational inequalities but also magnified the urgency of providing inclusive solutions to ensure continued learning for all.

Impact on Working Women

The COVID-19 pandemic has significantly intensified the challenges faced by working women, shedding light on enduring societal expectations and economic inequalities and adding more burden on women (Rani & Sharma, 2020). Despite the progress made in the twenty-first century, traditional gender roles persist, with domestic chores and childcare often designated as the primary responsibility of women. The pandemic has exacerbated this burden, as research indicates that a majority of women found themselves working even more than before (Del Boca et al., 2020). The impact on working women has been particularly pronounced in India and globally, with job losses, work-from-home arrangements, and disparities in the unorganized sector. Many women faced job terminations, while those in unorganized sectors experienced reduced salaries, and domestic workers, already vulnerable due to poverty and illiteracy, were compelled to accept meager wages, resulting in double marginalization (Kakkar & Roy, 2020). Women employed in corporate sectors encountered various challenges during the pandemic, including distractions at home, Internet and data issues, difficulties in maintaining regular schedules, resource shortages, limited communication with colleagues, familial responsibilities, and concerns about virus transmission (Jasrotia & Meena, 2021). The pandemic has underscored the urgent need for a more equitable and flexible approach to women's roles in the workforce, addressing systemic issues that perpetuate disparities.

Violence against Women

The COVID-19 pandemic has cast a distressing spotlight on the intensification of violence against women, revealing an alarming surge in various forms of abuse. Rooted in issues of dominance and power, violence against women encompasses domestic violence, sexual exploitation, sexual abuse, sexual harassment, and intimate partner violence. Throughout the pandemic, all these forms of violence witnessed a disturbing escalation, with domestic violence taking the forefront. Increased violence is the result of restrictions that have been imposed by the government to stop the spread of the virus (Krishnkumar & Verma, 2021). Further, UN women have termed this pandemic as a "shadow pandemic," drawing attention to the silent crisis faced by women globally (UN Women 2022). Shockingly, approximately one in three women, equivalent to around 243 million worldwide, has experienced sexual and physical violence at the hands of their intimate partners. In India, national lockdowns led to a doubling of domestic violence complaints, with an average of 25 daily calls to the police reporting violence cases and at least 40 cases registered from the lower strata of society (Vora et al., 2020, as cited in Krishnkumar & Verma, 2021; Kannan, 2020). This distressing escalation underscores the urgent need for comprehensive measures to address and combat the shadow pandemic within the pandemic.

Women in the Pandemic and the Role of Social Solidarity

The COVID-19 pandemic has significantly impacted women globally, unveiling multifaceted challenges that have highlighted the critical importance of social solidarity in mitigating adversity. The pandemic exacerbated existing gender disparities, placing an increased burden on women in terms of domestic responsibilities, economic vulnerabilities, and heightened risks of violence (Krishnkumar & Verma, 2021). In this context, social solidarity has emerged as a crucial mechanism for support and resilience. Communities, networks, and grassroots organizations have played a pivotal role in providing essential assistance to women, creating a collective spirit that transcends traditional boundaries (World Health Organization, 2020). The interconnectedness fostered through social solidarity has offered tangible support and provided a sense of belonging and shared responsibility, enabling women to navigate the challenges posed by the pandemic. This highlights the vital role of social solidarity in building more resilient and equitable societies where women can confront crises with dignity and communal support.

Welfare Schemes to Combat COVID-19 by the Indian Government

The Indian Government has implemented several welfare schemes to combat the challenges posed by the COVID-19 pandemic. One such initiative is the Prime Minister Street Vendor's Atma Nirbhar Nidhi Scheme, introduced in June 2020, offering working capital loans of up to INR 10,000 to urban street vendors impacted by the pandemic (National Portal of GOI, 2020a). Another significant effort is the Pradhan Mantri Garib Kalyan Yojana, launched in March 2020, which allocated a comprehensive relief package of Rs. 1.70 lakh crore to provide food and financial assistance to the poorest sections of society (National Portal of GOI, 2020b). Additionally, the Mission COVID Suraksha, led by the Department of Biotechnology, aims to develop and distribute a safe and effective COVID-19 vaccine to citizens, aligning with the vision of Atma Nirbhar Bharat (Ministry of Science and Technology GOI, 2020). The ESIC Covid-19 Relief Scheme, under the Ministry of Labour and

Economics, provides welfare provisions for insured persons and their dependents affected by the pandemic (Ministry of Labour and Economics GOI, 2020). Furthermore, the PM CARES for Children Scheme, initiated by the Ministry of Women and Child Development, focuses on supporting children who have lost parents or legal guardians due to COVID-19, aiming to provide comprehensive care and protection, education, health insurance, and financial assistance for self-sufficiency (Ministry of Women and Child Development, GOI, 2021). Collectively, these welfare schemes reflect the government's commitment to addressing the multifaceted impacts of the pandemic on various vulnerable segments of the population.

Conclusion and Suggestions

This chapter concludes that several women's lives became a challenge for themselves, and they face several challenges on a day-to-day basis, especially during this time. The burden on women intensified, encompassing increased workloads, psychological stress, family conflicts, and alarming rates of violence. Despite the government's initiatives to mitigate the impact of the pandemic through various schemes such as Pradhan Mantri Garib Kalyan Yojana/Package, Mission Covid Suraksha, Garib Kalyan Rojgar Yojna, and Pradhan Mantri Likhna Abhiyan, the ground reality reflects a grim picture, with girls dropping out of school and falling prey to child labor, unemployment, early marriage, and trafficking.

The parallel pandemic of violence against women is a stark reminder that there is much work to be done in creating a safer and more equitable environment for women. The adverse effects on education, employment, and overall well-being require immediate attention and comprehensive strategies. Further, promoting social solidarity is fundamental to collectively addressing the challenges faced by women. By fostering a sense of unity and shared responsibility, societies can better withstand the impact of crises, ensuring that no one is left behind. Social solidarity involves communities, institutions, and individuals coming together to support and uplift the vulnerable, creating a more resilient and compassionate society. Therefore, social media and other sources should promote empathetic understanding towards women. To cope with the challenges of this pandemic or any epidemic and pandemic, the government, family members, society, and employers can play crucial and responsible roles towards women. There is a need for a suitable policy framework for domestic workers to make them mainstream so that if a similar situation arises, they will not face a similar problem again in the future. For domestic violence, a savior community-based support system should be promoted. The policy should be framed in such a way that if a similar situation arises in the future where the government has the only option to lock down, food must be available in every community.

References

Anand, R. R., Sanu, S. K., Sharma, V. R., & Singh, M. (2022). Assessing sustainable development goals through social progress index for Bihar. *Gandhi Marg Quarterly, 44*(3), 359–384.

Anand, S., Sharma, V. R., & Sanu, S. K. (2023). Global pandemic Covid-19 as a health hazard: issues and challenges. In *A geographical exploration of urban risk and COVID-19: An innovative and systematic approach* (pp. 186–214). Cambridge Scholars Publishing. 1-5275-2940-1.

Del Boca, D., Oggero, N., Profeta, P., & Rossi, M. (2020). Women's and men's work, housework, and childcare, before and during COVID-19. *Review of Economics of the Household, 18*, 1001–1017. https://doi.org/10.1007/s11150-020-09502-1

Jasrotia, A., & Meena, J. (2021). Women, work, and pandemic: An impact study of COVID-19 lockdown on working women in India. *Asian Social Work and Policy Review, 15*(3), 282–291. https://doi.org/10.1111/aswp.12240

Kakkar, A., & Roy, C. (2020). Impact of COVID-19 on working women in India: A holistic approach. *Manpower Journal, 54*. Retrieved from http://iamrindia.gov.in/writereaddata/UploadFile/6.pdf

Kannan, S. (2020). *Domestic violence cases in Chennai are up, cops get 25 calls a day*. Times of India. Retrieved from https://timesofindia.indiatimes.com/city/chennai/domestic-violence-cases-in-chennai-up-cops-get-25-calls-a-day/articleshow/75153610.cms?from=mdr

Krishnakumar, A., & Verma, S. (2021). Understanding domestic violence in India during COVID-19: A routine activity approach. *Asian Journal of Criminology, 16*(1), 19–35. https://doi.org/10.1007/s11417-020-09340-1

Krishnkumar, A., & Verma, A. (2021). The shadow pandemic: Understanding the unseen crisis of domestic violence during COVID-19. *Frontiers in Global Women's Health, 2*, 614189.

Ministry of Labour and Employment, Government of India. (2020). *ESIC Covid 19 relief scheme*. Retrieved from https://labour.gov.in/

Ministry of Science & Technology, Government of India. (2020). *Mission COVID Suraksha*. Retrieved from https://dbtindia.gov.in/

Ministry of Women and Child Development, Government of India. (2021). *PM CARES for children scheme*. Retrieved from https://www.wcd.nic.in/

National Portal of India. (2020a). *Prime Minister street vendor's Atmanirbhar Nidhi scheme*. Retrieved from https://www.india.gov.in/

National Portal of India. (2020b). *Pradhan Mantri Garib Kalyan Yojana/Package*. Retrieved from https://www.india.gov.in/

Rana, D. (2021). *COVID-19, India, and girls' education*. Harvard Political Review. Retrieved from https://harvardpolitics.com/covid-19-girls/.

Rani, S., & Sharma, K. K. (2020). Gender and pandemic. *Journal of Exclusion Studies, Special Issue on Gender and Exclusion, 10*(2), 135–150. ISSN 22314547.

Sharma, P., Sanu, S. K., Sharma, V., & Anand, O. (2022). Impact assessment of Covid-19 on higher education: A case study of the University of Delhi, India. *Towards Excellence, 14*(3), 1343–1355. ISSN: 0974-035X.

Sharma, P., Pandey, H., Sanu, S. K., & Sharma, V. R. (2023a). Appraisal of Covid-19'S positive impact on India. *Research Journal (Arts), LI*, 135–158.

Sharma, P., Sanu, S. K., & Anand, R. R. (2023b). Indian female rights: A comparative analysis of political and social contributions of major Indian reformers and Dr. B R Ambedkar. *Sanshodhak, 4*, 2394–5990.

Thakur, G., Arora, A., Sikka, P., & Jain, V. (2022a). Impact of COVID-19 pandemic on severe maternal outcomes—An observational study from a referral institute in India. *Clin Epidemiol Global Health, 17*, 101–121. https://doi.org/10.1016/j.cegh.2022.101121

Thakur, G., Arora, A., Sikka, P., & Jain, V. (2022b). Impact of COVID-19 pandemic on severe maternal outcomes observational study from a referral institute of India. *Clinical Epidemiology and Global Health, 17*, 101121. https://doi.org/10.1016/j.cegh.2022.10112

The Hindu. (July 14, 2022). India ranks 135 out of 146 in the Global Gender Gap Index. Retrieved from https://www.thehindu.com/news/national/india-ranks-135-out-of-146-in-global-gender-gap-index/article65636167.ece

UN Women. (2022). *The shadow pandemic: Violence against women during COVID-19*. Retrieved from https://www.unwomen.org/en/news/in-focus/in-focus-gender-equality-in-covid-19-response/violence-against-women-during-covid-19

World Health Organization. (2020). *Social determinants of health*. Retrieved from https://www.who.int/westernpacific/health-topics/social-determinants-of-health.

Geography of Disaster, Coordination for Relief and Conflict Resolution: A Case of Kashmir Earthquake 2005

Ramanuj Kaushik

Introduction

The chapter delves into the possibility of cooperation and coordination between two estranged states, at the time of a humanitarian crisis caused by a natural disaster, and its consequence over the conflict resolution. It chose the earthquake that occurred in Kashmir on 8 October 2005. The epicenter of the earthquake was near Muzaffarabad, the capital of Pak-Occupied Kashmir, having a magnitude of 7.6 registered on the Richter scale (Fig. 1). As of November 8, the official death toll from the Pakistani Government stood at 73,276. Officials reported nearly 1400 deaths on the Indian side and 14 in Afghanistan. Most of the affected population resided in mountainous areas, where landslides blocked roads, resulting in approximately 3.3 million homeless people in Pakistan. The UN stated that more than four million people were directly impacted before winter snowfall began in the Himalayan region. The estimated damages exceeded US$5 billion (30 billion Pakistani rupees) (Wilkins, 2006).

The destruction resulting from such natural disasters imposes physical constraints on the ability of responders, including the state and its machinery, to mobilize and provide aid, communicate with each other, and coordinate joint responses with the affected community. This is because natural disasters damage physical and communication infrastructure. In the immediate aftermath of the October 2005 Kashmir earthquake, landslides triggered by the quake destroyed a kilometer-long section of mountain highway leading to the two main population centers in the Neelum and Jhelum valleys. The only way to deliver supplies to the affected population was via military-assisted helicopter lifts, which limited the amount of supplies that could be transported. As a result, the widespread destruction of infrastructure

R. Kaushik (✉)
Department of Geography, D.A.V. Post Graduate College, Jai Prakash University, Chapra, Siwan, Bihar, India

K. Arora, R. R. Anand (eds.), *The Solidarity Approach in Geography*,
https://doi.org/10.1007/978-3-031-86753-8_28

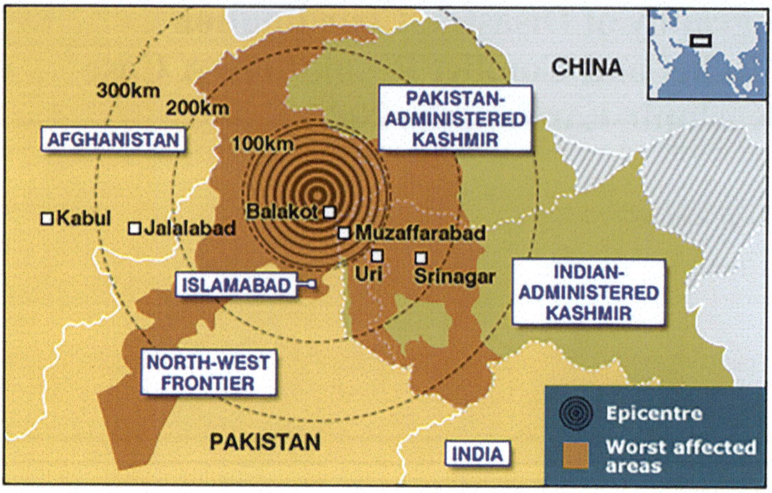

Fig. 1 Affected area of Kashmir earthquake, 2005. Sources: BBC News (2006)

created physical barriers, isolating the affected geographical area (Relief Web, 2006a). These unavoidable outcomes of natural disasters place physical constraints on the ability of national and international humanitarian assistance to respond promptly and deploy quickly in terms of logistics and coordination.

PoK Between Two Nations

The study focuses on the earthquake that occurred in Pak-Occupied Kashmir (PoK) and its adjoining Indian part, but with more emphasis on the first one on account of multiple reasons: *First*, more than 70,000 people perished in this earthquake, which is five times more than people killed in other parts, if we flock together.

Second, PoK has been a bone of contention between India and Pakistan. It is now under Pakistan as *Azad Kashmir*. However, India considers its legitimate right on the PoK, because it was ceded to India by the ruler of Kashmir Maharaja Hari Singh under the provision of the Independence of India Act of 1947. Pakistan claimed the whole of Kashmir on the ground that the state is Muslim dominated and signing the instrument of accession by the King of Kashmir was against the popular will of the Kashmir. Three wars were fought between two neighbors with many agreements that remained idle due to noncompliance and accusations and counter-accusations by both parties. Resultantly, the issue of Kashmir remained close to the heart of both countries and strained Indo-Pak relations at all levels.

Third, the geography of PoK has destined it as one of the most backward areas. The history of development marred by rugged mountainous terrain hindering necessary physical and infrastructural development for growth closed Islamic society with the increasing trend of Islamic militancy. Pakistan's policy made this region subordinate to the center with few human rights and freedom of economic and political choice. The historical matrix of PoK was consequently left with a lack of vibrant civil society, nondemocratic mobilization of the public, and opinion on the name of Islam and anti-Indian feeling.

Fourth, PoK has been a flash point due to the breeding ground of Islamic militancy. There, militants are the different shades of the *Jihadi* element that wage war against India in Kashmir as well as other parts of it. Various reports confirm that these *Jihadi* militants, which were subsequently declared as terrorist organizations due to the nature of the operation, belong to either banned Lashkar-e-Taiba (LeT) or Jaish-e-Mohammed (JeM). They, along with many such organizations, run camps to train the innocent and illiterate people of these areas and use them to spread terrorism not only to India but also to the Western hemisphere. The worst situation is that they are well under the patronage of the Pakistani military and receive all possible support from the administration. They are now so entrenched that the Government of Pakistan (GoP), despite being allies in the war against terrorism, is not able to have control over them. They spread their tentacles in the whole tribal-dominated NWFP and Balochistan area and have greater political clout. Terrorism is a major issue and an impediment in the Indo-Pakistan peace process. India calls them terrorists, while Pakistan recognizes them as militants meant to unshackle India's grip in Kashmir. The United States and its Western allies also recognize that they are more terrorist organizations than others elsewhere and have banned the LeT and JeM (MHA Report, 2004–2005). Therefore, analysis in those contexts in the post-disaster scenario will be interesting given the political dimension of this terrorist organization.

Coordination and Conflict Resolution

According to International Humanitarian Law, the primary responsibility for responding to natural disasters lies with the national government (ICRC, 2007). Bilateral or multilateral assistance can only be provided at the request of the affected government and must be coordinated through it. Because of the crucial role the national government plays in disaster relief, it fiercely protects its autonomy, particularly regarding domestic issues. Additionally, many national governments may lack the organizational or logistical capacity to manage disaster response effectively. The quality of disaster response and its coordination depends on the experience and administrative capabilities of the affected country's government. This creates inherent limitations on the effective deployment and coordination of the numerous international responders arriving at the disaster site (Katoch, 2006).

In addition to damaging physical infrastructure, natural disasters also affect local government officials and community leaders, who are responsible for leading the response. Even if these officials are not personally affected, their family members might be injured, and they could experience trauma from the loss of relatives and friends. Therefore, expecting local authorities to function normally, as the media and the outside world sometimes seem to do, is an unrealistic expectation.

In this context, coordination is essential in disaster relief to enhance the efficiency of the delivery system. It is the state's responsibility to coordinate the relief operations. Relief agencies often rush to the affected area and encounter various challenges. On the one hand, they need to operate in a foreign environment, while on the other, they must satisfy their donors.

The UN defines coordination as the systematic use of policy instruments to deliver humanitarian assistance cohesively and effectively. These instruments include strategic planning, data collection, information management, resource mobilization, and accountability, organizing a functional division of labor, negotiating and maintaining a workable framework with host political authorities, and providing leadership (Reindorp and Peter Wiles, 2001).

Highlighting the importance of coordination in conflict resolution, Comfort (2000) writes: "It is not the disaster, but the kind and mode of cooperation that is fostered among nation is an environment threatened or altered by severe destruction that creates the opportunity for change in relations among states previously in conflict." It implies that disaster can be proved to be a conflict shooter rather than monger, as has been believed if the kind and mode of cooperation are negotiated and implemented in such a way that "non-disaster issues" could also be assimilated.

Issues of Security vs. Sovereignty

According to the Hyogo Framework for Action, 2005–2015, coordinating disaster relief is the responsibility of the national government (United Nations, 2005). The developing nation, in the absence of a credible disaster relief mechanism, finds it very difficult to respond and coordinate operations. The disaster could be an opportunity for losing ground by its strategy towards the restructuring programs in this region. The concerned country, seeing the visible nature of a large disaster and global pressure to act swiftly, will always apprehend the interference of outsiders. In that scenario, the issue of sovereignty comes to the fore, as a power and the extent of power of the country to assist in this area. The state may feel the infringing sovereignty when international agencies for relief bypass the central authority. Sometimes, local authorities do not comply with the central sovereign order of relief operation. It is more conspicuous in a remote area with little cultural affinity with the center, center versus fringe area.

Response After Earthquake

The purpose of this research is to evaluate the political geography of the area where the disaster occurred. It can be better grasped through the proper framing of the research questions:

- Do the modalities of relief and rehabilitation give the opportunity to two estranged nations to sit together and resolve the problem?
- In what circumstance do the two nations sit together and coordinate?
- How does the geography of the area play a vital role in bringing states together in the pressing need for humanitarian relief?
- What type of coordination opens the avenue of more cooperation and thus finally leads to conflict resolution?

The compelling situation created after this great humanitarian crisis, however, had to reinforce the ongoing composite dialogue borne out in the joint statement on 6 January 2004 given by Atal Bihari Vajpayee and General Musharraf. The day of composite dialogue initiation was termed in the volatile history of India and Pakistan relations as a watershed for confidence-building measures (CBMs) between the two countries. Out of eight points in the dialogue, the issue of J&K was also included. A positive sentiment was created in the people of Kashmir when both parties aired the possibility of making the Line of Control (LoC) or the international border irrelevant, which has divided the families of both sides. Bus service started in April between two Kashmirs. Four days before the earthquake, both countries had agreed to resolve their dispute over troop withdrawals from the Siachen Glacier within 3 months (IPCS-Special-Report-12, 2006).

Against this backdrop, when the earthquake occurred on 8 October 2005, and killed both sides of people in Kashmir, it was expected that the CBM would be reinforced and entrenched under the low level of cooperation by giving way to multi-track diplomacy as a new urgency of post-disaster relief operation. The area of cooperation could be joint military relief operations, joint coordination of whole NGOs, working here for humanitarian relief, and finally evolving the mechanism that engaged from the top to local people that helped both sides to understand each other.

Following the earthquake, India and Pakistan worked together to ease Line of Control restrictions to support aid operations. On 19 October 2005, telephone links were re-established across the LoC, allowing families to communicate. In November, five locations along the LoC were opened over 9 days to enable the passage of relief supplies (Wilkins, 2006). At the donors' meeting, India promised $25 million of aid to Pakistan. In an interview with BBC on 21 October 2005, President Musharraf once again desired the LoC to be made irrelevant to handle the groundswell of human misery. He put forth the idea to identify what is Kashmir and then demilitarize the identified region, take all the soldiers out, and then give self-government to the people in the region (Relief Web, 2006a, b).

To assist helicopters in delivering relief to remote villages, India permitted Pakistan to operate in a no-fly zone along the heavily militarized Kashmir border. Pakistan's President Musharraf expressed his readiness to allow residents of Indian Kashmir to travel freely to the Pakistani side to aid in the reconstruction efforts—an offer India welcomed (Pasricha, 2005). Additionally, Indian Air Force planes transported medicines, blankets, and tents to Pakistan.

President Musharraf aimed to win back the hearts and minds in Kashmir, as the people were furious with the Pakistani Government and army for their slow response following the earthquake, which led to many deaths due to delayed aid. Meanwhile, India sought to present itself as a benevolent presence in Kashmir, where its military is despised by the populace due to years of brutal and indiscriminate suppression in response to the separatist movement. In response to India's gesture, the major Pakistani newspaper *The Dawn*, dated 16 October 2005, wrote: "India's offer to send relief goods this time has been timely not just for the earthquake affected people but also for the confidence building process between the two countries."

Geopolitics of Disaster Relief in PoK

The calamity followed with many opportunities for different actors involved in the operation. For Pakistan, it was an opportunity to show its benevolence to the Kashmir people on this side to convince them that they were not the colonial force who occupied the PoK illegally and treated them as second-class citizens. For Islamabad, it was a better hope to do more for the people of its side and convey the message to India Kashmir that Islamabad is more serious and prompter to the Kashmir people than the Government of India. For Pakistan, it was no less than a big chance to wipe out whole terrorist camps in NWFP and PoK, which were severely damaged by disaster. Pakistan, then, was the alliance partner in the war against terrorism. It has been very clear as time elapsed that the GOP and its military force are not in a position to control and close the deeply entrenched terrorist operation.

For the United States and its Western allies, the PoK disaster was looked upon from the perspective of current political development in the world arena. The projection of the United States and its allies as Christian forces clashing with the Muslim world and thus deteriorating their reputation as a crusader of Muslim civilization has substantially narrowed the sphere of influence in West Asia, Central Asia, and SE Asian communities. The so-called war against terrorism operations in Afghanistan and fringe areas of Pakistan as well as in other parts of the world contributed to projecting such an image. The Western forces did not consider it benign and viewed it as the part of neocolonialism. There has been a divergence of opinion regarding the West. The GoP on the one side cooperated with the Western forces in the war against terrorism, while the people of Pakistan including PoK were totally against such cooperation. There have been more than a hundred demonstrations against the United States in Pakistan since the 9/11 incident. The terrorist groups

Al-Qaeda LeT and Jaish-e-Mohammed, against which the Western force was waging the war, had a substantial grip on the people of Pakistan including PoK. In this context, the West hoped that direct cooperation with the people of this country would help to improve the image and change the perception towards the West. This would help to isolate the terrorist organization from the masses.

For international NGOs and the UN, it was a litmus test. For INGO, it was one of the few occasions where military forces were supposed to cooperate, while for the UN, the major challenge was to cooperate with Western forces and with INGO in the aftermath of the Volcker Report that highlighted the various irregularities in the UN's working system (UN Press Release, 2005). The new evolving cluster approach of the UN to carrying out R&R was also to be analyzed in the new context of the PoK's sociopolitical setup (Stoddard et al., 2007).

Role of National Government/Military

GoP/military's role may have two dimensions: first, its direct response to the bereaved people, and second, its coordination with the US and NATO forces and multiple local NGOs and INGOs. For the immediate response, government created a Federal Relief Commission (FRC) within days of the disaster to scale up coordinated action for rescue and relief operations. The government also established the Earthquake Rehabilitation and Reconstruction Authority (ERRA) to shore up medium- to long-term rebuilding efforts. It serves as the primary liaison between the government and international lending institutions, other international organizations, as well as national authorities and philanthropic organizations focused on rehabilitating the affected areas. The organization was led by the army's chief engineer, Lieutenant General Mohammad Zubair (Krick, 2008).

However, the reality was that after 2 weeks since the quake, rescue workers had still not managed to deliver supplies to about 15% of the affected area. The government solely entrusted military with this responsibility, which proved to be inefficient and unskilled and with lack of logistic support. The UN emergency relief chief Jan Egeland told the BBC that his organization had never encountered such a logistical nightmare, stating that it was nearly worse than the Asian Tsunami. Comparing the military's poor response to the 1970 cyclone that hit East Pakistan (now Bangladesh), a former head of the military's Inter-Services Intelligence (ISI) said that government insensitivity at the time led to the deaths of 300,000–500,000 people and fueled a separatist uprising (BBC News, 2005a).

The military's top-down approach significantly impacted the efficiency of the quick delivery system and revealed the weak structural setup. Forces were stuck in their camp for a week, awaiting orders from higher-ranking officials, which hindered swift action. Poor performance and forced removal of affected people from tents after April 2006 sparked protests. The Government of PoK was marginalized, and a large portion of the aid was being funneled through NGOs that were sponsored by Islamic political parties. In NWFP, the World Bank and Asian Development

Bank directed their efforts towards Peshawar (the capital of NWFP), while aid for Kashmir was routed not through Muzaffarabad but through Islamabad. This client relationship was further evidenced when substantial funds for reconstruction were spent without meaningful input from Kashmiri politicians (Qureshi and Bamforth, 2006). Rejecting criticism of the military's inept response, Education Minister and former ISI chief Lt. General Javed Ashraf Qazi shifted the blame, stressing that the army was not responsible for crisis management and disaster relief. "These functions are carried out by the civil administration," he told the National Assembly, the lower house of parliament. However, when international aid and relief organizations arrived, the army assumed control of every task (The Dawn, 2005).

The weakness of GoP/military did not mean that they loosened their grip over the foreign military or INGOs. After 3 days, Islamabad formerly requested the US military and NATO forces to help in the rescue operation. The foreign forces coordinated well with the Pakistani military with several Chinook helicopters and mobile hospitals. The local and international NGOs anchored in this area were directed to deliver the relief material at 43 points identified by military force rather than go alone. It was only to prevent the other actors of Pakistani civil society and political organizations from involving in this operation (NATO, 2010).

Virtually all organizations collaborated closely with the Pakistani military and other forces for logistical support. Respondents generally expressed great appreciation for the arm's role in the earthquake response. Initially, the UN found the military to be very supportive, with officials describing a level of cooperation from the government that was unprecedented in their experience, even stating that it had been a pleasure to work with them. The military's role was further legitimized when UN Secretary-General Kofi Annan invited them to participate in a donor conference in October 2006 (UN News, 2005).

The overly close working relationship established with the Pakistani army allowed President Musharraf to put a military face on the relief operation. This helped to restore the credibility lost due to the ill-planned and poorly executed emergency response and also to project the military internationally as Pakistan's sole organized and effective institution. Speaking to the media at the World Economic Forum in Davos, he said: "I can very proudly say that the world has learned how to handle a catastrophe because I think we have done much better than other calamities around the world. I take pride in being the leader of this army" (ICG Report, 2006).

The purpose of explaining the role of GoP/military was to explain how the sovereign right of ensuring coordinated R&R operation of the national government did not bear the desired fruit. The marked weakness and lack of vision on the part of a military government led to the slow demise of disaster diplomacy with India. President Musharraf himself declared that the disaster could serve as "an opportunity of a lifetime to end decade's old hostilities" (Relief Web, 2006b). One of the most significant developments after the Kashmir earthquake due to the military's callousness was the resurgence of Islamic militancy and a new focus on terrorism, which not only affected the peace process between two enemies but also created a new diplomatic dilemma for the West and their military and INGO who some way or other cooperated with these terrorists.

The Resurgence of Islamic Organization

The lackluster attitude of the military gave the Islamic organization and *Jihadi* group an opportunity to fill the gap and win local support. The quick response to these groups helped to garner support and acquire a major chunk of relief material and money from local and international donors. It was also feared that they succeeded in capturing a large number of child orphans who may be the potential *Jihadi* and shortly they will strengthen the organization and thus perpetuate the problems for its neighbor and the world. *Jamaat-ud-Dawa* (JD), the new name of *Lashkar-e-Taiba* banned by the government and listed as a terrorist group by the UN; *Al-Khair* trust of *Jamiat Ulema-e-Islam*; *Jamaat-e-Islami* and banned *Al-Rasheed* Trust; and renamed *Khuddamul Islam* of banned *Jaish-e-Mohammed* were the active Islamic organizations involved in the relief operation and that cooperated well with military force and various local and international NGOs (ICG Report, 2006).

Many reasons explain this development: *First*, more than 10,000 military personnel were killed in the earthquake. Consequently, they were not quick to respond. *Second*, most Islamist organizations have better knowledge of the local conditions and are well equipped for battling with harsh conditions. Some of the organizations that have linkages to the world terrorism network have better logistical support, mobile hospital satellite phones, and modes of transportation. *The Observer* reported (16 November 2005) that Pakistan-based Islamic militants, who spent the past decade fighting Indian rule in the region, have announced a "holy war" to help victims of the earthquake. Brussels-based International Crisis Group, meant for working to prevent conflict worldwide, reported (2006): "*Jihadi* and Islamist groups were the first to conduct rescue operation, establish initial medical emergency camp, surgical units, and dispensaries for earthquake survivors, and send assessment teams to isolated areas. They raised a volunteer army of thousands of *madrassa* students. *Jihadi* outfits and Islamic groups provided doctors' clinics, X-ray services, dental care, reconstruction material, ambulance service, burials, and mosque rebuilding. They also cared for orphans and displaced widows. They organized transport for relief goods to isolated areas, and commandeered lifting equipment and tents. In the reconstruction phase, these groups have established programs providing cheap reconstruction materials and subsidies to sawmills."

Third, the most important thing was the military cooperation with these Islamic organizations. "The government stressed that help from any quarter was welcome at a time of dire need when state institutions and agencies had been rendered nonfunctional by the magnitude of the disaster. However, the government further portrayed the *Jihadi*s as the prime contributors to relief work, ignoring the contribution of countless NGOs, charities, and volunteers. Interior minister Aftab Ahmad Khan

Sherpao, for instance, said that they (the *Jihadis*) are the lifeline of our rescue and relief work in Azad Kashmir right now" (ICG Report, 2006).

The question remains as to why Pakistan, despite being a partner of a global coalition of terrorism and projecting as a champion of moderate Islamic country, relied on these organizations. The possible answer may be that, given its inability to respond to disaster swiftly, Islamabad feared that there may be separatist movement that could emerge led by civil society or political parties, for they have better penetration in masses in these areas.

The *Jihadi* groups have been serving Pakistan's strategic interest in Kashmir and Afghanistan and served as an extension of the army. This time, the military government found it easier to cooperate with them as a bullwork against the civil society and moderate political parties. They helped in every aspect and channeled funds to tighten the grip over these groups who are impartial in thinking. In Pakistan, this strategy worked well. The people of Kashmir, while condemning the army, welcomed these Islamic organizations (Clark, 2006).

The fourth and final reason is related to Pakistan's domestic politics. Pakistan's main political opposition leaders Benazir Bhutto and Nawaz Sharif remained in exile. Yet, their joint political party meant for the restoration of democracy in Pakistan criticized the lackluster approach of the government. They also insisted upon the discussion in parliament before inviting the US and NATO forces to strategically sensitive areas. They were also charged with the corruption in R&R operations. Meanwhile, Musharraf's political party *Pakistan Muslim League-Quaid-e-Azam* (PML-Q) was struggling for both internal cohesion and popular support. As a consequence, the government relied upon minority Islamic parties such as the *Muttahida-Majlis-e-Amal* (MMA) coalition and to a lesser extent the *Muttahida Qaumi Movement* (MGM) to prop up its electoral power base. These Islamist political parties played a greater role in guiding decision-making after the earthquake and helped to funnel funds to its affiliated *Jihadi* organization, ignoring local civil society and the secular forces (Qureshi and Bamforth, 2006).

International Presence

While India shut down the door of cooperation with foreign agencies, Pakistan had to invite the US and NATO forces for the rescue operation. The doors were opened for INGOs as well to anchor in this area. The US and NATO military and the US Agency for International Development (USAID) arrived in Pakistan 3 days later on 10 October 2005 to help in relief efforts.

The 84-bed Mobile Army Surgical Hospital (MASH) arrived on 24 October 2005 (CRS Report for Congress, 2006). They worked under the command and control of Pakistan. The political party as well as the Islamist party including those in coalition government vehemently opposed these forces. However, on the other side, Western forces took it as an opportunity to improve their image in the oriental world.

Now Islamabad had twin problems: *first*, the deployment of foreign forces and INGOs for a better R&R program, and *second*, to voice down the agitating opposition and Islamic parties. The solution was found in deploying the forces along with the Pakistani army while keeping control of relief aid by directing all donors to report at 43 points chosen by the army. This helped to funnel most of the aid to these fundamentalist groups who were given more weight to show their presence in disaster relief rather than boycotting Western forces this time. Their close presence in masses was sufficient to outweigh the Western effect. The US urgency to work in this scenario led to the compromise of its anti-terrorism position. Brussels-based think tank International Crisis Group confirmed that virtually all international NGOs and USAID, OXFAM, etc. worked with these *Jihadi* groups (ICG Report, 2006). Journalist Clark of the BBC News also reported that UN agencies, to ensure that their cluster approach works well, cooperated with extremists which were banned by this organization itself. The internal rivalry between USAID and the UN led to liaisons with the extremist organizations which were made more powerful by the GoP at the cost of local political parties and civil societies.

The UN was entrusted with the responsibility to coordinate the whole nonmilitary relief and rehabilitation operation covering international and local NGOs and local civil society in order to ensure its cluster approach. The coordination and linkages were the main thrust of this approach, and due to that basic aspect, the approach could not gain much result. Action Aid International (2006), which analyzed the UN cluster approach in Kashmir, pointed out that the UN should have engaged with local NGOs, who have expertise in handling the situation. Rather, it chose extremist-sponsored trust and sidelined Pakistan's largest NGO, Edhi.

The problems also cropped up with the engagement of women in relief work. There were women from outside Pakistan working in this area under many INGOs, and this prompted local women to work with them in the relief operation. The Islamic force vehemently opposed it. There were *fatwa* against INGO, and all women including foreigners were told to either leave the area or work under *Burka*. On 6 May 2007, the *Awami* Action Form (AAF), a group with local support from the religious section, threatened to attack the vehicle of the UN and other NGOs, if they were found to carry women workers (Chandran et al., 2009).

The legitimacy this terrorist organization got in working with NGOs and military forces was now asserting against the foreign presence. The UN and NGO workers were attacked on some or other issues. In the first week of May, all the humanitarian groups here went on a token strike, suspending their work for a day. In the second week of May 2007, the UN announced closing all its operations and offices in the Bagh district (hardest hit by the earthquake) in PoK following the attack on its officials by a radical element (Chandran et al., 2009).

Discussion and Conclusion

The geopolitics of disaster relief in PoK highlights that contrary to the principle that society is the first responder in disaster relief, the distribution of aid in strategically sensitive areas is almost entirely controlled by state authorities. Although the state is responsible for coordinating relief efforts, it must ensure greater participation of grassroots governance to distribute aid efficiently and transparently. If local government and civil society were promoted, then there would have been fair change in which LoC could be made irrelevant. The people on either side of the line could move to each other's area to extend timely help. Many of the Baltistan areas could have been reached through the Indian side, but the terrorist organizations taking charge of the whole relief and rehabilitation program made the Indian side very skeptical.

The shared geography of Kashmir necessitates a collaborative approach to relief operations, involving local civil societies. This approach could foster more person-to-person contact and mutual understanding between the two sides of Kashmir. However, the top-down approach has hindered genuine civil society and NGOs from participating in relief efforts, instead allowing terrorist organizations to exploit the situation. The Pakistani military and state apparatus, lacking a deep understanding of the region's geography, found it more convenient to work with these groups. Pakistan has historically treated PoK as a colony, using it to advance anti-India campaigns by promoting various terrorist organizations. Over time, these groups have become deeply entrenched in Kashmiri society. Their extensive knowledge of the region's geography, honed through their infiltration activities into Indian-administered Kashmir, has further solidified their influence.

It is important to note that at the time, Pakistan was involved in the US-led war against terrorism and had initiated confidence-building measures (CBMs) with India in 2004. However, when the earthquake struck in 2005, the inadequacies of the state machinery at both the national and local levels provided an opportunity for terrorist organizations to take charge of relief operations. These organizations had a better understanding of the geography and society of PoK. Consequently, even international NGOs and UN agencies had to coordinate with these groups. Given the visible nature of the disaster and the pressure from the global community, state and international agencies were compelled to work with these organizations.

The emergence of a terrorist organization as a first responder, due to their in-depth understanding of the geography and sociology of Kashmir, India. The other relief agencies did not find space wherein they could show their presence, even facilitating more people-to-people contact if not sending military logistics in earthquake-affected area. Similarly, it became hard for the Pakistani side to get help from India, despite the proximity and easy accessibility through it, because of the open anti-India stance of terrorist organizations.

Thus, given the causality and the urgency the earthquake in Kashmir created, we can conclude that it was merely not because of a history of distrust between the two states but a lack of geographical understanding of the area that led to the emergence of new non-state actors (mostly terrorist organizations) as relief agencies, and their direct collaboration with international relief agencies (thus bypassing the state agency) hindered the two state agencies to collaborate. It also thwarted people-to-people contact. These two were keys for better coordination and conflict resolution. The geopolitics of disaster relief without an actual understanding of geography does not ensure peace and better understanding between two estranged parties.

References

Action Aid International. (2006). *The evolving UN cluster approach in the aftermath of the Pakistan earthquake: An NGO perspective*. Retrieved March 15, 2022, from https://inee.org/sites/default/files/resources/doc_1_91_ActionAid_Report_on_UN_Cluster_Approach_April_2006.pdf

BBC News. (10 Oct 2005a). *Quake is UN's worst nightmare*. Retrieved March 18, 2022, from http://news.bbc.co.uk/2/hi/south_asia/4358902.stm

BBC News. (05 Oct 2006). *South Asian quake*. Retrieved Feb 11, 2007, from http://news.bbc.co.uk/cbbcnews/hi/newsid_4320000/newsid_4327300/4327360.stm

Chandran, D. S., et al. (2009). India disaster relief diplomacy. *Indian Foreign Affair Journal., 4*(2), 63–80.

Clark, K. (2006). *UN quake aid went to extremists*. BBC News, North-west Pakistan. Retrieved Feb 11, 2007, from http://news.bbc.co.uk/go/pr/fr//2/hi/south_asia/5402756.stm

Comfort, L. K. (2000). *Disaster: Agent of diplomacy or change is international affairs*. Retrieved Feb 11, 2007, from http://www.igs.Berkeley.edu/publication/workingpaper/wp200018pdf

CRS Report for Congress. (2006). *The earthquake in South Asia: humanitarian assistance and relief operations*. Retrieved Feb 11, 2007, from http://digital.library.unt.edu/govdocs/crs/permalink/meta-crs-8257:1

Hyogo Framework for Action 2005–2015. *World conference on disaster reduction 18–22 January 2005, Kobe, Hyogo*. Japan, International Strategy for Disaster Reduction. Retrieved March 15, 2007, from https://www.unisdr.org/2005/wcdr/intergover/official-doc/L-docs/Hyogoframework-for-action-english.pdf

ICG Report. (15 March 2006). *Pakistan: Political impact of the earthquake*. Retrieved March 15, 2007, from http://efaidnbmnnnibpcajpcglclefindmkaj/; https://www.files.ethz.ch/isn/15458/b046_pakistan_political_impact_of_the_earthquake.pdf

International Committee of the Red Cross (ICRC). (2007). International humanitarian law and the challenges of contemporary armed conflicts. : ICRC. Retrieved March 1, 2010, from https://www.icrc.org/en/document/international-humanitarian-law-and-challenges-contemporary-armed-conflicts

IPCS-Special-Report-12. (2006). *Indo pak composite dialogue 2004–05: A profile*. sRetrieved March 15, 2007, from http://efaidnbmnnnibpcajpcglclefindmkaj/; https://www.files.ethz.ch/isn/100224/IPCS-Special-Report-12.pdf

Katoch, A. (2006). The responders' cauldron: The uniqueness of international disaster response. *Journal of International Affairs., 59*(2), 153–172.

Krick, J. (2008). *Building back better: Post-earthquake responses and educational challenges in Pakistan*. UNESDOC Digital Library. Retrieved Sept 12, 2022, from https://unesdoc.unesco.org/ark:/48223/pf0000163575

MHA Report. (2004–2005). *Ministry of Home Affairs, Government of India*. Retrieved March 15, 2007, from https://www.mha.gov.in/en/document/annual-report-ministry-of-home-affairs-2004-2005

NATO. (2010). Pakistan earthquake relief operation. Retrieved July 14, 2015, from https://www.nato.int/cps/en/natohq/topics_50070.htm

Pasricha A. (2005). *India, Pakistan call peace process irreversible, relief web*. Retrieved March 15, 2007, from https://reliefweb.int/report/india/india-pakistan-call-peace-process-irreversible

Qureshi and Bamforth. (2006). *Earthquake Jihad: The role of Jihadis and Islamist Groups after the October 2005 earthquake, International Crisis Grou*. Retrieved July 15, 2015, from http://www.crisisgroup.org/home/index.cfm?action=friend&d_id=25311&t_id=4270

Reindorp and Peter Wiles. (2001). *Humanitarian coordination: Lessons from recent field experience, office for the coordination of humanitarian affairs (OCHA)*. Retrieved July 15, 2015, from http://efaidnbmnnnibpcajpcglclefindmkaj/; https://cdn.odi.org/media/documents/4186.pdf

Relief Web. (2006a). Pakistan 2005 earthquake: Early recovery framework—with preliminary costs of proposed interventions. Retrieved 15 July 2015 http://www.reliefweb.int/rw/RWB.NSF/db900SID/RMOI6J89V9?OpenDocument

ReliefWeb. (29 June 2006b). *Pakistan: The role of the military in the Pakistan earthquake*. Retrieved Apr 2, 2007, from https://reliefweb.int/report/pakistan/pakistan-role-military-pakistan-earthquake

Stoddard, A, et al. (2007). *Cluster approach evaluation*. Retrieved Feb 11, 2007, from http://efaidnbmnnnibpcajpcglclefindmkaj/; https://www.humanitarianresponse.info/sites/www.humanitarianresponse.info/files/documents/files/Cluster%20Approach%20Evaluation%201.pdf

The Dawn. (16 Oct 2005). *Disaster Relief a civil job, says minister*. Retrieved Feb 11, 2007, from https://www.dawn.com/news/161439/disaster-relief-a-civil-job-says-minister

UN Press Release. (2005). *Security council SC/8492, independent inquiry finds maladministration, evidence of corruption in un 'oil-for-food' programme, security council told*. Retrieved Feb 11, 2007, from https://press.un.org/en/2005/sc8492.doc.htm

United Nations. (2005). *Hyogo framework for action 2005–2015: Building the resilience of nations and communities to disasters*. UNISDR. Retrieved Feb 11, 2007, from https://www.unisdr.org/2005/wcdr/intergover/official-doc/L-docs/Hyogo-framework-for-action-english.pdf

Wilkins, L. (2006). *Quake survivors still struggle*. BBC NEWS, Published online, 9th October. Retrieved Feb 11, 2007, from http://news.bbc.co.uk/go/pr/fr/-/2/hi/uk_news/5403014.stm

Part VIII
Building Solidarity for Natural Resource Management and Conservation

Commercialization of Water and Shift from Traditional to Modern Technologies: Evidence from Water Insecure Western Rajasthan

Anindita Sarkar

Introduction

An elderly man in the village in Jaisalmer exclaimed, "In olden days when unexpected guests arrived in our house, we served them buttermilk or milk instead of water. Only when they asked for it, we gave them a glass of water mixed with brackish water. We used to be conscious that if they asked for more water, how would we give it? Water is precious." (Translated from interview). This narrative brings out the essence of how scarcity of potable water in the deserts of Rajasthan makes drinking water so valuable that its judicious use is ingrained into the local cultures of everyday life. People walk miles to fetch drinking water and use each drop of it very carefully. Surface and subsurface water harnessing technology has evolved from historical times in Rajasthan based on village solidarity for water resource management and conservation (SAARC, 2009). This is one of the reasons why the challenges of living in an arid desert did not result in scarcity, making Rajasthan one of the most populated deserts in the world, and enabling it historically to play a dominant role in trade, culture, and politics (Mishra, 2001).

Gradually with the diffusion of technological interventions like piped water infrastructure and modern drilling technology, taps, hand pumps, and deep tubewells multiplied in the villages of western Rajasthan (Rathore, 2004; O'Reilly & Dhanju, 2014; Sarkar & Singh, 2020). As demand increased, informal water markets evolved to fill the gap in drinking water supply that public infrastructure could not serve. At present, several villages in western Rajasthan get water at their

A. Sarkar (✉)
Department of Ecology and Natural Resources Management, Center for Development Research, University of Bonn, Bonn, Germany

Department of Geography, Miranda House, University of Delhi, Delhi, India
e-mail: anindita@uni-bonn.de

K. Arora, R. R. Anand (eds.), *The Solidarity Approach in Geography*,
https://doi.org/10.1007/978-3-031-86753-8_29

doorsteps brought by the water tankers with payment per unit of water (Sarkar & Singh, 2020; Sarkar, 2023).

The study highlights geographic research on neoliberal interventions in the use of modern water technologies for drinking water in the heterogeneous rural communities of western Rajasthan. It demonstrates the ways that social norms, village water governance, and regulations of water marketing converged to create "new" ways of thinking about water access and payment that undermined the social goals and the existing physical infrastructure. The chapter tries to answer research questions like: has the introduction of modern water technologies eased the process of water access? How has the commercialization of water changed the waterscapes in the villages characterized by heterogeneous societies? What are the traditional water management systems and why are these systems based on village solidarity no longer widely accepted? Who suffers when these traditional systems die and degenerate?

Materials and Methods

For this study, western Rajasthan was selected for three reasons. First, this region has low precipitation and witness drought every 3–4 years. Second, climate variability is projected to increase unpredictable precipitation (Pandey et al., 2010), raising the average temperature and evapotranspiration (Mall et al., 2006) in Rajasthan. This will accelerate desertification and land degradation (Arya et al., 2009). Third, with successive years, the region has experienced water depletion and degradation resulting in drinking water shortages and economic losses impacting social and economic life (Sarkar & Singh, 2020).

Western Rajasthan comprises 12 districts, i.e., Sriganganagar, Hanumangarh, Churu, Sikar, Jhunjhunu, Nagaur, Bikaner, Jodhpur, Barmer, Jalore, Pali, and Jaisalmer. Six sample villages were purposively selected from the districts of Jaisalmer, Jodhpur, and Barmer that had heterogeneous caste compositions and different kinds of drinking water sources. A total of 190 households were surveyed and randomly selected across villages. Both structured and unstructured questionnaires were used to gather information on demographic and socio-economic factors of the inhabitants, availability, and state of water supply, and water-collection practices. This was supplemented by 14 focus group discussions, where homogeneous groups were formed based on caste, class, and gender so that participants could express their diverse views as freely as possible. 15 in-depth semi-structured interviews were conducted with women who were selected based on a stratified random selection procedure to ensure geographical spread, and balance in age, caste, and class to include all sections of society in the villages.

This chapter begins with a literature survey and theoretical and conceptual insights on hydro-solidarity and the political ecology of technological evolution and its effects on the socioecology of a region followed by a brief description of the study region. After elaborating on the traditional water technologies in the villages, the study looks at the trajectory of the transformation of the drinking water landscape from solidarity-based to commercially driven water technologies. All these changes are then discussed along with their impact on the village ecology and sociocultural processes. In the section on discussion, both the two technologies are critically evaluated in the context of achieving equitable conditions of water access.

Literature Review and Theoretical Background

The review of the literature encapsulates three aspects of the issue under study to give a theoretical background of the empirical findings. First, it tries to unpack how social solidarity is constructed in the management of natural resources like water. Second, it looks at how studies have demonstrated the use of technology and market influences on solidarity-based common resource management. Third, it shows how studies have proven that the use of water technology and its diffusion is not usually based on ethics and justice but is governed by power struggles between caste, class, and gender.

Social solidarity refers to fostering a sense of shared responsibility, cooperation, and support among individuals, communities, organizations, and governments to protect and sustainably manage natural resources (Boelens et al., 2005). Since competition over water resources is inherently local and context-specific in their manifestations, building solidarity for its management and conservation is also local (Falkenmark & Folke, 2002) based on a choice for collective arrangements (Keessen et al., 2016), especially when the resources are scarce. Literature on local knowledge in sustainability transformation research has established that local knowledge should be understood from people's perspective of their natural resource endowment and their use in the contexts of their social and cultural milieu. These may be in the form of technology social and economic practices or even behavior (Lam et al., 2020).

These three factors of social solidarity, traditional knowledge, and local context (in this case aridity and water shortages) typically characterize the traditional water management systems in Rajasthan (Rathore et al., 2020). There is comprehensive documentation of these traditional water harvesting systems which demonstrates the ingenuity and innovation of the local communities in developing cost-effective and appropriate technology using locally available resources for coping with drought and scarcity of drinking water (Agarwal & Narain, 1997).

A substantial amount of scholarly work grounded in Common Property Resources (CPR) theory has established that "scarcity" of resources could be the reason for initiating and enhancing cooperation between local community members largely seen as "building solidarity in communities" (Wade, 1988; Bromley & Cernea, 1989; Berkes et al., 1989; Ostrom, 1990). These scholars asserted that local people and institutions can effectively manage resources and their allocation at varying scales and diverse locations. Water control problems are ecological human-made or both. Invariably, it impacts outcomes locally that are especially influenced by existing social and political processes (Boelens et al., 2005). Thus, the potential of the hydro-solidarity concept as a successful tool to address scarcity and unequal distribution of water and reconcile daily conflicts over water access needs to be grounded in the social, cultural, and political dimensions of societies on which commonly accepted norms are based (Boelens et al., 2005).

The creation of social capital and transforming it into collective action for a common societal goal is not easy. It requires deliberate interventions, communication, and building of trust among the members (Menon, 2008). The studies that establish social solidarity as cooperation and association between local communities are often guided by characterizing it with universal norms of community participation like "social justice," "equality," and even "democracy," between the group members. These are the basic fundamental components of ethics that need to be recognized in a solidarity society because they have far-reaching implications on its sustainability and communities' well-being (Barkin & Lemus, 2014). However, there is no denying that in reality, different normative orders exist in heterogeneous societies that contribute to the conceptualizing of complex resource use and governance realities in communities (Boelens et al., 2005). Often the societal goals operate in the interest of the most powerful, who manipulate institutions and policies to sponsor their interests. This can even lead to exacerbating both environmental destruction and social disparities leading to a more unequal and unjust society in the future (Barkin & Lemus, 2014).

The above point is further reiterated by the ample scholarly work on community participation that established that community participation is rarely representative or democratic (Agarwal, 2001; Cornwall, 2003; O'Reilly, 2007). Participation should rather be understood as an ongoing process that undergoes a unique struggle for power (Cooke & Kothari, 2001). Cleaver (2001) even goes on to show that in the struggle for natural resources marginalized groups often opt out of participatory schemes. This is particularly true for women and lower caste groups (in mixed caste spaces) because they are already vulnerable due to their disadvantaged position in the social hierarchy. Social power relations in villages may also transform when users show uneven coping strategies to sustain access to water (Birkenholtz, 2010).

For this reason, several welfare programs and policy interventions that aim to reduce scarcity and unjust resource allocation not only fail but often create opportunities for future degradation and unjust accumulation that they are seeking to ameliorate (Ferguson, 1994; Escobar, 1995; Scott, 1998). This is probably because interrelationships between communities are likely to lead to social power struggles that ultimately subvert the new system and insist on staying with the status quo (Bauwens & Niaros, 2017). O'Reilly and Dhanju (2014) in their study in rural Rajasthan found that the supply of tap water that was intended to create greater equality across caste for access to drinking water among its users did not reach its intended goal. Unfortunately, the public taps that required payments for water use proved to be counter-technology because they suffered from elite capture by the upper caste and wealthy households in the villages. It seems this was even socially and culturally acceptable (O'Reilly & Dhanju, 2014). Thus, the introduction of any scarcity-reducing technologies becomes shrouded with existing social power relations, local knowledge, and the social acceptability of such practices (Mitchell, 2002).

Though caste-based inequalities in access to domestic water are well-documented in India (Agarwal, 2001), access to water is most often gendered (Sultana, 2009; Sarkar, 2023). Studies in Rajasthan (O'Reilly, 2006; Sarkar, 2023) and in other places (Sultana, 2009) have noted that the introduction of water-supply technologies has disrupted existing water-collection practices and social power relations leading to further marginalization of women. However, the new socioecology is created by unique relationships forged between the new technologies and existing technologies as they rework the existing relationships in unexpected ways that may be both complex and contradictory (Birkenholtz, 2013).

The practices of a social solidarity economy often become vulnerable to capitalism's "isomorphic pressures" (Mason, 2012). In other words, it is common for the market to put pressure and induce private enterprises to prioritize the market over social and environmental concerns acting under the influence of increasing demand (Dey, 2014; Mason, 2012). Desai and Dubey (2011) concluded that caste mattered the most in matters of resource access in Indian villages "even as market mechanisms take hold" (Desai & Dubey, 2011). In other words, the market mechanisms assisted dominant caste groups in seizing public resources. Even in other studies on water access with payment, caste-based discrimination persisted where caste groups operated to maintain their distinction (Joshi & Fawcett, 2006; Mehta, 2007; Sarkar, 2023). Water governance reforms impact everyday caste/class relations to the disadvantage of lower caste groups and neoliberal reforms can stake no claim to political neutrality (Loftus, 2006).

Neoclassical economists believe that technology is the reason why the modern economy has broken through the constraints of natural resources and achieved rapid growth (Van Vliet et al., 2021). Technological changes in the water resources sector similarly can break through the constraints of the water crisis in terms of long-term economic growth (Liu et al., 2011) that includes technological innovation, dissemination, and adoption (Van Vliet et al., 2021). Typically, solutions for drinking water shortages focus on the use of clean water technologies, such as desalinated water use and treated wastewater reuse (Elimelech & Phillip, 2011; Gude, 2017; Jones et al., 2019). However, the real issue of water technology is the rate of diffusion and adoption of innovation (O'Callaghan et al., 2020).

While the scientific principles that are used in making technology may be value-neutral, the decision-making and actual use are always embedded in historical, aesthetic, political, and cultural meanings (Arisalea, 2001). This is because technology is not merely an enhanced manipulation of tools but is a "form of action" expressing a cultural and political preference through the means of material production (Pretty, 1995). Thus, technology never exists independent of social context and is not culturally neutral (Feenberg, 1991). A growing number of political ecologists are drawing on political economy and science and technology studies (STS) approaches to investigate the relationship between water, social power, and water-supply technologies in urban areas (Swyngedouw, 1999; Birkenholtz, 2010; Phadke, 2011; Sarkar, 2022).

From the literature, we can find two research gaps. First, the concept of hydro-solidarity stresses much on the positive effects of cooperation rather than the effects of competition and power struggles to maneuver water access and use. It ignores the inherently local and context-specific manifestations of water-sharing practices embedded in historical and cultural backgrounds by assuming that a universal set of commonly accepted norms and rules (may) exist. Consequently, the hydro-solidarity concept remains a "buzzword" that has not yet been able to capture the complex realities of water storing and conserving practices. Secondly, the political economy and science and technology studies that have looked at the issues of water supply and social power are concentrated in the context of urban water technologies. This study fills these two research gaps by looking at the complex relationships between the adoption of water technology, social solidarity, and social power relations that undergo enduring transformation with time.

Study Area

Rajasthan is the largest state in India covering 10.5% of the country's geographical area but having only 1.15% of its water resources. It is one of the lowest rainfall areas in the world with 60–75% of its area designated as arid or semiarid. In the last 50 years, a threefold increase in the human population and a doubling of the livestock populations have put tremendous pressure on its fragile water and land resources. The estimated annual, per capita water availability is expected to be 439 m^3 by the year 2050, against the national average of 1140 m^3 by 2050. Groundwater is overexploited with very low water tables and several reserves are either saline or fluoride contaminated.

Rural communities in Rajasthan are heterogeneous and highly differentiated along the lines of caste, class, gender, and age. There are multiple castes in each study village. Most of the upper castes consist of the Rajputs and Brahmins, and some backward classes[1] like Kumhars, scheduled tribes like Bhil, and lower caste and ex-untouchable groups included Nai, Meghwals, and Chamars (Scheduled Caste [SC[2]]).

[1] OBCs (Other backward castes) are a bureaucratic category selected on the basis of "social and educational backwardness."

[2] SCs are a government-defined group of castes that face "extreme social, education and economic backwardness arising out of the traditional practice of untouchability." Members of ex-untouchable groups interchangeably referred to as SC, Harijan or *Dalit*.

Traditional Water Technologies

Due to scattered settlements in the villages, water sources have always been decentralized in Rajasthan. In most parts of western Rajasthan, there is a layer of gypsum at varying depths starting from Northwest to Southwest districts. Below the gypsum layer, the groundwater is deep and mostly brackish of varying intensity (Mishra, 2001). The traditional drinking water sources are designed to overcome these limitations taking into account the availability of water from rainfall, subsurface runoff, geological conditions like soil and rock types, and types of settlements (SAARC, 2009; Mishra, 2001). Water harvesting structures have dotted the villages for centuries whose technological knowledge is passed through generations (Rathore et al., 2020). However, it is important to point out that in the deserts of Rajasthan different caste groups usually have their water sources that remain privy to that particular social group. The social norms for civilized behavior, *inter alia*, encouraged the community members to maintain these facilities, conserve and protect water from pollution, and ensure its equitable and fair distribution (Rathore et al., 2020).

Although there are numerous traditional water harvesting/storing and water conservation structures in Rajasthan, this chapter will focus on only the structures that were found in the study villages.

1. *Kund or kundi* is a circular underground shallow well dug at about a depth of 45–55 feet to tap the rainwater (subsurface water) before it percolates to the salty water table. It has a saucer-shaped catchment area (called Agor) that gently slopes toward the center where the well is situated (CGWB, 2011). A wire mesh is put across the water inlets to prevent debris from falling into the well and is covered with a dome-shaped lid. The sides of the well pit are covered with (disinfectant) lime and ash. It was observed that the *kunds* are mostly privately owned by people with money to invest and land in which to construct them. The water in the *kunds* can meet the needs of a few households or single households in the years of normal rainfall. When rainwater is exhausted, people transport water from a distance and fill these structures.

   For the poor, large public *kunds* were built for water harvesting that had an adequately large catchment area (has to be selected or artificially prepared) to produce adequate runoff to meet the storage requirements. A 2-ha catchment area, having a 2–3% slope on heavy textured soil free from vegetation, is generally sufficient for a *kund* of 200 cum capacity (Saxena, 2017). Before the onset of the monsoon, the catchment area is cleaned, and the bottom of the wells are desalinized by community labor of all the beneficiaries. The use of water from the common *kunds* was monitored by the community.

   The success of *kunds* depended on the selection of the site, particularly its catchment characteristics and clearly, with the increase in population in the village it was difficult to find such sites and adequately meet drinking water demand. In the focus group discussion, villagers pointed out that the common *kunds* remained dry and were not sufficient to meet the demand for drinking water. Hence people have stopped taking water from them. At the time of the survey, the public *kunds* were dirty and dry with cattle grazing at its entry. Villagers confirmed that it is no longer customary to fill these *kunds* with transported water, and they do not put any labor into its maintenance.

2. *Kuis/Beris* are 10–12 m deep pits with very narrow mouths dug near tanks to collect the seepage and also harvest rainwater (CGWB, 2011). The pit gets wider as it goes deeper under the ground so that water can seep into a large surface area and at the same time prevent evaporation of the collected water. These were mostly constructed above the gypsum layer, so the water does not turn brackish (Saxena, 2017). The openings of these structures are generally covered with planks of wood or put under lock and key. Some *beris* were used by the low-caste villagers.
3. *Jhalara* is a local name given to step-wells essentially meant for community use and religious rites and not for drinking water. Often rectangular in design, *jhalaras* have stepped on three or four sides. Some are community step-wells that are shallower and have beautiful arches along their full height. They were secular structures from which everyone could draw water. They were traditionally built by the nobility usually for strategic and/or philanthropic reasons. With the introduction of piped water-supply schemes and the dumping of waste material in the upstream lakes, these water sources have deteriorated. Most of them are defunct today while the beautiful structures remain.
4. *Baoris/Bers* are community wells used mainly for drinking purposes and many villagers were still using them. *Baoris* can hold water for a long time because of almost negligible water evaporation when compared to other water bodies. Cost-benefit ratios as stated by Narain et al. (2005) indicate that these structures need public funds and repeated revival for perennial use. *Baoris* and *jhalaras* were essentially constructed at the site of sweet water aquifers getting a regular and heavy recharge from subterranean seepage of a talab, or lake located upstream.
5. *Nadi* or dug-out village pond is the oldest and still the most prevalent storage structure for rainwater harvesting. The location of the *nadi* had a strong bearing on its storage capacity due to the size and runoff characteristics of the catchment area (CGWB, 2011). Some *nadis* have stone walls on one or two sides to enhance capacity and water retention. The water stored in a *nadi* is generally used for drinking by both livestock and human beings. Water availability from the *nadi* would generally range from 2 months to a year after the rains (Narain et al., 2005). With proper care in repair, maintenance, and desilting, a *nadi* may function for a much longer duration than the designed life. Poorly maintained *Nadi* suffers high water losses through evaporation, seepage, and biotic interference resulting in rapid siltation and pollution (Narain et al., 2005).

   In the study villages, the *nadis* were used by the livestock and mostly the lower caste people fetched water from them when all other sources were dry. So, in the focus group discussions, the *Dalit* women said, that in years of poor rainfall when these *nadis* dry up in their villages, they have to walk miles to the nearby villages in search of water. Also, since the water was also used by the livestock, the high-caste houses preferred to buy water from the tankers rather than fetch water from them.

6. *Talabs* are water reservoirs situated in valleys and natural depressions. In old talabs, only the slope side was provided with strong parapet walls to hold the rainwater. Other sides were naturally supported by outcrops of hillocks or elevated rocky formations. On the parapet side, steps were provided to reach the water. Once these talabs served as the main source of drinking water for the local inhabitants, but they are now used more for cattle and irrigation purposes. When their water dries up, the beds are used for agricultural purposes.
7. *Tanka* is a local name given to underground cisterns generally made of masonry or concrete for the collection and storage of surface runoff from the roofs. They were the most common source of drinking water in my study villages. It is constructed in a circular or rectangular shape, normally located in a room, or a hall on bare ground in the courtyard. The water is generally withdrawn manually from its rooftop or sometimes through a pump. The size of the *tanka* is in proportion to the size of the roof and also considering the family water requirement. To ensure its purity, no one goes on the rooftop wearing any footwear. A traditional tanka constructed with lime plaster and thatched with bushes has a life span of 3–4 years (CGWB, 2011). The life span of *tankas* can be more than 20 years if they can be kept free of pollutants (Narain et al., 2005).

People in the villages are now storing their private *tankas* with water from the tankers and keeping them locked. However, it is not easy to build private *tankas* at home as it entails heavy initial investment for its construction with fewer subsidies available. Moreover, maintenance costs were also high which required special technical skills.

On deeper investigation, several constraints were revealed by the villagers that deterred them from using and maintaining these systems. The first and the most significant reason given by the villagers was the deterioration of water quality. Local people believed that revival of the traditional sources would be futile since there are more water losses in Indigenous structures due to rise in high temperature and insufficient rainfall leading to recurrent droughts and the absence of good quality groundwater. All these perceptions may have arisen due to the effects of climate change or due to an increase in demand with the rise in population. However, greater in-depth empirical research and scientific data are needed to validate these perceptions that are beyond the scope of this study.

The second reason was a shift in water usage after the advent of the piped water supply. The respondents said, "When the pipelines and the taps came, they (municipality officials) told us not to use the *talab* or *nadi* for our drinking water as this was contaminated. They said we would be backward if we did so. We wanted to be modern. But [now] water does not come from the pipe and the *talab* has gone dry." With water at the doorsteps by the tankers, the dominant castes who also happen to be the patrons take lesser or no interest in maintenance activities of the community sources. It has become more convenient to rely on buying water from tankers rather than to engage in such community tasks. Respondents confided, "We have not worked in the *talab* for maintenance since the pipeline came"; "Now we buy water from the tankers and only our livestock drinks water from the *nadi*." De Beer and

Koster (2009) observed that solidarity can be affected by a lack of community spirit, perceived abuse, and free-rider behavior. The obligation to assist is morally commendable but not binding. However, it is interesting to note that all the dominant castes continue to use these traditional water sources for their livestock but refrain from taking responsibility for their maintenance.

Ethical commitment to a community is an important component of social solidarity that addresses the needs of all people in the community and makes provision for the well-being of future generations (Barkin & Lemus, 2014). However, in reality, one does not show solidarity with everyone else, but only with the other members of the community to which one believes one's self to belong (Bayertz, 1999). The dominant caste groups ranted, "Those who continue to rely on local sources must allocate more of their labor to maintain those sources." The local leadership was ineffective in maintaining the traditional sources as they buy water. For them, it is both convenient and cheaper than to take the initiative to rejuvenate and reform these traditional structures (D'Silva & Pai, 2003).

Similar to the findings of some other studies (Harris, 2008; Birkenholtz, 2013), this study also found that in the hope of getting assured piped water, rural communities have neglected the management of traditional systems. At present, intermittent and inadequate water supply from the pipeline has resulted in water scarcity, especially among people who cannot afford to buy tanker water. This has served to subjugate water users, while further differentiating them along existing and new axes of socio-economic differences (Birkenholtz, 2013). A deeper investigation is needed to see if modern water technologies deter community solidarity in natural resource management.

Even though the traditional sources were sustained for the common good, the knowledge about the nature, type, and specifications of these assets associated with their construction, maintenance, and use remain undocumented, private, or known to specific groups and communities (SAARC, 2009). The ingenuity of these systems and their constant experimentation and evolution were transferred orally from one generation to another through a process of socialization and were internalized by the communities as part of their lifestyle activities. Unfortunately, with the gradual shift to other modes of water sources, it is natural that this knowledge is lost since they no longer remain a way of life. It is not surprising that when the villagers were asked if they knew the technique of traditional methods of water harvesting and maintenance, the answer was very dismal (Table 1).

Table 1 Do they know about the traditional system of water storage and conservation?

	Jaisalmer	Barmer	Jodhpur
Yes	24	22	30
no	66	78	70
Somewhat	10	0	0
	100	100	100

Source: Data calculated from primary data collected from the field

In the last decade, there has been growing recognition of these traditional sources of water and their revival for addressing scarcity. In this context, Mitchell (2002) points out that the introduction of modern water technologies by the state and the recent calls by government technocrats (and development practitioners) for resurrecting local systems of water management represents "a concentration and reorganization of knowledge rather than an introduction of expertise where none had been before" (page 41). The local people in this study area lacked skill, and it seems the technology of traditional management systems did not evolve with need. With the earlier patrons gone, and new patrons deciding to buy water from tankers, the hopes for the revival of traditional structures seem grim without external help.

Modern Water Technologies

Piped water is supposed to be the most reliable, safest, and easiest source of drinking water available. Over the last decade, coverage of piped water has increased to 73% but in terms of access at the household level, only 21.6% of the households have access to piped water within premises (Sarkar & Singh, 2020) and 20% reported inadequacy in its supply (Reddy, 2010). Many households have access to water from groundwater sources like hand pumps, wells, tube-wells, and boreholes. There are about 25.4 million hand pumps installed in rural areas. About 23% of these pumps are either unfit or unusable sources, as these are located in areas where there is no water (12%) or with very poor-quality water (11%).

According to the latest available data, only 33% of the villages in Rajasthan are fully covered by rural drinking water. In terms of supply provision, 88.5% of the total villages and habitations receive <40 lpcd (L/capita/day) of water. Among these, 8% receive <10 lpcd, 33% between 10 and 20 lpcd and 41% between 20 and 30 lpcd. The demand-supply gap is quite substantial if we take net supplies, after accounting for leakages (40–50%) in the case of piped water supplies (Rathore, 2004). Around 28% of the villages in the state are not covered with safe drinking water due to poor quality of water. Water quality has emerged as a growing concern in Rajasthan both due to arsenic and fluoride contamination of groundwater and bacteriological contamination of surface water (Government of India, 2012).

There is a major difficulty in reaching universal coverage in rural Rajasthan as there is a growing problem of "slip back" which is estimated at 14%, with habitations suffering a fall in the water table and water quality, especially given the growing dependence on groundwater. Infrastructure, in terms of overhead tanks and distribution lines, was constructed to provide piped water supplies to rural areas. Almost all these schemes depend on groundwater. In the majority of cases, these water-supply schemes have become defunct due to the drying up of groundwater aquifers. The deterioration of the traditional water bodies has also decreased recharge and increased runoff. At present, the huge infrastructure (physical capital) is lying waste and dilapidated depending on the groundwater resources. It has greatly aggravated the availability of drinking water and has also deteriorated the quality.

In my study village, most of the high-caste households had their sources of drinking water that were within their premises (Table 2). It could be a tap connection at home a tube-well or a well in their courtyard. Having an individual underground storage cistern or *tankas* was a universal phenomenon of all high-caste households and water sharing from their private cistern was rare. The low-caste households (both SC and ST) mostly accessed water from the common sources, which included both the public communal taps and the local traditional sources Like *Nadi*, *Talab*, *Kui,* and *Beri* depending on the quantity and quality of the water and the distance of the source from their houses.

Just like the proliferation of the informal urban water markets due to the gap in demand and supply of drinking water in the unserved and inadequately unserved urban spaces, villages in water-scarce Rajasthan are thriving with private water tanker markets. At the state level, 4% of the rural households in Rajasthan are buying water, which is the highest in the country (Reddy, 2010). Though macro-level aggregate data is not available at present, it was clear from the field that a greater number of people in the villages are buying water (Table 3). More than 60% of the households are currently buying water in the sample villages as the taps are dry and hand pumps do not provide safe water anymore due to contamination. Though water quality tests were not done in the study, the villagers complained about water being salty and unfit for drinking. Most of the villagers, especially children, had fluorosis in their teeth, indicating fluoride contamination.

The water tankers in the villages mostly acquire water from the government canals, local streams, water taps, and local water bodies. Equipped with pumps, these tankers siphon water from well-maintained water bodies and sell it to people in villages that do not have well-maintained sources. Of course, common users of these water bodies defend these sources, but tankers are typically able to exploit lapses in oversight or individual integrity. On average, one tanker of water is sold for Rs. 400 to Rs. 500 depending on the water quality, availability, and the distance from the source of the water to the villages. Though domestic water consumption varied across the households depending on the number of family members and their water usage behavior, most of them reported buying tanker water twice to thrice in a month spending an average of Rs. 800 to Rs.1000.

To understand the impact of modern technologies on access to drinking water, the respondents were asked about their perceptions of the changing situation. The perception looks inconclusive as only around half of the residents thought modern technologies have improved their access to drinking water (Table 4) mostly by the introduction of piped tap water supply and tube-wells (Table 5). Tanker water supply was popular in high-caste households as they got water at their doorsteps and at a convenient time (Table 3), but it was still not perceived as an improvement in water access as people had to buy water which was both against their cultural norm and became a permanent component of their household budget. However, proponents of the tanker technology voiced one concern that the tap water through free was intermittent, inadequate, and unreliable. They believed that due to the

Table 2 Source of drinking water used by different caste groups in the study villages (Percentages)

	Jaisalmer				Jodhpur				Barmer				
	General	SC	ST	Total	General	SC	ST	Total	General	OBC	SC	ST	Total
Own	39 (83)	19 (56)	2 (29)	60 (68)	10 (71)	15 (56)	11 (61)	36 61)	6 (50)	0 (0)	12 (60)	5 (71)	23 (53)
Common	8 (17)	13 (38)	5 (71)	26 (30)	4 (29)	11 (41)	7 (39)	22 (37)	6 (50)	4 (100)	8	2 (29)	20 (47)
Neighbor	0 (0)	2 (6)	0 (0)	2 (2)	0 (0)	1 (4)	0 (0)	1 (2)	0 (0)	0 (0)	0 (0)	0 (0)	0 (0)
Total (N)	47	34	7	88	14	27	18	59	12	4	20	7	43

Source: Data calculated from primary data collected from the field
Note: The figure in parentheses shows the percentage of households in each caste group

Table 3 Caste-wise prevalence of households buying water from private tankers

Caste	Jaisalmer	Barmer	Jodhpur
General	46 (98)	10 (91)	11 (92)
SC	18 (53)	17 (43)	16 (47)
ST	5 (61)	11 (61)	6 (66)
Total	69 (78)	38 (64)	33 (77)

Source: Data calculated from primary data collected from the field
Note: The figure in parentheses shows the percentage of households in each caste group

Table 4 Whether modern water technologies have improved access to drinking water

Percentage of households	Jaisalmer	Barmer	Jodhpur
Have improved	44 (50)	36 (61)	20 (47)
Not improved	31 (35)	16 (27)	15 (35)
Remained same	5 (6)	2 (3)	1 (2)
Cannot say	8 (9)	5 (8)	7 (16)
	88 (100)	59 (100)	43 (100)

Source: Data calculated from primary data collected from the field
Note: The figure in parentheses shows the percentage of households

Table 5 The water technologies that have improved access to drinking water

Percentage of households	*Jaisalmer*	*Barmer*	*Jodhpur*
Tanker	14	25	35
Tap water supply	41	64	40
Tube-well	45	11	25

Source: Data calculated from primary data collected from the field

unreliability of tap water, many houses have and will be shifting to tanker water in the future.

Social Power Relations and Adoption of Water Technology

Caste and Access to Water from Taps and Tankers

Even though state-based institutions prohibit water-based discrimination, the "higher" castes still insist on discriminatory practices. Caste groups might mix at a public tap but separate taps for upper caste and ex-untouchable groups were the norm in the villages as also noted in other studies (Mehta, 2007; O'Reilly & Dhanju, 2014). The separate geographical location of taps allowed us to maintain a clear distinction between caste-based access and water supply.[3] For example, chamas

[3] In rural India, caste also has a geography. Upper caste and/or dominant caste groups usually live at the village center, while SCs occupy the village outskirts.

(SC) used taps in the chamar neighborhood. The villagers also referred to the public community taps by the names of castes, "Rajput taps" or "Meghwal taps" signifying public taps having caste-based, locational association, and even branded ones. Even though the water tankers did not confirm geography as they were not fixed infrastructure, they were also confined to the norms of differential access. The tankers had castes written on them since they were private and thus it was the prerogative of their owner to legally brand them. The presence of water tankers in the villages was a material reminder of caste power.

While paying for water played a role in equalizing access to water, access to private water was not always guaranteed with payment where caste and kinship relationships played an upper hand. The fact that water buying was most prevalent in the high-caste households (Table 3), testifies to that. A few respondents asserted that "All castes can buy from tankers, except the *Dalits* (ex-untouchables) who have their own designated tankers or designated days in which water is sold to them." For this man, and others, lower caste and higher caste groups might share the tankers, but the ex-untouchables must have their own specific designated turns or tankers. It is not surprising that the tankers selling water mostly to the upper caste groups would seldom refuse to sell water to the *Dalits* as their upper caste fixed customers would refrain from buying from them. An elderly *Harijan* woman exclaimed, "We pay for water but when the tankers are owned by higher caste people how can we claim their water. It is their wish if they will sell us water." She stated that the water in the *talab* and the *nadi* is "our water" (community water) but what can we say for private water. Access to water was governed by rules of "purity and pollution" dictating whose water can be drunk, whose should be avoided, and who should fetch the water.

When there was no water, the lower caste groups asked the tankers to sell them water. Even if not regularly, around half of the low-caste households who could afford to buy water have also started buying water from the tankers. Thus, the rules and restrictions based on "purity and pollution" are bent or even totally dropped under certain circumstances. Some sellers sell water to anyone irrespective of their caste provided they can pay. Villagers also mentioned that during drought periods, "higher" castes do not hesitate to drink water from Dalit wells. Thus, even caste-based institutions display a certain degree of flexibility during times of drought. High-caste villagers explain this differently. According to them, sub-terrain water is the same everywhere; it becomes differentiated only when it acquires the attributes of the user (Mehta, 2007). Thus, according to village logic, water in a well-used by Dalits is not impure, but the water in a Dalit's house is. Similarly, when the tankers enter the courtyard of a *Dalit* house, they become impure and cannot sell water to a higher caste till it is cleaned and refilled with water.

Commercialization of water is sometimes viewed as a freedom to buy water from the tankers that have given a few lower caste groups an opportunity for equal access, no matter how few of them. High-caste groups also understood that payment put lower caste groups on an equal footing where access to water was concerned. The common voices of *Dalits* were "All people pay to buy water and if any high caste tells us, we can tell them we too pay for water like others do, so we have a full water right," "There are no caste issues, at least not in this village. We know our rights." However, this also bothered some upper caste people as they felt the caste system was being undermined by the commercialization of water by the tankers. One Rajput woman confided, "The *Dalits* buy water from the same tanker from where we buy water. This tanker system has spoiled the [caste] system." She and her Rajput neighbors felt the combined pinch of *Dalit* women's assertion of their right to water in the commercialized water systems.

Caste, Water Storage Infrastructure, and Access to Water

A house connection enabled convenient access to water when water was flowing (2 h a day) and could fill a *tanka* to enable long-term storage for when water was not flowing. Most of the high-caste families (who were also relatively wealthy) now have *tankas* in their compounds (also noted by O'Reilly & Dhanju, 2012; Mehta, 2007). There were no tap connections inside the houses of the *Dalits*. One Meghwal (SC) boy exclaimed, "There were house connections everywhere, except for the poor Meghwal, Chamar, and Nayak (SC) families. That's because they can't afford the pipe and tap for the house connection and *tankas* for water storage." In a focus group discussion, participants said "People who do not have *tankas,* have to go to the *nadis* and *beris* when there is no water."

Water was never sold in small quantities and only people who had *tankas* could fill them up with tanker water. Several *Dalits* lamented that they had the money to buy water, but they could not afford a hefty one-time cost to build a *tanka* inside their house. Most of the houses in the *Dalit* locality were closely spaced and did not have the space and means to accommodate these underground storage cisterns. All high-caste households compulsorily had private *tankas* with locks. The water that was bought was inadvertently precious and needed extra protection. So, both factors of caste and class are intertwined to ensure differential water supply and access.

A *Dalit* woman confided that she borrows water from her high-caste neighbor when water was not available at the community tap or night. This was probably an exception as all other *Dalits* in the interviews exclaimed, that they were not allowed to fetch water from Brahmin and Jat *tankas* when no water was there in the taps in their neighborhoods. So, in most cases, they walked miles to other villages or to local communal water sources to fetch water. *Dalits*, when asked about their difficulty in lugging water from far-off places often remarked, "Water is so expensive, but, if need be, it had to be paid." 'Sometimes we pay and sometimes we walk and get it. We decide as per our budget." The *Dalit* women stated that they get money for working in the government minimum daily wage scheme (MGREGA[4]). So, they

[4]The Mahatma Gandhi National Rural Employment Guarantee Act (MGNREGA) programs in my study villages employed local labor for integrated watershed management programs. Women at the time of my survey were working on rejuvenation of the *khadins*, an ingenious construction designed to harvest surface runoff water for agriculture.

had to decide if they would buy water and spend money or go to work to earn daily wages.

Caste, Gender, and Access to Water

After tankers have arrived, Rajput and Brahmin women do not go out to fetch water at all. A Rajput woman commented, "We have always had a *tanka* in our house for generations. Earlier we used to fill up water from the rains, now we fill it from the pipeline mostly, and in the dry summer months we buy water from the tanker-wallah." "This gives us more time to do other things like reading and knitting." Ironically in the same village, the low-caste women and girls spent vast amounts of time on water collection. Invariantly, they get less or no time to pursue other empowering activities that high-caste women could do. In the drier months when the water scarcity is at its peak, these women have to walk 1–3 km to fetch water and spend time regularly searching for community wet sources (that can be a tap, tube-well, or a *Kui*). A Meghwal woman commented, "These days it is not safe for the girls to walk far and collect water from the other villages. Moreover, there is not enough time to do anything else." For the *Dalit* women who work on the farms or as daily wage laborers, fetching water is like a triple burden as they have to manage their domestic chores, paid work, and also fetch water from far-off places.

There is differential access to water between women stratified by village. Thus, lived experiences of water scarcity are observed at both intervillage and intragender levels. It shows that technologies of water access may minimize the stress of physical accessibility of water but "lived scarcities" of water are created by village-level institutions. Technologies become more accessible by dominant caste groups whose actions increase water scarcity along caste lines. Interestingly, even when water was sold at a price, dominant caste groups were still able to re-assert their dominance through their control of the tankers and the stored water in their tankas. Mehta (2007) in her work on arid Gujarat argues that decentralization of water governance has played a role in the reproduction of caste inequality. Similarly, this work shows how water commercialization through water tankers reproduces caste inequality in the arid villages of western Rajasthan. In other words, it seems the new water technology has rediscovered caste identities. Jeffrey (2001) observed that the modern water-supply network was enmeshed and reproduced persistent caste and class power struggles.

Results and Discussion

Since water scarcity is usually portrayed in the physical availability of water and is calculated in volumetric terms, it becomes imperative to perceive water-accessing technology as a panacea to solve the problem of water access. However, to solve the problem of water access, one needs to delve deeper into the complexities of how the "problem" of scarcity is constructed, manifested, and manipulated. The basic thing to understand this is to disaggregate water users and their entitlements. Thus, to undertake the task of finding a solution, priority needs to be given to understanding the politics of "technology choice" vs a vi the constraints of "technology accesses" both from a justice lens and resource governance.

For centuries, the socio-ecological system of the arid region of western Rajasthan has been described as drought-resilient because local communities have thrived in the desert with traditional infrastructure for harvesting and conserving every drop of rain with great ingenuity and caution. Unfortunately, the modern state with its technology, manpower, and wealth has failed to provide universal safe drinking water in the villages of Rajasthan. At present, there is the dual dilemma of dilapidated traditional water sources and drying up of the public source or breakdown in pipeline due to physical water shortages. This has culminated in both intervillage and intravillage water scarcity on the one hand and the proliferation of a usurious system of private water tankers on the other.

The findings of this study, like the previous studies, also show the contradictory effects on access to water by the introduction of new technologies. This is because socio-economic inequality and water scarcity are intertwined, and it is heavily influenced by social systems and historical conditions. Modern water-supply technologies rework these existing systems and create a new unique water-supply landscape. This goes on to validate the fact that socioecological histories are not displaced by new technologies; rather, they hybridize and create a more complex one. Therefore, the potential for its transformation lies not directly with the water-supply network per se but with the way these technologies are introduced and how they interact with historical social power relations.

Though the most preferable way to provide universal safe water coverage in any area is through a piped water system, it has major financial implications. In the case of rural Rajasthan, there are added dilemmas of increase in population and depletion and deterioration of freshwater reserves, especially groundwater. Additionally, there is the universal problem of poor water services resulting in intermittent and unreliable water supply. At present, the alternative to public taps is private water tankers whose demand will continue to grow by the day. It is important to mention here that the author is not opposed to modern technologies of water supply, especially piped water systems, but the lack of physical availability of water in the pipes gives sufficient reason for thinking through the politics and capacities that technologies have on socioecological changes. At the same time, it nudges to recognize that the "enthusiasm" generated by the new interest in traditional water management systems as an alternative for solving the rural water crisis, needs to be justified against their actual potential.[5] Though these traditional systems need to be applauded for their rich local knowledge and solidarity-based sociocultural practices, without sci-

[5]There are modern technologies of rainwater harvesting and groundwater recharge in Rajasthan such as anicut, percolation tank, subsurface barrier, and pond with infiltration wells (Narain et al., 2005). But such structures were absent in my study villages and hence not mentioned.

entific upscaling of their numbers, they cannot keep pace with the growing water demand that increases over time.

Conclusion

It was envisaged by development practitioners that modern technology of water supply by a centralized system and networked distribution would displace the traditional community water structures. The traditional or old, local, water systems came to become synonymous as "backward" and "unsafe". However, in the context of climate change adaptation and growing awareness to foster community resilience to climate change impacts, there is a renewed interest in going back to the "traditional" and "local." This narrative aligns with the mandate of rejuvenation of the traditional knowledge and within that traditional knowledge of water harvesting and local community water structures. It calls for reinventing and strengthening institutional frameworks that support the harvesting of rainwater in arid areas. Demand-driven water governance that propagates "mindful" and "wise" water usage is also associated with the re-establishment of water sharing based on community solidarity. However, it seems at present that traditional and modern water technologies are competing with each other theoretically and in practice. Perhaps it is time to reconceptualize a hybridized form of water system that harmonizes these two systems together. This could serve as an alternative water-supply management landscape. In practice, it could be a combination of a publicly funded centralized water-supply system accountable to local populations and a state-monitored institutional mechanism that promotes community solidarity for investing and maintaining local water bodies with enhanced local participation. Both these water systems have their own positive and negative externalities, but they can complement each other to solve the perennial problems of drinking water shortages in arid western Rajasthan.

References

Agarwal, B. (2001). Participatory exclusions, community forestry, and gender: An analysis for South Asia and conceptual framework. *World Development, 29*, 1623–1648.

Agarwal, A., & Narain, S. (1997). Dying wisdom: The decline and revival of traditional water harvesting systems in India. *The Ecologist, 27*(3), 112–117.

Arisalea, Y. (2001). Women carrying water: At the crossroads of technology and critical theory. In J. Paris & W. Wilkinson (Eds.), *Essays on liberation, new critical theory*. Rowman and Littlefield.

Arya, A. S., Dhinwa, P. S., Pathan, S. K., & Raj, K. G. (2009). Desertification/land degradation status mapping of India. *Current Science, 97*, 1478–1483.

Barkin, D., & Lemus, B. (2014). Rethinking the social and solidarity society in light of community practice. *Sustainability, 6*(9), 6432–6445.

Bauwens, M., & Niaros, V. (2017). *Value in the commons economy: Developments in open and contributory value accounting*. Heinrich Böll Stiftung, P2P Foundation.

Bayertz, K. (Ed.). (1999). *Solidarity* (Vol. 5). Springer Science & Business Media.
Berkes, F., Feeny, D., McCay, B. J., & Acheson, J. M. (1989). The benefits of commons resources. *Nature, 340*, 91–93.
Birkenholtz, T. (2010). 'Full cost recovery': Producing differentiated water collection practices and responses to centralized water networks in Jaipur, India. *Environment and Planning A, 42*, 2238–2253.
Birkenholtz, T. (2013). "On the network, off the map": Developing intervillage and intragender differentiation in rural water supply. *Environment and Planning D: Society and Space, 31*(2), 354–371.
Boelens, R., Zwarteveen, M., & Roth, D. (2005). Legal complexity in the analysis of water rights and water resources management. In D. Roth, R. Boelens, & M. Zwarteveen (Eds.), *Liquid relations: Contested water rights and legal complexity* (pp. 1–19). Rutgers University Press.
Bromley, D., & Cernea, M. (1989). *The management of common property natural resources: Some conceptual and operational fallacies*. The World Bank.
Central Groundwater Board (CGWB). (2011). *Select case studies rain harvesting and artificial recharge: 9–11*. Ministry of Water Resources.
Cleaver, F. (2001). Institutions, agency and the limitations of participatory approaches to development. In B. Cooke & U. Kothari (Eds.), *Participation: The new tyranny?* Zed.
Cooke, B., & Kothari, U. (Eds.). (2001). *Participation: The new tyranny?* Zed.
Cornwall, A. (2003). Whose voices? Whose choices? Reflections on gender and participatory development. *World Development, 31*, 1325–1342.
D'Silva, E., & Pai, S. (2003). Social capital and collective action: Development outcomes in forest protection and watershed development. *Economic and Political Weekly, 38*(14), 1404–1415.
De Beer, P., & Koster, F. (2009). *Sticking together or falling apart? Solidarity in an era of individualization and globalization*. Amsterdam University Press.
Desai, S., & Dubey, A. (2011). Caste in 21st century India: Competing narratives. *Economic and Political Weekly, 46*, 40–49.
Dey, P. (2014). Governing the social through 'social entrepreneurship': A Foucauldian view of 'the art of governing in advanced liberalism. In *Social entrepreneurship and enterprise: Concepts in context* (pp. 55–72). Tilde University Press.
Elimelech, M., & Phillip, W. A. (2011). The future of seawater desalination: Energy, technology, and the environment. *Science, 333*(6043), 712–717.
Escobar, A. (1995). *Encountering development: The making and unmaking of the third world*. Princeton University Press.
Falkenmark, M., & Folke, C. (2002). The ethics of socio-ecohydrological catchment management: Towards hydrosolidarity. *Hydrology and Earth System Sciences, 6*(1), 1–9.
Feenberg, A. (1991). *Critical theory of technology* (Vol. 5). Oxford University Press.
Ferguson, J. (1994). *The anti-politics machine: 'Development', depoliticization, and bureaucratic power in Lesotho*. University of Minnesota Press.
Government of India (2012). *Rural drinking water and sanitation, twelfth five year plan (2012–2017)*, Economic Sectors Vol. II, Planning Commission Government of India.
Gude, V. G. (2017). Desalination and water reuse to address global water scarcity. *Reviews in Environmental Science and Bio/Technology, 16*(4), 591–609.
Harris, L. M. (2008). Water rich, resource-poor: Intersections of gender, poverty, and vulnerability in newly irrigated areas of Southeastern Turkey. *World Development, 36*, 2643–2662.
Jeffrey, C. (2001). 'A fist is stronger than five fingers': Caste and dominance in rural north India. *Transactions of the Institute of British Geographers, 25*, 1–30.
Jones, E., Qadir, M., van Vliet, M. T., Smakhtin, V., & Kang, S. M. (2019). The state of desalination and brine production: A global outlook. *Science of the Total Environment, 657*, 1343–1356.
Joshi, D., & Fawcett, B. (2006). Water, Hindu mythology and an unequal social order in India. In T. Tvedt & T. Oestigaard (Eds.), *A history of water* (The world of water) (Vol. 3). IB Tauris.
Keessen, A., Vink, M. J., Wiering, M., Boezeman, D., Ernst, W., Mees, H., & Van Eerd, M. C. (2016). Solidarity in water management. *Ecology and Society, 21*(4), 35–41.

Lam, D. P., Hinz, E., Lang, D., Tengö, M., von Wehrden, H., & Martín-López, B. (2020). Indigenous and local knowledge in sustainability transformations research: A literature review. *Ecology and Society, 25*(1). https://doi.org/10.5751/ES-11305-250103

Liu, Y., Yang, X., Zhou, R., Duan, Y., & Yao, C. (2011). A comparative study on the "growth tail effect" of soil and water resources in the economic growth of the central region. *Resource Science, 9*, 1781–1787.

Loftus, A. (2006). Reification and the dictatorship of the water meter. *Antipode, 38*(5), 1023–1045.

Mall, R. K., Gupta, A., Singh, R. S., & Rathore, L. S. (2006). Water resources and climate change: An Indian perspective. *Current Science, 90*, 1610–1626.

Mason, C. (2012). Isomorphism, social enterprise and the pressure to maximize social benefit. *Journal of Social Entrepreneurship, 3*(1), 74–95.

Mehta, L. (2007). Whose scarcity? Whose property? The case of water in western India. *Land Use Policy, 24*(4), 654–663.

Menon, S. (2008). Social capital, community participation and panchayats: Study of Aapni Yojnain Rajasthan. *The ICFAI University Journal of Public Administration, 4*(28), 26–37.

Mishra, A. (2001). *The Radiant Raindrops of Rajasthan*. Technology and Ecology, New Delhi.

Mitchell, T. (2002). *Rule of experts: Egypt, techno-politics, modernity*. University of California Press.

Narain, P., Khan, M. A., & Singh, G. (2005). *Potential for water conservation and harvesting against drought in Rajasthan, India*. Working Paper 104. International Water Management Institute.

O'Callaghan, P., Adapa, L. M., & Buisman, C. (2020). How can innovation theories be applied to water technology innovation? *Journal of Cleaner Production, 7*, 122910.

O'Reilly, K. (2006). 'Traditional' women, 'modern' water: Linking gender and commodification in Rajasthan, India. *Geoforum, 37*, 958–972.

O'Reilly, K. (2007). Where the knots of narrative are tied and untied: The dialogic production of gendered development spaces in Rajasthan, India. *Annals of the Association of American Geographers, 97*, 613–634.

O'Reilly, K., & Dhanju, R. (2012). Hybrid drinking water governance: Community participation and ongoing neoliberal reforms in rural Rajasthan. *Geoforum, 43*, 623–633.

O'Reilly, K., & Dhanju, R. (2014). Public taps and private connections: The production of caste distinction and common sense in a Rajasthan drinking water supply project. *Transactions of the Institute of British Geographers, 39*(3), 373–386.

Ostrom, E. (1990). *Governing the commons: The evolution of institutions for collective action*. Cambridge University Press.

Pandey, D. N., Gupta, A. K., & Ravindranath, N. H. (2010). *Climate change impacts, mitigation and adaptation: Science for generating policy options in Rajasthan, India*. Rajasthan State Pollution Control Board.

Phadke, R. (2011). Reclaiming the technological imagination: Water, power, and place in India. In M. J. Goldman, P. Nadasdy, & M. D. Turner (Eds.), *Knowing nature: Conversations at the intersection of political ecology and science studies*. University of Chicago Press.

Pretty, J. N. (1995). *Regenerating agriculture*. Joseph Henry Press.

Rathore, M. S. (2004). *Water and sanitation in Rajasthan: Achievements against MDGs*. Institute of Development Studies.

Rathore, V. S., Kantamneni, J., Jamadar, A., Anupam, S., & Kachchap, D. (2020). Manifestations of traditional knowledge in water systems: The cases of the Kuchaman Fort and Rathi Haveli, Rajasthan. *Journal of Traditional Building, Architecture and Urbanism, 1*, 421–433.

Reddy, V. R. (2010). Water sector performance under scarcity conditions: A case study of Rajasthan, India. *Water Policy, 12*(5), 761–778.

SAARC. (2009). Indigenous knowledge of the vulnerable communities living in the drought prone areas of Rajasthan, India. In *Indigenous knowledge for disaster risk reduction in South Asia, New Delhi*.

Sarkar, A. (2022). How 'smart' are we with smart technology: Comparative cases of water ATMs in Nairobi and Delhi. *Water Practice and Technology, 17*(10), 2160–2170.
Sarkar, A. (2023). Women, water and technology: Changing nature of water transactions in Western Rajasthan. *Journal of Water, Sanitation and Hygiene for Development, 13*(1), 30–38.
Sarkar, A., & Singh, S. (2020). *"Waters" of Western Rajasthan: Myths, traditions, life and livelihoods*. New Delhi.
Saxena, D. (2017). Water conservation: Traditional rainwater harvesting systems in Rajasthan. *International Journal of Engineering Trends and Technology, 52*(2), 91–98.
Scott, J. C. (1998). *Seeing like a state: How certain schemes to improve the human condition have failed*. Yale University Press.
Sultana, F. (2009). Fluid lives: Subjectivities, gender and water in rural Bangladesh gender. *Place and Culture, 16*, 427–444.
Swyngedouw, E. (1999). Modernity and hybridity: Nature, Regeneracionismo, and the production of the Spanish waterscape, 1890–1930. *Annals of the Association of American Geographers, 89*, 443–465.
Van Vliet, M. T., Jones, E. R., Flörke, M., Franssen, W. H., Hanasaki, N., Wada, Y., & Yearsley, J. R. (2021). Global water scarcity including surface water quality and expansions of clean water technologies. *Environmental Research Letters, 16*(2), 024020.
Wade, R. (1988). *Village republics: Economic conditions for collective action in South India*. Cambridge University Press.

Prevention of Hydrological Disasters Centered on Hydro-Solidarity: Estimation of Sediment Yield in Bhagirathi Basin and Prioritization

Bratati Dey, Juveriya Naaz, and Arushi Jha

Introduction

The process of soil erosion by water comprises the stages of detachment, transportation, and deposition of sediment through the impact of raindrops and the movement of flowing water. The primary forces driving these processes originate from both the impact of raindrops and the runoff of water. Soil erosion mechanisms come into play when water from areas experiencing sheet flow converges under specific conditions, forming small rills that give rise to minor channels (Julien, 1998; Wischmeier & Smith, 1978). As the flow concentrates into these channels, it leads to the erosion of soil and less resistant lithological formations, facilitating the transportation of a considerable amount of material within these pathways. While soil erosion may not be readily apparent on exposed soil surfaces, even in cases where raindrops are causing substantial sediment erosion, the effects can be striking in instances where concentrated flows result in the development of extensive rill and gully systems (Kim, 2006).

The rapid loss of soil and water due to human activities and natural elements poses a significant threat to land and water resources and the ecological environment. The examination of soil erosion and the movement of sediment are crucial because soil erosion plays a central role in source pollution. Analyzing soil erosion is fundamental for determining sediment yield and understanding sediment

B. Dey (✉)
ICSSR Post Doctoral Research Fellow, Shaheed Bhagat Singh College, University of Delhi, New Delhi, India

J. Naaz
Swastik Edustart-Institute of GIS and Remote Sensing, New Delhi, India

A. Jha
University School of Environmental Management (USEM), GGS Indraprastha University, Dwarka, New Delhi, India

K. Arora, R. R. Anand (eds.), *The Solidarity Approach in Geography*,
https://doi.org/10.1007/978-3-031-86753-8_30

transport, constituting essential groundwork in environmental studies (Gassman et al., 2007). The initial step toward effective water management and environmental restoration involves thoroughly comprehending the watershed's hydrological processes. Factors such as climate fluctuations, alterations in land cover (such as deforestation), and excessive grazing have collectively played a role in causing land degradation and soil erosion (Haregeweyn et al., 2013).

Sediment yield modeling is a critical aspect of watershed management, aiming to understand and predict the dynamics of sediment transport in river systems. The Soil and Water Assessment Tool (SWAT) Model has gained prominence as an effective tool for simulating sediment yield in various environmental settings (Yesuf et al., 2015; Wolde & Bogale, 2017; Setegn et al., 2010; Tesfahunegn et al., 2012; Mengistu & Sorteberg, 2012; Tadele & Forch, 2007). It is a comprehensive, process-based hydrological model developed to simulate the impact of land management practices on water, sediment, and agricultural chemicals in large complex watersheds. SWAT incorporates various components such as weather, soil, land use, and management practices, making it well-suited for assessing sediment yield. The literature reviewed underscores the significance of the SWAT model in sediment yield modeling across various watersheds (Duru et al., 2017; Arnold et al. 2012a, b; Jeong et al., 2010; Pandey et al., 2016). Its versatility, when coupled with proper calibration, validation, and integration with other technologies, positions SWAT as a valuable tool for understanding and managing sediment transport in a changing environment.

India, with its diverse geographical and climatic conditions, faces significant challenges related to soil erosion and sediment transport in its watersheds. The modeling of sediment yield in Indian watersheds holds immense significance for several reasons as agricultural productivity, water resource management, flash flooding and erosion control, infrastructure planning, biodiversity conservation, climate change resilience, soil fertility health, and community livelihoods (Pandey et al., 2008; Kaini et al., 2012; Issaka & Ashraf, 2017). India is primarily an agrarian economy, and soil erosion adversely impacts agricultural productivity. Sediment-laden runoff can contribute to flash floods during monsoon. Sedimentation in rivers and reservoirs is a critical concern for water resources management. The deposition of sediments reduces the storage capacity of reservoirs impacting water availability for irrigation, domestic use, and hydropower generation. Sediment yield modeling helps in understanding erosion patterns, enabling the implementation of soil conservation measures to sustain and enhance agricultural yields (Singh et al., 2012; Sanjay et al., 2010; Gull & Shah, 2022). Accurate modeling provides insights for efficient reservoir management (Pandey et al., 2016; Prabhanjan et al., 2015). Modeling sediment yield helps in evaluating the impact of erosion on soil fertility. This is crucial for the implementation of sustainable land management practices and for maintaining the long-term health of agricultural soils (Uniyal et al., 2020; Verma & Jha, 2015). In conclusion, sediment yield modeling in Indian watersheds is crucial for sustainable development, environmental conservation, and the well-being of communities. It provides the necessary tools to anticipate mitigate and adapt to the challenges posed by soil erosion and sediment transport in a country with diverse and dynamic landscapes.

Study Area

The Bhagirathi basin lies in the North-western part of Uttarakhand, India. The basin stretches from 30° 16′ to 31°10′North and 78°11′ to 79°17′ East; it covers an area of 7590 sq km and spreads over districts of Uttarkashi, Pauri Gharwal, and Tehri Gharwal. The Bhagirathi basin lies over the middle and lower Himalayan ranges which constitute low- and high-altitude mountains, the elevation of the basin ranges from 296 m to 7801 m above the mean sea level (Fig. 1). The climate of the basin falls under the temperate monsoonal zone, the maximum and minimum temperature of the basin range between 30°–36° and 0°–6°, respectively, and annual rainfall received is about 150–25 (Tripathi & Kumar, 2014). The runoff in the region is governed by summer and winter precipitation (snow and rainfall). Southwest monsoon carriers rain in during summer (June–September) while western disturbances are responsible for winter precipitation (Dimri et al., 2023). Thus snowmelt and rainfall both are responsible for potential runoff and sediment yield in the Bhagirathi basin. The study area is dominated by vegetation and range land with some patches of agriculture and built-up in the south with open bare and snow-covered mountains in the north with range grasses, also India's highest dam, i.e., Tehri Dam is located on this Bhagirathi basin at an elevation of 713 m above MSL.

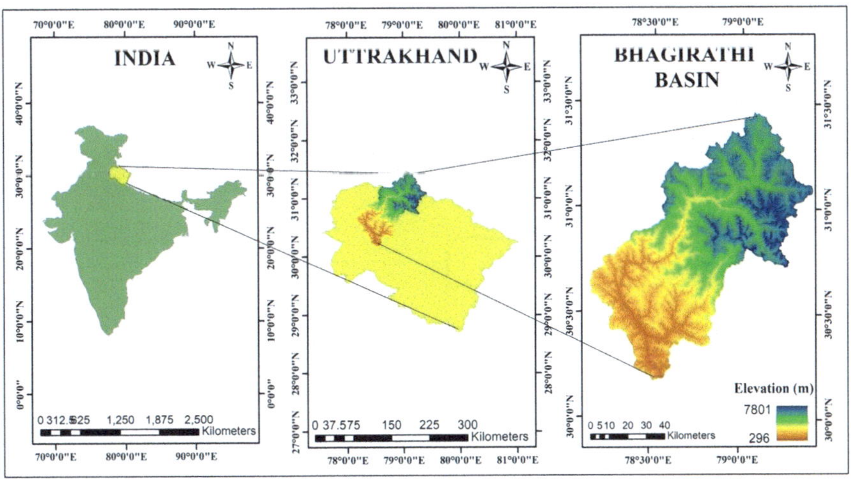

Fig. 1 Location map, Bhagirathi basin

Materials and Methods

To set up the model and simulate the hydrological process of basin SWAT requires various spatial data. The datasets needed to run the SWAT model are the digital elevation model (DEM), land use land cover (LULC), soil data, slope, and basic climate parameters such as temperature and precipitation sources of data and procedure involved in preparing model input are discussed below (Fig. 2).

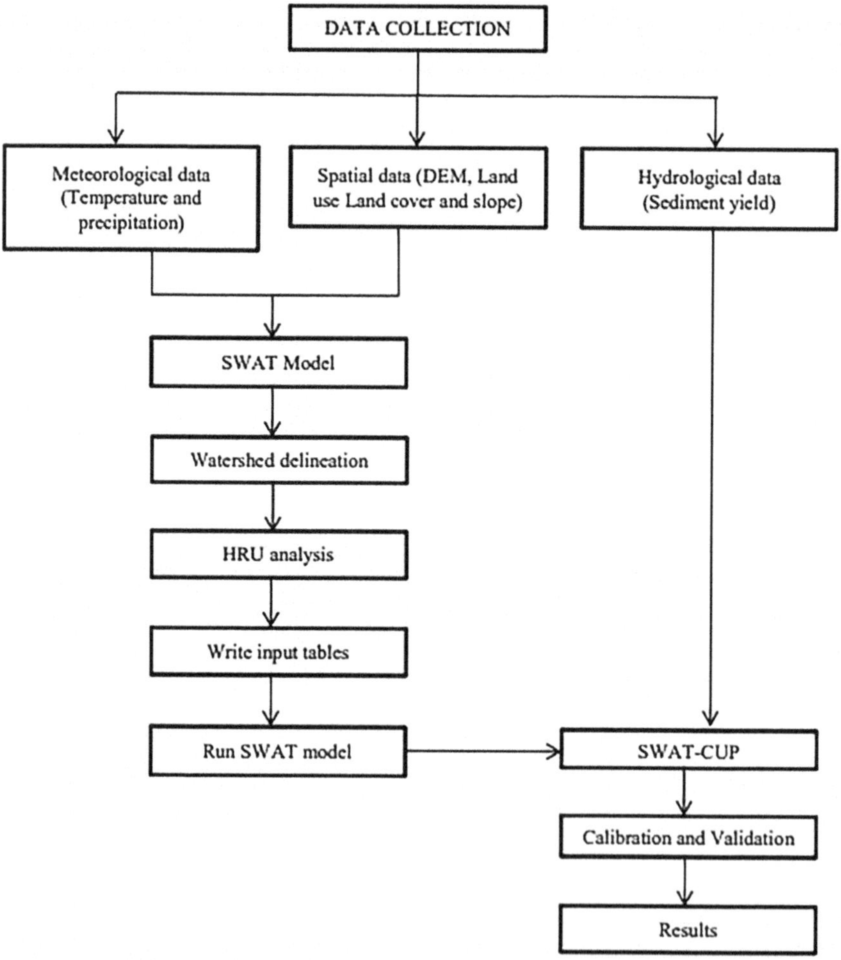

Fig. 2 Methodology cart

Digital Elevation Model

DEM is a prerequisite in the model, in this study Shuttle Radar Topography Mission (SRTM) DEM is used, this high-resolution data of 1-arc second (30 m) is provided at the United States Geological Survey (USGS) website. DEM is used to delineate basins and further demarcate the sub-basins, and our study area has 9 sub-basins (Fig. 3). Further, a slope map was also developed which then divided the Bhagirathi basin into many Hydrological Response units (HRUs). HRU is delineated based on

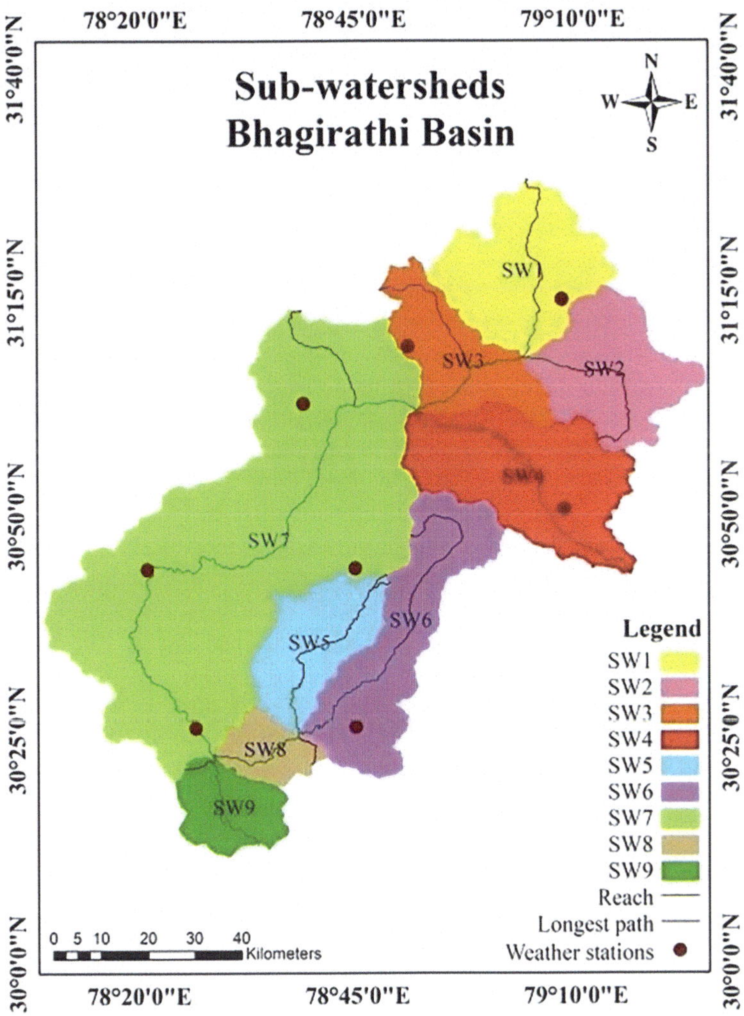

Fig. 3 Sub-watershed

uniformity of land use, soil type, and slope. The entire basin was divided into 118 HRUs.

Land Use Land Cover Data

LULC is an essential element of the watershed determining runoff, soil loss, sediment yield, and other hydrological processes of the watershed. For model input, the LULC data provided by Esri 10 m annual land use land cover map is used (Fig. 4).

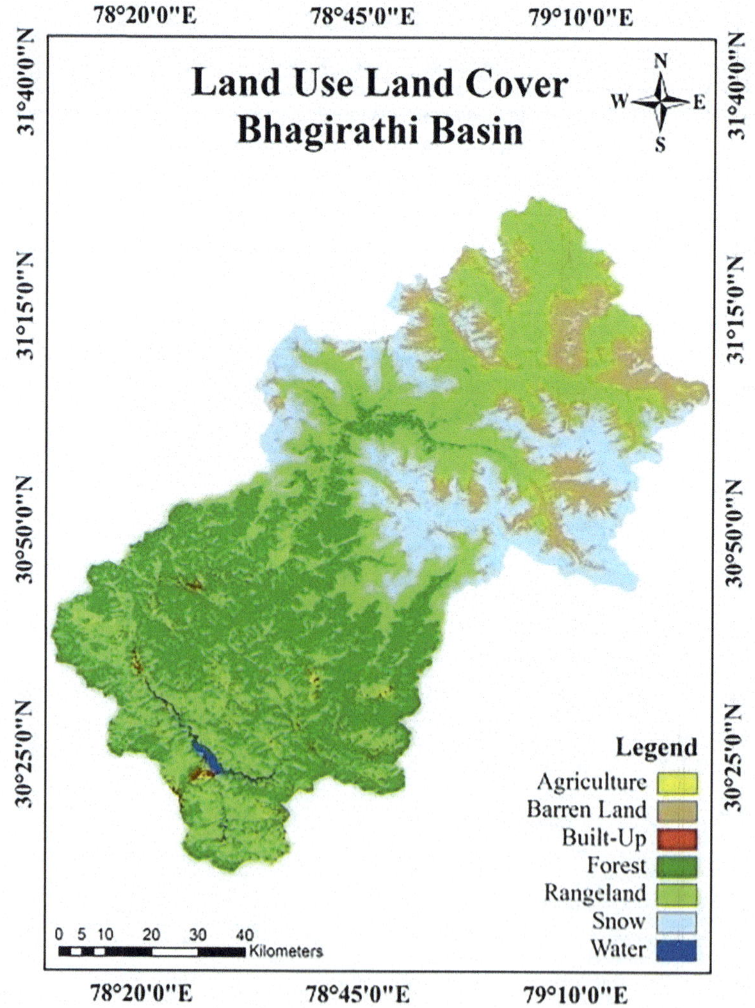

Fig. 4 Land use land cover

Table 1 Land use land cover, Bhagirathi basin

Land use classes	Area (km^2)	Area (%)
Agriculture	29.44	0.38
Barren land	1110.61	14.63
Built-up	70.13	0.92
Forest	2366.31	31.77
Rangeland	2547.16	33.55
Snow	1412.64	18.61
Water	53.61	0.70

This dataset is generated by Impact Observatory, which used billions of human-labeled pixels (curated by the National Geographic Society) to train a learning model for land classification. The LULC map is obtained from ESA Sentinal-2 imagery at 10 m resolution. The LULC maps are assessed to have an average accuracy of 75%. The Bhagirathi basin is classified into seven broad categories such as agriculture, barren land, built-up, forest, rangeland, snow, and water (Table 1).

Soil Data

The soil map of the catchment area is derived from the Global soil map fostered by the Food and Agriculture Organization (FAO) of the United Nations (Fig. 5). It is observed that the study area comprises four types of soil: dystric cambisols (clay loam), gleysols (clay), humic cambisols (sandy clay loam), and xerosols (loamy sand) (Table 2). The basin is largely covered by cambisols.

Slope

The slope of a watershed demonstrates the rate of change of elevation respective to distance along the principal path of flow (Indian Council of Agriculture Research). The slope map of the basin is developed in the SWAT model by dividing the DEM of the basin into five classes of slope percent rise, 0–10, 10–20, 20–30, 30–40, and above 40 (Table 3) (Fig. 6). It is observed that the slope of the catchment varies greatly, a small portion of land is under a gentle slope and a large area of the basin has a steep slope.

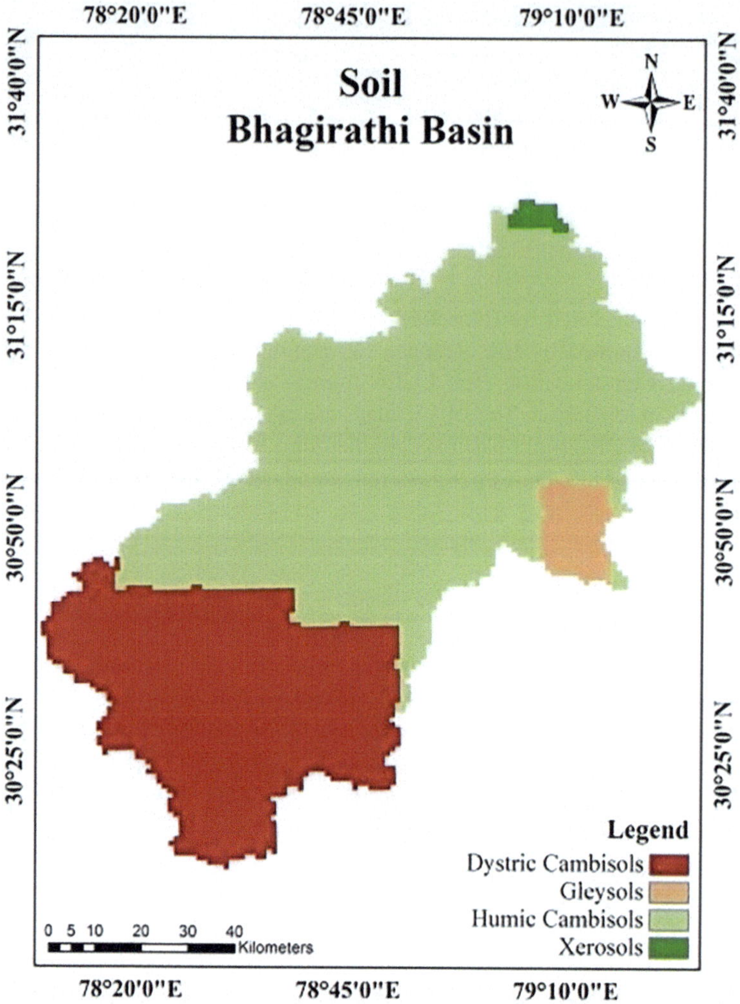

Fig. 5 Soil types

Table 2 Soil types, Bhagirathi basin

Soil class	Texture	Area (sq. km)	Area (%)
Dystric Cambisols	Clay loam	2500.55	33.3
Gleysols	Clay	243.74	3.21
Humic Cambisols	Sandy clay loam	4782.6	63.12
Xerosols	Loamy sand	48.99	0.64

Table 3 Slope classes, Bhagirathi basin

Slope class (%)	Area (sq. km)	Area (%)
0–10	201.20	2.65
10–20	448.76	5.91
20–30	651.73	8.59
30–40	888.54	11.71
>40	5399.78	71.14

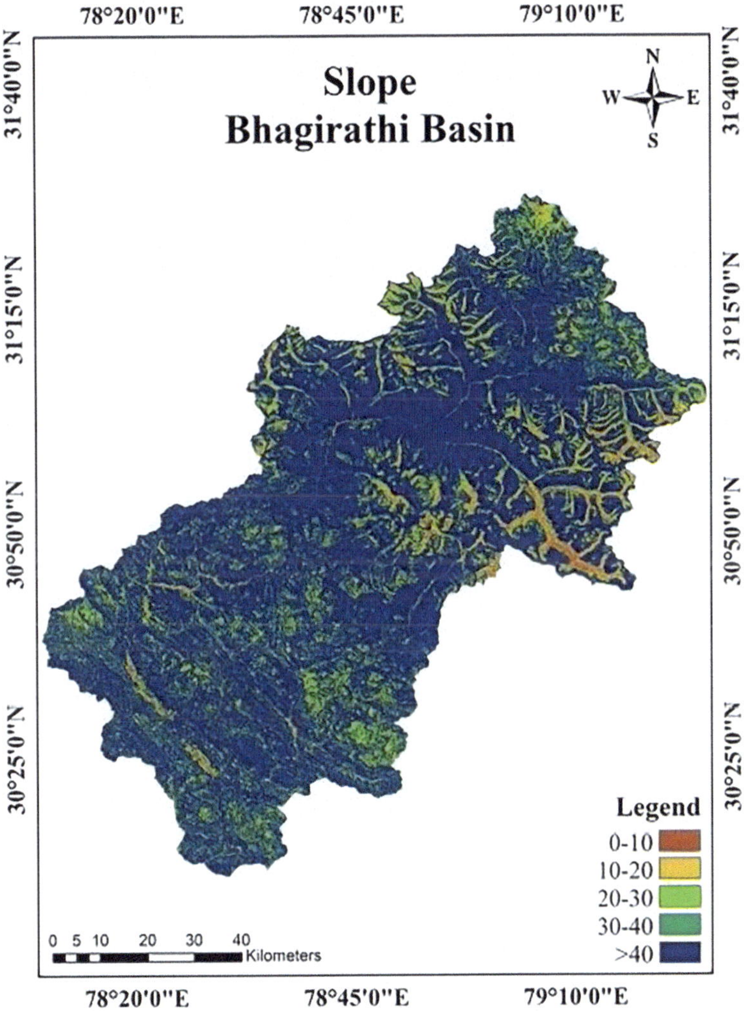

Fig. 6 Slope classes

Table 4 Weather stations

Stations	Latitude	Longitude	Elevation (m)
1	31.25	79.15	5316
2	31.15	78.85	5140
3	31.05	78.65	4400
4	30.85	79.15	5891
5	30.75	78.35	1463
6	30.75	78.75	3406
7	30.45	78.45	1248
8	30.45	78.75	1991

Hydro-Meterorological Data

Climate data such as temperature and precipitation are essential to run the SWAT model. Daily maximum and minimum temperature and precipitation data extracted from global weather data provided by SWAT at https://swat.tamu.edu/data/ in SWAT-supported format is given as weather data input to the model. Eight prominent weather stations (Table 4) (Fig. 2) with their location and elevation are listed. Inventory data for calibration and validation are collected from India-WRIS.

SWAT Model Description

The USDA Agriculture Research Service (USDA-ARS) developed the semi-distributed SWAT model, which works on watershed level and continuous time. The Simulator for Water Resources in Rural Basins (SWRRB) model, the Routing Outputs to Outlet (ROTO) model's river course calculation module, and the Enhanced Stream Water Quality model's inland river power module make up the SWAT model process. According to Arnold et al. (2012a, b), the SWRRB model combines the Environmental Policy Integrated Climate (EPIC) model, the Chemicals, Runoff, and Erosion from Agricultural Management Systems (CREAMS) model, and the Groundwater Loading Effects of Agricultural Management Systems (GLEAMS) model. Early in the 1990s, SWAT formally released the initial SWAT 94.2 version, which has since been updated and improved. SWAT 2012 is the most recent version (https://swat.tamu.edu/). The SWAT model's source code is freely accessible online. The daily time step SWAT model is used to forecast how land use and management would affect water, sediment, and agricultural chemical yields in unmeasured watersheds.

The entire simulation process of the SWAT works in two parts: the land surface (runoff and slope confluence) and the water surface (concentration of channel). The first part controls the input of water, sediment, nutrients, and chemical substances in the main channel of each sub-watershed. The latter determines the process of transferring the movement of water, sediment, and other substances from the river network to the basin and the calculation of the load.

Sub-watershed, reservoir, and river calculus comprise the bulk of the SWAT model's framework. There are multiple sub-watersheds in each module. For instance, hydrology, weather, sediment, soil temperature, plant growth, nutrients, pesticides, and agricultural management are all part of the water cycle process in sub-watersheds (Arnold et al., 1998; Douglas-Mankin et al., 2010).

According to the principle of water circulation, the water balance expression used in the SWAT model is given by Eq. (1).

$$\mathrm{SW}_t = \mathrm{SW}_0 + \sum_{i=1}^{t} \left(R_{\mathrm{day}} - Q_{\mathrm{surf}} - E_{\mathrm{a}} - W_{\mathrm{seep}} - Q_{\mathrm{gw}} \right) \tag{1}$$

where SW_t is the final water content of the soil, SW_0 is the water content in the early stage of the soil, t is time in days, R_{day} is the daily precipitation, Q_{surf} is the daily surface runoff, E_{a} is the daily evapotranspiration (ET), W_{seep} is the daily percolation, Q_{gw} is the daily return flow, all units are in mm. SWAT calculates soil loss using RUSLE given by. Sediment yield is predicted separately at the HRU scale using Eq.(2)

$$S_{\mathrm{y}} = 11.8\left(Q_{\mathrm{surf}} \times q_{\mathrm{peak}} \times \mathrm{A}_{\mathrm{hru}} \right)^{0.56} \times K_{\mathrm{USLE}} \times \mathrm{LS}_{\mathrm{USLE}} \times C_{\mathrm{USLE}} \times P_{\mathrm{USLE}} \times \mathrm{CFR} \tag{2}$$

where S_{y} is sediment yield of the day (Mg), Q_{surf} is surface runoff (mm/ha), peak is peak rate of runoff (m^3/s), A_{hru} is area of HRU in hectare, K_{USLE} is soil erodibility factor of universal soil loss equation (USLE), $\mathrm{LS}_{\mathrm{USLE}}$ topographic factor of USLE, C_{USLE} cover management factor of USLE, P_{USLE} is conservation practice factor of USLE and CFRG coarse fragment factor.

In the process of applying the SWAT model, due to different land use types and soil types, the sub-watersheds are divided into multiple hydrological response units (HRUs) to improve the simulation accuracy. The original data required include watershed Digital Elevation Model (DEM), land use map and index table, soil type and soil attribute table, meteorological data, observation runoff and management measures, reservoirs, wetlands, etc. The ungauged watersheds can use the SWAT weather generators to simulate daily wind, solar radiation, and relative humidity.

Results and Discussion Sensitivity Analysis

Model calibration is done by adjusting the range of important parameters affecting the sediment yield of the catchment, till satisfactory agreement is achieved between observed and simulated variables. Where a large number of parameters are controlling the processes, it becomes essential to sort the most sensitive parameter. In this study, global sensitivity analysis is performed at monthly time steps in the SWAT-CUP model with seven hundred simulations. The significance of each parameter is measured based on two indicators, viz, *p*-value and *t*-stat. Parameters with smaller *p*-values and high *t*-stat are considered more sensitive ones. Nine parameters were selected of which Curve number (CN2), Base flow alpha factor (ALPHA_BF), Groundwater delays (GW_DELAY), and Threshold depth of water GWQMN are sensitive for all variables (Abbaspour et al., 2007). Channel re-entrained linear parameter (SPCON), Channel re-entrained exponent parameter (SPEXP), Channel erodibility factor (CH_COV1), Channel cover factor (CH_COV2), USLE equation soil erodibility (K) factor (USEL_K) are sensitive to sediment. Out of the selected parameters curve number appeared to be the most sensitive parameter. It indicates the runoff response characteristic of the basin and thus curve number plays an important role in sediment study. Rank-wise parameters are listed in Table 5

Model Calibration for Sediment Yield

The processes of calibration and validation are carried out using the Sequential Uncertainty Fitting (SUFI-2) algorithm in the SWAT Calibration and Uncertainty Program (SWAT-CUP). To do so most sensitive parameters (Table 5) were selected and adjusted till a satisfactory relation was achieved with observed data. It is seen that the observed and simulated sediment yield of sub-basins into the reach are at proximity during the calibration period 2005–2014 (Fig. 7). The overall performance of the calibration model was evaluated based on different indicators such as the scatter plot between observed and simulated sediment yield (Fig. 8) gives the value of coefficient of determination (R^2) to be 0.94, also the model prediction potential has been assessed through Nash–Sutcliffe Efficiency (NS) value which is 0.88.

The values of the *P*-Factor and *R*-Factor for calibration are 0.72 and 0.71, respectively. Results achieved were up to the required standard (Table 6).

Table 5 Sensitivity analysis

Rank	Parameter name	Description	*t*-Stat	*p*-Value
1	CN2	Curve number	−13.16	0.00
2	ALPHA_BF	Base flow alpha factor	−13.46	0.00
3	GW_DELAY	Groundwater delays	−0.79	0.42
4	GWQMN	Threshold depth of water	0.00	0.99
5	SPCON	Channel re-entrained linear parameter	−1.98	0.04
6	SPEXP	Channel re-entrained exponent parameter	0.23	0.81
7	CH_COV1	Channel credibility factor	−0.30	0.75
8	CH_COV2	Channel cover factor	0.59	0.55
9	USLE_K	USLE equation soil erodibility (K) factor	−1.59	0.11

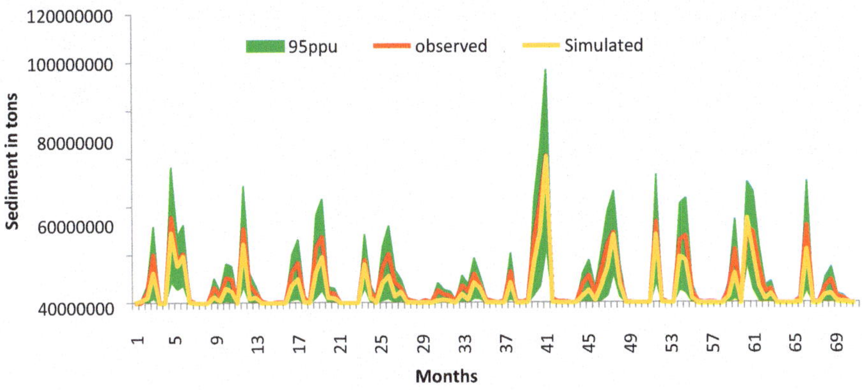

Fig. 7 Model calibration for sediment yield at monthly time step (2005–2014)

Fig. 8 Scatter plot for calibration

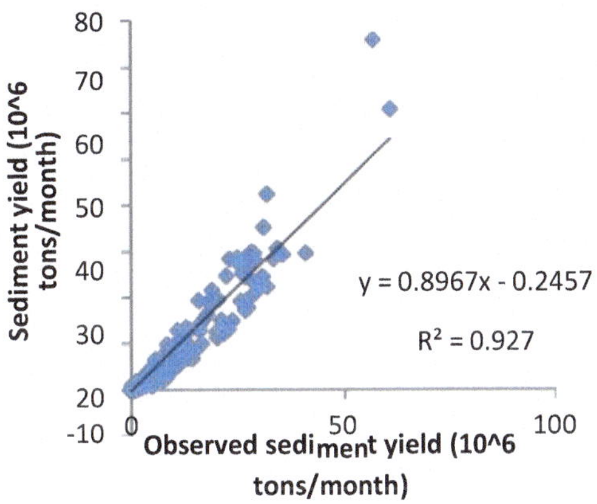

Model Validation for Sediment Yield

Validation is done without further adjusting the range of parameters and a good enough correlation is found between the observed and simulated sediment yield during the validation period 2015–2019 (Fig. 9). This assessment is established based on satisfactory values achieved for multiple indicators of validation performance. The value for R^2 is 0.81 as represented by a scatter plot (Fig. 10). The capability of the model to predict is sustained based on the value acquired for NSE which is 0.79. Values of *P*-Factor and *R*-Factor are 0.60 and 0.80, respectively (Table 7). The overall assessment is based on the 95 ppu band, most of the simulated values lie under the range of the 95 ppu band which indicates that the model is capable of simulating sediment yield satisfactorily even under parameter uncertainty (Panda et al., 2021)

Table 6 Summary statistics of model performance indicator for calibration

Method	R^2	NSE	*P*-Factor	*R*-Factor
SUFI-2	0.94	0.88	0.72	0.71

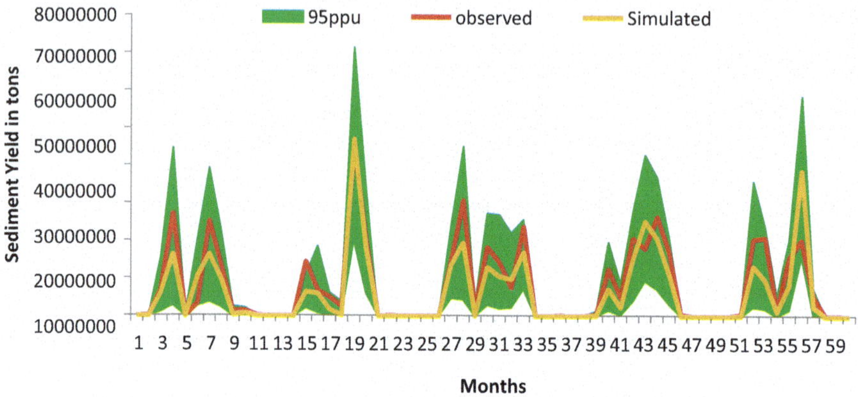

Fig. 9 Model validation for sediment yield at monthly time step (2015–2019)

Prioritization

Effective soil and water conservation methods need the recognition of core locations with high sediment output at the sub-basin level (Debela, 2022). In this study, average sediment yield from each sub-basin in the Bhagirathi basin is calculated for the entire simulated, calibrated, and validated period (2005–2019). Average sediment yield from sub-basins into the reach ranged from 0–90 × 10^6 tons, thus all the nine sub-basins are categorized into three classes—low risk (0–30 × 10^6 tons), moderate risk (30 × 10^6–60 × 10^6 tons), and high risk (60 × 10^6–90 × 10^6 tons) (Fig. 11). It is found that SW4 covering 11.88% (902 sq km) of the total area is at high risk that alone counts for an average of 90.77 × 10^6 tons of sediment output into the reach. This catchment is categorized by a steep slope, dominated by two soil types: humic cambisols (sandy clay loam) and gleysols (clay). The major area in this sub-basin is covered by glaciers, and lower elevation is categorized by barren land, range land, and little forest along the river. SW5 and SW6 lie under the moderate risk category with averages of 41.05 × 10^6 and 61.22 × 10^6 tons sediment yield, respectively. This moderate risk zone comprises 16.86% (1280 sq km) of the total area and is dominated by dystric cambisols (clay loam) and humic cambisols (sandy clay loam). Most of the area here is covered by forest with some patches of built-up and little of range land, barren land, and glacier. A large area of Bhagirathi basin, namely SW1, SW2, SW3, SW7, SW8, and SW9 is under low risk with average sediment yield of 10.85 × 10^6, 29.07 × 10^6, 0, 2.07 × 10^6, 0, and 2.49 × 10^6 tons, respectively. The low-risk zone covers 71.23% (5407 sq km) of the total area; this zone is diverse with three types of soils, namely dystric cambisols (clay loam) and humic

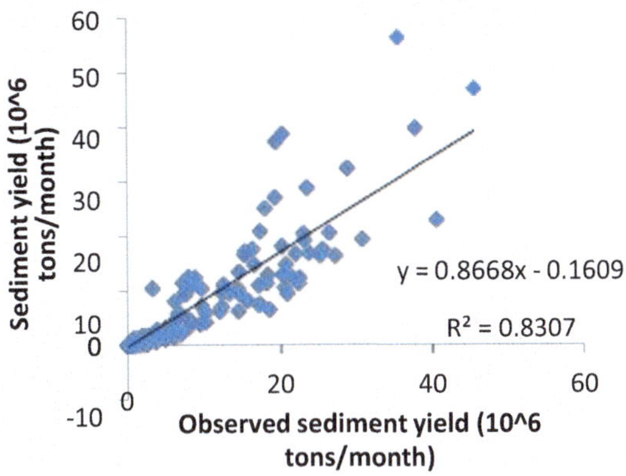

Fig. 10 Scatter plot for validation

Table 7 Summary statistics of model performance indicators for validation

Method	*R*	NSE	*P*-Factor	*R*-Factor
SUFI-2	0.81	0.79	0.60	0.80

cambisols (sandy clay loam) and aerosols (loamy sand). Mixed land use land cover is found here with forest dominating the southern with small patches of built-up, agriculture, and range land while the northern region is largely covered by range land and barren land with ice cove in high altitude. The risk zone assessment of the Bhagirathi basin can be well understood by Table 8.

Conclusion

Hydrological disaster has become a nightmare for the Himalayan region as it is fed by many major rivers. Proper planning and management are urgent requirements but planning needs an effective base specifying the region of concern.

Remote sensing and GIS tools and techniques have proved to be a great help in planning and activities as they give the facilities to work with various parameters together spatially. In this chapter, an effort was made to prioritize the area with maximum sediment yield in the Bhagirathi basin which eventually will create flood havoc. The work was done using the SWAT model which works based on various parameters.

From the study, it was found that the northeastern part of Bhagirathi is more susceptible to hydrological disasters as it was categorized under maximum priority. This can be used as a base for drafting mitigation strategies.

The SWAT model works based on different parameters which are essential for hydro-solidarity. The term "hydro-solidarity" describes an increasingly integrated approach to managing water resources that relies not only on the integration of different physical parameters but also on participation and coordination among

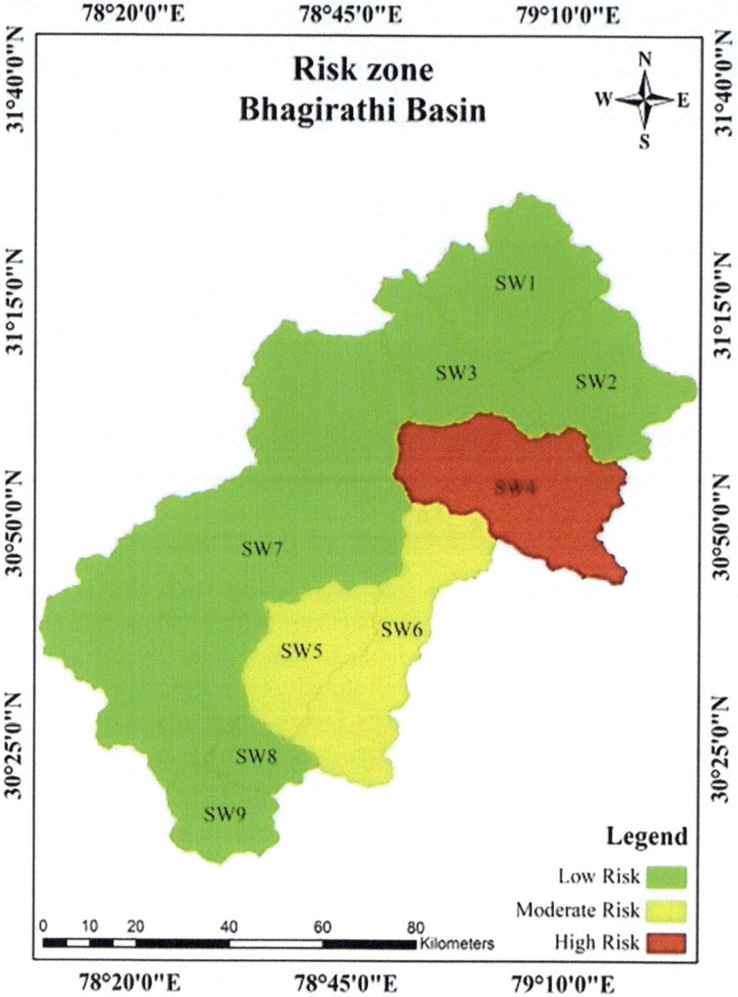

Fig. 11 Prioritization of risk zone map

Table 8 Prioritization of risk zone, Bhagirathi basin

Risk zone	Sediment yield (10^6 tons)	Sub-watershed	Priority level	Area (km²)	% Cover
Low risk	0–30	SW1, SW2, SW3, SW7, SW8, SW9	III	5407	71.23
Moderate risk	30–60	SW5, SW6	II	1280	16.86
High risk	60–90	SW4	I	902	11.88

community stakeholders, water-related management agencies, and local, state, and national governments. A fundamental goal of hydro-solidarity is the cooperative, unified management of shared water resources, whether at the national or international level.

The Bhagirathi River originates from the Gangotri glacier regions of the Himalayas (Garhwal) and results in high sediment yields causing flood hazards downstream of the Ganga River and high sediment flux to the Bay of Bengal. Differences in sediment yield between climate zones suggest potential linkages between climate and both fluvial processes and stream morphology. The analysis has shown that predicting the sedimentation pattern is a prerequisite for developing optimal sediment evacuation strategies, involving, for instance, flushing operations. Sound modeling of sediment transport and deposition in reservoirs is highlighted as being highly beneficial for supporting sediment management with the perspective of achieving reservoir sustainability and helping to prevent potential disasters.

The general problems due to soil erosion and deposition are worldwide problems and are not restricted to India. However, for sustainable development, it will be essential to thoroughly understand and solve these problems. For this purpose, a detailed screening analysis should first be carried out to identify areas of severe erosion and sedimentation sites, SWAT model can be very helpful in this. A greater number of streams will have to be gauged to determine their sediment loads. Also, detailed reservoir surveys will have to be carried out. Next, priority basins and reservoirs may be selected for experimentation to study the effectiveness of soil conservation measures and reservoir operation policies. Basin's dynamics originate from hydrological realities and cannot be neglected if foreseeable problems are to be avoided. All water-related activities in a basin have to be orchestrated for compatibility: water-impacting land and water use, water-dependent land and water uses and ecosystems, as well as water-consumptive terrestrial ecosystems and water-dependent aquatic ecosystems. Successful basin management depends on the involvement of all stakeholder groups and the proper application of the hydro-solidarity approach.

References

Abbaspour, K. C., Yang, J., Maximov, I., Siber, R., Bogner, K., Mieleitner, J., & Srinivasan, R. (2007). Modeling hydrology and water quality in the pre-alpine/alpine Thur watershed using SWAT. *Journal of Hydrology, 333*(2-4), 413–430.

Arnold, J. G., Srinivasan, R., Muttiah, R. S., & Williams, J. R. (1998). Large-area hydrologic modeling and assessment: Part I. Model development. *American Water Resources Association, 34*, 73–89.

Arnold, J. G., Moriasi, D. N., Gassman, P. W., Abbaspour, K. C., White, M. J., Srinivasan, R., Santhi, C., Harmel, R. D., Van Griensven, A., Van Liew, M. W., Kannan, N., & Jha, M. K. (2012a). SWAT: Model use, calibration, and validation. *Transactions of the ASABE, 55*(4), 1491–1508.

Arnold, J. G., Kiniry, J. R., Srinivasan, R., Williams, J. R., Haney, E. B., & Neitsch, S. L. (2012b). *Soil and water assessment tool input/output documentation version 2012. Texas Water*

Resources Institute Technical Report 436. Texas A & M University System. http://swat.tamu.edu/documentation/2012-io/. (Accessed 17 December 2014)

Debela, S. K. (2022). Prioritization of sediment yield at a sub-watershed level using swat model in Finca'aa watershed, Abay Basin, Ethiopia. *Journal of Sedimentary Environments, 7*(2), 305–315.

Dimri, T., Ahmad, S., & Sharif, M. (2023). Hydrological modeling of Bhagirathi River basin using HEC-HMS. *Journal of Applied Water Engineering and Research, 11*(2), 249–261.

Duru, U., Mazdak, A., & Wohl, E. E. (2017). Modeling streamflow and sediment yield using the SWAT model: A case study of Ankara River basin, Turkey. *Physical Geography, 39*(3), 264–289.

Gassman, P. W., Reyes, M. R., Green, C. H., & Arnold, J. G. (2007). The soil and water assessment tool: Historical development, applications, and future research directions. *Transactions of the ASABE, 50*, 1211–1250.

Gull, S., & Shah, S. R. (2022). Hydrological modeling for streamflow and sediment yield simulation using the SWAT model in a forest-dominated watershed of the north-eastern Himalayas of Kashmir Valley, India. *Journal of Hydroinformatics*, jh2022042.

Haregeweyn, N., Poesen, J., Verstraeten, G., Govers, G., De Vente, J., Nyssen, J., Deckers, J., & Moeyersons, J. (2013). Assessing the performance of a spatially distributed soil erosion and sediment delivery model (WATEM/SEDEM) in northern Ethiopia. *Land Degradation and Development, 24*, 188–204.

Issaka, S., & Ashraf, M. A. (2017). Impact of soil erosion and degradation on water quality: A review. *Geology, Ecology, and Landscapes, 1*(1), 1–11.

Jeong, J., Narayanan, K., Jeff, A., Glick, R., Gosselink, L., & Raghavan, S. (2010). Development and integration of sub-hourly rainfall–Runoff modeling capability within a watershed model. *Water Resources Management, 24*, 4505–4527.

Julien, P. Y. (1998). *Erosion and sedimentation*. Cambridge University Press.

Kaini, P., Artita, K., & Nicklow, J. W. (2012). Optimizing structural best management practices using SWAT and genetic algorithms to improve water quality goals. *Water Resources Management, 26*(7), 1827–1845.

Kim, H. S. (2006). *Soil erosion modeling using RUSLE and GIS on the Imha watershed, South Korea.*, Master's D. thesis. Colorado State University Press.

Mengistu, D. T., & Sorteberg, A. (2012). Sensitivity of SWAT simulated streamflow to climatic changes within the Eastern Nile River basin. *Hydrology and Earth System Sciences, 16*(2), 391–407.

Panda, C., Das, D. M., Raul, S. K., & Sahoo, B. C. (2021). Sediment yield prediction and prioritization of sub-watersheds in the Upper Subarnarekha basin (India) using SWAT. *Arabian Journal of Geosciences, 14*, 1–19.

Pandey, A., Chowdary, V. M., Mal, B. C., & Billib, M. (2008). Runoff and sediment yield modeling from a small agricultural watershed in India using the WEPP model. *Journal of Hydrology, 348*(3–4), 305–319.

Pandey, A., Himanshu, S. K., Mishra, S., & Singh, V. P. (2016). Physically based soil erosion and sediment yield models revisited. *Catena, 147*, 595–620.

Prabhanjan, A., Rao, E. P., & Eldho, T. I. (2015). Application of SWAT model and geospatial techniques for sediment-yield modeling in ungauged watersheds. *Journal of Hydrologic Engineering, 20*(6), C6014005.

Sanjay, K. J., Jaivir, T., & Vishal, S. (2010). Simulation of runoff and sediment yield for a Himalayan watershed using SWAT model. *Journal of Water Resource and Protection, 2*(3), 267–281.

Setegn, S. G., Dargahi, B., Srinivasan, R., & Melesse, A. M. (2010). Modeling of sediment yield from Anjeni gauged watershed, Ethiopia using the SWAT model. *JAWRA Journal of the American Water Resources Association, 46*(3), 514–526.

Singh, A., Imtiyaz, M., Isaac, R. K., & Denis, D. M. (2012). Comparison of soil and water assessment tool (SWAT) and multilayer perceptron (MLP) artificial neural network for predicting sediment yield in the Nagwa agricultural watershed in Jharkhand, India. *Agricultural Water Management, 104*, 113–120.

Tadele, K., & Forch, G. (2007). *Impact of land use/land cover change on streamflow: A case study of Hare Watershed, Ethiopia, 6* (pp. 80–84). FWU Water Resource Publication. ISSN No. 1613-1045.

Tesfahunegn, G. B., Vlek, P. L. G., & Tamene, L. (2012). Management strategies for reducing soil degradation through modeling in a GIS environment in northern Ethiopia catchment. *Nutrient Cycling in Agroecosystems, 92*, 255–272.

Tripathi, D. K., & Kumar, M. (2014). *Monitoring of snow cover in Bhagirathi Watershed, Garhwal* (pp. 289–300).

Uniyal, B., Jha, M. K., Verma, A. K., & Anebagilu, P. K. (2020). Identification of critical areas and evaluation of best management practices using SWAT for sustainable watershed management. *Science of the Total Environment, 744*, 140737.

Verma, A. K., & Jha, M. K. (2015). Evaluation of a GIS-based watershed model for streamflow and sediment-yield simulation in the upper Baitarani river basin of Eastern India. *Journal of Hydrologic Engineering, 20*(6), 1–12.

Wischmeier, W. H., & Smith, D. D. (1978). *Predicting rainfall losses: A guide to conservation planning. Agricultural Handbook No. 537*. USDA. Journal of Soil and Water Conservation.

Wolde, K., & Bogale, G. (2017). Effect of land use land cover dynamics on the hydrological response of watershed: A case study of Tekeze Dam watershed, northern Ethiopia. *International Soil and Water Conservation Research, 5*(2017), 1–16.

Yesuf, H. M., Assen, M., Alamirew, T., & Melesse, A. M. (2015). Modeling of sediment yield in Maybar gauged watershed using SWAT northeast Ethiopia. *Catena, 127*, 191–205.

Ecology-Social and Solidarity Economy Model (ESSE) for Sandstone Mining in Mirzapur, Uttar Pradesh

Archana Verma

Introduction

The importance of mineral resources in shaping the economy of any nation cannot be denied in today's perspective. In the competitiveness of industries and for the well-being of the population with the continuous growth of the economy, mineral resources, and the mining industry can play a crucial role and be the backbone of local as well as national economic development. India has been extremely blessed with this gift of nature. Raniganj was the first site where an English company was granted permission for the mining of coal in the year 1774 as per the records. Since Indian independence, growth in the mining sector was the result of successive Five-Year Plans that was a visionary act by the then government that proved to be the right decision with time. There were ambitious plans to increase the production of minerals during the 8th Five-Year Plan and thereafter. The lower Vindhyan sediments of Sonbhadra contain deposits of cement-grade limestone, flux-grade dolomites, building stone, and coal while the Upper Vindhyan Sandstones are suitable for making tiles and decorative items. India is abundant in terms of quantity and quality of a variety of minerals available throughout the country. More than 27% of the total stone production in the world is produced by India and India shares more than 11% of stone exports in the world (CDoS). The Ahraura and Marihan tehsils of Mirzapur are surrounded by the Vindhya Range, a group of discontinuous chains of hill ranges, mountain ridges, highlands, and plateau escarpments extending from the eastern part of Gujarat, Rajasthan, Madhya Pradesh, and the southern region of Uttar Pradesh. These hill ranges in the Mirzapur and Sonbhadra districts in Uttar Pradesh are significantly mined for sandstone extraction. The 1960s was the decade when mining on a grand level took place here. The lush green forests were destroyed for mining. Approximately 250 mines are currently given on lease in the district.

A. Verma (✉)
Remote Sensing Applications Centre, Lucknow, Uttar Pradesh, India

K. Arora, R. R. Anand (eds.), *The Solidarity Approach in Geography*,
https://doi.org/10.1007/978-3-031-86753-8_31

Some sandstones are resistant to weathering and still are easy to work. Durability is one of the qualities that makes sandstone a common building and paving material including asphalt concrete. Because of the hardness of individual grains, uniformity of grain size, and being fragile in their structure, some types of sandstone are excellent materials to make grindstones, for sharpening blades and other implements. Whether it the construction of the Ram Temple in Ayodhya or the Vishwanath Corridor in Varanasi, or the monuments lead park devoted to political leaders in Lucknow, it is hard to ignore the contribution to the coveted pink sandstone from Mirzapur district.

The mining industry in India contributes remarkably to the Indian economy. The gross domestic product (GDP) contribution of the mining sector stretches from 2.2% to 2.5% only but going by the GDP of the total industrial sector, it contributes around 10–11%. Mining done even on a modest scale contributes 6% to the entire cost of mineral production. Recently, Union minister Pralhad Joshi urged all stakeholders from the mining industry to take the sector's contribution to GDP to 2.5% by 2026–27 and said that the government will come up with investor- and industry-friendly norms. He pitched for the use of technology, global best practices, and artificial intelligence in exploration and mining. The Indian mining industry provides around 700,000 job opportunities to individuals. The traditional and old techniques of mining are time- and money-consuming; however, in recent times, modern exploration techniques like photo-geology, remote sensing, non-coring drilling, geophysical surveys, etc. are being increasingly used for detailed and precise exploration.

Study Area

Mirzapur

Mirzapur is a city in Uttar Pradesh, India. It is located at 23.52 and 25.32 North latitude and 82.7 and 83.33 East longitude (Fig. 1). It is best known for its carpets and brassware industries, along with the tradition of Kajari, Birha Music and Tourism. Other than its historic identity this city also has a geographic identity that this city is surrounded by several hills of the Maikal Range. This geographic identity now is becoming a threat to the region as due to several minerals in this city, mining has been a boon that has been converting into a curse of the town for more than a decade. To understand the ground situation, a primary survey was carried out in the Ahraura town, municipal board, and Marihan tehsils of Mirzapur district (Fig. 2), which are well known for sandstone mining.

Fig. 1 Location of Mirzapur district. Source: Author (2023)

Fig. 2 Mining plants. Source: Google Earth Bird's eye view of mining plants in Mirzapur and Ahraura

Methodology

To analyze the value of ecological and social solidarity in mining-based economy and mineral exploration in the study areas, a primary survey was carried out from 19th December 2023 to 24th December 2023. The total respondents were 40 and they were from different age groups. Schedules, questionnaires, and random sampling methods are used for the collection of first-hand data in the form of public opinion. The interviews of representatives of different organizations have also been conducted, recorded, and used in the analysis along with the satellite data. The secondary data and information have been also collected from different sources.

During the survey, it was observed that the majority of the villagers (all age groups) were not in favor of the expansion of mining near their residential areas. Opinions and complaints of the villagers are included in this study as generalized statements.

To assess the various aspects such as physiographic and demographic changes with ecological changes, cadastral maps, satellite images, census data, and remote sensing and GIS techniques have been used. The satellite images have been used to interpret the temporal changes and expansion of the area of mining. Satellite images visualize the physiographic changes in the area. Continuous years of blasting have turned the area into a wasteland, some hillocks have disappeared altogether, as a meteorite hit them, leaving behind a crater full of blue and green groundwater.

As far as other environmental changes are concerned, a clear improvement was visible in 1980s and afterward when once deep forests in these two tehsils were linked to the Chandraprabha Wildlife Sanctuary which is situated in the adjoining Chandauli district, that too can be identified through the remote sensing data. Remote sensing data in mining involves using satellite imagery to monitor and manage mineral exploration and production.

This technology improves efficiency and sustainability in the mining sector. A Ground Truth visit has been done for the confirmation of information retrieved from the satellite data and other information.

For analyzing the social change schedule, a questionnaire and visual interpretation of satellite images have been used which identifies the changes in habitats and loss of greenery which is an unnoticed fact of mining that forced locals to not domesticate animals for their livelihood other than being laborers in mines which indirectly played a role in the migration of locals from surrounding areas.

Problem and Prospects of Sandstone Mining in Mirzapur

When mining started in Mirzapur, it provided tremendous employment opportunities, but later turned out to be a curse for the local environment as it ravaged topsoil, destroyed vegetation by altering physicochemical properties of soil, and disorganizing nearby ecosystems, which directly ruined the existing landscape and environment. Deposition of stone dust on soil and crops with the wind and runoff leads to degradation of soil quality and crop production. Lack of proper management and waste treatment daily is one of the major issues associated with quarrying and requires be sensitive resolution. The unrestrained mining activity has rendered the hillocks over roughly 10 sq km in the Sonpur-Ahraura area of Mirzapur district bald and desolate. Pink sandstones have been mined for more than three decades from these hills but in the last 2 years, the recurrence of mining operations—which include "blasting" and powerful explosions on the hillside—has increased. At a time, 300 kg of gunpowder is set off for the blast, if the same amount of gunpowder is set off in the village, it would turn the village into ashes in a few minutes (Fig. 3). Living with a layer of dust is common for the villagers of Sonpur rather can be said a part of their lifestyle. Every second person is suffering from lung disease here, some with serious and painful symptoms. Repeated explosions are the reason why, cracks have formed on the walls of every house in this village. Smoking a cigarette per day.

Reduces one's lifespan by 6 years. What about them? They breathe in stone dust 24 h a day. Even the children born here are not secure and start breathing in air pollution immediately after their birth. Rural inhabitants with diseases are slowly moving towards death with the baggage of breathing disorders like asthma, silicosis, and tuberculosis.

Approximately 50,000 rural inhabitants in these two tehsils are compelled to eat and breathe dust in their daily lives. Not a single household could be found that doesn't have a family member suffering from a respiratory disease. A couple suffering from silicosis, a lung disease caused by breathing in tiny bits of silica, a common mineral found in sand, quartz, and many other types of rock, works in the same Sonpur mine. This incident depicts a horrifying future for rural inhabitants living there. Corruption in mining activity by mine contractors forces villagers to sell their lands for mining. As a result, the mining area in the district exceeds the legal lease allotted by the government.

Mining is a dangerous activity as it adversely impacts the environment and villagers' lives. But the entire system runs on breach of trust, there's nobody to consider the safety of the villagers living here. A large pool of groundwater was exposed due to the blasting and mining of the hills and nobody was paying any heed towards it. It is unsafe for children to play here, as a deep mine was dug there! The children could fall inside the mine and die. Quarrying is an activity that affects all aspects of the environment. The air, water, soil, and the constant explosions of gunpowder and other methods used to break the rocks make it very difficult to live there with a manmade disaster called earthquake-like conditions with air and sound pollution. The constant noise of heavy machinery operating in the quarry is also irritating, and

(a) (b)

Fig. 3 (**a**) Stone Thresher plant near settlement. (**b**) A blast. Source: Google Earth

it can lead to mental and emotional traumas in the long run. Water, air, and noise pollution is everywhere.

According to the Directorate General of Mines Safety, all blasting activity in the mines and the quarries is directed to be carried out at least 300 m away from the settlement area. But houses and huts in Sonpur village are <100 m away from the mine and quarries site. In some cases, this distance reduces to 50–60 m. They fear the mines will one day swallow up their homes too. The mining activities continue day and night. A normal day is disturbed by loud explosions that split the air, and the nights are disrupted by heavy machinery as it shreds the blasted rocks into smaller pieces and it makes it nearly impossible for villagers to take a nap even. With such mining activity, the topsoil, the land surface over a significant area is destroyed, and it is left behind with unstable land, causing landslides, erosion, siltation, and polluted water. Pits and artificial hills formed during mining operations such as the deposition of overburden are a major problem in the area; it makes that area or overburden land useless after the mining ends and may cause environmental problems even after the mine has been closed.

There were around 200 patients from Marihan and Rajgarh going through TB treatment at this health facility. But these patients are just the tip of the iceberg and both villagers and local activists claim the actual number of patients suffering from respiratory ailments was much higher as a large chunk of the rural residents were unable to access healthcare facilities due to financial constraints or paucity of time needed for the treatment.

As a result of all the digging and mining work, the groundwater used by the villagers from hand pumps is also contaminated. The water they drink is often muddy. It has been nearly 15 years since they have had the pleasure of sipping clean water. Sonpur Pahadi is an area that has no electricity either. The residue from the mines, if not treated properly, flows down the rivers and results in sedimentation which raises the levels of the rivers and makes flooding more severe and disastrous. The improper and unsuitable quarrying practices are severely affecting the nearby environment.

Ecology and Temporal Changes in Mining Areas: Marihan and Ahraura

Minerals are an immensely important and principal resource for the core sectors of any economy. With the rapid growth in urbanization and the targeted growth of concrete jungle and in the multiple sectors mainly in the manufacturing sector, minerals stand out with their high demand. The mining industry has long been a base of global resource extraction and economic growth. But in the last two decades, this industry has been on its boom as the political and social environment has been continuously changing and developing with every passing year. With the initiatives like Make in India and Vocal for Local the demand for minerals is likely to shoot up. Extraction and management of minerals have to be guided by long-term national goals and perspectives and integrated into the overall strategy of the country's economic development. Mining technology should be upgraded with the trend and time to ensure uprooting and exertion of the entire Run-of-Mines. But the reality check of the pink sandstone extraction industry is that it has pushed many things on the borderline.

The minerals extraction from nature repeatedly creates an imbalance that adversely affects the environment in all aspects and forms. The key environmental impacts of mining are on wildlife and fishery habitats, the water balance, local climates and the pattern of rainfall, sedimentation, exhaustion of forests, and the disruption of the ecology. Through the technologies like remote sensing and other methods like ISR mining comes under the radar. These techniques are used to provide satellite images to identify the temporal changes in the study area and surroundings that indicate the increase in the number of mining sites and expansion in the area of existing sites along with the number of stone crushers and cutting plants. These images from the years 2016 and 2023 mark the changes and indicate the diminishing pattern of agricultural land and deforestation in nearby areas (Fig. 4).

Ecological, Social, and Solidarity Economy Model

The word solidarity itself consists of a concept of oneness or altogether. The term indicates the support of one group of people for another as they agree with each other's aims. Ecological, Social, and Solidarity Economy (ESSE) belongs to a wide concept of economic activities that aim to align social profitability instead of only monetary benefits (Fig. 5). Another aspect is the emphasis placed on the governance system. It is an emergence which cast around the ecological social and economic quality of life of a region or community based on solidarity. It mainly embraces the activities organized to address and alter the exploitation of ecology and human power under different economic activities led by any large-corporation, large-shareholder-dominated economy and can include diverse activities. One ESSE approach focuses mainly on making the current Ecological, Social, and Economic System sustainable. Its objective is the creation of enterprises that serve its members

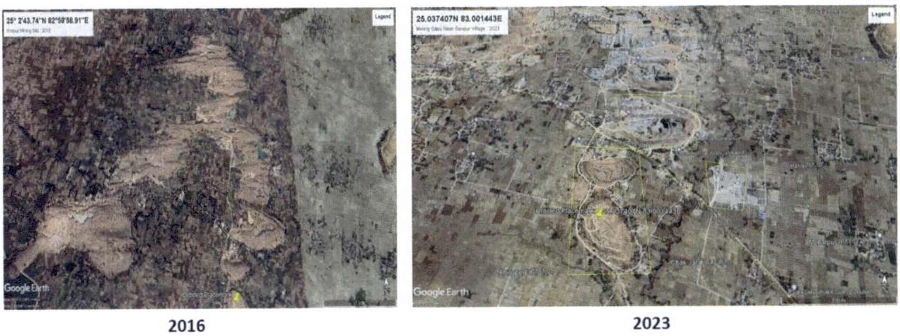

Fig. 4 Expansion in area and increment in number of mining and Stone Thresher Plants and Loss of Greenery from 2016 to 2023

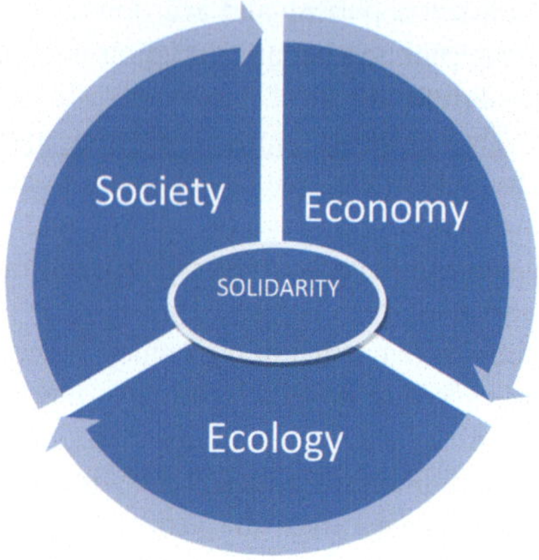

Fig. 5 Solidarity linkages with society, ecology, and economy

or the community along with nature and ecology, instead of simply striving for financial profit by prioritizing people and work over capital in the distribution of revenue. It's a concept that centralizes human beings and human needs rather than commercial and economic benefits. Environmental consideration in economic development, also known as sustainable development, is a kind of economic and social progression that safeguards living standards without destroying the ecosystem and is designed to the needs of each sustained region to become both humanistic and sustainable over time. In this regard, prevention and mitigation of adverse environmental effects due to mining by the latest scientific norms and modern afforestation shall be formed as an integral part of the mine development strategy. Mining activities should not generally be taken up in ecologically fragile and biologically rich areas. The mining regulating agencies should identify critically fragile areas in terms of ecology and take appropriate actions. To realize a better guesstimate between mineral-based development and the environment, there should be a will to develop an Exclusive Mining Zone (EMZ). Before starting any mining projects prior in-principle statutory clearances were demarcated for the mineralized zone to sidestep conflicts of interest and to avoid delay in the beginning of mining operations. To lessen pollution, carbon footprint, and operational costs, the usage of renewable sources of energy at mining sites should be promoted through apposite inducements. Proper exposure and training about environmental matters can also be given to all stakeholders involved in mining operations.

In 2015–16, India's mining sector contributed about 2.6% of GDP. According to a 2017 report titled Socio-Economic Impact Study of Mining and Mining Policies on the Livelihoods of Local Population in the Vindhyan Region of Uttar Pradesh, 23 % of the working population in Mirzapur district was employed as laborers in mines. It was also found that in total, 81 % of the local population directly or indirectly depended on mining for their livelihood. So far out of this 81%, about 27% were involved as laborers, 38% as transporters, 10% as mine owners, and 6% were dependent on mining in various ways, either involved as a middleman or selling, etc. (Dubey, 2017) Dehradun-based Indian Council of Forestry Research and

Education published another project report in 2017, which was funded by NITI (National Institute for Transforming India) Aayog, which mentioned that practically all the mines in the area are open cast, involving digging of pits and excavating hillsides, and are responsible for developing an undulating unsafe landscape that's a threat to locals as it may cause harm to their toddlers and sometimes adults too.

In 2021, the first consignment of pink sandstone for the Ram Temple was sent from Mirzapur to Ayodhya. Rs.797 crores for the beautification of Ram Temple are allocated by the Government, which will ultimately generate employment and boost the level of economy in which the sandstone from Mirzapur is also playing a major role. Sandstone from these mines was also used in the building (2003–2008) of Dr. Bhimrao Ambedkar Memorial Park in Lucknow, built by the Mayawati government worth Rs. 700 cr. which is highly controversial yet provided jobs to 2700 workers then. For building the park, the pink sandstone from Mirzapur district was transported to Rajasthan for cutting and sewing and then further transported to Lucknow for the construction.

Ram Temple in Ayodhya and the Kashi-Vishwanath project in Varanasi will use these stones. The construction of Ram Mandir will not only strengthen the spiritual and religious beliefs but will also boost the economy of the state as this will increase tourism by ten times. The government is making other efforts also to create infinite possibilities in the area of religious-spiritual tourism in the state.

Vindhyan and the Bundelkhand are the areas where significant mineralization has taken place. Mineral-based industries are the major contributors to the overall economy of these regions. Important minerals include coal, pink sandstone (*laal ya gulabi balua patthar*), magnesite, sulfur, silica sand and limestone, pyrophyllite, etc. Several industries using these minerals are already in operation, including fertilizers, aluminum, alumina, caustic soda, Ferro-alloys, refractory and ceramics, sulfuric acid, and oil refineries. In addition, a few other minerals are also found in the state. These include marble in Mirzapur and Sonebhadra, non-plastic fireclay in Bansi and Makri-Khoh area of Mirzapur district, and Uranium found in the Lalitpur district. Barytes and Andalusite are found in the Mirzapur and Sonebhadra districts. Mining cannot be banned abruptly here as it is the second-highest contributor to the national GDP. Mining gives employment to many villagers, and it is the only source of income for less educated people in these tehsils. A complete ban on mining will ultimately lead to migration to the nearby cities and eventually to the distant cities which will result in a complete imbalance from village level to cities. Migration not only leads to physical mass movement of the people but also creates a void in families as mostly male and young members of the family migrate which also adversely affects the social aspect of villages and somehow creates certain imbalances on different levels in cities too. By adopting the Ecological, Social Solidarity in Economy, the people and their needs can be brought in the center of the economy, which will ultimately lead to a sustainable economy with the better life of miners. ESSE emphasizes solidarity by placing more importance on the present and future human population and their needs rather than the generation of only economic profit.

Role of Remote Sensing and GIS Techniques to Achieve ESSE

Technological advancement plays an important role in all economic sectors. In this context, the mining sector has also been evolving through techniques and technologies. In the earliest days of mining, miners used primitive digging tools made of stone. Eventually, the industry progressed from picks and hammers to fire for clearing tunnels. In the Middle Ages, significant advances in mining technologies occurred as miners began to use explosive black powder to dismantle large rocks, which were eventually replaced with dynamite. During the nineteenth century, the mining industry also saw developments in motorized mining tools, such as lifts, steam-powered pumps, and drills. The mining industry has come so far from the ancient practice of picks and hammers to the modern-day practice of explosives and gunpowder. This shows the role of technology which eventually increased the production and optimization of minerals and technologies also minimized environmental harm. In this regard, the application of remote sensing and GIS technology can be a game changer. In the generation of technologies and techniques taking over our day-to-day life to imagining our social cultural and economic life, remote sensing and GIS are techniques that can be used in mining that involves using satellite imagery to track and manage mineral exploration and production. Prominent applications include identifying mineral deposits, assessing environmental impact, monitoring infrastructure conditions, and detecting illegal activities. This technology improves efficiency and sustainability in the mining sector. From site selection to geological mapping to targeted exploration site safety and risk management to monitoring land reclamation and post-mining site assessment, remote sensing can be a master stroke. In this order, GIS techniques like creating thematic maps and other maps can also be used by mining companies to achieve the ESSE. Numerical data can be hectic to be studied by anyone. So the planners or miners can use GIS maps in this regard for a better mining experience from its initial stage to the post-mining stage. GIS technique can solve multiple issues at the same time. Different maps can be layered so one can toggle between them to see different representations of the numerical data. The maps generated by this technique like georeferenced maps can be used to locate its exact geographic location with coordinates. With the exact geographic location, multiple tasks can be performed at once. With the help of data or maps generated through remote sensing and GIS better mine plan can be created. Maps with coordination can help to point out the site to assemble machinery and to pave the road for to and fro transportation of extracted material. Other than these benefits of technology, these maps can give a better view and idea to implement rules and make decisions about economic and social aspects connected with mining and mining laborers. Through line feature, drainage patterns can be better drawn or planned so that it doesn't affect nearby places and groundwater levels. A pre-planned way can better create a best-from-worst kind of effect related to mine slurry or dumping areas. With the help of the point feature in GIS, a map can be created that can show the rehabilitation points for villagers or pinpoints where hand pumps, wells, schools, PHCs, or many other things can be constructed while keeping all the aspects i in mind like distance from mines and crusher plants and what should be the distance from agricultural lands and residential areas too. Polygon and buffer can be used to delineate a particular distance that should be maintained from a source of water or any other water body and fertile land so mine dust doesn't affect them adversely.

Polygon also can calculate the surface area and this data may also be used for other than mining.

To provide possible solutions to the aforementioned problems, proper assessment of the environmental and health impacts of unsustainable quarrying practices is the first step, then these measures can be taken

- Mandating Corporate Social Responsibility can be an initial yet important step towards building solidarity in ecological, social, and economic aspects. Authorities will have to brief and implement the CSR to the companies and make them understand their moral duty towards ecological and environmental protection too rather than focusing on commercial profits only. CSR has been a part of Indian tradition and culture in many different forms for ages. In current time frames, some of the countries do the same with a different name which is the Mine Closure Program. It is done in the form of social and ecological welfare by providing schools and medical facilities near the mining sites for the families of workers drilling boreholes for drinkable water, providing lighting facilities, and paving roads for better transportation.
- Other than CSR, taking proper measures like using small fountains across the quarry to spew water routinely to keep the dust down when the mining is underway is needed and other guidelines mentioned by the government should also be strictly followed.
- Curbing corruption from various steps of mining is also one of the most neglected yet to be more concerned about the topic, so corruption in lease and other areas of mining needs to be cured by strict rules and regulations.
- To ensure the imposition of Ecological and Social Solidarity in mining plans, the State Directorates of Mining and Geology and the Indian Bureau of Mines (IBM) should have power and willingness with appropriate manpower, tools and equipment, and expertise to upgrade the mining plans.
- Environmental Clearances should be simple yet efficient with inflexible, transparent, accountable, and punctual procedures.
- There shall be an ample and effective law and regulations promoting no residue or zero-waste mining policies as the ultimate aim and a responsibility to prevent flawed and absurd mining, eco-friendly and efficient modes should be applied. Ecological, Social Solidarity in the Economy is a frame that covers the perspective of bringing people together for each other's growth so that they can conserve the earth's ecology and environment along with their economic interests.
- A broad ESSE perspective is needed to make the efforts truly transformative and impact all spheres of society. Some state and climate-specific policies are also required to make a difference on a major level (Fig. 6).

Fig. 6 Requirements for effective ESSE. Source: Author (2024)

Conclusions

The overview of the ESSE in the mining sector of the Ahraura and Marihan tehsils of Mirzapur exhibits that strong political as well as social desire with long-term vision is the key to manifesting the Solidarity on Socio-Ecological and Economic level which will lead to better life of mine workers and sustainably protect ecology. Without securing the future of marginalized workers and residents no long-term benefits of mining can be enjoyed by the government and companies who are the major players in this sector as in the long run nature will eventually show its power if things are changed against nature. Many examples of mine collapse and other natural and manmade disasters have been witnessed by many of us. In achieving solidarity in any field, governance on the local and national levels needs to work together as neither of them cannot make things work progressively alone. A determined vision and efficient approach have to be taken appropriately and compulsorily so that some of the important Sustainable Development Goals that are automatically included under the broad spectrum of Ecological Social Solidarity in the Economy can be achieved. Sustainable Development Goals should be taken as a parameter and a step towards achieving ESSE. SDGs promote growth and sustainable benefits with sustainable habits for everyone. So human civilization could enjoy the benefits of natural resources in a better manner without harming nature in any aspect. ESSE falls under the umbrella of SDGs. Some of the main SDGs that are automatically linked with this chapter are SDGs **3** which encourages healthy lives and well-being of all at all ages, SDG **6** which ensures access to drinkable water and proper sanitation, improves water quality by reducing pollution, eliminating dumping and minimizing release of hazardous chemicals and materials, halving the proportion of untreated wastewater (wastewater treatment) and substantially increasing recycling and safe reuse globally.

SDG **8** promotes sustainable economic growth, full and productive employment, and satisfactory work for all, promotes youth employment, education and training, and a safe working environment. SDG **11** advocates making cities safe, strong, and sustainable for living, and it also promotes to protect the cultural and natural heritage. SDG **13** is to limit and adapt to climate change and its impact, improve education, raising awareness and human and institutional capacity on climate change reduction, adaptation, impact reduction, and anticipatory action and most importantly SDG **15** assists to protect, restore, and promote sustainable use of terrestrial or earthbound ecosystems, viably manage forests, tackle desertification, and draw up and reverse land degradation and stop biodiversity loss that are critically entangled with mining sector too. It also advocates taking urgent and significant action to reduce the degradation of natural habitats, halt the loss of biodiversity, and integrate ecosystem and biodiversity values into national and local planning. All the measures and major actions have been taken on national and international levels to achieve SDGs properly. The above-mentioned SDGs are an integral part of ESSE. As all the areas of society are now taking a step ahead with the help of techniques and available technologies to achieve their respective goals, similarly an advancement in tools and technologies with the complete determination of companies working in this sector and of course the political will and approach is required. Eventually, technological advancement with a gradual and collective approach is the key to achieving an Ecological Social Solidarity Economy in the mining sector in the current scenario.

References

Dubey, K. (2017). *Socio-economic impact study of mining and mining policies on the livelihoods of local population in the Vindhyan region of Uttar Pradesh, Project Report NITI Aayog, New Delhi*. Centre for Social Forestry and Eco-Rehabilitation.

Further Reading

http://www.springer.com/series/7377
https://www.academia.edu/36283720/Applied_Mining_Geology_Modern_Approaches_in_Solid_Earth_Sciences
https://thewire.in/health

https://www.gaonconnection.com/read/mirzapur-stone-mining-uttar-pradesh-elections-air-water-noise-hazard-pollution-asthma-tuberculosis-silicosis-health-rural-disease-50483
https://www.sciencedirect.com/book/9780323956086/geospatial-analysis-applied-to-mineral-exploration,GeospatialAnalysis2
https://timesofindia.indiatimes.com/city/varanasi/chunar-sandstone-second-natural-good-to-get-gi-tag/articleshow/68568263.cms
https://upmines.upsdc.gov.in/major_mineral.aspx
https://mines.gov.in/admin/storage/app/uploads/64352887bcfa41681205383.pdf
https://pib.gov.in/newsite/PrintRelease.aspx?relid=155093
https://mines.gov.in/webportal/nationalmineralscenario
https://www.spatialpost.com/application-of-remote-sensing-in-mining/
https://mirzapur.kvk4.in/district-profile.html
https://www.outlookindia.com/magazine/story/india-news-the-cost-of-piety-a-mirzapur-dirge
https://www.iisd.org/articles/insight/seven-ways-new-technology-will-impact-mining-sector
https://www.researchgate.net/publication/344876280_Environmental_Impacts_of_Sandstone_Quarrying_and_Its_Waste_A_Case_Study_of_Jodhpur_India
https://www.researchgate.net/publication/354208628_Application_of_Remote_Sensing_in_Mining

Enhancing Solidarity through a Comparative Analysis of NDVI and SAVI Vegetation Analysis: A Geospatial Approach for Sustainable Land Use and Environmental Management

Surender Kumar and Ripudaman Singh

Introduction

The twenty-first century has witnessed an unprecedented transformation in our understanding of environmental sustainability and the crucial role it plays in ensuring the well-being of our planet and its inhabitants. The pressing challenges of climate change, deforestation, habitat loss, and resource scarcity have necessitated innovative approaches to land use and environmental management. One key aspect of this endeavor is the effective monitoring and assessment of vegetation cover and health, which are indispensable indicators of ecosystem vitality and resilience. Vegetation is the most important factor in energy exchange processes for the hydrological cycle, biogeochemical cycle, and earth's surface (Xu et al., 2017). Vegetation cover change analysis has great significance for estimating and evaluating surface vegetation variance and micro-level ecological quality (Wardle et al., 2012; Nowacki & Abrams, 2015). NDVI and SAVI vegetation indices are effective methods for vegetation health analysis which is very helpful in studying climatic change, landscape ecology, and land degradation processes (Gray et al., 2011; Li et al., 2010). The geospatial technique is an efficient tool for obtaining better information about vegetation health (Ahmadi & Nusrath, 2012). Geospatial techniques are applied in vast areas such as land use and land cover classification, vegetation index calculation, soil moisture assessment (Kumar & Kumar, 2019), and snow cover (Chouhan & Rao, 2004).

S. Kumar (✉)
Department of Geography, Dayanand College, Hisar, Haryana, India
e-mail: surenderkumar@dnc.ac.in

R. Singh
Department of Geography, Institute of Science, Banaras Hindu University, Varanasi, India

K. Arora, R. R. Anand (eds.), *The Solidarity Approach in Geography*,
https://doi.org/10.1007/978-3-031-86753-8_32

Geospatial Technique and Vegetation Index

Vegetation analysis, facilitated by remote sensing technologies, has emerged as a powerful tool in this context (Arvind et al., 2022). Among the multitude of indices and algorithms developed for vegetation assessment, the Normalized Difference Vegetation Index (NDVI) and the Soil-Adjusted Vegetation Index (SAVI) have garnered considerable attention. NDVI is widely employed due to its simplicity and effectiveness in capturing vegetation dynamics, while SAVI offers improved accuracy in arid and semi-arid regions by accounting for soil brightness. Remote sensors obtain the spectral reflectance response from various phenomena, objects, and areas without any contact (Gandhi et al., 2015; Karaburan, 2010). A very large area is covered by remote sensing methods which are economically feasible to measure the vegetation health (Vani & Mandla, 2017). The higher spectral resolution provides more information (Honkavaara et al., 2013) which makes it easy to interpret any satellite data. Many researchers have analyzed the spatial-temporal change in vegetation with NDVI and SAVI indices (Bannari et al., 2002). Based on the visible and near-infrared bands, vegetation change analysis is done by scholars at the local to international levels (Rondeaux et al., 1996)

$$\text{NDVI} = \frac{\text{NIR} - R}{\text{NIR} + R} \tag{1}$$

Where NDVI stands for Normalized Difference Vegetation index, NIR and R stands for Near Infra-Red and Red band of electromagnetic spectrum. But soil noise and leaves area index impact the NDVI results (Epiphanio and Huete, 1995; Rondeaux et al., 1996). To overcome this issue, SAVI index was developed (equation-2) and soil adjustment factor was used to solve the issue (Huete, 1988).

$$\text{SAVI} = \frac{\text{NIR} - R}{\text{NIR} + R + L}(1 + L) \tag{2}$$

where SAVI is soil-adjusted vegetation index, NIR and Red stand for reflectance band in the electromagnetic spectrum. L stands for a constant and is normally used as 0.5. The optimal value for soil-adjusted factor remains varies from 0 to 1 which means a lower value stands for dense vegetation (Huete, 1988). Ren (2014) used NDVI and SAVI indices for the Sunitezuo Banner of arid grasslands in Mongolia. Results showed that a negative SAVI factor provides satisfactory soil noise reduction. The comparative analysis of NDVI and SAVI has been the subject of numerous studies, with findings suggesting the suitability of each index varies depending on the ecological context. For example, studies in arid regions have demonstrated SAVI's superiority in capturing vegetation dynamics over NDVI (Pettorelli et al., 2005). Conversely, in more humid regions, NDVI has shown competitive performance (Thenkabail et al., 2004). These varying results underscore the importance of understanding which index is most appropriate for specific environmental conditions. A comparative study of NDVI and SAVI was done by Vani and Mandla (2017) in the Anantapur District of Andhra Pradesh. LANDSAT data were used for the study which resulted in the SAVI index being the best vegetation index. Land use and land cover were classified by Da Silva (2019) in Campo Belo do Sul, Brazil by using NDVI and SAVI index. The result showed that the SAVI vegetation index has good measurements. Effective environmental management and sustainable land use are critical for addressing global challenges such as climate change, biodiversity loss, and food security. Vegetation analysis using remote sensing tools plays a pivotal role in achieving these goals by providing valuable data for informed decision-making (Foley et al., 2005). Accurate vegetation assessments, as influenced by the choice of vegetation index, are central to identifying areas in need of conservation, monitoring reforestation efforts, and evaluating the impact of land use changes on ecosystems. In recent years, there has been a growing recognition of the need for global solidarity in environmental conservation efforts. Climate change, deforestation, and other environmental issues transcend national borders, necessitating collaborative solutions. The United Nations Sustainable Development Goals (SDGs) highlight the importance of international cooperation in achieving a sustainable and equitable future for all (United Nations, 2015). The major objective of the study is to analyze the spatial-temporal change detection for vegetation cover in the selected area. To fulfill this purpose, NDVI and SAVI index were used.

This research paper seeks to contribute to the discourse on solidarity in environmental conservation by examining the comparative performance of NDVI and SAVI in various ecological settings. The findings of this study will provide valuable insights for researchers, policymakers, and land managers, fostering a sense of global solidarity in the pursuit of sustainable land use and environmental management.

Study Area

Central Haryana mainly includes the area of Rohtak, Sonipat, Charkhi Dadri, Jind, Bhiwani, Hisar, and Jhajjar. The study area lies between 28°21′ to 29°27′ N and 75°55′ to 77°00′ E. The whole area is bounded by Jind, Delhi, Sonipat, and Gurugram districts in the east; Karnal and Jind districts in the North; Hisar, Charkhi Dadri, and Bhiwani districts in the west and Rewari districts in the south. Central Haryana covered an area of 7715.57 km^2. Geomorphologically, most of the study area lies in the Ghaggar-Yamuna plain. The study area has an elevation of 201–250 m with 251–300 m elevation in the southern part. The whole part of the study area has four geomorphological units, namely Older Alluvial—Aeolian Plain, Sand Dune, and Dune Complex. The average annual rainfall received by the targeted area is 30–110 cm and 24.60–25.0 °C mean annual temperature. Most of the agricultural area is irrigated by Western Yamuna, Jui Canal, and Sahibi Nadi. Solemnized and Sierozem are the major soil types of the study area. The main vegetation types of the study area are subtropical dry deciduous. The study area has groundwater quality in the safe, critical, semi-critical, and over-exploits category (Central Ground Water Board, 2014–15; HARSAC, n.d.-digital Maps). The study area is an agricultural-dominated region of Haryana state. According to the census 2011, the study area has a population of 39, 99,019. Figure 1 displays the location of Central Haryana.

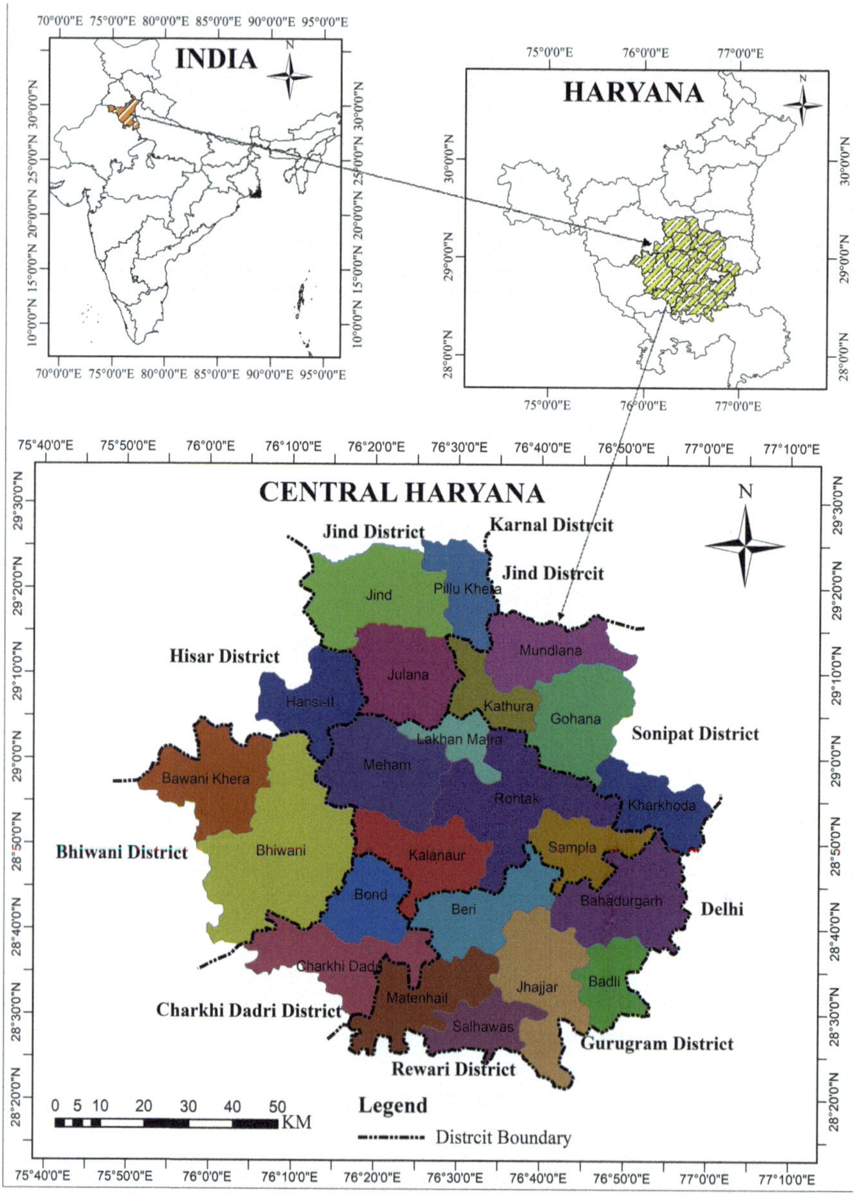

Fig. 1 Location map of study area

Methodology

The secondary source of data was used for the study. LANDSAT series data (MSS to OLI) for the years 1975, 1979–80, 1989–90, 2000, 2010, and 2019–20 of pre- and post-monsoon were used for NDVI and SAVI calculation. Year-wise satellite band combination details are present in Table 1. USGS Earth Explorer is used for Landsat data download. All required data was available in Geo Tiff format having WGS (World Geodetic System) coordinates system (84 Universe Transverse Mercator-Zone 43N). LANDSAT 2, 3, 5, and 8 were geometric and radiometric corrected. LANDSAT 7 data for the year 2010 were affected by strip gaps. This strip gap was filled by the LANDSAT Toolbox. tbx. Image correction (Radiometric Correction) of LANDSAT 7 data for the year 2010 is done using LANDSAT Toolbox. (Tech-Tutor with Fitsum, 2020). The toolbox was added to the Arc toolbox in GIS software. After that LANDSAT Toolbox. tbx. opened and completed the process with the Fix Landsat 7 Scan Line Errors option. The complete dataset has a different resolution as 60 m and 30 m. The resampling method was applied to get uniformity in all data resolutions. The imagery of 1975, 1979–80 of both periods as pre- and post-monsoon was resampled in ERDAS IMAGINE 2015 by applying the nearest neighbor method. After that study area was clipped from each band of all imageries of the selected period. Image processing was done by using ERDAS IMAGINE 2015 and Arc GIS 10.2 software. Detailed methodology is shown in Fig. 2.

Result and Discussion

NDVI and SAVI results are classified mainly into four main categories. Detailed information is given below:

Table 1 Satellite data and band combination information

Sr. no.	Satellite series	Year (pre- and post-monsoon)	NDVI band combination	SAVI band combination
1.	Landsat 2	1975	NIR = B-7; Red = B-5	NIR = B-7; Red = B-5
2.	Landsat 3	1979–80	NIR = B-7; Red = B-5	NIR = B-7; Red = B-5
3.	Landsat 5	1989–90	NIR = B-4; Red = B-3	NIR = B-4; Red = B-3
4.	Landsat 7	2000–10	NIR = B-4; Red = B-3	NIR = B-4; Red = B-3
5.	Landsat 8	2019–20	NIR = B-5; Red = B-4	NIR = B-5; Red = B-4

Source: various literature reviews

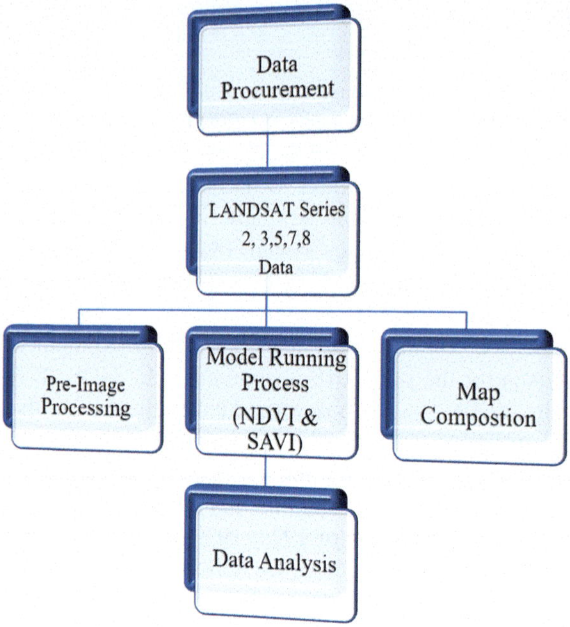

Fig. 2 Flow chart methodology

NDVI

In NDVI calculation, ranges were varying and were categorized into four classes—1–0 (No vegetation), 0–0.3 (Built-up/Fallow/Scrubs), 0.3–0.6 (Thin vegetation), and above 0.6 (Dense vegetation). The area of each class was also calculated and analyzed. Figures 3 and 4 displayed the NDVI-based map of 1975, 1979–80, 1989–90, 2000, 2010, and 2019–20 for pre- and post-monsoon period. Tables 2 and 3 show the NDVI area calculation in a hectare of each class and the NDVI area (ha) change analysis from pre-to-post monsoon.

SAVI

For NDVI and SAVI, same categories were identified based on the −1 to 1 SAVI range. The spatial-temporal change was analyzed in 1975, 1979–80, 1989–90, 2000, 2010, and 2019–20 for pre- and post-monsoon. Comparatively, the change in vegetation cover in the study area was also analyzed from 1975 to 2020 of the pre- and post-monsoon period. Figures 5 and 6 present the SAVI-based map of 1975, 1979–80, 1989–90, 2000, 2010, and 2019–20 for the pre- and post-monsoon

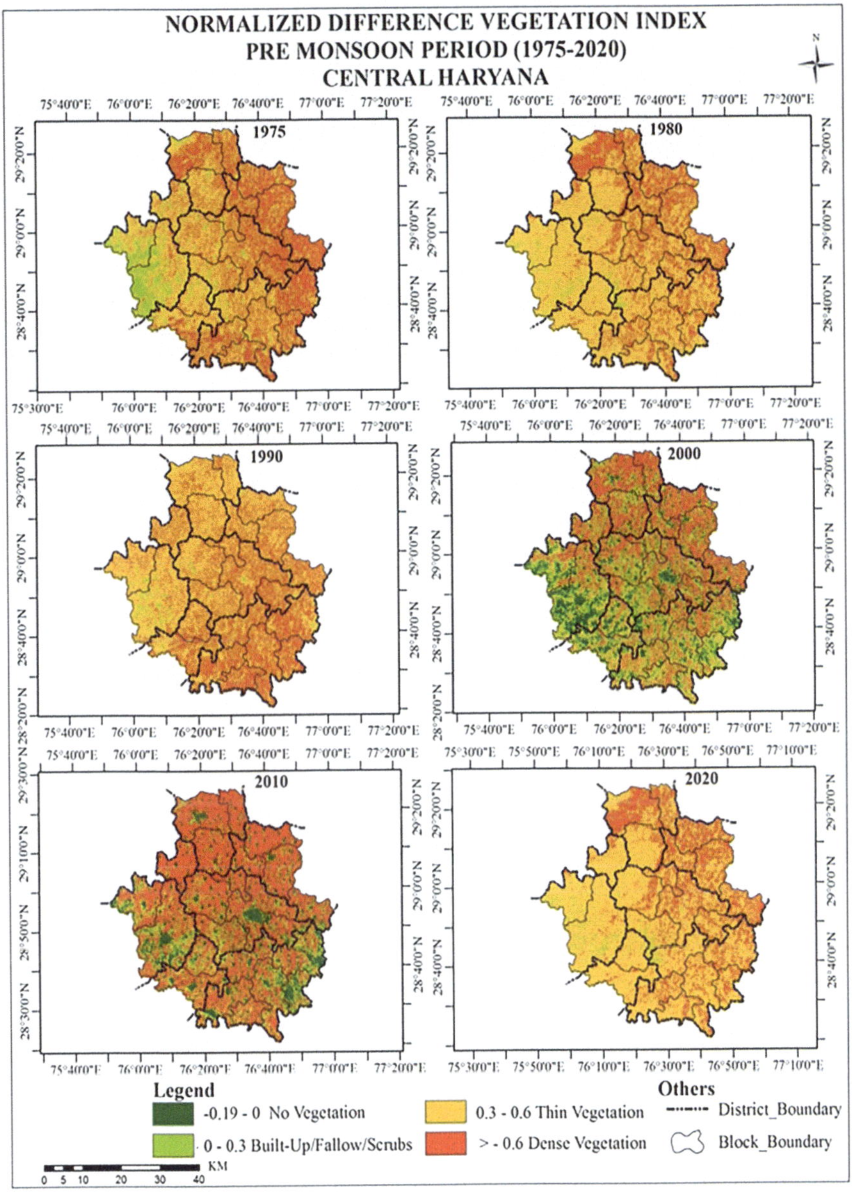

Fig. 3 NDVI Calculation pre-monsoon 1975–2020

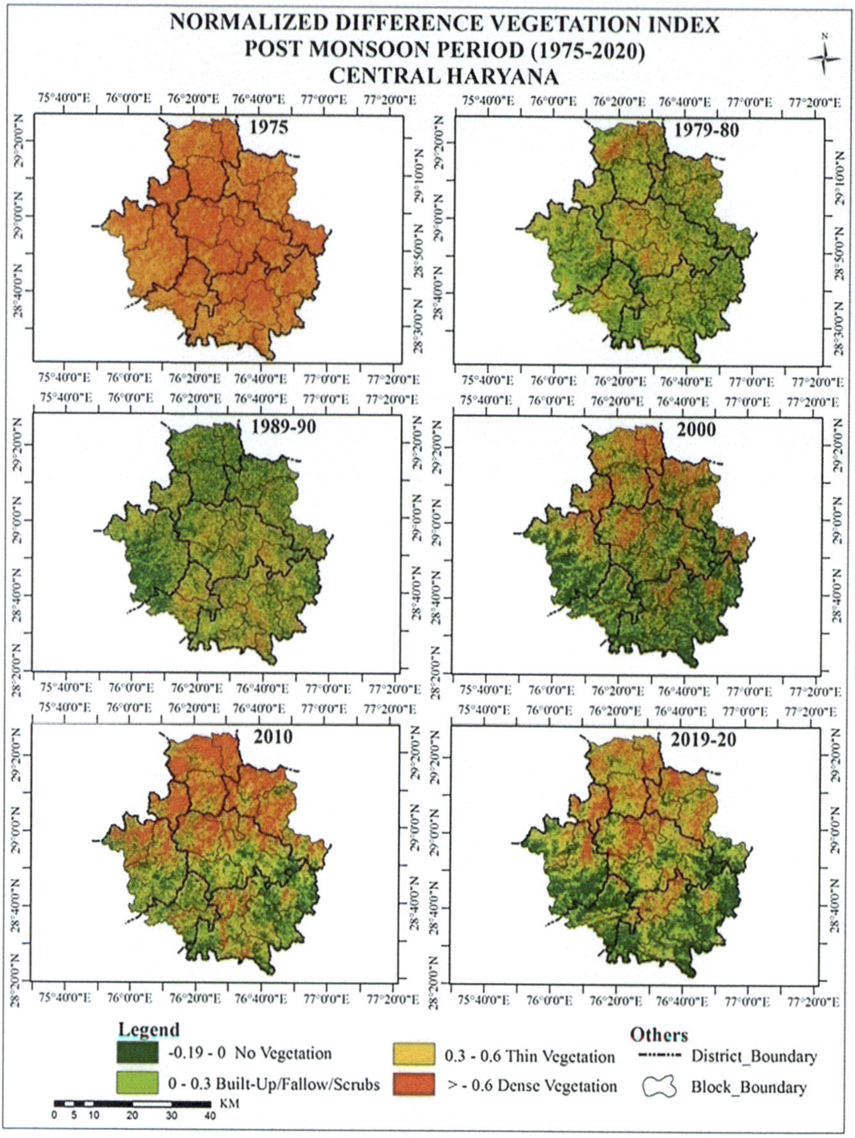

Fig. 4 NDVI Calculation post-monsoon 1975–2020

periods. Tables 4 and 5 showed the SAVI area calculation in a hectare of each class and the SAVI area (ha) change analysis from pre-to-post monsoon.

Table 6 and Fig. 7 showed the comparative results of NDVI and SAVI indices of 1975, 1979–80, 1989–90, 2000, 2010, and 2019–20 for the pre- and post-monsoon period of the study area.

Table 2 NDVI calculation from 1975 to 2020 pre- and post-monsoon

Years/duration		NDVI Classes No vegetation −1 to 0	Built-up/fallow/scrubs 0–0.3	Thin vegetation 0.3–0.6	Dense vegetation >–0.6
1975	Pre-monsoon	87,554	449,103	226,758	8085
	Post-monsoon	54,989	310,760	290,478	115,273
1979–80	Pre-monsoon	105,925	128,308	387,199	150,068
	Post-monsoon	116,867	318,221	193,405	143,007
1989–90	Pre-monsoon	220,203	258,145	213,066	80,086
	Post-monsoon	261,756	267,936	164,892	76,916
2000	Pre-monsoon	102,356	120,808	270,189	278,147
	Post-monsoon	292,229	194,341	146,502	138,428
2010	Pre-monsoon	103,714	105,815	191,044	370,927
	Post-monsoon	152,523	199,747	198,106	221,124
2019–20	Pre-monsoon	123,543	116,833	240,587	290,537
	Post-monsoon	218,725	209,234	216,127	127,414

Source: NDVI-based software area calculation

Table 3 NDVI change calculation from 1975 to 2020 of pre-to-post monsoon

Years	NDVI Classes No vegetation −1 to 0	Built-up/fallow/scrubs 0–0.3	Thin vegetation 0.3–0.6	Dense vegetation >–0.6
1975	−32,565	−138,343	63,720	107,188
1979–80	10,942	189,913	−193,794	−7061
1989–90	41,553	9791	−48,174	−3170
2000	189,873	73,533	−123,687	−139,719
2010	48,809	93,932	7062	−149,803
2019–20	95,182	92,401	−24,460	−163,123
1975–2020	62,617	−45942	39,260	−55,935

Source: NDVI-based software area calculation

Validation of Results

NDVI and SAVI indices values were extracted by sampling technique to establish the relationship between these two indices. Near about randomly 100 samples were selected to get the average value for every class pixel. Average values were compared with the already verified map. SAVI values were very close to the map. It

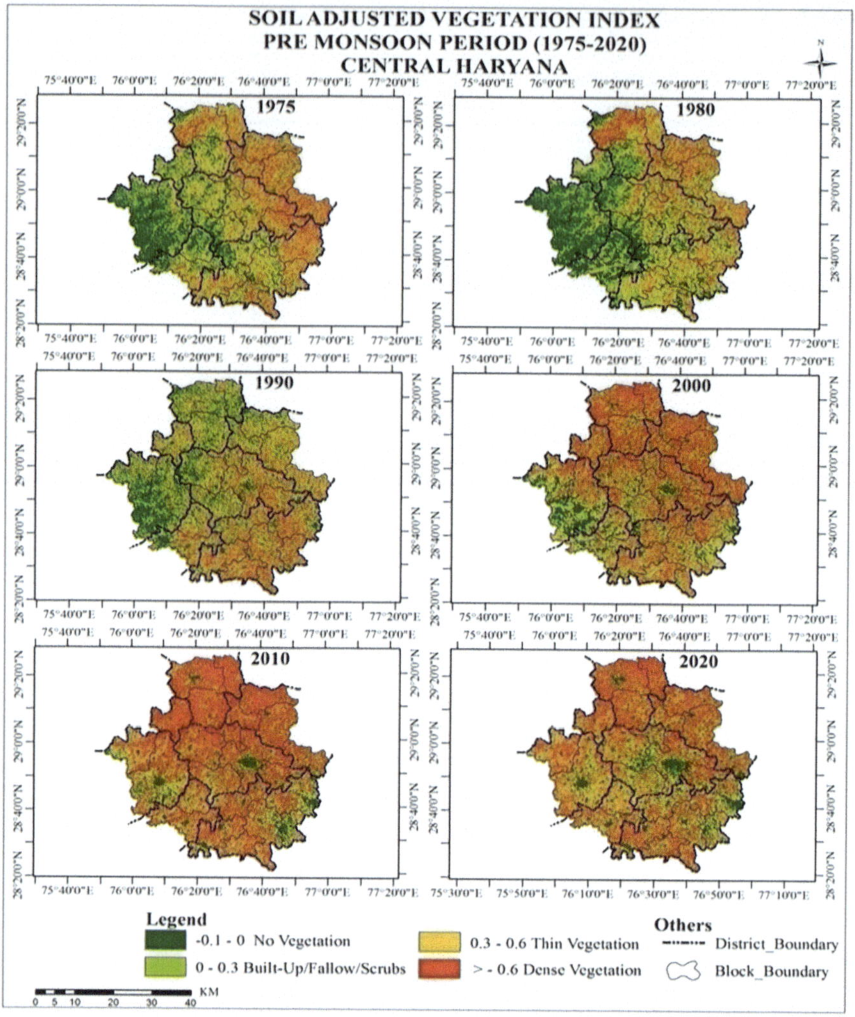

Fig. 5 SAVI Calculation pre-monsoon 1975–2020

proves that SAVI is a better index for micro-level analysis. A similar result is obtained by Vani V. (2017) for the Anantpur district.

Accuracy Assessment

After analyzing all temporal data, the effectiveness of NDVI and SAVI index for waterlogged area change was evaluated by performing an accuracy assessment. Comparative change for the years 1989–90, 2000, 2010, and 2019–20 for both

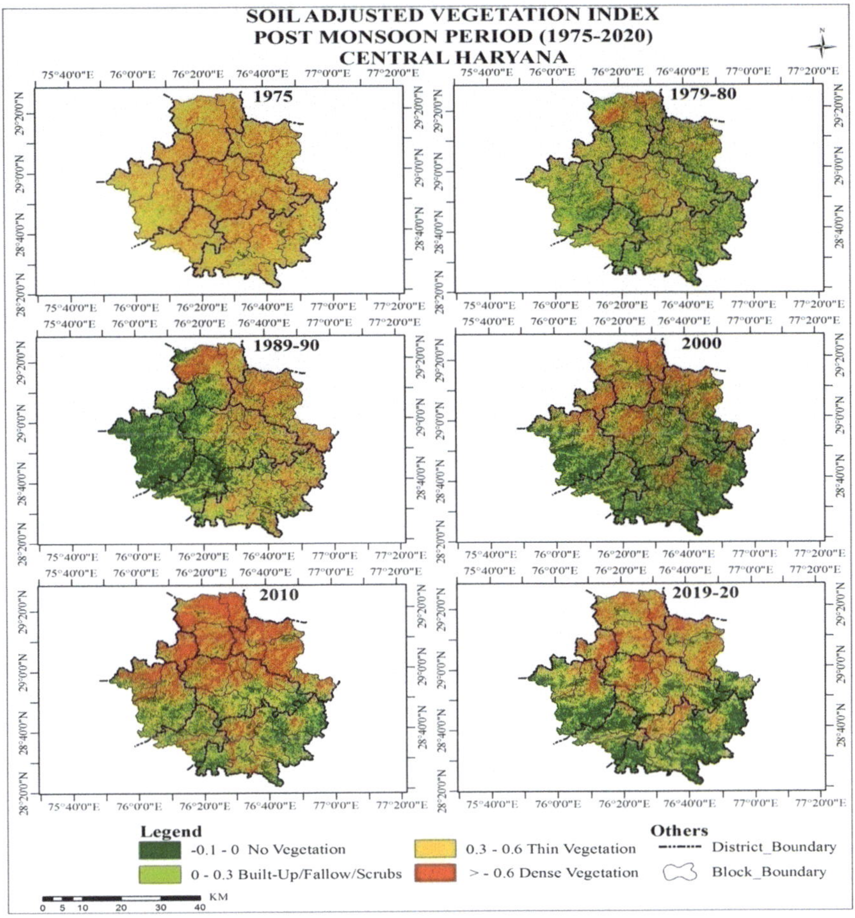

Fig. 6 SAVI Calculation post-monsoon 1975–2020

monsoon periods was evaluated by calculating the Kappa coefficient, overall accuracy, producer accuracy, and overall accuracy. For user accuracy assessment, randomly a set of 40 points was created and compared with Google Earth (producer accuracy). After that overall accuracy and kappa coefficient were calculated. All calculation is given in the table.

The above table describes the NDVI area calculation (ha) of Central Haryana from 1975, 1979–80, 1989–90, 2000, 2010, and 2019–20. Each class has a fluctuating trend. No vegetation cover area was 87554 and 54989 ha in 1975 for pre- and post-monsoon, but the area was increased to 123543 and 218725 ha in 2019–20 for pre- and post-monsoon. But built-up/fallow/scrubs have a negative trend from 1975 to 2019–20. The thin vegetation class area increased in the pre-monsoon period of 1975 to 2019–20 but decreased in the post-monsoon period. Drastic change (8085 ha to 290,537 ha) has been seen in the dense vegetation class of the monsoon period of

Table 4 SAVI area calculation (ha) from 1975 to 2020 of pre- and post-monsoon

Years/duration		SAVI Classes No vegetation −1 to 0	Built-up/fallow/ scrubs 0–0.3	Thin vegetation 0.3–0.6	Dense vegetation >−0.6
1975	Pre-monsoon	187,685	219,071	222,035	142,709
	Post-monsoon	52,200	263,001	310,033	146,266
1979–1980	Pre-monsoon	222,164	240,229	195,138	113,969
	Post-monsoon	199,915	303,642	193,637	74,306
1989–1990	Pre-monsoon	205,289	215,414	203,200	147,597
	Post-monsoon	282,525	255,906	160,821	72,248
2000	Pre-monsoon	144,302	134,765	222,511	269,922
	Post-monsoon	291,394	194,790	146,885	138,431
2010	Pre-monsoon	108,204	102,423	200,950	359,923
	Post-monsoon	154,530	207,109	199,097	210,764
2019–2020	Pre-monsoon	124,290	119,836	247,597	279,777
	Post-monsoon	217,959	209,996	216,130	127,415

Source: SAVI-based software area calculation

Table 5 SAVI area (ha) change calculation from 1975 to 2020 of pre-to-post monsoon

Years	SAVI Classes No vegetation −1 to 0	Built-up/fallow/scrubs 0–0.3	Thin vegetation 0.3–0.6	Dense vegetation >−0.6
1975	−135,485	43,930	87,998	3557
1979–80	−22,249	63,413	−1501	−39,663
1989–90	77,236	40,492	−42,379	−75,349
2000	147,092	60,025	−75,626	−131,491
2010	46,326	104,686	−1853	−149,159
2019–20	93,669	90,160	−31,467	−152,362
1975–20	−41,816	46,230	56,531	−148,805

Source: SAVI-based software area calculation

1975 to 2019–20. In post-monsoon, increasing trend is showing. Increasing and decreasing trends can also be seen in the years 1979–80, 1989–90, 1990, 2000, 2010, and 2019–20. Thus, the NDVI area calculation table analyzes the LULC change of the selected area. The fallow land and scrub class is transformed into other classes.

Table 6 Comparative area (ha) analysis of NDVI and SAVI of pre-to-post monsoon

Year/duration		NDVI and SAVI Classes No vegetation −1 to 0	Built-up/fallow/scrubs 0–0.3	Thin vegetation 0.3–0.6	Dense vegetation >–0.6
1975	1* to 1*	100,131	−230,032	−4723	134,624
	2* to 2*	−2789	−47,759	19,555	30,993
1979–80	1* to 1*	116,239	111,921	−192,061	−36099
	2* to 2*	83,048	−14,579	232	−68,701
1989–90	1* to 1*	−14,914	−42,731	−9866	67,511
	2* to 2*	20,769	−12,030	−4071	−4668
2000	1* to 1*	41,946	13,957	−47,678	−8225
	2* to 2*	−835	449	383	3
2010	1* to 1*	4490	−3392	9906	−11,004
	2* to 2*	2007	7362	991	−10,360
2019–20	1* to 1*	747	3003	7010	−10,760
	2* to 2*	−766	762	3	1

(1* to 1* = pre-to-pre monsoon change; 2* to 2* change=post-to-post monsoon change)
Source: NDVI and SAVI-based software area calculation

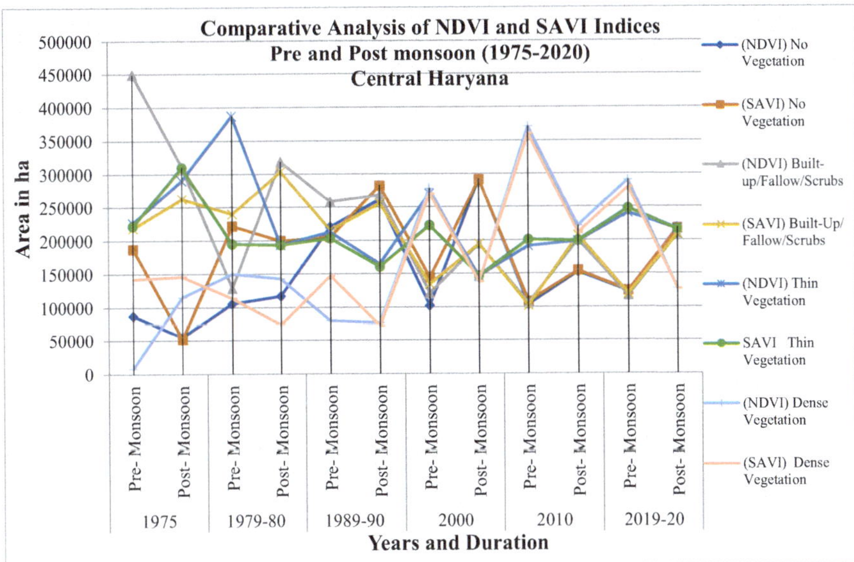

Fig. 7 Comparison of NDVI and SAVI indices

Table 3 shows the NDVI area change calculation of pre-to-post monsoon periods of 1975, 1979–80, 1989–90, 2000, 2010, and 2019–2020 of the study area. In 1975, the area of no vegetation class decreased but after that increased. Built-up/fallow/scrubs area decreased in 1975 after that area of this class increased. The area under the thin vegetation class has an ups and downs trend from 1975 to 2020. Dense vegetation class has increased trend in 1975 but decreased after that. The dense vegetation class area is continuously decreasing after 1975 and its area transformed into the thin vegetation class. In between 1975 and 2019–20, a fluctuation trend can be seen. Overall change from 1975 to 2020 has variation in every class. From 1975 to 2020, the No vegetation class increased to 62,617 ha, the built-up/fallow/scrubs class decreased −45,942 ha, the thin vegetation class increased to 39,260 ha, and the dense vegetation class area declined to −55,935 ha.

The above table displays the SAVI area calculation (ha) of Central Haryana from 1975, 1979–80, 1989–90, 2000, 2010, and 2019–20. Each class has a fluctuating trend. No vegetation cover area was 187,685 and 52,200 ha in 1975 for pre- and post-monsoon but area was increased as 124290 and 217,959 ha in 2019–20 for pre- and post-monsoon. Built-up/fallow/scrubs have a negative trend from 1975 to 2019–20. Thin vegetation and dense vegetation class areas increased in the monsoon period of 1975 to 2019–20 but decreased in the monsoon period. In every decade of 1980, 1990, 2000, 2010, and 2020, the area under every class has been transferred to another class in the pre- and post-monsoon period. Thus, the SAVI area calculation table analyzes the land use and land cover change in Central Haryana. It showed that area in which class increased or decreased.

Table 5 shows the SAVI area change calculation of pre-to-post monsoon periods of 1975, 1979–80, 1989–90, 2000, 2010, and 2019–2020 of the study area. In 1975 and 1979–80, the area of no vegetation class decreased but after that increased. The built-up/fallow/scrub area was increased. The area under the thin vegetation class increased in 1975 but after that decreased. Dense vegetation class has increased trend in 1975 but decreased after that. The dense vegetation class area continuously decreased after 1975 and its area transformed into thin vegetation class. Increasing and decreasing trend is seen in the pre-to-post monsoon periods of each class such as no vegetation, built-up/fallow/dune, thin vegetation, and dense vegetation in the decades of 1980, 1990, 2000, 2010, and 2020. Overall change from 1975 to 2020 has variation in every class. From 1975 to 2020, the No vegetation class decreased to −41,816 ha, the built-up/fallow/scrubs class increased to 46,230 ha, the thin vegetation class increased to 56,531 ha, and the dense vegetation class area declined to −148,805 ha.

Table 6 and Fig. 7 present the NDVI and SAVI change of pre-to-pre and post-to-post monsoon period for the years 1975, 1979–80, 1989–90, 2000, 2010, and 2020. In 1975, the difference in the area under the No vegetation class of pre-to-pre monsoon of NDVI to SAVI is 10,031 ha but in post-to-post monsoon, the area is decreased as -2789 ha. The area of the built-up/fallow/scrub class was decreased to -230032 and -47759 ha in pre-to-pre and post-to-post monsoon. Area

under thin vegetation class is decreased as −4723 ha in pre-to-pre monsoon and increased as 19555 ha in post-to-post monsoon. The dense vegetation class area is increased as 134624 and 30993 ha in pre-to-pre and post-to-post monsoon. Similarly, the difference of pre-to-pre and post-to-post monsoon periods is also seen in the years 1979–80, 1989–90, 2000, 2010, and 2019–20. The difference between NDVI and SAVI results is lower in the last three decades rather than 1980 and 1990 decade.

Discussion

A powerful geospatial strategy that aids sustainable land use and environmental management (Kumar & Singh, 2021) is the comparison of NDVI and SAVI vegetation analyses. These indices play a critical role in solving global concerns such as climate change and food security by monitoring vegetation health, identifying environmental threats, and building unity via informed decision-making. Collaboration, resource allocation, community participation, and evidence-based policies based on NDVI and SAVI data are critical for increasing community solidarity and attaining a more sustainable and equitable future. The present study has been showing vegetation indices as NDVI and SAVI from LANDSAT series data of different resolutions of Central Haryana. Analysis was done for the years of 1975, 1979–80, 1989–90, 2000, 2010, and 2020 of pre- and post-monsoon period. NDVI vegetation has errors related to soil noise. However, SAVI has overcome this error by using an adjustment factor. This comparative study analyzes the study area by categorizing it into four classes. Due to development activity, every class has negative and positive changes in every decade. From 1975 to 2020, most areas declined in vegetation class. Sample value comparison of NDVI and SAVI indices with the original map proves SAVI is the best vegetation index in both of them because of having good adjustment factors for thematic mapping or classes.

NDVI is an appropriate index for studying vegetation globally. A regional scale analysis results in higher resolution datasets for analysis, which means that background effects can become more prominent and affect index values. As a result, more indicators are developed to mitigate these effects. In SAVI, soil reflectance is taken into account rather than soil texture. This difference is because SAVI uses soil-adjustment factors. In addition to manual adjustments, the L functions reduce soil backdrop variations, increasing vegetation dynamic range. A forest cover, for example, should result in an L that approaches 0, similar to the NDVI, while a plantation should result in an L that approaches 1, as opposed to the NDVI.

Conclusion

The sustainable management of land resources and the environment is a global challenge that requires interdisciplinary approaches and the utilization of advanced technologies. In this context, geospatial analysis, particularly the use of vegetation indices like the Normalized Difference Vegetation Index (NDVI) and the Soil-Adjusted Vegetation Index (SAVI), plays a crucial role in assessing and monitoring changes in vegetation cover. This comparative analysis of NDVI and SAVI vegetation analysis seeks to explore how these indices can enhance solidarity among communities by contributing to sustainable land use and environmental management. This discussion delves into the theoretical foundations of NDVI and SAVI, their applications, and the potential for fostering community solidarity through informed decision-making. NDVI and SAVI are crucial for monitoring vegetation health and helping detect changes such as deforestation, land degradation, and the spread of invasive species. By identifying areas with declining vegetation, authorities can take proactive measures to prevent environmental degradation.

References

Ahmadi, H., & Nusrath, A. (2012). Vegetation change detection of Neka River in Iran by using remote sensing and GIS. *Journal of Geography and Geology, 2*(1), 58–67.

Arvind, S., Veer, V., Kumar, S., & Singh, R. (2022). GIS-based review for monitoring the spatial distribution of COVID-19: A case study of Haryana. *International Journal of Research in Applied Science and Engineering Technology (IJRASET), 10*(II), 1093–1099.

Bannari, A., Asalhi, H., & Teillet, P. M. (2002). Transformed difference vegetation index for vegetation cover mapping. *IEEE International Geoscience and Remote Sensing Symposium.* https://doi.org/10.1109/igarss.2002.1026867

Central Ground Water Board (2014–15, 2015–16 and 2016–17). "Groundwater yearbook of Haryana state", report by North Western Region Chandigarh, pp. (19, 21); (12, 20); (23,25)

Chouhan, R., & Rao, N. (2004). Vegetation detection in multispectral remote sensing images: Protective role analysis of coastal vegetation in 2004. *Indian Ocean Tsunami, Geoinformatics for Disaster Management, Procedia Technology, 6*, 612–621.

Epiphanio, J. N., & Huete, A. R. (1995). Dependence of NDVI and SAVI on sun/sensor geometry and its effect on fAPAR relationship in Alfalfa. *Remote Sensing of Environment, 51*(3), 351–360. https://doi.org/10.1016/0034-4257(94)00110-9

Gandhi, G. M., Parthiban, S., Thummalu, N., & Christy, A. (2015). NDVI: Vegetation change detection using remote sensing and GIS-a case study of Vellore District. *Procedia Computer Science, 57*, 1199–1210. https://doi.org/10.1016/j.procs.2015.07.415

Gray, L. K., Nylander, T., Mbogga, M. S., Chen, P. Y., & Hamann, A. (2011). Assisted migration to address climate change: Recommendations for aspen reforestation in western Canada. *Ecological Applications, 21*(5), 1591–1603. https://doi.org/10.1890/10-1054.1

HARSAC (n.d.). http://hrsdi.in/link/maps.html

Honkavaara, E., Saari, H., Kaivosoja, J., Polonen, L., Hakala, T., Litkey, P., Makynen, J., & Pesonen, L. (2013). Processing and assessment of spectrometric; spectrometric imagery collected using a lightweight UAV spectral camera for precision agriculture. *Remote Sensing, 5*, 5006–5039. https://doi.org/10.3390/rs5105006

Huete, A. R. (1988). A soil adjusted vegetation index-SAVI. *Remote Sensing Environment, 25*, 295–309.

Karaburan, A. A. K. B. (2010). Estimation of C factor for soil erosion modeling using NDVI in Buyukcekmece watershed. *Ozezan Journal of Applied Science, 3*, 77–85.

Kumar, S., & Kumar, G. (2019). Role of geospatial application in waterlogging and salinity assessment: A review. *Our Heritage, 67*(1), 1–10.

Kumar, S., & Singh, R. (2021). Geospatial applications in land use/land cover change detection for sustainable regional development: The case of Central Haryana. *India. Geomatics and Environmental Engineering, 15*(3), 81–98. https://doi.org/10.7494/geom.2021.15.3.81

Li, J., Zaho, W., & Zhang, X. (2010). The application of remote sensing data to assess soil erosion. In *Proceeding of the International Conference on Multimedia Technology (ICMT 10)* (pp. 1–4). IEEE. https://doi.org/10.1109/ICMUT.2010.5629726

Nowacki, G. J., & Abrams, M. D. (2015). Is climate an important driver of post-European vegetation change in the Eastern United States? *Global Change Biology, 21*(1), 314–334. https://doi.org/10.1111/gcb.12663

Pettorelli, N., Vik, J. O., Mysterud, A., Gaillard, J. M., Tucker, C. J., & Stenseth, N. C. (2005). Satellite-derived NDVI and antler size in Finnish brown bears Ursus arctos. *Wildlife Biology, 11*(3), 219–224.

Rondeaux, G., Steven, M., & Baret, F. (1996). Optimization of soil-adjusted vegetation indices. *Remote Sensing of Environment, 55*(2), 95–107. https://doi.org/10.1016/0034-4257(95)00186-7

Tech-Tutor with Fitsum (2020). https://drive.google.com/file/d/1FKc-G1vMVtWXi66hh15VKoj-zk3UPWgX/view

Thenkabail, P. S., Smith, R. B., De Pauw, E., & Gillies, R. R. (2004). Global irrigated area map (GIAM), derived from remote sensing, for the end of the last millennium. *International Journal of Remote Sensing, 25*(21), 4091–4137.

United Nations. (2015). *Transforming our world: The 2030 agenda for sustainable development.* United Nations General Assembly.

Vani, V., & Mandla, V. R. (2017). Comparative study of NDVI and SAVI vegetation indices in Anantpur District Semi-Arid Areas. *International Journal of Civil Engineering and Technology, 8*(4), 559–566.

Wardle, D. A., Jonsson, M., Bansal, S., Bardgett, R. D., Gundale, M. J., & Metcalfe, D. B. (2012). Linking vegetation change, carbon sequestration and biodiversity: Insights from island ecosystems in a long-term natural experiment. *Journal of Ecology, 100*(1), 16–30. https://doi.org/10.1111/j.1365-2745.2011.01907.x

Xu, D., Chen, R., Xing, X., & Lin, W. (2017). Detection of decreasing vegetation cover based on empirical orthogonal function and temporal unmixing analysis. *Hindwai-Mathematical Problems in Engineering, 2017*, 1–11. https://doi.org/10.1155/2017/5032091

Ren, Z., Zheng, H., He, X., Zhang, D., Yu, X., & Shen, G. (2014). Spatial estimation of urban forest structures with Landsat TM data and field measurements. *Urban Forestry & Urban Greening, 14*(2), 336–344. https://doi.org/10.1016/j.ufug.2015.03.008

Da-Silva, S. V., Salami, G. and da-Silva, O. I. M., (2019). Emanuel Araújo Silva, José Jorge Monteiro Junior & Elisiane Alba (2020) Methodological evaluation of vegetation indexes in land use and land cover (LULC) classification, Geol. Ecol. *Landscapes*, 4(2):159–169, https://doi.org/10.1080/24749508.2019.1608409

Foley JA, Defries R, Asner GP, Barford C, Bonan G, Carpenter SR, Chapin FS, Coe MT, Daily GC, Gibbs HK, Helkowski JH, Holloway T, Howard EA, Kucharik CJ, Monfreda C, Patz JA, Prentice IC, Ramankutty N, Snyder PK. (2005). Global consequences of land use. *Science.* 309(5734):570–4. https://doi.org/10.1126/science.1111772. PMID: 16040698.

Fostering Environmental Solidarity through the Assessment of Hydrogeomorphic Problems of the Shiddhi Hill Region for the Sustainable Development of Shirur Town in Pune District, Maharashtra

Atul M. Jethe

Introduction

During the last two to three decades, the basement water problem has been very serious because of industrialization and urbanization. With the increase in population, people construct basements at the toe of the hill which causes property loss, furniture loss as well as money loss, the gravity water problem is not local but it's a very dangerous problem (Holmes, 1996). It's a worldwide problem. There is more property loss by the given problem. At the world level more than 50 cr. loss, at the national level more than 2.5 cr. loss, and at the state level more than 80 lakh property loss (Sinha 2014; Murthy, 1998).

As the growth of population increases, people construct basements at the toe of hills, therefore changing the land and so land prices increase more (Shajahan, 2004; Reddy & Sundresha, 2010).

Aim

Management and planning to control the problem of basement water.

A. M. Jethe (✉)
Department of Geography, C.T. Bora College of Arts, Commerce and Science, Shirur, Pune, Maharashtra, India

K. Arora, R. R. Anand (eds.), *The Solidarity Approach in Geography*, https://doi.org/10.1007/978-3-031-86753-8_33

Objective

1. To manage the gravity, water in the basement of the building to reduce property loss.
2. To find out the suitable engineering techniques to control the problem of basement water.

Study Area

The area selected for this investigation is located along the Shiddhi hillside region of Shirur city. The total study area of this region is about 7 km. It is located between 18^{0}49'00" North Latitude to 18^{0}50'30" North Latitude to 74^{0}21'30" East Longitude to 74^{0}23'30" East Longitude (Fig. 1). The slope of the Shiddhi Hill region is more than 45°.

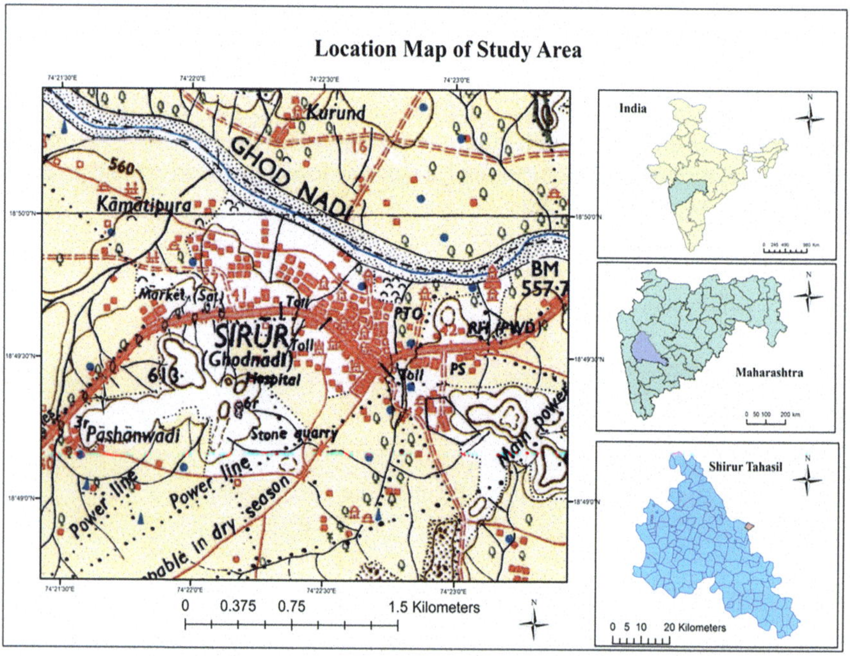

Fig. 1 Location map of study area

Research Method

In the present investigation, the following methods have been adopted:

Data Acquisition

(1) Toposheet—SOI sheet no. 47 J/4 and quadrant B3
(2) Google Earth Images—Google Earth programming
(3) Rainfall Data—Hydrological department of Shirur

Field Work and Lab Work

A further detail of methodology is divided into three components.

Fieldwork

To study the physical and chemical properties of soil, collect some soil samples. In fieldwork, the water level is measured by an infiltrometer tool. In the GPS survey, important points of the study area are tracked using GPS.

Lab Work Component

In lab work, this study used Google Earth Images, i.e., Program 2008–2015. As a secondary sources articles related to lab work, research paper and newspaper clippings also used to write this research paper.

Map Work Component

In map work component SOI topographical map on the scale of 1:25000 has been used as a base map. The present study area lies in SOI sheet no 47 J/5 of quadrant B3.

Preparation of Thematic Maps

Various thematic maps have been prepared in the Surfer software using techniques of spatial interpolation.

Result and Analysis

Paleohydrological Survey, Infiltration, and Hydraulic Conductivity

In the given full research, I have used a paleohydrological study which consists of historical watermarks present on the wall of the basement of affected buildings due to increasing water levels by infiltration of water from the slope of the region.

I have selected seven affected buildings in which the infiltration rate is high (Fig. 2).

Table 1 shows infiltration rate and hydraulic conductivity of the study area

In the riparian region of Shiddhi hill, the ground surface in a large number of permeable rocks and aquifers the water infiltrates into the ground more amount of water penetrates the region behind the Hudako colony at which streams are blocked

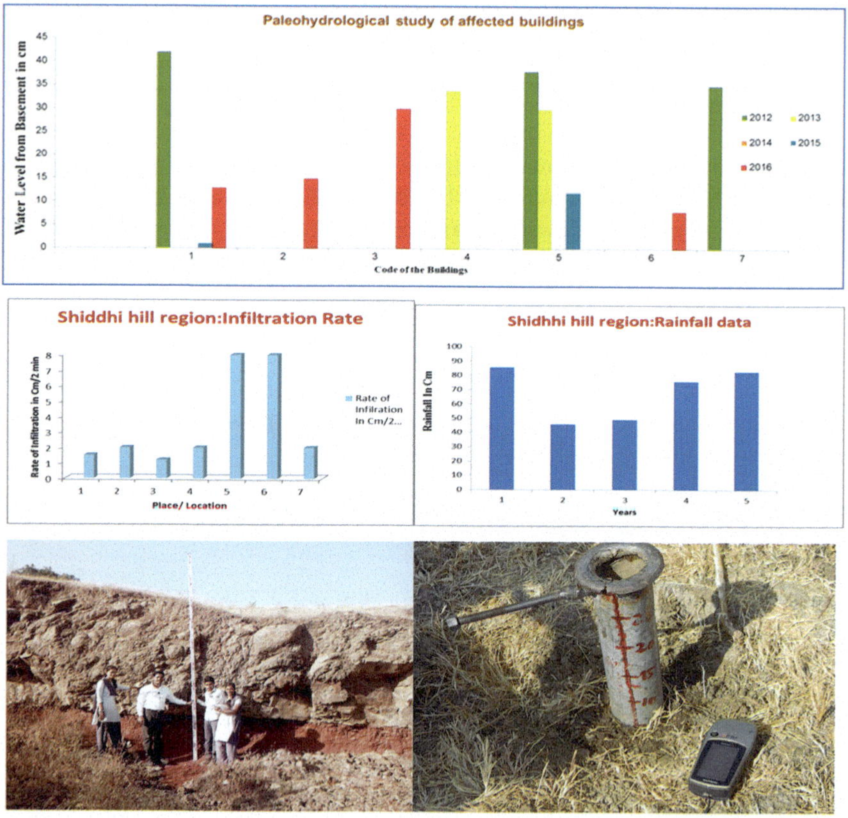

Fig. 2 Paleohydrological survey of the basement of affected buildings. (**a**) Infiltration rate (**b**) Rainfall. (**c**) Profile observation. (**d**) Infiltrometer

Table 1 Infiltration rate and hydraulic conductivity of the study area

Sr. no.	Sample code	GPS location of point		Height in meters	Infiltration (cm/2 min)	H.C.
		Lat. ext. X	Long. ext. Y			
1	C.T. Bora College ground	18°49′16.5″N	74°22′17.5″ E	580 m	1.5 cm	6 min
2	The Upper part of C.T. Bora Stream	18°49′17.8″ N	74°22′09.0″ E	581 m	2 cm	1.30 min
3	Hillside slope along Baburao Nagar	18°49′16.2″ N	74°21′59.8″ E	586 m	1.2 cm	2 min
4	Dargah stream near Pritam Prakash nagar	18°49′24.7″ N	74°21′55″ E	583 m	2 cm	1.30 min
5	Near the Renuka Mata temple along the hillside	18°49′23.9″ N	74°21′36.4″ E	573 m	8 cm	3.5 min
6	Pabal phata stream	18°49′25.9″ N	74°22′2.7″ E	585 m	8 cm	50 s
7	Kukadi colony	18°49′21.5″ N	74°22′21.0″ E	573 m	2 cm	2 min

and also at the region of stone query because of water stores for a long time so it has more time seeps into the ground. Infiltrated or penetrated, gravity water enters in subsurface layers and flows along the slope as a through flow and comes into the basement (Davie, 2019).

In the following map (Fig. 3), red color indicates high infiltration zone while blue color express moderate zone and green color shows less infiltration zone.

Some suggestions are explained by the following research.

Suggestions

(1) Remodification of the basement of affected buildings using total cement coating.
(2) Make the basement using steel sheet piles.
(3) Total waterproofing of the basement.
(4) Make the pumping well along the slope direction.
(5) Grade the area of high infiltration so that gravity water level comes into the basement.
(6) Make the cement water outlet pipeline.

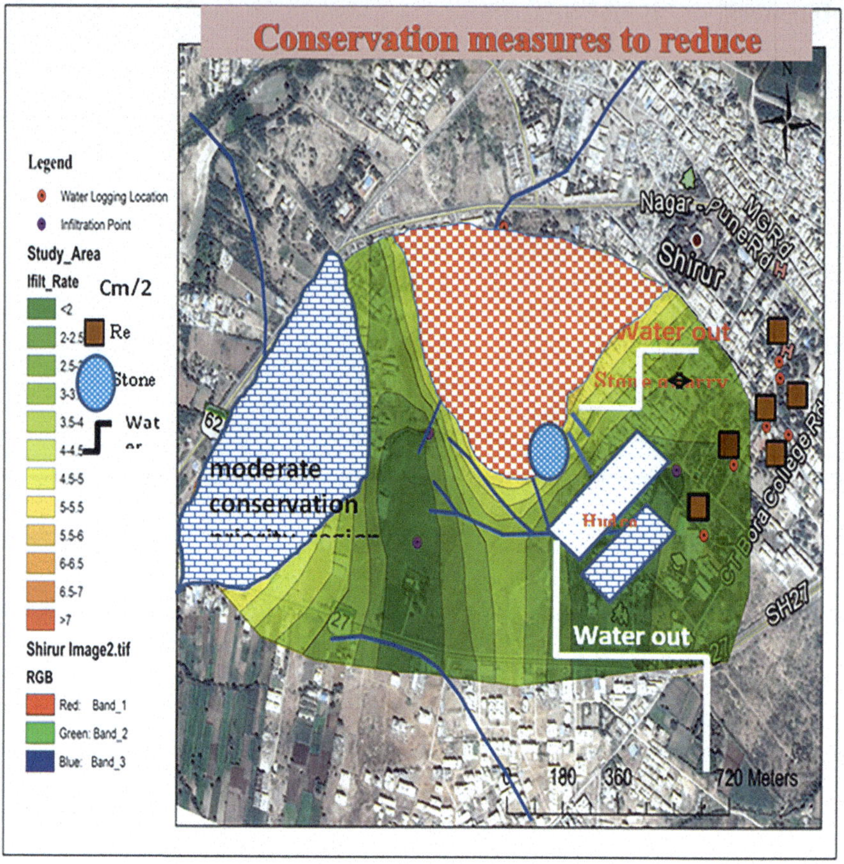

Fig. 3 Conservation measures suggested to reduce the severity and priority of the problem (Source author 2022)

Conclusion

Thus, for the most urbanized and densely populated area where the settlement located in the riparian region faces basement water can be solved with good planning and by using suitable engineering techniques. Wealth loss in the region can reduce and the basement can be used for commercial purposes in a better way. Hence, fostering environmental solidarity and implementing comprehensive strategies, Shirur town can address the hydrogeomorphic problems of the Shiddhi hill region. Sustainable development efforts will protect the natural resources and also enhance the well-being and resilience of the local community.

References

Davie, T. (2019). *Fundamentals of hydrology*. Rutledge.

Roy Holmes (1996). Introduction to civil engineering construction collage of estate management.

Murthy, J. V. S. (1998). *Watershed management: Principles parameters and programs*. New Age International.

Reddy, J.N., & Sundresha, B.S.M. (Eds.) (2010). Geospatial technologies for sustainable urban and rural development. IGI Global

Shajahan, K. M. (2004). *Environmental management in India: Issues and concern*. APH Publishing Corporation.

Sinha, R. K. (2014). *Hydrogeomorphic hazards: Risks, vulnerability and mitigation measures*. Springer.

Conclusion

Solidarity in the Indian Subcontinent: A Geographical Perspective

Solidarity, the expression of common interests encompasses various dimensions including political, economic, social, and cultural cooperation among countries. Incorporating a solidarity approach into geography helps appreciate the complex ways people come together to support common interests and face shared challenges based on their geographic contexts. This interplay shapes social cohesion, political dynamics, and economic development across various scales, from local communities to global regions.

A solidarity approach is particularly useful for geographers as it aligns with core geographical concerns such as spatial justice, interconnectedness, and sustainability. Here are few key areas in which a solidarity approach may benefits geographers:

Addressing Spatial Inequality

Geographers are often concerned with the unequal distribution of resources and opportunities across different regions. A solidarity approach emphasizes the need for collective action to address these spatial inequalities, promoting policies and practices that ensure more equitable development. The Indian subcontinent has many dimensions of spatial inequality, such as urban-rural disparity, east-west divide, regional differences in human development, among which the most challenging is rural-urban disparities, which lead to massive rural migration and haphazard unplanned growth of urban areas. For managing this solidarity approach can help in achieving spatial Justice through urban-rural linkages. Urban areas, despite their density, might experience weaker local solidarity compared to rural areas due to higher anonymity and mobility. In contrast, rural areas often exhibit stronger

K. Arora, R. R. Anand (eds.), *The Solidarity Approach in Geography*, https://doi.org/10.1007/978-3-031-86753-8

community bonds due to long-term residence and fewer social options. The strong connections between urban and rural areas, encourage balanced development and reduce disparities between these regions.

It has been seen that geographic regions with strong identities develop movements aimed at promoting regional interests, which can enhance solidarity among residents. This is evident in Indian federal systems where states or regions have specific representation. Geographically large and diverse countries might promote solidarity through decentralization, giving more autonomy to regions to manage their affairs, which can strengthen regional solidarity while maintaining national unity.

There are various examples of regional movements in the Indian subcontinent based on the diverse linguistic, ethnic, and cultural landscape of the region with each one having its unique context, motivations, and aspirations. It's important to note that these movements have evolved, and their current status and demands may vary. Time and again region has experienced movements for separate statehood based on language or resources. Some of the examples are:

- **Telangana Movement**: The decades-long struggle for a separate Telangana state, carved out of Andhra Pradesh, based on linguistic and cultural identity, culminated in the formation of Telangana in 2014.
- **Gorkhaland Movement**: The demand for a separate Gorkhaland state in the Darjeeling hills of West Bengal, based on the distinct Gorkha identity and cultural aspirations, has been ongoing for several decades.
- **Bodoland Movement**: The Bodo people in Assam have sought a separate Bodoland state to safeguard their cultural identity and address developmental concerns.
- **Bundelkhand Movement**: The demand for a separate Bundelkhand state, encompassing parts of Uttar Pradesh and Madhya Pradesh, has been driven by concerns over underdevelopment and neglect of the region.
- **Vidarbha Movement**: The Vidarbha region in Maharashtra has seen demands for separate statehood due to perceived neglect in terms of development and representation.

Some movements arose for autonomy and greater representation such as

- **Dravidian Movement**: The Dravidian movement in Tamil Nadu advocated for greater autonomy and representation for the Dravidian people, leading to significant political and social changes in the state.
- **Jharkhand Movement**: The movement for a separate Jharkhand state, carved out of Bihar, was driven by demands for greater autonomy and resource development for the tribal communities in the region.

There are some movements which has ethnic and linguistic underpinnings

- **Kashmir Conflict**: The ongoing conflict in Kashmir involves complex issues of identity, autonomy, and political aspirations, with various groups seeking different outcomes, including independence or greater autonomy within India or Pakistan.
- **Naga Insurgency**: The Naga insurgency in Northeast India has sought greater autonomy and self-determination for the Naga people, based on their unique ethnic and cultural identity.

These movements show the greater need for the application of a solidarity approach to achieve the common goals of development and well-being of all residents of the Indian subcontinent.

Promoting Environmental Sustainability

A solidarity approach encourages cooperative efforts among communities, regions, and nations to address environmental challenges and Collective Environmental Action. The Chipko Movement in Garhwal Himalayas is an important testament to how solidarity between local communities and global networks created lasting change and strengthened the connection between global environmental concerns and local livelihoods. The Chipko Movement protected the Himalayan forests but also empowered local communities, particularly women, to take charge of their environmental destiny. It demonstrated the power of grassroots activism and global solidarity in achieving environmental and social justice. The movement's legacy continues to inspire environmental movements worldwide, highlighting the interconnectedness of local and global struggles for a sustainable future.

Environmental issues such as climate change, deforestation, and pollution require collective action across different scales. The Indian subcontinent is particularly vulnerable to climate change impacts often fostering solidarity around environmental initiatives and policies aimed at mitigation and adaptation. The region is highly vulnerable to climate change impacts, such as rising temperatures, erratic rainfall patterns, extreme weather events (heat waves, floods, droughts), and sea-level rise. Intense and prolonged heat waves have become more frequent and severe, particularly in North and Central India. The 2022 heat wave, for instance, saw temperatures soar to record highs, causing heatstroke deaths, agricultural losses, and water scarcity. Changing rainfall patterns, coupled with inadequate infrastructure, have led to devastating floods in various parts of the subcontinent. The 2022 Assam floods displaced millions, destroyed crops, and submerged vast areas, highlighting the vulnerability of the region.

The Himalayan region is witnessing an increase in GLOFs due to accelerated glacial melting. These sudden floods can cause catastrophic damage downstream, as seen in the 2021 Uttarakhand disaster, which resulted in numerous deaths and destruction of infrastructure. Recently, Sikkim experienced severe flooding and landslides, particularly in October 2023 and July 2024. Heavy rains caused the glacial South Lhonak Lake to overflow, leading to a devastating flash flood. The flood destroyed the Teesta III Dam and caused widespread damage downstream, including the loss of lives and the destruction of infrastructure like bridges.

Torrential rains triggered multiple landslides in the Mangan district, resulting in fatalities and leaving numerous tourists stranded. Additionally, the main highway connecting Sikkim to the rest of the country (NH10) was severely disrupted.

Likewise, the region has experienced increased intensity of tropical cyclones. The intensity and frequency of cyclones in the Arabian Sea and Bay of Bengal have increased, causing widespread damage to coastal communities. Cyclone Amphan in

2020, one of the strongest cyclones ever recorded in the Bay of Bengal, devastated parts of West Bengal and Bangladesh.

These changes threaten agriculture, water resources, coastal communities, and infrastructure, impacting livelihoods and exacerbating existing inequalities.

These incidents highlight the region's vulnerability to climate change-induced disasters, especially with the increasing frequency and intensity of extreme weather events. By promoting shared responsibility and mutual support, solidarity helps geographers advocate for sustainable development practices that consider the needs of both current and future generations.

Enhancing Social Cohesion and Integration

Geographers studying social dynamics also use a solidarity approach to promote social cohesion and integration among communities. This can be particularly important in multicultural and multi-ethnic contexts. Solidarity involves standing with marginalized and disadvantaged groups, which aligns with geographers' interest in social justice and equity. This can lead to more inclusive and supportive policies and practices. The various initiatives that contributed to enhancing social cohesion and integration in the Indian subcontinent are:

There are many examples of the many SHGs working across India to empower women, promote social cohesion, and drive economic development. Each group has its unique approach and focus, but they all share a common goal of improving the lives of marginalized communities through collective action and solidarity. SEWA (Self-Employed Women's Association) established in 1972, is one of the pioneering SHGs in India. It focuses on empowering women working in the informal sector through microfinance, skill development, healthcare, and legal aid.

Likewise, in India where caste is an important composition of society, several caste-based organizations and movements have emerged to promote solidarity and uplift marginalized communities. While caste-based discrimination remains a persistent challenge in India, several organizations and movements have emerged to address this issue and promote social integration.

The Ambedkarite Movement is an example of a caste-based movement that has played a crucial role in promoting solidarity and social justice in India. They have challenged the traditional caste hierarchy, advocated for the rights of marginalized communities, and fostered a sense of unity and empowerment among Dalits and other oppressed groups.

These movements have worked towards empowering marginalized castes, providing them with educational and economic opportunities, and advocating for their rights. They have also challenged discriminatory practices and fostered dialogue. These organizations have contributed to greater social cohesion and equity.

India with its diverse ethnicity, languages, vibrant cultural landscape, and festivities provides opportunities for social cohesion and integration. These festivals bring together people from different backgrounds, fostering a sense of shared identity and belonging. Most of the time these festivals are celebrated across religious boundaries and promote unity among communities. This enhances the social cohesion and integration which is crucial in the diverse and complex social fabric of the Indian subcontinent.

Strengthening Global-Local Connections

A solidarity approach recognizes the interconnectedness of local and global processes and global interdependence. In smaller geographic areas, such as neighborhoods or villages, geographic proximity facilitates stronger social ties and a higher degree of mutual support and solidarity. People are more likely to know each other and engage in community activities. Globally also shared cultural and historical backgrounds often foster a sense of solidarity among people within a specific geographic region. This is seen in diasporic communities across the world.

Geographers can use this perspective to study and address global issues such as migration, trade, and climate change, understanding how local actions impact global outcomes and vice versa. Cross-border cooperation also fosters international cooperation and partnerships, which are essential for addressing transnational issues and promoting global well-being. There are few examples of cross-border cooperation in the field of infrastructure, energy, disaster management, and trade facilitation in Indian subcontinent. This enhances solidarity in the region.

Infrastructure:

- **BBIN Motor Vehicle Agreement (MVA)**: This agreement between Bangladesh, Bhutan, India, and Nepal aims to facilitate the seamless movement of passenger, personal, and cargo vehicles across their borders. This initiative is expected to boost trade and tourism in the region.
- **India-Bangladesh Railway Connectivity**: Several railway projects connecting India and Bangladesh are underway, including the Agartala-Akhaura rail link, which is expected to significantly reduce travel time and improve trade between the two countries.

Energy:

- **India-Bhutan Hydropower Cooperation**: Bhutan has immense hydropower potential, and India has invested in several hydropower projects in Bhutan. This cooperation provides a clean energy source for India and revenue for Bhutan.
- **South Asia Regional Initiative for Energy Integration (SARI/EI)**: This initiative aims to promote regional energy cooperation in South Asia by developing cross-border electricity transmission infrastructure and facilitating electricity trade among the member countries.

Disaster Management:

- **BIMSTEC Centre for Weather and Climate**: This center, established under the Bay of Bengal Initiative for Multi-Sectoral Technical and Economic Cooperation (BIMSTEC), focuses on regional cooperation in weather forecasting, climate modeling, and disaster risk reduction.

- **SAARC Disaster Management Centre**: This center, based in New Delhi, promotes regional cooperation in disaster risk reduction, preparedness, and response among the South Asian Association for Regional Cooperation (SAARC) member countries.

Trade Facilitation:

- **Framework Agreement on the Facilitation of Cross-Border Paperless Trade in Asia and the Pacific**: This United Nations Economic and Social Commission for Asia and the Pacific (UNESCAP) agreement, signed by several countries in the region, including India, aims to streamline trade procedures and reduce the reliance on paper-based documentation.
- **South Asian Free Trade Area (SAFTA)**: This agreement aims to reduce tariffs and other trade barriers among the SAARC member countries, promoting regional trade and economic integration.

Cross-border cooperation and solidarity in the Indian subcontinent are deeply intertwined. There are challenges to the relationship between cross-border cooperation and solidarity in the Indian subcontinent which is essential for the region's development and stability. By addressing the existing tensions and prioritizing cooperation, the countries in the region are building a more prosperous, peaceful, and resilient future together.

Participatory Research and Action with Community Engagement

Geographers often engage in participatory research that involves working closely with communities. By prioritizing community engagement, co-creation of knowledge, and mutual respect, they are not only generating valuable insights but also empowering communities to address their challenges and achieve sustainable development goals. Some of the examples of participatory research and actions are as follows:

ParticipatoryGIS(PGIS) in Land Rights Struggles: In tribal areas of Odisha and Jharkhand state of India, geographers are using PGIS to map community land use and customary rights. By involving local communities in the mapping process, they empower them with spatial data and knowledge, which can be used to advocate for their land rights and challenge encroachments. This collaborative approach strengthens community solidarity and agency in resource management.

Community-Based Disaster Risk Reduction (CBDRR): In flood-prone regions like Assam and Bihar state in India, geographers are working with communities to develop CBDRR plans. This involves participatory vulnerability assessments, risk mapping, and co-creation of disaster preparedness and response strategies. By engaging communities in the process, geographers foster ownership and local knowledge integration, leading to more effective and sustainable disaster management.

Action Research on Urban Slum Upgrading: In Indian cities like Mumbai and Delhi, geographers are using action research to understand the complex challenges faced by slum dwellers and co-create solutions with them. This involves participatory mapping, focus group discussions, and collaborative planning for slum upgrading projects. This approach ensures that development interventions are contextually relevant and address the specific needs and aspirations of the community.

Participatory Water Management: In drought-prone regions like Rajasthan and Gujarat state of India, geographers are working with communities to develop participatory water management systems. This involves mapping traditional water sources, co-creating water conservation strategies, and establishing community-led water governance structures. By empowering communities to manage their water resources, geographers foster solidarity and resilience in the face of water scarcity.

Community-Based Tourism: In ecologically sensitive regions like the Western Ghats and the Himalayas, geographers are promoting community-based tourism initiatives. This involves working with local communities to develop tourism products that showcase their unique cultural heritage and natural landscapes while ensuring that the benefits of tourism are shared equitably and that the environment is protected.

It also facilitates Interdisciplinary Collaboration. Many geographical issues are complex and require interdisciplinary and holistic understanding. Solidarity promotes collaboration across disciplines, encouraging geographers to work with sociologists, economists, environmental scientists, and others to address multifaceted problems. It aligns diverse stakeholders around common goals, facilitating coordinated efforts to address shared challenges.

This collaborative approach strengthens social bonds, fosters collective action, and contributes to a more equitable and just society. It emphasizes mutual respect, collaboration, and co-creation of knowledge, leading to more effective and meaningful research outcomes. By fostering solidarity, geographers can empower communities to take collective action and advocate for their own needs and interests.

Policy Development and Policy Advocacy

Geographers can use insights from a solidarity approach to advocate for policies that promote social and spatial justice, environmental sustainability, and equitable development. This can influence local, national, and international policy agendas.

Geographic disparities in the regions with the least development require an approach of regional economic solidarity for economic development. Differences in physical geography, resource availability, historical development, and government policies call for solidarity in the form of redistributive policies. Less developed regions may seek support from wealthier areas within a country or economic union. North-South Divide, Coastal vs. Inland Areas, Urban vs. Rural Areas, Himalayan

Region, North Eastern region, tribal region, arid and semi-arid region, addressing these disparities requires targeted policies and investments in infrastructure, education, healthcare, and livelihoodopportunities, particularly in disadvantaged regions. It also necessitates greater attention to environmental sustainability and climate changeadaptation, given the region's vulnerability to natural disasters and changing climatic conditions.

Constraints for Using the Solidarity Approach

Geographers using a solidarity approach might feel constrained by ethical dilemmas. Geographers may struggle with questions about representation, ownership of knowledge, and the potential for research findings. Traditional research methods may not be suitable to apply solidarity approach. Geographers may need to develop new, participatory methods that empower the community and prioritize their knowledge and perspectives which can balance research objectives with the needs and desires of the community.

Maintaining long-term solidarity relationships can be challenging. Geographers may face constraints related to funding, time, and institutional support. Ensuring that the benefits of the research are sustained within the community can also be a concern. Academic institutions may not always be supportive of a solidarity approach. Geographers may face pressure to adhere to traditional research practices, publish in specific journals, or focus on topics that are deemed more "important" by the academic community. Solidarity research sometimes involves working with marginalized communities. Geographers may feel constrained by their positionality and the inherent power imbalances in these relationships. This can be time-consuming and require additional resources.

Solidarity work can be emotionally demanding. Geographers may witness first-hand the struggles and injustices faced by the communities they work with. This can lead to feelings of guilt, frustration, or burnout.

Despite these constraints, the approach is highly useful for geographers as it aligns with their core values and research interests. It helps address spatial inequalities, promote environmental sustainability, enhance social cohesion, strengthen global-local connections, support participatory research, facilitate interdisciplinary collaboration, and contribute to policy development. By adopting a solidarity approach, geographers can analyze more effectively and address the complex and interconnected challenges which the world is facing today. It is essential for addressing social and environmental injustices. By acknowledging and navigating these challenges, geographers can contribute to meaningful change and build stronger relationships with the communities they serve.

Solidarity in the Indian subcontinent is essential for regional stability, economic growth, social development, and environmental sustainability. By fostering political, economic, social, cultural, and security cooperation, the countries in this region can address common challenges more effectively and work towards a more integrated and harmonious future. Despite existing challenges, ongoing efforts and initiatives continue to promote solidarity and cooperation in this diverse and dynamic region.

Solidarity while holding diverse meanings in different contexts, is a fundamental pillar for achieving peace, stability, and development in the Indian subcontinent. The Indian subcontinent, with its rich tapestry of cultures, religions, and political landscapes, faces unique challenges that can only be effectively addressed through collective action and mutual support.

Fostering solidarity in political, economic, social, cultural, and security spheres allows countries in the subcontinent to pool resources, share expertise, and collaborate on solutions for common problems. This can range from addressing shared environmental concerns like climate change and water scarcity to combating transboundary threats such as terrorism and human trafficking. By recognizing their interconnectedness and working together, nations can build a more resilient and prosperous future for the entire region. By applying the Solidarity approach in geography, geographers can contribute to policy and practice, community-based planning and advocacy, and activism. They can work with communities to develop participatory planning processes that empower residents and promote social justice. This involves facilitating dialogue, building trust, and negotiating competing interests. They can use their research and expertise to advocate for policies and practices that promote solidarity, social justice, and environmental sustainability. This involves engaging with policymakers, civil society organizations, and the public. This approach is dominant paradigms, amplifies marginalized voices, and promotes collective action to address social and environmental challenges. Moreover, solidarity is deeply ingrained in the cultural and spiritual traditions of the Indian subcontinent. The concept of "Sarve bhavantu sukhinaḥ," a Sanskrit phrase meaning "may all beings be happy," reflects a profound commitment to universal well-being. Similarly, the concept of "Vasudhaiva Kutumbakam," which translates to "the world is one family," emphasizes the importance of interconnectedness and shared responsibility for the welfare of all. These principles, deeply rooted in major religions provide a moral compass for fostering solidarity and cooperation.

Geographers, with their unique understanding of the spatial dimensions of human interactions and the environment, can play a crucial role in promoting human solidarity with its physical environment. It also promotes interpersonal, interstate solidarity in the changing global order. In practice, a solidarity-based policy approach encourages participatory governance, cross-border cooperation, and attention to marginalized voices. It foregrounds policies that support climate justice, urban equity, transnational migration rights, and post-disaster recovery, recognizing that spatial inequalities are often rooted in historical injustices and global power asymmetries.Through research and analysis, geographers can highlight the interconnectedness of issues, identify areas of potential collaboration, and advocate for policies that prioritize collective well-being. By engaging with communities, policymakers, and other stakeholders, geographers can contribute to informed decision-making that takes into account the diverse perspectives and needs of the region. By integrating geographical insights with a solidarity ethic, policymakers can move toward frameworks that are not only efficient but also just, resilient, and inclusive.

The solidarity approach in geography provides a powerful lens to understand how spatial relationships, socio-political processes, and collective identities intersect in shaping the human experience. It emphasizes interdependence, mutual support, and ethical responsibility across space and scale. This approach challenges individualistic, neoliberal, and exclusionary spatial logics. From David Harvey's spatial justice to Doreen Massey's relational spatiality, the solidarity framework reveals how places are not isolated but interwoven through histories, power dynamics, and shared futures. Geographers using this lens contribute to more inclusive knowledge, drawing attention to grassroots movements, transnational alliances, environmental justice, and humanitarian developmental geographies. Ultimately, the solidarity approach not only deepens our understanding of spatial inequality but also offers transformative pathways for resistance, resilience, and regeneration in the face of global crises—from climate change to displacement and systemic violence. It calls for a geography that not only explains the world but seeks to change it through collective agency and ethical spatial practices. In this way, solidarity serves as a unifying force, bridging differences and fostering a sense of shared purpose. By embracing solidarity as a guiding principle, the Indian subcontinent can overcome its challenges and create a more just, equitable, and sustainable future for all its inhabitants. In a world facing interconnected crises, from pandemics to climate emergencies, the solidarity approach in geography and policy offers a critical pathway to foster collective well-being, sustainable futures, and spatial justice.

Index

K. Arora, R. R. Anand (eds.), *The Solidarity Approach in Geography*,
https://doi.org/10.1007/978-3-031-86753-8

The manufacturer's authorised representative in the EU is Springer Nature Customer Service Centre GmbH, Europaplatz 3, 69115 Heidelberg, Germany. If you have any concerns regarding our products, please contact ProductSafety@springernature.com

Printed and bound by CPI Group (UK) Ltd, Croydon, CR0 4YY
05/10/2026
02228127-0017